HYPERCHOLESTEROLEMIA,
HYPOCHOLESTEROLEMIA,
HYPERTRIGLYCERIDEMIA,
IN VIVO KINETICS

ADVANCES IN EXPERIMENTAL MEDICINE AND BIOLOGY

Recent Volumes in this Series

A Continuation Order Plan is available for this series. A continuation order will bring delivery of each new volume
immediately upon publication. Volumes are billed only upon actual shipment. For further information please contact
the publisher.

HYPERCHOLESTEROLEMIA, HYPOCHOLESTEROLEMIA, HYPERTRIGLYCERIDEMIA, IN VIVO KINETICS

Edited by

Claude L. Malmendier

Research Foundation on Atherosclerosis
Brussels, Belgium

P. Alaupovic

Oklahoma Medical Research Foundation
Oklahoma City, Oklahoma

and

H. Bryan Brewer, Jr.

National Heart, Lung, and Blood Institute
National Institutes of Health
Bethesda, Maryland

PLENUM PRESS • NEW YORK AND LONDON

Library of Congress Cataloging-in-Publication Data

International Colloquium on Atherosclerosis (5th : 1990 : Brussels,
  Belgium)
    Hypercholesterolemia, hypocholesterolemia, hypertriglyceridemia,
  in vivo kinetics / edited by Claude L. Malmendier, P. Alaupovic, and
  H. Bryan Brewer, Jr.
       p.   cm. -- (Advances in experimental medicine and biology ; v.
  285)
    "Proceedings of the Fifth International Colloquium on
  Atherosclerosis, held March 14-16, 1990, in Brussels, Belgium"--T.p.
  verso.
    Includes bibliographical references and index.
    ISBN-13: 978-1-4684-5906-7
    1. Atherosclerosis--Pathophysiology--Congresses.
  2. Hypercholesteremia--Congresses. 3. Hypocholesteremia-
  -Congresses. 4. Hypertriglyceridemia--Congresses.
  5. Atherosclerosis--Molecular aspects--Congresses.   I. Malmendier,
  Claude L. II. Alaupovic, P. III. Brewer, H. Bryan. IV. Title.
  V. Series.
    [DNLM: 1. Atherosclerosis--etiology--congresses. 2. Cholesterol-
  -blood--congresses. 3. Hypercholesterolemia--metabolism-
  -congresses. 4. Hypertriglyceridemia--metabolism--congresses.   W1
  AD559 v. 285 / WG 550 I6033h 1990]
  RC692.I4674 1990
  616.1'36071--dc20
  DNLM/DLC
  for Library of Congress   91-2557
                                                              CIP

Proceedings of the Fifth International Colloquium on Atherosclerosis,
held March 14-16, 1990, in Brussels, Belgium

ISBN-13: 978-1-4684-5906-7   e-ISBN-13: 978-1-4684-5904-3
DOI: 10.1007/ 978-1-4684-5904-3

© 1991 Plenum Press, New York
Softcover reprint of the hardcover 1st edition 1991
A Division of Plenum Publishing Corporation
233 Spring Street, New York, N.Y. 10013

PREFACE

The past two decades have seen steady progress in our understanding of the pathogenesis of atherosclerosis. The role of low density lipoprotein (LDL) increase and of LDL receptor deficiency or malfunctions in familial hypercholesterolemia has been largely enlightened by the works of Brown and Goldstein. These authors postulated also that modification of LDL to a form recognized by the scavenger or acetyl-LDL receptor may be required for lipid loading of macrophage-derived foam cells in the lesions. A growing body of evidence suggests that oxidative modification of LDL could enhance its atherogenicity by its implication as a factor in the generation of foam cells.

Thus, if the role of LDL in the pathogenesis of hypercholesterolemia was well established a great deal of information appears currently on new approaches such as the mechanisms leading to the accumulation of foam cells, the impact of LDL structural alterations, notably oxidation and the role of gene mutations of apolipoprotein B and/or LDL receptor

The opening topic is devoted to these new avenues outlined in the field of hypercholesterolemia. The first part concerns the genetic aspects of atherosclerosis: mainly the genetics of apolipoproteins, their transcriptional regulation, the amino acid mutations of the apo B gene and of the LDL receptor gene, the structural domains and the acylation sites of apoprotein B. The second part of the topic is devoted more directly to cholesterol: the mechanisms regulating its distribution between lipoproteins in relation to the activity of the cholesteryl ester transfer protein, the role of its precursors in the hepatic lipase expression, the LCAT-mediated formation of its esters, the role of HDL receptor, of apolipoproteins A-I, A-II, A-IV and of CETP in its efflux from cells. The relation of structural characteristics (for instance amphipathic helices) and alterations with the function of LDL and HDL is also documented in this section.

A short second topic is dedicated to hypocholesterolemia. Surprisingly, while hypercholesterolemia has been the object of thousands of very competent papers, hypocholesterolemia which may be, when chronic and severe, harmful for the maintenance of normal membrane integrity and cell function and for adequate synthesis of steroid hormones has been often left out. It is worth remembering that a relationship has been long since suspected between hypocholesterolemia and cancer.

The third topic honors Mones Berman's contribution to kinetics. Whereas in vitro studies were often at the origin of basic discoveries, in vivo kinetic experiments made difficult by the very high number of variables appeared essential to be realized for our understanding of metabolism. Most "static" studies demonstrate the changes from normal of many parameters without furnishing any view on their dynamics. In vivo kinetics is also submitted to a number of problems: ethical but essentially methodological. The use of tracers, the validity of their use as "physiological" markers have given rise to much controversy since many years, but despite these snags such studies, difficult of access, are irreplaceable. It is sure that Mones Berman a physicist of the NIH with a particularly high ability to apprehend physiology was a pioneer in the concept of compartmental modeling of physiological data, and of its application for identifying and quantifying metabolic routes in normal state and in a variety of pathologies, and also for assessing the prognosis, the mechanism of action of dietary conditions, of other risk factors and of hypolipidemic drugs. The fact that this approach is surely one of the most tedious has limited the number of teams and of studies devoted to these problems. Numerous examples are given showing the complexity of the subject, aggravated by the recent introduction of the concept of heterogeneity partly related to the new lipoprotein particle's theory.

The potential risk accompanying hypertriglyceridemia has not been until now well defined. Many contradictory studies have not allowed to result in a definite conclusion whether it is an additional or dependent risk factor. The fourth part of this volume deals with this question, particularly the role of genetic variation of the lipoprotein lipase gene, the factors regulating lipoprotein lipase, the effect of lipoprotein lipase and apoprotein C-II deficiency, the role of apoproteins in modulating the metabolism and uptake of triglyceride-rich particles, the relation with LDL composition, the potential role of Lp(a) pathogenicity, the influence of lipoprotein particles' apoprotein composition, the binding and uptake of triglyceride-rich lipoprotein remnants.

Finally the closing topics concerns some new approaches of atherosclerosis: -the cytotoxicity of triglyceride-rich remnants, the role of oxidized LDL and their preferential uptake by macrophages, the effect of certain drugs or vitamins on these modifications; -the relation between plasma factors and structural constituents of the arterial wall as the proteoglycans; -the contribution of immunocompetent cells to the atheromatous lesion and the formation of autoantibodies against endothelial cells, modified LDL, and circulating immune complexes.

If it is certain that atherosclerosis represents one entity the intricacy of the various mechanisms corresponding to many origins, isolated or combined, gives way to many further studies.

It is also highly desirable that these extraordinary strides in fundamental research will find a quick application in the diagnosis, prevention and treatment of atherosclerotic manifestations in routine clinics.

C. L. Malmendier

CONTENTS

CHOLESTEROL METABOLISM

PATHOGENESIS OF ATHEROSCLEROSIS

CHOLESTEROL METABOLISM

TRANSCRIPTIONAL REGULATION OF THE HUMAN APOLIPOPROTEIN GENES

Vassilis I. Zannis, Dimitris Kardassis, Kinya Ogami
Margarita Hadzopoulou-Cladaras, and Christos Cladaras

Section of Molecular Genetics, Cardiovascular Institute
Departments of Medicine and Biochemistry
Boston University Medical Center, Boston, MA 02118

I.   INTRODUCTION

The transcription of eukaryotic genes is controlled by
the interaction of regulatory gene sequences (promoter
elements) with specific nuclear proteins (transcription
factors) (1-3). The interaction of the transcription
factors with the promoter elements controls: a) tissue
specific gene expression (4-6); b) gene expression during
differentiation and development (7,8); and c) gene expression in response to intracellular and extracellular stimuli
such as hormones and metabolites (9-12).

A.   Methodologies Used to Study the Regulation of Gene Transcription

Several experimental advances have facilitated the
study of eukaryotic promoters and has led to the identification and characterization of several eukaryotic transcription factors. These include:

a)   Definition of the promoter region necessary for
gene transcription. For this analysis a promoterless gene,
usually chloramphenicol acetyl transferase (CAT), is placed
under the control of the promoter to be studied in an
expression plasmid (13). Following transfection of cells
with this plasmid, cell extracts are prepared and analyzed
for their ability to convert ^{14}C-chloramphenicol to the
mono- and diacetylated forms.

b)   Identification of the different factors which bind
to a specific promoter region and of their precise binding
domains. For this purpose a variety of techniques are used
including DNase I footprinting, gel electrophoretic mobility
shift assays, methylation interference assays and in vitro
mutagenesis. The DNase I footprinting assays show the
ability of nuclear extracts to protect regions of DNA from
degradation by DNase I (7,14). Degradation is demonstrated
by a series of bands of specific size following gel analysis
and autoradiography. Protection is demonstrated by disappearance of the bands in the presence of nuclear extracts,

Hypercholesterolemia, Hypocholesterolemia, Hypertriglyceridemia
Edited by C.L. Malmendier *et al.*, Plenum Press, New York, 1990

thus giving a clear area on the autoradiogram. The gel electrophoretic mobility shift assays test the binding of specific DNA sequences to factors present in nuclear extracts (15,16), since specific binding of these proteins to the DNA will retard the migration of the labeled DNA in native polyacrylamide gels. DNA methylation interference assays show the effect of partial methylation of critical G residues on the binding of nuclear factors (7). The unmethylated and methylated DNA species can be distinguished following cleavage at the G residues and analysis by a Maxam and Gilbert sequencing method (16,17).

Finally, in vitro mutagenesis of the promoter region can be used to assess the importance of specific residues for transcription in vivo (CAT assays) (13) and in vitro (4) and to assess the ability of the mutated sequence to bind nuclear factors as described above.

Progress towards purification of transcription activation factors has been limited primarily due to the relatively small concentrations in nuclear extracts. However, several laboratories have recently demonstrated the feasibility of isolating such factors from mammalian cells and tissues and the ability of the purified factors to bind DNA and activate transcription in vitro (18-22). A key step in these purifications is DNA specific affinity chromatography with concatamers of the DNA binding site of the factor as ligand. Two main approaches have also been employed recently for the isolation of cDNAs encoding mammalian transcription factors. The first involves screening of cDNA libraries with oligonucleotide probes corresponding to a partial protein sequence of the factor (23,24). The second approach involves screening of expression cDNA libraries with ^{32}P-labeled synthetic double stranded oligonucleotides corresponding to the DNA binding site of the corresponding factor or with appropriate antibodies (25). Isolation and expression of cloned factors provides the biological material required to study specific mechanisms which are responsible for transcriptional activation of eukaryotic genes.

B. Apolipoprotein Synthesis and its Regulation

The sites of apolipoprotein synthesis were originally determined by protein analysis of different tissues and cell lines (26-30). Additional information on the sites of apolipoprotein synthesis was obtained from blotting analysis of RNA isolated from different tissues (31-34). The RNA methodologies also provided an additional tool to study the regulation of apolipoprotein synthesis and the transcriptional regulation of the apolipoprotein genes. These analyses showed tissue and species specific apolipoprotein synthesis and also suggested hormonal and environmental regulation of expression of the apolipoprotein genes (35-38). A summary of the sites of apolipoprotein synthesis and its regulation is provided in Tables 1 and 2, respectively.

TABLE I.   TISSUE DISTRIBUTION OF APOLIPOPROTEIN mRNA*

Apolipoprotein	Major Sites In Humans	Minor Sites In Humans	Minor Sites Other Species
A-I	Liver, Intestine	Gonads, Kidney, Adrenals	Skeletal Muscle
A-II	Liver	Intestine	
A-IV	Intestine	Liver	Spleen, Lung
B-100	Liver, Intestine	Placenta	Kidney
B-48	Intestine	Liver	
CI,CII,CIII	Liver	Intestine	
D	Kidney, Pancreas Adrenal	Spleen, Intestine Liver	
E	Liver, Adrenal, Brain, Kidney	The Majority of Peripheral Tissues	

*The information is based on references 26-34.

TABLE II.   REGULATION OF APOLIPOPRTOEIN SYNTHESIS*

	Tissue Specific	Developmental	Hormonal
A-I	Yes	Yes	Insulin Dexamethazone 17b estradiol
A-IV	Yes	Yes	Insulin Dexamethazone
B	Yes	---	Cholesterol 17b-estradiol
CII	---	---	17b-estradiol
CIII	Yes	Yes	---
E	No	Yes	Cholesterol 25-OH Cholesterol 17b-estradiol

*The information is based on references 35-38.

C.   Progress in Understanding the Regulation of Expression of the Human Apolipoprotein Genes

Recent work by different laboratories has shown that positive as well as negative 5' regulatory elements control expression of human apoAI (39,40), apoB (41,42), apoCIII (43,44) and apoE (45,46). Multiple promoter elements for human apoA-1, apoA-II, apoB, apoCIII and apoE have been identified by us (47-50) and others (45,46,51) and are summarized in Figure 1.

In sections II and III we are reviewing and discussing recent findings in our laboratory on the promoter elements and factors involved in the regulation of human apolipoprotein B (apoB) and apolipoprotein CIII (apoCIII) genes.

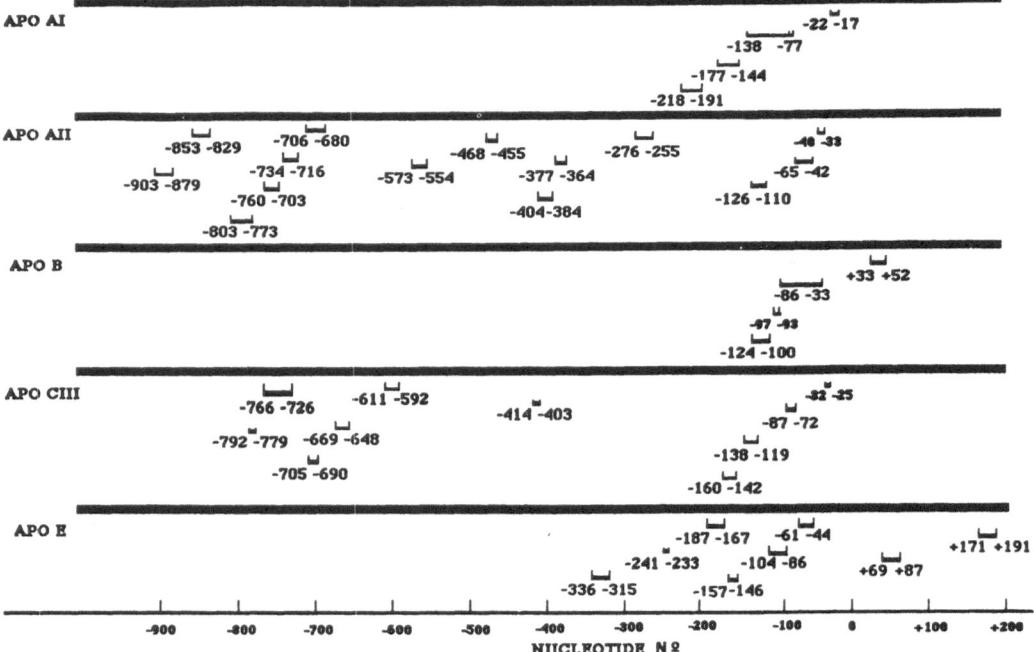

Figure 1. Schematic representation of the promoter elements of human apolipoprotein genes as determined by DNase I footprinting analysis. The information on apoE is based on references 45 and 46. The information on A-I, apoB, apoCIII and apoA-II is based on references 47-51.

D.   ApoB

Human apoB is the ligand for the cellular recognition and catabolism of LDL by the LDL receptor (52,53). The LDL receptor-apoB interaction and subsequent catabolism mediates the clearance of LDL from plasma and regulates cellular cholesterol biosynthesis (52,53). Thus, it is believed that apoB plays a major role in the maintenance of cellular cholesterol homeostasis as well as in the pathogenesis of atherosclerosis. The primary structure of human apoB-100 has been determined by derivation of cDNA (54-57), gene (58,59) and protein sequences (60). The mature protein contains 4,536 amino acids and has a calculated molecular weight of 513,000, implying that there is one apoB molecule per LDL particle (10-13). ApoB is synthesized by the liver, the intestine, and the placenta (Table I) (28,54,61). Numerous population studies indicated that elevated LDL cholesterol or apoB levels are associated with increased risk of coronary artery disease (62-65). In contrast, moderately decreased LDL cholesterol levels (66) and hypo-betalipoproteinemia (67,68) are associated with longevity. These observations indicate the importance of mutations

4

which affect the function of apoB or alter apoB gene expression are very important.

E.   ApoCIII

Human apoCIII is a 79 amino acid protein of known primary structure (69,70) and is a major component of VLDL and a minor component of HDL (71). The gene and cDNA sequences encoding the human apoCIII have been isolated and characterized (72-74). ApoCIII has been implicated in the catabolism of triglyceride rich lipoproteins (75-79) and thus may play some role in the development of hypertriglyceridemia. Furthermore, the apoCIII gene displays a stringent tissue specific expression. In humans and mammalian species apoCIII mRNA and protein have been detected in the liver and to a lesser extent in the intestine (Table I) (31,33,35).

II. PROMOTER ELEMENTS AND FACTORS REGULATING THE EXPRESSION OF THE HUMAN APOB GENE

A.   Effect of Deletion Mutations on the Transcriptional Activity of the ApoB Promoter

To study the promoter elements present in the 5' upstream region of the human apoB gene, we isolated two genomic lambda clones extending approximately 12 and 5 kb upstream from the transcription initiation site. Initially, two apoB promoter fragments, -1800 to +124 and -899 to +124, were cloned into a derivative of the CAT expression vector, placing the expression of the promoterless CAT gene under control of different portions of the apoB promoter (13,47). The sequence of the promoter region extending to nucleotide -899 is shown in Figure 2.

APOB PROMOTER SEQUENCE

```
          PvuII
 -899 CTGGGCAGAGGCAGCGAGCGCTGGCTGAAGTTTCCGGTGGGAAATGGGCAGTGCCTAGAA
                         NalI
 -839 GAGAGGGAAACGATGCATGAGGAGGTTCCAGATGTCTATGAGGAACATGACGTGTCCTGT
 -779 CCACTACTCTGCTTTTCCTCGTCCGCCTCCCCACCACTGGAGGAAACCTAGAAGCTGGTGC
 -718 AGGAAATCCTCCTCTCAACAACCCAAGAACACTTTGCACAAGAGGGGTGCGCCCCTCGGA
                    StuI
 -659 GGTTGCTCTTCCCCAGAGGCCTCTCCTCGCTGGGGTTTCTTGAAGACAGATACTTGGACTCC
 -597 TGCTGGGACCAGGCAGGCCACCCATCCTCAGGGGCAGTGACTGGTCACTCACCAGACCTC
 -537 CCTGCATCCCCCTTCTCTCTTCTCCCCCAGCACGGGCTGAACCCCCGCAGCCACAGATTCTG
 -476 ACCAGGATTAGGGTGTGGGTGCAAATCCAAGGTCCACCAAAATGGAAAAGAAGTAACCGA
 -416 TGGGAACACGTCTCCACCAAGACAGCGCTCAGGACTGGTTCTCCTCGTGGCTCCCAATTCAG
 -354 TCCAGGAGAAGCAGAGATTTTGTCCCCATGGTGGGTCATCTGAAGAAGGCACCCCTGGTCA
                    NaeI
 -293 GGGCAGGCTTCTCAGACCCTGAGGCGCCGGCCATGGCCCCACTGAGACACAGGAAGGGCCGC
 -231 GCCAGAGCACTGAAGACGCTTGGGGAGGGAACCCACCTGGGACCCAGCCCCTGGTGGCTG
                    ApaI
 -170 CGGCTGCATCCCAGGTGGGCCCCCTCCCCGAGGCTCTTCAAGGGCTCAAAGAGAAGCCAGTGT
                    SmaI
 -108 AGAAAAGCAAACAGGTCAGGCCCGGGAGGCGCCCTTTGGACCTTTTGCAATCCTGGCGCTCT
                    NaeI
  -46 TGCAGCCTGGGCTTCCTATAAATGGGGTGCGGGCGCCGGCCGCGCATTCCCACCGGGAACCTGC
 +18 GGGGCTGAGTGCCCTTCTCGGTTGCTGCCGCTGAGGAGCCCGCCCAGCCAGCCAGGGCCGCG
                    PvuII
 +80 AGGCCGAGGCCAGGCCGCAGCCCCAGGAGCCGCCCCACCGCAG
```

Figure 2. Nucleotide sequence of a 5' upstream PvuII to PvuII fragment containing the promoter region of human apoB.

5

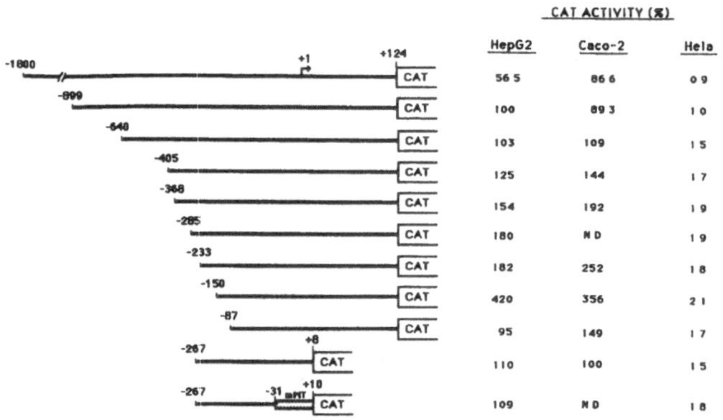

	CAT ACTIVITY (%)		
	HepG2	Caco-2	Hela
-1800 / +1 +124 CAT	56 5	86 6	0 9
-899 CAT	100	89 3	1 0
-640 CAT	103	109	1 5
-405 CAT	125	144	1 7
-368 CAT	154	192	1 9
-285 CAT	180	N D	1 9
-233 CAT	182	252	1 8
-150 CAT	420	356	2 1
-87 CAT	95	149	1 7
-267 +8 CAT	110	100	1 5
-267 -31 +10 mut CAT	109	N D	1 8

Figure 3. A schematic representation of apoB-CAT deletion mutants. The numbers indicate the position of nucleotides with respect to the transcription initiation transcription site. The CAT activities were measured in HepG2, Caco-2, and HeLa cells and were expressed relative to the activity of the apoB-CAT -899 to +124 construct.

Progressive deletions of the 5' upstream apoB region -1800 to +124 in the CAT vector generated the constructs shown in Figure 3. The ability of the truncated promoter regions to promote transcription of the CAT gene in HepG2, Caco-2 and HeLa cells is also shown in Figure 3. This analysis showed that the apoB promoter regions direct the transcription of the CAT gene in HepG2 and Caco-2 but not in HeLa cells. In HepG2 and Caco-2 cells there was an increase in CAT expression as the deletions progress from the 5' to 3' direction. The highest promoter activity is achieved with the deletion of the region spanning NT's -1800 to -150. Further deletions to nucleotide -87 result in decreased promoter activity.

Analysis of the -267/+8 and -267/-31 deletion mutants showed that both resulted in CAT activity similar to the -87 to +124 construct but 1.5-fold lower than that of the -285 to +124 construct. These experiments suggest that the region from nucleotides -150 to -31 contain promoter elements important for the liver and intestine specific expression of the apoB gene. The increase and/or decrease in promoter activity observed as the 5' upstream and/or 3' downstream sequences were deleted may be explained by the existence of other promoter elements that may modulate transcription in a negative or positive way, respectively.

B. Definition of the Proximal ApoB Promoter Elements by DNase I Footprinting

The observed reduction in the transcriptional activity of the CAT gene is pressumed to result from elimination of important promoter elements which constitute the binding sites of transcriptional factors. To investigate the binding domains of the factors which recognize the -150 to -31 promoter region we initially performed DNase I footprinting (7,14). For this analysis, DNA fragments spanning the apoB promoter region from -150 to +8 and -267 to +8 were end labeled and incubated with nuclear extracts from rat

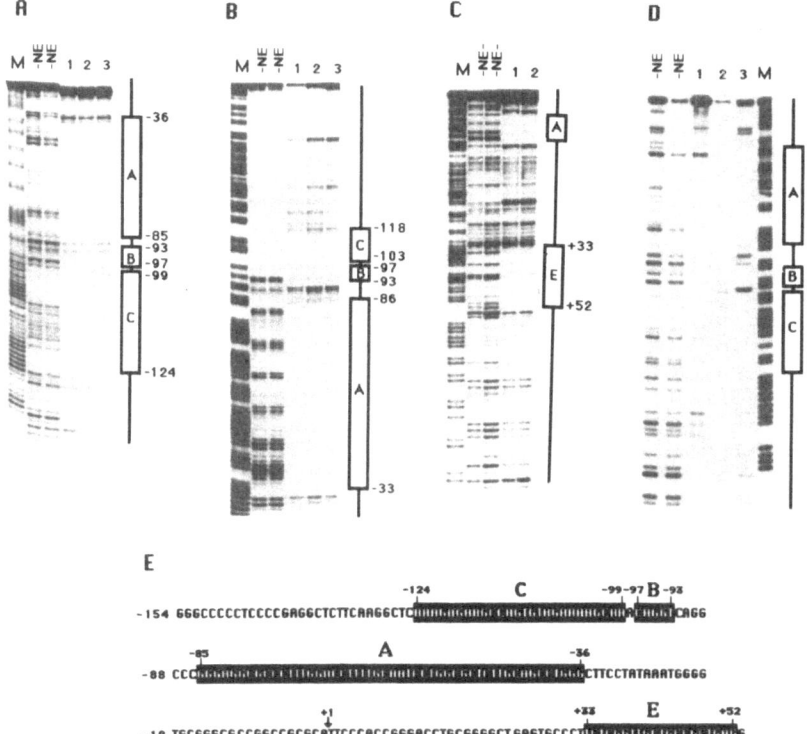

Figure 4. DNase I footprinting analysis of apoB promoter region -150 to +124 with rat liver nuclear extracts. The fragments used were: -150 to -8 labeled at nucleotide -124 (panels A and D), -267 to +8 labeled at nucleotide +8 (panel B), -85 to +124 labeled at nucleotide +124 (panel C). DNA fragments were labeled with gamma ^{32}P-ATP and T4 polynucleotide kinase. The amounts of rat liver nuclear extracts used were: Panel A, lanes 1, 2, 3: 25, 35, and 45 ug, respectively; Panel B, lanes 1, 2, 3: 35, 45 and 55 ug, respectively; Panel C, lanes 1, 2: 45 and 55 ug, respectively; Panel D, lanes 1, 2, 3: 40 ug extracts from HeLa cells, HepG2 cells and rat liver nuclei, respectively.

Lanes -NE represent reactions performed in the absence of nuclear extracts. Lane M represents G+A chemical cleavage of the same DNA fragment used in footprint analysis. Boxes indicate the regions protected from DNase I digestion; numbers refer to the position relative to the transcription initiation site. Panel E. Schematic representation of the apoB promoter areas which are protected from DNase I digestion by nuclear extracts. The protected areas are enclosed in boxes.

liver, human HepG2 and HeLa cells. The analysis with
extracts from rat liver (Figure 4 A-C) or HepG2 cells
(Figure 4D) showed three protected areas designated footprints A, B and C spanning nucleotides -33 to -86, -93 to
-97 and -100 to -124, respectively (Figure 3A-C and E). In
contrast, analysis with HeLa extracts provided only one
small protected area within footprint A between nucleotides
-74 to -53 (Figure 4D). This analysis indicates that a
different set of nuclear factors exists in HepG2 cells which
express apoB as compared to HeLa cells which do not.

C.   In Vitro Mutagenesis of the ApoB Promoter Elements Defined by Footprinting Analysis

    The importance of the proximal nucleotide -150 to -31
region in promoting apoB transcription was analyzed in
detail with in vitro mutagenesis. Eight sets of nucleotide
substitutions were introduced with cassettes of synthetic
oligonucleotides (47). The location of these changes and
their effect in promoting CAT transcription in HepG2 and
Caco-2 cells is shown in Figure 5. Mutations in the area
-105 to -91 of the apoB promoter cause a five fold reduction
and mutations in the region -77 to -68 abolish the transcription of the CAT gene. Mutations in the region -61 to
-60 and -57 to -53 are also associated with 8 and 2-fold
reduction, respectively, in CAT transcription both in HepG2
and Caco-2 cells. This analysis indicated the presence of
at least three elements within the apoB promoter region -119
to -41 important for the transcriptional regulation of the
apoB gene in both hepatic and intestinal cells.

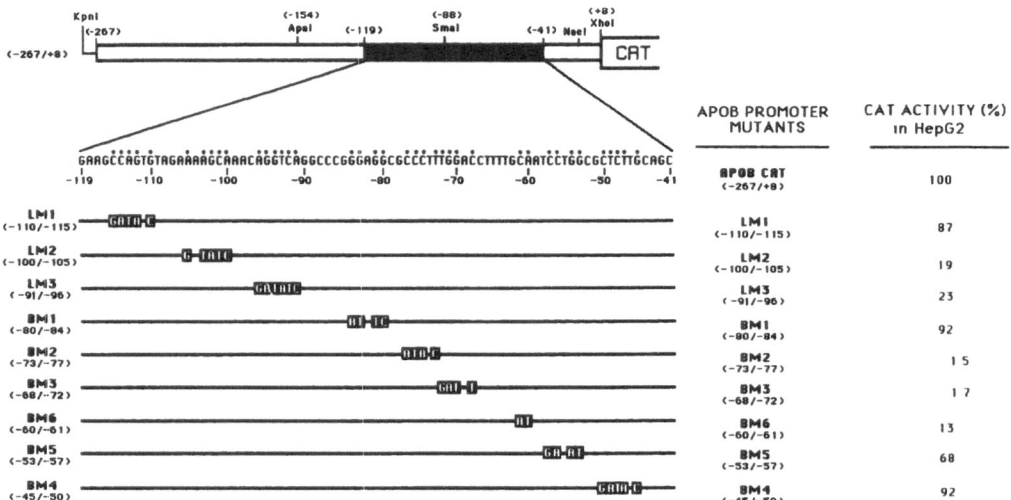

Figure 5. Structure and expression of the apoB-CAT nucleotide substitution mutants. The relationship of the substitution mutations to the -267/+8 apoB-CAT construct is
indicated. Dots indicate nucleotides that have been altered. The altered nucleotide sequence on each mutant is
enclosed in dark boxes. The CAT activities were measured in
HepG2 and Caco-2 cells and are normalized relative to the
activity of the apoB-CAT -267/+8 expression plasmid.

Another element may exist in the region -53 to -41 which has a much lower activation potential. We did not observe an increase over control transcription in any of the substitution mutations. This indicates the absence of a negative cis-acting element operating in this region as was previously suggested (42). The nucleotide substitution mutagenesis of Figure 5 is compatible with the footprinting analysis of Figure 4A-E. Thus, mutations of nucleotides -105 to -100 and -96 to -91 which reduce transcription five-fold are located within footprints C and B, respectively. Similarly, mutations in the regions -77 to -53 which abolish or reduce transcription 2- to 8-fold are within footprint area A. DNase I footprinting analysis of the fragment -85 to +124 identified one protected area located in the coding region (footprint E, nucleotides +52 to +33). This region may account for the 1.7 fold increase in the promoter activity observed with the -285 to +124 as compared with the -285 to +8 CAT deletion mutants of Figure 3. The deletion and oligonucleotide substitution analysis of the promoter region combined with the footprinting patterns indicated the presence of tissue specific nuclear regulatory factors which determine the transcription of the apoB gene in hepatic and intestinal cells. Furthermore, the observed footprinting pattern suggested a complex organization of the regulatory sequences in the apoB promoter region.

D.   Binding of Hepatic Nuclear Factors to the Regulatory Elements Defined by Footprint A

The binding of hepatic nuclear factors to the putative promoter elements was further tested with DNA binding gel electrophoretic assays (15.16) using double stranded synthetic oligonucleotides corresponding to the footprint area A or segments of it. Utilization of synthetic oligonucleotides corresponding to footprint areas eliminates artifacts arising from non specific protein binding to DNA sequences. The oligonucleotides used were designated BA (-88 to -36), BA1 (-88 to -61), BA2 (-61 to -36) and BA3 (-78 to -48). Incubation of ^{32}P-labeled oligonucleotide BA with rat liver nuclear extracts showed the presence of at least 9 DNA-protein complexes which could be competed out with a 100 or 200 fold excess of unlabeled oligonucleotides (Figure 6A). Similar analysis with oligonucleotide BA3 gave five DNA-protein complexes designated a to e. Finally, analysis with oligonucleotide BA2 gave fewer slow migrating complexes and analysis with oligonucleotide BA1 gave one complex with intermediate electrophoretic mobility (Figure 6B). A diagram of the oligonucleotides used is shown in Figure 6C.

E.   A Single Nuclear Factor Binds to the Regulatory Region -78 to -68 of ApoB

The factor which binds to the oligonucleotide BA1 was further tested for binding to oligonucleotides having similar alterations in the -88 to -62 region. Figure 6 D and F show that substitution of residues -77 to -75 (BM2) or -71, -70 and -68 (BM3) abolished binding of the factor to the mutated synthetic oligonucleotide. In contrast, substitutions of oligonucleotides -69 and -67 (DA2) partially affected binding and substitution of residues -79 and -77

(DA1) had no effect on the binding of the factor. The BM2 and BM3 mutations which completely abolish the transcriptional activity of the apoB promoter in hepatic and intestinal cells (Figure 5) also abolish binding of the hepatic nuclear factor (compare Figure 5 with 6D). We conclude that this factor is essential and acts positively in the transcription of the apoB gene.

The DNA binding domain of the transcription factor was further defined by methylation interference analysis (7). The methylation interference analysis of the sequence -88 to -61 showed that methylation of the G residues at positions -78, -71 and -70 of the coding and -76, -75 and -68 of the noncoding strand, respectively, interferes with the binding of the hepatic nuclear factor (Figure 7). The combined data of Figures 5, 6D and 7 suggest that the binding site of this factor resides within the sequences defined by residues -79 to -63.

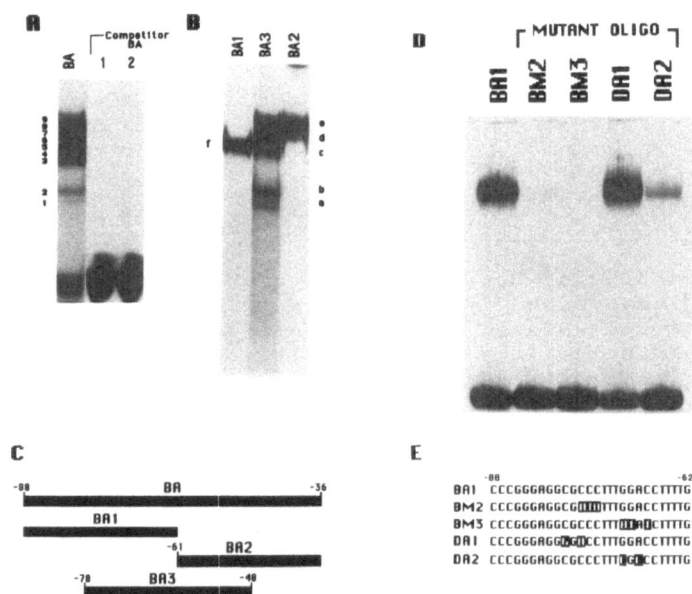

Figure 6. DNA binding gel eletrophoretic assays of radiolabeled double stranded oligonucleotides with rat liver nuclear extracts.
Panel A. Lanes designated BA1 and BA2 represent assays of oligonucleotide BA competed by 100- (lane 1) and 200- (lane 2) fold excess of unlabeled oligonucleotide BA, respectively.
Panel B. BA1, BA2, and BA3 represent assays with the corresponding oligonucleotide fragments. Specific DNA-protein complexes are represented by numbers and letters.
Panel C. Schematic representation of oligonucleotides BA, BA1, BA2 and BA3 used in panels A and B. Panel D. Assays of radiolabeled BA1 (-88 to -61) and mutant oligonucleotides. Panel E. Oligonucleotides used for binding in panel D. The mutations are enclosed in dark boxes.

The sequence -79 to -63 contains the motif CGCCCTT-
TGGACCTTT, which has sequence homology with the regulatory
sequences of human alpha1-antitrypsin (-124 to -104), human
apoAI (-212 to -192) and human apoA-II (-736 to -715)
(80-82). DNA binding competition experiments using synthetic oligonucleotides as competitors showed that the human
apoAI sequence (-212 to -192) competes as effectively as the
homologous apoB sequence whereas the human apoA-II (-736 to
-716) and alpha1-antitrypsin (-124 to -104) compete less
effectively. The observation that mutations in the binding
site of this factor affect similarly the intestinal and
hepatic transcription of apoB gene suggests that this factor
may be present both in the liver and the intestine. The
factor binding to the regulatory region -79 to -63 of apoB
has been purified to apparent homogeneity by a combination
of ion exchange and affinity chromatography. The purified
factor acts as a positive regulator of transcription in
vitro transcription systems and binds to the regulatory
regions of apoA-I (-212 to -191), apoAII (-740 to -719) and
apoCIII (-87 to -63).

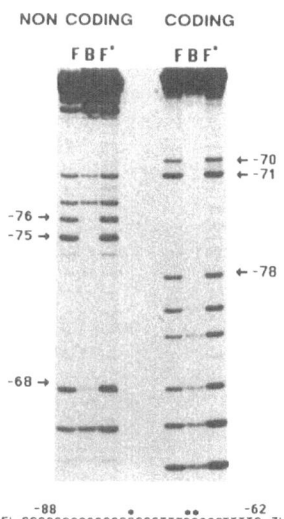

Figure 7. Methylation interference pattern of the DNA-
protein complex obtained with the apoB region -88 to -62 and
liver nuclear extracts. The non coding strand and coding
strand are indicated; F: free DNA before binding; F*: free
DNA recovered after binding; B: bound DNA recovered from the
DNA-protein complex. The arrows (and asterisks of the
bottom panel) indicate the position of guanine residues at
which methylation interferes with the binding. The numbers
indicate the position of nucleotides with respect to the
transcription initiation site.

III. PROMOTER ELEMENTS AND FACTORS REGULATING THE EXPRESSION
OF THE HUMAN APOCIII GENE

A.   Identification of the Promoter Elements Required for
Hepatic Transcription of the Human ApoCIII Gene by Deletion
and DNase I Footprinting Analysis

The promoter elements required for hepatic and intestinal transcription of the human apoCIII were localized and
their importance for transcription were elucidated with
approaches similar to those employed for the apoB promoter.
Sequential 5' deletions of the apoB promoter region (-1411
to +24) localized the elements required for hepatic and
intestinal transcription downstream of nucleotide -871
(Figure 8).   Deletion of the region -871 to -755 reduced
transcription by approximately 50% whereas deletion of the
region -871 to -686 abolished transcription (Figure 8A).
Sequential deletions between nucleotides -686 to -99 gave a
residual promoter activity in HepG2 cells and were not
informative. To assess the importance of putative regulatory
elements within the -686 to -99 region we initially generated internal deletions within this region and assessed
their effect on hepatic and intestinal transcription (Figure
8B).   The CAT vectors used for construction of the sequential and internal deletions are shown in Figure 8C.   The
internal deletion analysis showed that deletions of the -686
to -553 region approximately doubled intestinal transcription 2-fold and decreased hepatic transcription 89%.   In
addition, internal deletions in the region -408 to -163
decreased hepatic transcription 50-75% without affecting the
intestinal transcription.   The exact location of the regulatory elements required for hepatic transcription were
further identified by DNase I footprint analysis of the
region -890 to +24 with rat liver nuclear extracts.   This
analysis identified 10 protected regions as follows: A -32
to -25, B -87 to -72, C -138 to -119, D -160 to -142, E -414
to -403, F -611 to -592, G -669 to -648, H -705 to -690, I
-766 to -726 and J -792 to -779 (Figure 9).   The footprinting analysis gave a rational explanation of the reduction of the CAT activity in response to the various mutations.   Thus, the deletion of the sequence -871 to -755
eliminates the footprint J (-792 to -779) and a portion of
the footprint I (-766 to -726).   Further deletion of the
region -755 to -686 eliminates the footprints I and H (-705
to -690).   Regulatory factors which bind to the footprint
areas J, I and H are necessary but not sufficient for
hepatic transcription.   Thus elimination of the footprint
area G (-669 to -648) and F (-611 to -592) reduced by 9-fold
the hepatic transcription (compare Figures 8B and 9).
Disruption of the region -413 to -408 with a synthetic
heptanucleotide reduced hepatic transcription approximately
50%.   This reduction is associated with disruption of the
footprint area E (-414 to -403).   Further sequential deletions extending from -413 to -163 had no addition effect on
hepatic transcription suggesting the absence of additional
regulatory elements in this region.

B.   Hepatic and Intestinal Transcription are Controlled by
Different Sets of Regulatory Sequences

Similar to hepatic transcription, intestinal transcrip-

12

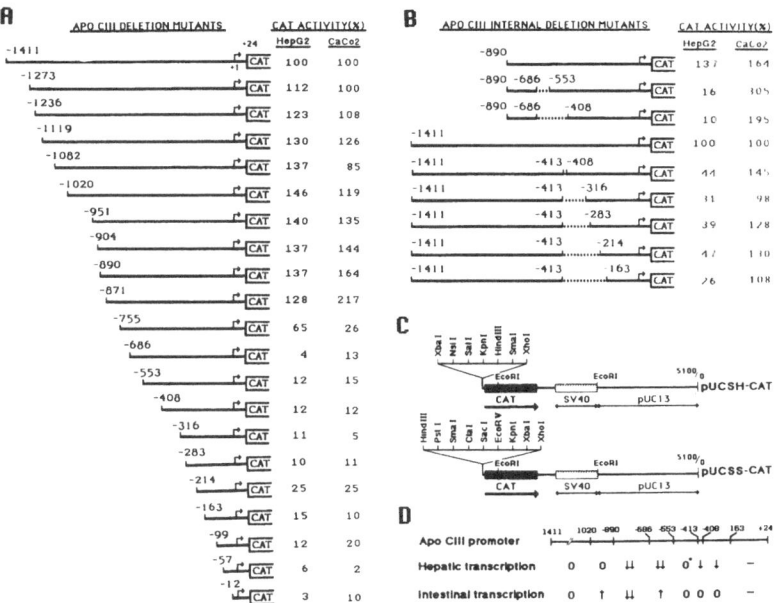

Figure 8. Structure and expression of apoCIII-CAT constructs. Panel A shows a schematic representation of apoCIII-CAT deletion mutants. The right panel shows the effect of deletion of the apoCIII promoter region in the transcription of the promoterless CAT gene in HepG2 and Caco-2 cells. Panel B: The left side shows internal deletion mutants and the right side CAT activity in HepG2 and Caco-2 cells. Activities are expressed relative to those achieved with the -1411 to +24 construct. Panel C: Schematic representation of the pUCSH-CAT and pUCSS-CAT vectors and their corresponding polylinker regions. Panel D: Schematic representation of the promoter regions of human apoCIII and their importance for intestinal and hepatic transcription. Symbols indicate increases or decreases in transcription associated with deletion of the region as follows: ↑↑ or ↓↓ increases or decreases, respectively 9-fold. ↑ or ↓ increases or decreases, respectively, 2- to 3-fold. 0 changes in transcription of 0.75- to 1.5-fold. The symbol * indicates that the importance of this region in gene transcription is inferred from the footprint analysis. The symbol - indicates that this region is important in apoCIII transcription but its precise role has not been determined.

tion as assessed by CAT assays in Caco-2 cells decreases 17-fold after deletion of the region -871 to -686 (Figure 8A) suggesting that this region may contain positive regulatory elements. However, the regions upstream of nucleotide -871 and downstream of nucleotide -686 affect hepatic and intestinal transcription differently. Thus, deletion of the region -1020 to -871 has no effect on hepatic transcription while it approximately doubles intestinal transcription (Figure 8A). More importantly, as mentioned, internal deletions of the region -686 to -553 decreased the hepatic transcription by 89%, and approximately doubled intestinal transcription (Figure 8B). The disruption of the region -413 to -408 and the internal deletions in the region -408 to -163, which reduced hepatic transcription 50-75%, had only marginal effects on intestinal transcription. Finally, the findings are consistent with the hypothesis that the region -890 to -686 is recognized by tissue specific nuclear factors which promote both intestinal and hepatic transcription. In contrast, the region -686 to -553 is recognized by nuclear factors which promote only hepatic transcription and may represent a strict liver specific regulatory region. The effect of sequential and internal deletions on intestinal and hepatic transcription is shown in Figure 8D.

The precise role of the other regulatory regions for intestinal and hepatic transcription cannot be assessed with this level of analysis. The small increases (2-fold) in intestinal transcription by deletion of the regions -1020 to -871 and -686 to -553 may indicate binding of intestinal factors which inhibit transcription to these regions. Alternatively, it may merely reflect position effects, i.e., stronger interactions between the factors binding to the distal and proximal promoter regions. Liver specific regulatory elements (4-7,83-88) and transcriptional factors (6,84,87-89) have been identified in several laboratories. However, information on regulatory elements and factors involved in intestinal transcription is lacking.

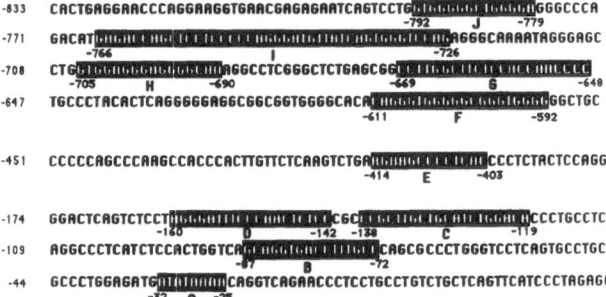

SUMMARY OF FOOTPRINTS OF APO CIII PROMOTER REGION

Figure 9. Schematic representation of apoCIII promoter regions A through J which are protected from degradation by DNase I in the presence of rat liver nuclear extracts.

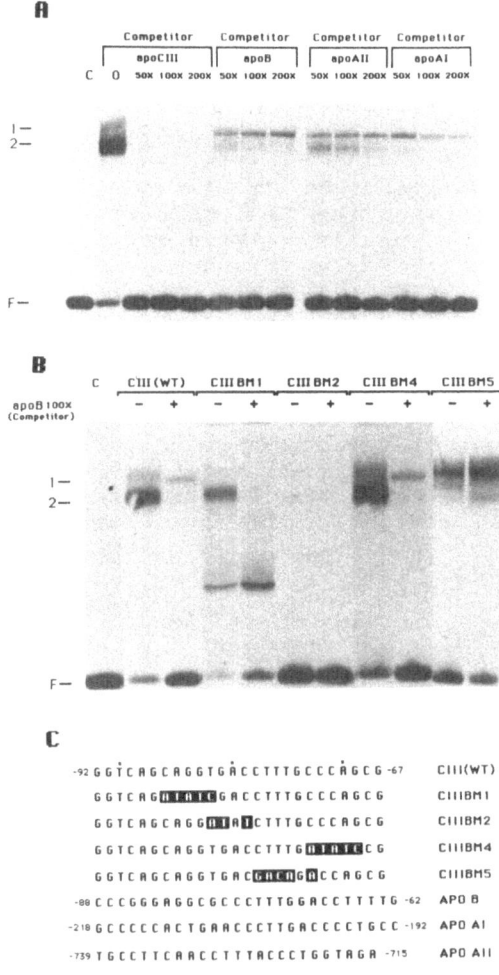

Figure 10. Competition gel electrophoresis mobility shift
assays of the normal and mutated apoCIII region −92 to −67
and competition experiments with regulatory regions of other
apolipoprotein genes. The double stranded apoCIII oligonucleotides were labeled with [alpha ^{32}P] dGTP and [alpha ^{32}P]
dCTP and Klenow polymerase. Panel A: Normal −92 to −67
sequence competed synthetic oligonucleotides corresponding
to regulatory elements of apoAI (−218 to −192), apoAII (−739
to −715), and apoB (−88 to −62). The competitor oligonucleotides were used at 50, 100 and 200 fold excess with respect
to the the apoCIII oligonucleotide.

Panel B shows gel electrophoresis mobility shift assays
of the mutated oligonucleotides CIIIBM1 to CIIIBM5 in the
presence or absence of 100-fold excess of apoB sequence −88
to −62 as indicated on the figure. In both panels lane C
contains ^{32}P-labeled double stranded apoCIII oligonucleotide
−92 to −67. Bands 1 and 2 represent the DNA-protein complexes involving factors CIIIB1 and CIIIB2, respectively,
and band F represents the free oligonucleotide. The two
bands in lanes 4 and 5 of panel B represent a consistent
artifact.

Panel C. Sequences of normal and mutated regulatory
elements used in Panels A and B.

C.   The Importance of the Proximal ApoCIII Promoter Region.
Two Mutually Exclusive FActors Bind to the -86 to -74
ApoCIII Promoter Region and Modulate Transcription

As discussed, footprinting analysis of the proximal region identified four additional protected areas in the regions -160 to -142, -138 to -119, -87 to -72 and -32 to -25 designated footprints D, C, B, and A, respectively.   The last footprint contains the TATA box (Figure 9).   The importance of the regulatory region defined by footprint B and the factors which recognize this regulatory element were analyzed in detail.   This included general DNA binding and competition experiments as well as correlation of the effects of specific nucleotide substitutions within this region on CAT transcription with the binding of the factors to the mutated sequences.   Competition gel electrophoretic mobility shift assays indicated that this region (-86 to -74) is recognized by two distinct factors, CIIIB2 and CIIIB1 (Figure 10A).   The binding of factor CIIIB2 is competed out by synthetic oligonucleotides corresponding to the regulatory elements of other liver specific genes

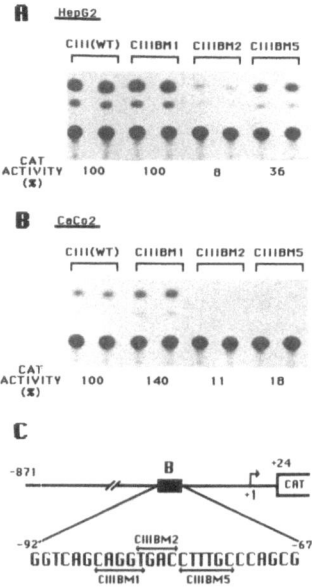

Figure 11. Effect of nucleotide substitution mutation of the apoCIII promoter on the transcription of the CAT gene. Panels A and B: Transcription in HepG2 and Caco-2 cells. respectively.   The designations CIIIBM1 to CIIIBM5 are explained in Figure 10C.   Panel C: Schematic representation of the location of the nucleotide substitution mutants CIIIBM1 to CIIIBM5 in the pUCSH-CAT plasmid.   The mutated sequences are shown in Figure 10C.

(35,51,81,82), whereas the latter may represent a new transcription factor. In addition, oligonucleotide substitution mutagenesis showed that mutations in the region -86 to -82 did not affect transcription whereas mutation in the regions -82 to -79 and -78 to -73 reduced hepatic transcription 12- and 2.5-fold and the intestinal transcription 9- and 5-fold, respectively (Figure 10 B and C). The effect of nucleotide substitution on hepatic and intestinal transcription is shown in Figure 11 A-C. Comparison of results of Figures 10B and 11 A and B indicates that the -82 to -79 mutation which reduced hepatic transcription 92% abolishes the binding of both factors. In contrast, the mutation in the region -86 to -82 and -78 to -73 which are associated with normal and 40% reduced hepatic transcription abolish the binding of factors CIIIB1 and CIIIB2, respectively. The findings define the binding sites of factor CIIIB1 and CIIIB2 between residues -86 to -79 and -82 to -73, respectively, and indicate that the binding of the two factors is mutually exclusive. Methylation interference analysis indicated that residues -86 and -79 of the noncoding strand and residues -81, -83, and -84 of the coding strand are involved in the binding of factor CIIIB2 (Figure 12) and confirm that the binding domain of this factor is within the region -86 to -79.

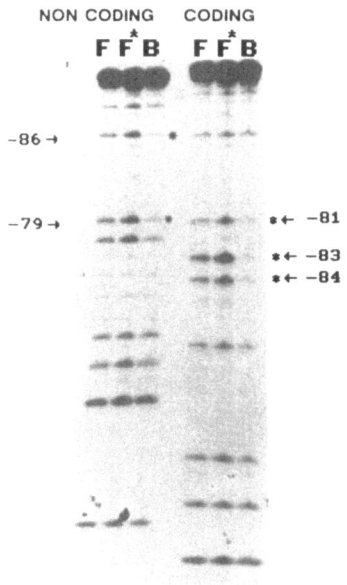

Figure 12. Methylation interference pattern of the DNA-protein complex obtained with the apoCIII region -92 to -67 and rat liver nuclear extract in the presence of 100-fold excess of apoB nucleotide -88 to -62. The non-coding strand and coding strands are indicated. F: free DNA before binding; F*: free DNA recovered after binding; B: bound DNA recovered from the DNA-protein complex. The arrows and asterisks indicate the position of guanine residues involved in binding. The numbers indicate the position of nucleotides with respect to the transcription initiation site.

Factor CIIIB1 which binds in the -86 to -79 region has been purified by a combination of ion exchange and affinity chromatography (93). In addition to apoCIII region -87 to -72, the purified factor recognized several regulatory elements of the human apoAII gene (49).

It has been noted previously that sequence -149 to -160 encompassing footprint regions C and D is complementary to the beta interferon sequence -65 to -54 (90) and therefore may play a negative role in apoCIII gene regulation. Furthermore, it was reported (44) that the region -86 to -74 of apoCIII has sequence homology with apolipoprotein (6,39, 47,48,51,81,91) and other liver specific genes (92), and that it represents the binding site of a positive hepatic regulatory factor, designated AP-1 (44). Figure 11 shows that this region is occupied by two (rather than one) mutually exclusive factors. Furthermore, the present data do not allow us to determine whether the footprint areas C and D act as positive or negative regulatory elements for hepatic transcription. Such information can be obtained by functional analysis following mutagenesis of each of the footprint areas. Our data are consistent with the hypothesis that hepatic transcription of the human apoCIII gene is modulated by the interaction of transcriptional factors which bind in the distal regions defined by footprints J through E with other factors which bind in the proximal (footprints A through D) regions. The identity of these factors and their precise role in the transcriptional activation of the apoCIII gene will require extensive new studies.

SUMMARY

A. ApoB Gene Regulation

1.   The regulatory elements of Human apoB are local-ized in four domains between nucleotides -124 to +52.

2.   Several transcription factors bind in these domains and regulate hepatic and intestinal transcription (Figure 13A).

3.   A single factor binds to the apoB region -78 to -68 and positively regulates hepatic transcription.

B. ApoCIII Gene Regulation

1.   The regulatory elements of human apoCIII are localized in ten domains between nucleotides -792 and -25.
2.   The intestinal and hepatic transcription is controlled by both common as well as unique regulatory elements.
3.   Elements which affect hepatic and intestinal transcription differently are localized in the -686/-553 region.
4.   The region -86 to -76 is recognized by two mutu-ally exclusive factors with overlapping domains designated CIIIB1 and CIIIB2. Binding of CIIIB2 is associated with normal transcription whereas binding of CIIIB1 is associated with reduced transcription. The transcription factors which recognize the proximal apoCIII promoter region are illus-trated in Figure 13B.

C. Underline Conclusions

The regulation of transcription of eukaryotic genes represents a complex biological process which involves several nuclear proteins.

The challenges of future research are:

1. To characterize (by protein and DNA techniques) the regulatory proteins involved in different tissues.

2. To determine the mechanism by which the interaction of a set of tissue specific regulatory proteins with a specific promoter enhances or suppresses the transcription of the corresponding gene.

A **PROXIMAL APOB PROMOTER ELEMENTS**

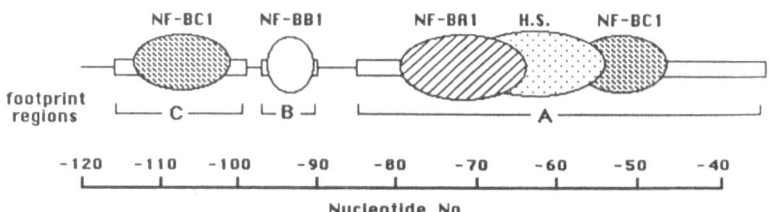

B **PROXIMAL APOCIII PROMOTER ELEMENTS**

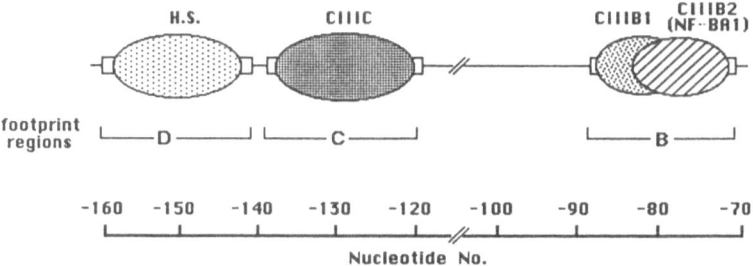

Figure 13. Panels A and B. Schematic representation of the proximal promoter elements of the human apoB (Panel A) and apoCIII (Panel B) genes. This tentative diagram is based on the information presented in this manuscript and references 47, 48, 82, and 93. Factors derive their names from the promoter region they protect. For instance, BA1 designates factor 1, which binds to the the footprint region A1 of the apoB gene. H.S. indicates heat stable factor(s).

ACKNOWLEDGEMENTS

This work was supported by a grants from the National Institutes of Health (HL33952 and HL43909), the Charles H. Hood Foundation, and the March of Dimes Birth Defects Foundation (1-1141). Kinya Ogami was supported by Kirin Brewery Co. Ltd. Dr. Christos Cladaras is an Established Investigator of the American Heart Association. Dr. Margarita Hadzopoulou-Cladaras is supported by an NIH training grant. The research was performed in part at the Housman Medical Research Center of Boston University Medical Center. We would like to thank Elizabeth Walsh, Gayle Forbes, Anne Minnich, and Philippe Cardot for their expert assistance.

REFERENCE

1. Maniatis, T., S. Goodbourn, J.A. Fischer. 1987. Science 236:1237-1245.
2. Briggs, M.R., Kadonaga, J.T., Bell, S.P., and Tjian, R. (1986) Science 234, 47-52.
3. Jones, N.C., Rigby, P.W.J., and Ziff, E.B. (1988) Genes & Development 2, 267-281.
4. Gorski, K., Carneiro, M., and Schibler, U (1986) Cell 27, 767-776.
5. Cereghini, S., Raymondjean, M., Carranca, A.J., Herbomel, P., and Yaniv, M. (1987) Cell 50, 627-638.
6. Monaci, P., Nicosia, A., and Cortese, R. (1988) EMBO J. 7, 2075-2087.
7. Treisman, R. (1986) Cell 46, 567-574.
8. Atchison, M.L., and Perry, R.P. (1987) Cell 48, 121-128.
9. Lenardo, M., Pierce, J.W., and Baltimore, D. (1987) Science 236, 1573-1577.
10. Montminy, M.R., and Bilezikjian, L.M. (1987) Nature 328, 175-178.
11. Hurst, H.C., and Jones, N.C. (1987) Genes Dev. 1, 1132.
12. Sudhof, T.C., Russell, D.W., Brown, M.S., and Goldstein, J.L. (1987) Cell 48, 1061-1069.
13. Gorman, C.M., Moffat, L.F., Howard, B.H. (1982) Mol. Cell. Biol. 2, 1044-151.
14. Staudt, L.M., Singh, H., Sen, R., Sirth, T., Sharp, P.A., and Baltimore, D. (1986) Nature 323, 640-643.
15. Fried, M., and Crothers, D.M. (1981) Nucl. Acids Res. 9, 6505-6525.
16. Strauss, F., and Varshavsky, A. (1984) Cell 37, 889-901.
17. Maxam, A.M., and Gilbert, W. (1977) Proc. Natl. Acad. Sci. USA 74, 560-564.
18. Kadonaga, J.T., and Tjian, R. (1986) Proc. Natl. Acad. Sci. USA 83, 5889-5893.
19. Briggs, M.R., Kadonaga, J.T., Bell, S.P., and Tjian, R. (1986) Science 234, 47.
20. Jones, K.A., Kadonaga, J.T., Rosenfeld, P.J., Kelly, T.J., and Tjian, R.T. (1987) Cell 48, 79-89.
21. Rosenfeld, P.J., Kelly, T.J. (1986) J. Biol. Chem. 261, 1398-1408.
22. Lee, W., Mitchell, P., and Tjian, R. (1987) Cell 49, 741-752.
23. Kadonaga, J.T., Carner, K.R., Masiarz, S.R., and Tjain, R. (1987) Cell 51, 1079-1909.

24. Santoro, C., Mermod, N., Andrews, P.C., and Tjian, R. (1988) Nature 334, 218-224.
25. Singh, H., LeBowitz, J.H., Baldwin, A.S., Jr., and Sharp, P.A. (1988) Cell 52, 415-423.
26. Windmueller, H.G., Herbert, P.N., and Levy, R.I. (1973) J. Lipid Res. 14, 215-223.
27. Felker, T.E., Fainaru, M., Hamilton, R.L., and Havel, R.J. (1977) J. Lipid Res. 18, 465-473.
28. Wu, A.L., and Windmueller, H.G. (1979) J. Biol. Chem. 254, 7316-7322.
29. Zannis, V.I., Breslow, J.L., SanGiacomo, T.R., Aden, D.P., and Knowles, B.B. (1981) Biochemistry 20, 7089-7096.
30. Zannis, V.I., Kurnit, D., and Breslow, J.L. (1982) J. Biol. Chem. 257, 536-544.
31. Zannis, V.I., Cole, S.F., Jackson, C., Kurnit, D.M., and Karathanasis, S.K. (1985) Biochemistry 24, 4450-4455.
32. Newman, T.C., Dawson, P.A., Rudel, L.L., and Williams, D.L. (1985) J. Biol. Chem. 260, 2452-2457.
33. Lenich, C., Brecher, P., Makrides, S., Chobanian, A., and Zannis, V.I. (1988) J. Lipid Res. 29, 755-764.
34. Thomas, M.S., Prack, M.M., Dashti, N., Johnson, F., Rudel, L.L., and Williams, D.L. (1988) J. Biol. Chem. 263, 5183-5189.
35. Haddad, I.A., Ordovas, J.M., Fitzpatrick, T., and Karathanasis, S.K. (1986) J. Biol. Chem. 261, 13268-13277.
36. Elshourbagy, N.A., Boguski, M.S., Liao,, W.S.L., Jefferson, L.S., Gordon, J.I., and Taylor, J.M. (1985) Proc. Natl. Acad. Sci. USA 82, 8242-8246.
37. Tam, S.P., Archer, T.K., and Deeley, R.G. (1986) Proc. Natl. Acad. Sci. USA 83, 3111-3115.
38. Sorci-Thomas, M., Prack, M.M., Dashti, N., Johnson, P., Rudel, L.L., and Williams, D.L. (1989) J. Lipid Res. 30, 1397-1403.
39. Sastry, K.N., Seedorf, Udo, Karathanasis, S.K. (1988) Mol. Cell. Bio.. 8, 605-614.
40. Higuchi, K., Law, S.W., Hoeg, J.M., Schumacher, U.K., Meglin, N., and Brewer, H.B., Jr. (1988) 263, 18530-18536.
41. Das, H.K., T. Leff, and J.L. Breslow. 1988. J. Biol. Chem. 263:11452-11458.
42. Carlsson, P., and Bjursell, G. (1989) Gene 77, 113-121.
43. Reue, K., Leff, T., and Breslow, J.L. (1988) J. Biol. Chem. 263, 6857-6864.
44. Leff, T., Reue, K., Melian, A., Culver, H., and Breslow, J.L. (1989) J. Biol. Chem. 264, 16132-16137.
45. Smith, J.D., Melian, A., Leff, T., and Breslow, J.L. (1988) J. Biol. Chem. 263, 8300-8308.
46. Paik, Y.K., Chang, D.J., Reardon, C.A., Walker, M.D., Taxman, E., and Taylor, J.M. (1988) J. Biol. Chem. 263, 13340-13349.
47. Kardassis, D., Hadzopoulou-Cladaras, M., Ramji, D.P., Cortese, R., Zannis, V.I., and Cladaras, C. (1990) Mol. Cell. Biol. 10, 2653-2659.
48. Ogami, K., Hadzopoulou-Cladaras, M., Cladaras, C., and Zannis, V.I. (1990) J. Biol. Chem. In Press.
49. Chambaz, J., Cardot, P., Pastier, D., Roghani, A., Zannis, V., and Cladaras, C. (1990) Arteriosclerosis. In Press.

50. Papazafiri, P., Ramji, D., Ogami, K., Cladaras, C., Cortese, R., and Zannis, V.I. (1990) Arteriosclerosis. In Press.
51. Lucero, M.A., Sanchez, D., Ochoa, A.R., Brunel, F., Cohen, G.N., Baralle, F.E., and Zakin, M.M. (1989) Nucleic Acids Res. 17, 2283-2300.
52. Goldstein, J.L., and Brown, M.S. (1977) Ann. Rev. Biochem. 46:897-930.
53. Goldstein, J.L., and Brown, M.S. (1982) In: The Metabolic Basis of Inherited Disease, 5th ed. (J.B. Stanbury, J.B. Wyngaarden, D.S. Fredrickson, J.L. Goldstein, and M.S. Brown, eds.), pp. 672-712, McGraw-Hill, New York.
54. Cladaras, C., Hadzopoulou-Cladaras, M., Nolte, R.T., Atkinson, D., and Zannis, V.I. (1986) The EMBO J. 5:3495-3507.
55. Law, S.W., Grant, S.M., Higuchi, K., Hospattankar, A., Lackner, K., Lee, N., and Brewer, H.B., Jr. (1986) Proc. Natl. Acad. Sci. USA 83, 8142-8146.
56. Yang, C.Y., Chen, S.H., Gianturco, S.H., Bradley, W.A., Sparrow, J.T., Tanimura, M., Li, W.L., Sparrow, D.A., DeLoof, H., Rosseneu, M., Lee, F.S., Gu, Z.W., Gotto, A.M., Jr., and Chan, L. (1986) Nature 323, 738-742.
57. Knott, T.J., Pease, R.J., Powell, L.M., Wallis, S.C., Rall, S.C., Jr., Innerarity, T.L.. Blackhart, B., Taylor, W.H., Marcel, Y., Milne, R., Johnson, D., Fuller, M., Lusis, A.J., McCarthy, B.J., Mahley, R.W., Levy-Wilson, B., and Scott, J. (1986) Nature 323, 734-739.
58. Blackhart, B.D., Ludwig, E.M., Pierotti, V.R., Caiati, L., Onasch, M.A., Wallis, S.C., Powell, L., Pease, R., Knott, T.J., Chu, M.L., Mahley, R.W., Scott, J., McCarthy, B.J., and Levy-Wilson, B. (1986) J. Biol. Chem. 261, 15364-15367.
59. Ludwig, E.H., Blackhart, B.D., Pierotti, V.R., Caiata, L., Fortier, C., Knott, T., Scott, J., Mahley, R.W., Levy-Wilson, B., and McCarthy, B.J. (1987) DNA 6, 363-372.
60. Yang, C.Y., Gu, Z.W., Weng, S.Q., Li, W.H., Gotto, A.M., Jr., and Chan, L. (1989) Arteriosclerosis 9, 96-108.
61. Demmer, L.A., Levin, M.S., Elovson, J., Reuben, M.A., Lusis. A.J., and Gordon, J.I. (1986) Proc. Natl. Acad. Sci. USA 83, 8102-8106.
62. Kannel, W.B., Castelli, W.P., and Gordon, T. (1979) Ann. Intern. Med. 90, 85-91.
63. Heiss, G., and Tyroler, H. (1982) Proceedings of the Workshop on Apolipoprotein Quantification (U.S. Dept. of Health and Human Services, National Institutes of Health, Bethesda, MD), NIH Publ. No. 83-1266, 7-24.
64. Brunzell, J.D., Sniderman, A.D., Albers, J.J., and Kwiterovich, P.O., Jr. (1984) Arteriosclerosis 4:79-83.
65. Sniderman, A.. Shapiro, S., Marpole, D., Skinner, B., Teng, B., and Kwiterovich, P.O., Jr. (1980) Proc. Natl. Acad. Sci. USA 77, 601-603.
66. Siervogel, R.M., Morrison, J.A., Kelly, K., Meliles, M., Gartside, P., and Glueck, C. (1980) J. Clin. Genet. 17, 13.
67. Steinberg, D., Grundy, S.M., Mok, H.Y.I., Turner, J.D., Weinstein, J.J., Brown, W.V., and Albers, J.J. (1979) J. Clin. Invest. 64, 292-301.

68. Young, S.G., Peralta, F.P., Dubois, B.W., Curtiss, L.K., Boyles, J.K., and Witztum, J.L. (1987) J. Biol. Chem. 262, 16604-16611.
69. Brewer, H.B., Jr., Shulman, R., Herbert, P., Ronan, R., and Wehrly, K. (1974) J. Biol. Chem. 249, 4975-4984.
70. Shulman, R. S., Herbert, P. N., Fredrickson, D. S., Wehrly, K., and Brewer, H. B., Jr. (1974) J. Biol. Chem. 249, 4969-4974.
71. Herbert, P. N., Assmann, G., Gotto, A. M., Jr., and Fredrickson, D. S. (1982) In: The Metabolic Basis of Inherited Diseases (Stanbury, J.B., Wyngaarden, J.B., Fredrickson, D. S., Goldstein, J.L., and Brown, M.S., eds.), pp. 589-651, McGraw-Hill, New York.
72. Protter, A. A., Levy-Wilson, B., Miller, J., Bencen, G., White, T., and Seilhamer, J. (1984) DNA 3, 449-456.
73. Sharpe, C.R., Sidoli, A., Shelley, C.S., Lucero, M.A., Shoulders, C.C., and Baralle, F.E. (1984) Nucleic Acids Res. 12, 3917-3932.
74. Karathanasis, S.K., Zannis, V.I., and Breslow, J.L. (1985) J. Lipid Res. 26, 451-456.
75. Brown, W.V., and Baginsky, M.L. (1972) Biochem. Biophys. Res. Commun. 46, 375-382.
76. Krauss, R.M., Herbert, P.M., Levy, R.I., and Fredrickson, D. S. (1973) Circ. Res. 33, 403-411.
77. Windler, E., Chao, Y., and Havel, R.J. (1980) J. Biol. Chem. 255, 5475-5480.
78. Shelburne, F., Hanks, J., Meyers, W., and Quarfordt, S. (1980) J. Clin. Invest. 64, 652-658.
79. Quarfordt, S.H., Michalopoulos, G., and Schirmer, B. (1982) J. Biol. Chem. 257, 14642-14647.
80. DeSimone, V., Ciliberto, G., HArdon, E.M., Paonessa, G., Palla, F., Lundberg, L., and Cortese, R. (1987) EMBO J. 6, 2759-2766.
81. Makrides, S.C., Ruiz-Opazo, N., Hayden, M., Nussbaum, A.L., Breslow, J.L., and Zannis, V.I. (1988) Eur. J. Biochem. 173, 465-471.
82. Kardassis, D., Zannis, V., and Cladaras, C. (1990) Arteriosclerosis. In Press.
83. Parker, C.S., and Topol, J. (1984) Cell 37, 273-283.
84. Hardon, E.M., Frain, M., Paonessa, G., and Cortese, C. (1988) EMBO J. 7:1711-1719.
85. Babiss, L.E., Herbst, R.S., Bennett, A.L., and Darnell, J.E. (1987) Gene Dev. 1, 256-267.
86. Courtois, G., Morgan, J.G., Campbell, L.A., Fourel, G., and Crabtree, G.R. (1987) Science 238, 688-692.
87. Courtois, G., Baumhueter, S., and Crabtree, G.R. (1988) Proc. Natl. Acad. Sci. USA 85, 7937-7941.
88. Lichtsteiner, S., and Schibler, U. (1989) Cell 57, 1179-1187.
89. Frain, M., Swart, G., Monaci, P., Nicosia, A., Stampfli, S., Frank, R., and Cortese, R. (1989) Cell 59, 145-157.
90. Goodburn, S., Burstein, H., and Maniatis, T. (1986) Cell 45, 601-610.
91. Elshourbagy, N.A., Walker, D.W., Paik, Y.K., Boguski, M.S., Freeman, M., Gordon, J.L., and Taylor, J.M. (1987) J. Biol. Chem. 262, 7963-7981.
92. Ciliberto, G., Dente, L., and Cortese, R. (1985) Cell 41, 531-540.
93. Ogami, K., Kardassis, D., Cladaras, C., and Zannis, V.I. (1990) Arteriosclerosis. In Press.

MUTATIONS AND VARIANTS OF APOLIPOPROTEIN B

THAT AFFECT PLASMA CHOLESTEROL LEVELS

Thomas L. Innerarity and Kristina Boström

Gladstone Foundation Laboratories for Cardiovascular Disease
Cardiovascular Research Institute, University of California, San Francisco
San Francisco, California 94140-0608

ABSTRACT

Apolipoprotein (apo-) B100 is the exclusive apolipoprotein of low density lipoproteins (LDL), which transport most of the plasma cholesterol in humans. Mutations in apo-B100 can cause either hypocholesterolemia or hypercholesterolemia. Familial hypobetalipoproteinemia, which leads to hypocholesterolemia, has been shown to be caused by defects in the apo-B gene that terminate translation prematurely and result in the production of truncated proteins. The mutations responsible for the hypocholesterolemia have been either single nucleotide substitutions or deletions. Familial defective apo-B100, which leads to hypercholesterolemia, is caused by a point mutation in the receptor-binding domain of apo-B100. The mutation disrupts the binding of LDL to the LDL receptor, thereby disrupting LDL receptor-mediated catabolism and resulting in hypercholesterolemia. A variant form of apo-B, apo-B48, is also critical for lipoprotein metabolism. Apolipoprotein B48 is obligatory for the secretion of chylomicrons. It is formed from an RNA-edited apo-B mRNA in which codon 2153 has been converted from a CAA (glutamine) codon to a premature UAA (stop) codon. The first cytosine in this codon is deaminated to form uracil. The minimum nucleotide recognition sequence for the editing mechanism has been reported to be between 26 and more than 63 nucleotides surrounding codon 2153. The apo-B mRNA editing mechanism, which appears to be a cytosine deaminase, and its regulation are being actively investigated.

INTRODUCTION

In the United States and other Western societies, heart disease is the leading cause of death. Numerous epidemiological studies have demonstrated that increased levels of plasma cholesterol, particularly low density lipoprotein (LDL) cholesterol, are associated with atherosclerosis (1, 2). One of the main areas of research on atherosclerosis is the identification of genetic factors in humans that increase the probability of developing premature atherosclerosis. Certainly, one of the dominant genes codes for the LDL

Hypercholesterolemia, Hypocholesterolemia, Hypertriglyceridemia
Edited by C.L. Malmendier *et al.*, Plenum Press, New York, 1990

25

receptor. Individuals who are homozygous for familial hypercholesterolemia possess few if any functional LDL receptors, have a six- to eightfold elevation of LDL cholesterol, and usually die of heart disease before the age of 20 (3). Three other genes that have been shown to be important for lipoprotein metabolism are the apo-B, the apo-E, and the lipoprotein(a) genes (4-6).

This brief review will focus on the importance of the apo-B gene in controlling plasma cholesterol levels and on the role of mutations in apo-B in causing either hypocholesterolemia or hypercholesterolemia. Apolipoprotein B is an essential part of a number of plasma lipoproteins: LDL, intermediate density lipoproteins (IDL), very low density lipoproteins (VLDL), chylomicrons, and chylomicron remnants. About two-thirds of plasma cholesterol in humans is transported in plasma LDL, high levels of which are positively correlated with coronary heart disease. In humans, apo-B100 is synthesized by the liver and is required for the assembly and secretion of VLDL. Very low density lipoproteins are converted by lipoprotein lipase to IDL and then to LDL. The LDL are cleared from the circulation by the LDL receptor, with apo-B100 serving as the ligand.

Apolipoprotein B is also of interest because of a unique mRNA editing process that enables the apo-B gene to make two structurally related but different-size apo-B proteins that have different functions: apo-B100 (4536 amino acids) and apo-B48 (2152 amino acids). In humans, apo-B48 is synthesized in the intestine. It plays a role analogous to that of apo-B100 in the liver, in that it is necessary for the assembly and secretion of another major class of lipoproteins, chylomicrons. Chylomicrons transport dietary triglycerides absorbed from the intestinal lumen and are converted to chylomicron remnants by lipoprotein lipase. The remnants are rapidly cleared from the circulation by the liver, with apo-E serving as the ligand. Because apo-B48 contains only the amino-terminal 2152 amino acids of apo-B100, it lacks the carboxy-terminal receptor-binding domain and therefore does not bind to LDL receptors. In addition, apo-B48-containing lipoproteins are not converted to LDL but are completely cleared from the plasma. Plasma from fasted subjects contains very little apo-B48 (for a review of apo-B, see Refs. 7 and 8).

STRUCTURE OF APOLIPOPROTEIN B

A number of laboratories determined the structure of apo-B100 by cloning and sequencing its cDNA. Human apo-B mRNA is 14.5 kilobases (kb) in length and codes for a mature protein consisting of 4536 amino acids that have a molecular weight of about 550,000 (about 10% of which is accounted for by carbohydrates) (7) (Fig. 1) The apo-B gene is 43 kb in length, contains 28 introns and 29 exons, and is located in the short arm of chromosome 2. The distribution of the introns within the gene is unusual in that 24 of the 28 introns occur in the 5'-terminal one-third of the gene. Over half of the protein is encoded by the extremely long exon 26, whose 7572 base pairs make it one of the largest exons yet reported in the human genome (13).

APOLIPOPROTEIN B MUTATIONS

The two genetic abnormalities attributed to mutations in the apo-B gene are familial hypobetalipoproteinemia, which is associated with low plasma cholesterol levels, and familial defective apo-B100 (FDB), which is characterized by high plasma cholesterol levels.

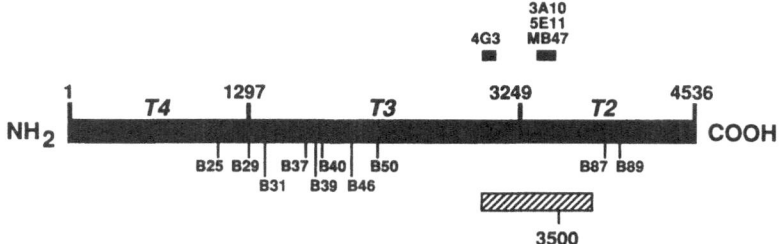

Fig. 1. Schematic structure of human apo-B100. Thrombin cleaves apo-B100 at residues 1297 and 3249, resulting in thrombolytic fragments T4, T3, and T2. Monoclonal antibodies 4G3, 3A10, 5E11, and MB47 have been mapped to the regions shown (9, 10). Each of these antibodies completely inhibits the binding of LDL to the LDL receptor. 3500 denotes the point mutation (CGG → CAG, which causes an Arg → Gln substitution) that disrupts the binding of LDL to the LDL receptor. The cross-hatched rectangle designates a best estimate of the receptor-binding region based on the evidence from the natural mutation that disrupts receptor binding (11) and from the monoclonal antibody studies (9, 10). The B25 through B89 notations denote the points of truncation of apo-B species associated with familial hypobetalipoproteinemia. (Reproduced, with permission, from Innerarity (12).)

Work from several laboratories has demonstrated that hypobetalipoproteinemia can be caused by mutations in the coding region of the apo-B gene. As a consequence, affected subjects who are heterozygous for this gene have only 25% to 50% of the normal levels of VLDL and LDL, and subjects with the homozygous phenotype have few if any apo-B100-containing lipoproteins in their plasma (7, 8, 13). Investigations using either apo-B gene-associated DNA polymorphisms or protein polymorphisms in association with antibody polymorphisms demonstrated that the disorder is linked to the apo-B gene (14, 15). More recently, a large number of mutations in the apo-B gene that cause a premature termination of translation have been identified (for a review, see Refs. 8 and 13). Examination of apo-B-containing lipoproteins from these subjects revealed truncated forms of apo-B100. From the analysis of several of these mutations, it appears that a minimum length of apo-B is necessary before any truncated apo-B-containing lipoproteins are detected in the plasma. Apparently, full-length apo-B is required for normal VLDL secretion (16).

The second lipoprotein disorder due to mutations of apo-B is FDB. This was first detected by an *in vitro* assay that measured the ability of LDL from hypercholesterolemic subjects to compete with normal ^{125}I-labeled LDL for binding to the LDL receptor. The LDL from subjects with this disorder had about 32% of normal receptor-binding activity (17). To identify the mutation responsible for the defect in apo-B100's ability to bind to its receptor, the mutant apo-B100 allele of the original proband was sequenced between nucleotides 7500 and 11916. This region, which codes for amino acids 2488 to 3901, includes the receptor-binding domain. Only one mutation was found in this domain: CGG was changed to CAG in codon 3500, causing a glutamine-for-arginine substitution at this site (11).

Because a single copy of apo-B100 is present on each LDL particle, LDL from heterozygotes with FDB are a mixture of normal and defective-binding LDL. The defective-binding LDL have very little receptor-binding activity. Thus, a single amino acid substitution at residue 3500 virtually abolishes the receptor binding of these LDL. Interestingly, this mutation is at or very near the epitope of MB47, a monoclonal antibody that effectively inhibits the receptor binding of LDL. MB47 binds to the LDL from various mammals, suggesting that it binds to an evolutionarily conserved sequence of apo-B100 (18). In addition, MB47 binds to the LDL from subjects with FDB with a higher affinity than to LDL from normal individuals (19).

The main clinical consequence of FDB is hypercholesterolemia. To determine the impact of this mutation on plasma cholesterol levels, we screened 1100 subjects for familial defective apo-B100 and uncovered 11 probands with this disorder. Family studies identified another 30 individuals, for a total of 41 heterozygotes for this mutation (20). In our study, FDB heterozygotes had an average plasma cholesterol level 81 mg/dl higher than age- and sex-matched controls. In two other studies the investigators found even higher levels of plasma LDL. Tybjaerg-Hansen et al. (21) found that 10 FDB heterozygotes had plasma cholesterol levels 163 mg/dl higher than the 50th percentile of the Lipid Research Clinic's age- and sex-matched controls. Schuster et al. (22) identified 18 subjects with this disorder and found they had levels of total cholesterol 134 mg/dl higher than the control subjects. No other known genetic mutation, with the exception of mutations in the LDL receptor gene, causes such a large increase in LDL plasma cholesterol (20).

APOLIPOPROTEIN B48

A unique physiological process produces a variant form of apo-B known as apo-B48. The complete amino acid sequence of apo-B48 is identical to the 2152 amino-terminal amino acids of apo-B100, and the protein possesses approximately 48% (240 kDa) of the molecular mass of apo-B100 (23-26). Both apo-B100 and apo-B48 are produced from the same gene; the latter is produced by an mRNA editing process that converts nucleotide 6666, a cytosine, to a uracil. The editing changes codon 2153 from a CAA (glutamine) codon to a premature UAA (stop) codon, which terminates the translation of apo-B48 mRNA (23, 25, 26).

Thus far, this apo-B mRNA editing has been found in intestine from the human, rabbit, and rat (23, 27). It has also been found in the rat liver (27) and in the cell lines CaCo-2 (human intestinal adenocarcinoma) (28), BNL CL.2 (mouse liver) (29), and McArdle 7777 (rat hepatoma) (30), all of which secrete apo-B48. Thus, the apo-B mRNA editing appears to be tissue-specific. However, recently we have found evidence for apo-B48 editing activity in a number of tissue-culture cell lines that do not synthesize apo-B (29).

The target for the apo-B mRNA editing, cytosine 6666, is part of a 26-nucleotide sequence that is completely conserved across several species: mice, rats, rabbits, and humans. This conserved sequence is believed either to be a part of or to constitute the recognition sequence for the editing machinery. At this point, the exact recognition sequence is not known. Using a series of deletion mutants around cytosine 6666 that were transfected into McArdle 7777 cells, Davies et al. found that as little as the 26 conserved nucleotides sufficed for efficient editing (30). However, when Driscoll et al. used an in vitro system prepared from McArdle 7777 cells, synthetic RNA of 55, 483, and 2383 nucleotides, but not 26 nucleotides, was edited (31).

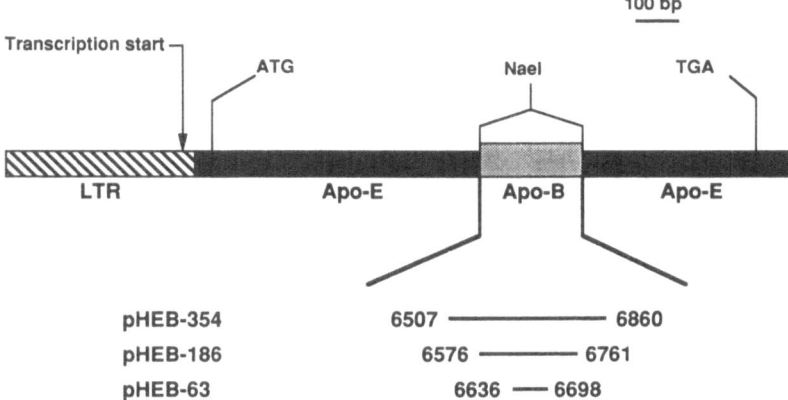

Fig. 2A. Schematic representation of expression vectors for chimeric apo-EB proteins. Apolipoprotein B nucleotide sequences of 63, 186, and 354 bp, all centered around the target base for the mRNA editing (cytosine 6666), were inserted into an apo-E expression vector. The resulting pHEB vectors were transfected into CaCo-2 cells, a human intestinal adenocarcinoma cell line that secretes apo-B48. LTR, Moloney murine leukemia virus long terminal repeat. ATG, translational start site; TGA, translational stop site.

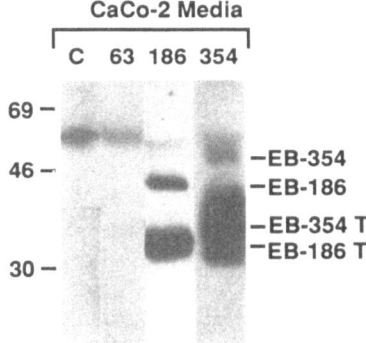

Fig. 2B. Immunoblot of chimeric apo-EB proteins secreted from CaCo-2 cells transfected with pHEB vectors. From CaCo-2 cells transfected with pHEB-354 or pHEB-186, both full-length (EB-354 and EB-186) and truncated proteins (EB-354T and EB-186T) were secreted and then detected on immunoblots with antipeptide 2140-2151. This apo-B antibody has a very high affinity for the carboxy terminus of apo-B48 and a very low affinity for apo-B100. The full-length EB-63 is not detected by antipeptide 2140-2151, but is readily detected with anti-apo-E antibodies (data not shown). No truncated EB-63 could be detected. This suggests that only the EB-354 mRNA and the EB-186 mRNA are recognized by the mRNA editing mechanism in CaCo-2 cells.

We inserted apo-B sequences of 354, 186, and 63 nucleotides into an apo-E expression vector and transfected the resulting vectors into CaCo-2, BNL CL.2, and McArdle 7777 cells. The chimeric apo-EB mRNA containing 354 and 186 nucleotides of apo-B mRNA was edited as efficiently as the endogenous apo-B mRNA in all three cell lines. By mutation analysis, Chen *et al.* (32) showed that the recognition mechanism might be somewhat tolerant: 20 mutations immediately flanking the cytosine 6666 had only a marginal effect on editing, whereas two other mutations abolished editing.

The apo-B48 mRNA editing seems to be regulated both on a developmental level and on a hormonal level. For example, during fetal life there is a progressive conversion in the intestine from the exclusive synthesis of apo-B100 to the exclusive synthesis of apo-B48 (33). Furthermore, Davidson *et al.* treated rats with thyroxine and found that the liver of the hyperthyroid rats produced only apo-B48 (34). The regulatory mechanisms for these developmental and hormonal changes are not known.

REFERENCES

1. Grundy, S.M., Greenland, P., Herd, A., Huebsch, J.A., Jones, R.J., Mitchell, J.H., and Schlant, R.C. (1987) Circulation 75: 1340A-1362A
2. Davignon, J., and Little, J.A. (1988) Can. J. Cardiol. 4, Suppl. A: 1A-38A
3. Goldstein, J.L., and Brown, M.S. (1989) In: *The Metabolic Basis of Inherited Disease,* 6th edition (Scriver, C.R., Beaudet, A.L., Sly, W.S., and Valle, D., eds.). McGraw-Hill, New York, pp. 1215-1250
4. Lusis, A.J. (1988) J. Lipid Res. 29: 397-429
5. Breslow, J.L. (1988) Physiol. Rev. 68: 85-132
6. Utermann, G. (1989) Science 246: 904-910
7. Kane, J.P., and Havel, R.J. (1989) In: *The Metabolic Basis of Inherited Disease,* 6th edition (Scriver, C.R., Beaudet, A.L., Sly, W.S., and Valle, D., eds.). McGraw-Hill, New York, pp. 1139-1164
8. Young, S.G. (1990) Circulation. In press.
9. Milne, R., Théolis Jr., R., Maurice, R., Pease, R.J., Weech, P.K., Rassart, E., Fruchart, J.-C., Scott, J., and Marcel, Y.L. (1989) J. Biol. Chem. 264: 19754-19760
10. Pease, R.J., Milne, R.W., Jessup, W.K., Law, A., Provost, P., Furchant, J.-C., Dean, R.T., Marcel, Y.L., and Scott, J. (1990) J. Biol. Chem. 265: 553-568.
11. Soria, L.F., Ludwig, E.H., Clarke, H.R.G., Vega, G.L., Grundy, S.M., and McCarthy, B.J. (1989) Proc. Natl. Acad. Sci. USA 86: 587-591
12. Innerarity, T.L. (1990) Curr. Opinion Lipidol. 1: 104-109
13. Blackhart, B.D., Ludwig, E.H., Pierotti, V.R., Caiati, L., Onasch, M.A., Wallis, S.C., Powell, L., Pease, R., Knott, T.J., Chu, M.-L., Mahley, R.W., Scott, J., McCarthy, B.J., and Levy-Wilson, B. (1986) J. Biol. Chem. 261: 15364-15367
14. Leppert, M., Breslow, J.L., Wu, L., Hasstedt, S., O'Connell, P., Lathrop, M., Williams, R.R., White, R., and Lalouel, J.-M. (1988) J. Clin. Invest. 82: 847-851
15. Young, S.G., Bertics, S.J., Curtiss, L.K., Dubois, B.W., and Witztum, J.L. (1987) J. Clin. Invest. 79: 1842-1851
16. Young, S.G., Hubl, S.T., Smith, R.S., Snyder, S.M., and Terdiman, J.F. (1990) J. Clin. Invest. 85: 933-942
17. Innerarity, T.L., Weisgraber, K.H., Arnold, K.S., Mahley, R.W., Krauss, R.M., Vega, G.L., and Grundy, S.M. (1987) Proc. Natl. Acad. Sci. USA 84: 6919-6923
18. Young, S.G., Witztum, J.L., Casal, D.C., Curtiss, L.K., and Bernstein, S. (1986) Arteriosclerosis 6: 178-188
19. Weisgraber, K.H., Innerarity, T.L., Newhouse, Y.M., Young, S.G., Arnold, K.S.,

Krauss, R.M., Vega, G.L., Grundy, S.M., and Mahley, R.W. (1988) Proc. Natl. Acad. Sci. USA 85: 9758-9762

20. Innerarity, T.L., Mahley, R.W., Weisgraber, K.H., Bersot, T.P., Krauss, R.M., Vega, G.L., Grundy, S.M., Friedl, W., Davignon, J., and McCarthy, B.J. (1990) J. Lipid Res. In press.

21. Tybjaerg-Hansen, A., Gallagher, J., Vincent, J., Houlston, R., Talmud, P., Dunning, A.M., Seed, M., Hamsten, A., Humphries, S.E., and Myant, N.B. (1990) Atherosclerosis 80: 235-242.

22. Schuster, H., Rauh, G., Kormann, B., Hepp, T., Humphries, S., Keller, C., Wolfram, G., and Zöllner, N. (1990) Arteriosclerosis. In press.

23. Powell, L.M., Wallis, S.C., Pease, R.J., Edwards, Y.H., Knott, T.J., and Scott, J. (1987) Cell 50: 831-840

24. Kane, J.P. (1983) Annu. Rev. Physiol. 45: 637-650

25. Chen, S.-H., Habib, G., Yang, C.-Y., Gu, Z.-W., Lee, B.R., Weng, S.-A., Silberman, S.R., Cai, S.-J., Deslypere, J.P., Rosseneu, M., Gotto, A.M., Jr., Li, W.-H., and Chan, L. (1987) Science 238: 363-366

26. Hospattankar, A.V., Higuchi, K., Law, S.W., Meglin, N., and Brewer, H.B., Jr. (1987) Biochem. Biophys. Res. Comm. 148: 279-285

27. Tennyson, G.E., Sabatos, C.A., Higuchi, K., Meglin, N., and Brewer, H.B., Jr. (1989) Proc. Natl. Acad. Sci. USA 86: 500-504

28. Boström, K., Lauer, S.J., Poksay, K.S., Garcia, Z., Taylor, J.M., and Innerarity, T.L. (1989) J. Biol. Chem. 264: 15701-15708

29. Boström, K. Unpublished observation.

30. Davies, M.S., Wallis, S.C., Driscoll, D.M., Wynne, J.K., Williams, G.W., Powell, L.M., and Scott, J. (1989) J. Biol. Chem. 264: 13395-13398

31. Driscoll, D.M., Wynne, J.K., Wallis, S.C., and Scott, J. (1989) Cell 58: 519-525

32. Chen, S.-H., Xiaoxia, L., Warren, S.L.L., June, H.W., and Chan, L. (1990) J. Biol. Chem. 265: 6811-6816

33. Glickman, R.M., Rogers, M., and Glickman, J.N. (1986) Proc. Natl. Acad. Sci. USA 83: 5296-5300

34. Davidson, N.O., Powell, L.M., Wallis, S.C., and Scott, J. (1988) J. Biol. Chem. 263: 13482-13485

MOLECULAR GENETICS OF FAMILIAL HYPERCHOLESTEROLEMIA

Katriina Aalto-Setälä and Kimmo Kontula

Institute of Biotechnology, University of Helsinki

Valimotie 7, SF-00380 Helsinki, Finland

INTRODUCTION

More than fifty years years ago first reports of an inherited disease with cholesterol clustering in tendons and the presence of coronary heart disease at an early age were reported[1]. This disease, called familial hypercholesterolemia (FH), has been found to be a single-gene, autosomally dominantly inherited disease[2]. The cause of FH has been demonstrated to be a defect in the amount or functioning of low density lipoprotein (LDL) receptors[2]. The function of these receptors is to carry cholesterol-rich particles from the bloodstream into hepatocytes and peripheral cells for synthesis of cell membranes and steroid hormones[3].

FH is characterized by a gene dosage effect[2]: those who are carrying one mutant and one normal LDL receptor allele have the milder (heterozygous) form of the disease, whereas those with two mutant LDL receptor alleles (the homozygotes) have a more severe form of the disease with the onset of cardiovascular symptoms already in the childhood. Serum LDL cholesterol concentration is elevated from birth on. Compared to the normal level, it is 2-3 times higher in heterozygotes and 5-8 times higher in homozygotes. Clinical manifestations of the disease include cholesterol accumulation in extensor tendons and arterial wall and, most importantly, a greatly increased risk of myocardial infarction. Typically, heterozygotes get their first myocardial infarctions at the age of 40-50 years and homozygotes during their two first decades of life.

At the cellular level FH was found to be a heterogeneous disease and the existence of four different types of defective LDL receptor alleles were demonstrated: null alleles, transport-deficient allelles, binding-deficient aleles and internalization-defective alleles[2, 3]. This favored the idea that multiple types of DNA alterations would ultimately be detected to explain the heterogeneity of FH.

STRUCTURE OF THE NORMAL LDL RECEPTOR AND ITS GENE

The LDL receptor gene is located on chromosome 19 and comprises about 45 000 base pairs[4]. The gene contains 18 exons, 13 of which share homology to other known genes. Molecular cloning of cDNA for the LDL receptor revealed that the LDL receptor mRNA is about 5 300 nucleotides in length and the LDL receptor contains five functional domains[5]. The first domain is negatively charged with a number of loop-forming cysteine residues. The function of this domain is to bind the positively charged ligand (apolipoprotein B of the LDL particle). The second domain, homologous to epidermal growth factor precursor, has been found to be

Hypercholesterolemia, Hypocholesterolemia, Hypertriglyceridemia
Edited by C.L. Malmendier *et al.*, Plenum Press, New York, 1990

essential for the normal recycling of the receptors and for the normal dissociation of the receptor from its ligand at the acidic pH in an endosome[6]. The third domain contains multiple carbohydrate chains and is proposed to increase the stability of LDL receptors[7]. The fourth domain contains 22 hydrophobic amino acids and spans the cell membrane anchoring the receptor to the cell. The fifth domain, the cytoplasmic part of the receptor, is needed for the clustering of the LDL receptors into coated pits along the cell membrane and for the normal internalization of the receptors into cells.

MOLECULAR BASIS OF FH

A vast number of mutations of the LDL receptor gene, ranging from single base changes to large deletions or insertions, have been reported[2, 3]. Only 2-6% of these mutations have been found to involve major gene rearrangements that are readily detectable by routine Southern blot analysis[8, 9]. Some of the mutations have been found to result in the production of mRNA molecules with an altered molecular size[10]. The LDL receptor has been shown to contain a large number of DNA polymorphisms. These apparently harmless alterations in functional terms provide a valuable tool to examine the inheritance of FH in affected families[11].

ENRICHMENT OF SPECIFIC LDL RECEPTOR GENE MUTATIONS IN GENETICALLY HOMOGENEOUS POPULATIONS

In most populations, the frequency of the heterozygous form of FH is approximately 1 in 500 whereas only 1 in 1 million suffer from the homozygous form of the disease[2]. However, there are several examples of populations in which FH has been enriched by an apparent founder gene effect, including the Lebanese, French Canadians and South Africans. In each case, one or two mutations alone seem to explain the increased prevalence of the disease. A fourth example of populations with a characteristic LDL receptor mutation is Finland.

The Finnish population forms a genetic isolate with a curious panel of inherited diseases and a virtual lack of genetic diseases common in other parts of the world, such as cystic fibrosis and phenylketonuria[12]. When DNA samples from Finnish patients with FH were examined, about 40% of them were found to carry the same type of LDL receptor mutation[13, 14]. Subsequent to the cloning of a portion of the mutant receptor gene and the corresponding cDNA[14, 15], the mutation was defined as a 9.5 kb deletion in the 3' coding region of the LDL receptor gene. The deletion involves exons 16 and 17 and a part of exon 18 (Fig. 1). These exons normally encode the carboxy terminus, i.e. the transmembrane and the cytoplasmic domains of the receptor. The carboxyterminal part of the mutant LDL receptor is encoded by 163 bases at the 5' end of intron 15 and thus shares no homology to the corresponding domain of the normal LDL receptor[14]. LDL binding studies with fibroblasts from FH patients with this gene deletion revealed an internalization-defective phenotype of FH[14]. When lipid and lipoprotein levels were compared in FH patients carrying the deleted allele and those Finnish FH patients with yet an unknown LDL receptor mutation, no significant differences were found[14]. This mutation of the LDL receptor gene, very common in the Finnish population and designated as FH-Helsinki according to the residence of the first proband examined in detail, has not been reported in other populations. In the lack of exact epidemiological studies we do not know yet whether the enrichment of the FH-Helsinki allele in the Finnish population also signifies an increased prevalence of FH in Finland.

In Lebanon the frequency of FH is high, i.e. about one patient in 100 subjects. Molecular genetic studies of the LDL receptor alleles from Lebanese FH patients revealed that more than 90% of these patients carry an identical LDL receptor mutation[16]. This mutation is characterized by a single nucleotide change at the codon for amino acid 660, causing the appearance of a premature inframe stop codon (Fig. 1). The resulted truncated LDL receptor is thus normal up to the last portion of the domain homologous to the epidermal growth factor precursor. This mutant receptor is not matured normally in the Golgi complex but it is degraded rapidly intracellularly.

A specific mutation of the LDL receptor gene explaines about two-thirds of FH cases in the French speaking population around Montreal area[17, 18]. This mutation is characterized by a large deletion of more than 10 kb in size eliminating the promotor region and the first exon of the LDL receptor gene (Fig. 1). Apparently this mutant gene is not transcribed at all and accordingly patients homozygotes for this type of LDL receptor mutation do not synthesize any LDL receptors. The other French Canadian mutations (one deletion and three point mutations) characterized so far account together less than 20% of the defective LDL receptor alleles in this population. The frequency of FH has been estimated to be two or three times higher among French Canadians than in an average Caucasian population[18].

Similar to Lebanon, the prevalence of FH is very high among South African Africaners[19]. Two LDL receptor mutations alone explain the presence of FH in about 95% of the South African FH patients[19]. Both mutations involve single nucleotide alterations causing an inframe amino acid change. One mutation affects the codon for amino acid 206 causing a substitution of glutamine for asparagine (Fig. 1). The change of the amino acid at position 206 impairs the maturation of the resultant LDL receptors which are rapidly degraded. Previous studies suggest that abnormal number or spacing of the cysteine residues may affect the normal maturation of LDL receptor[20-22]. It appears, however, that the South African LDL receptor mutation does not disturb the normal number or spacing of the amino acids in the first domain. The other type of South African mutation affects the codon for amino acid 408, resulting in a substitution of methionine

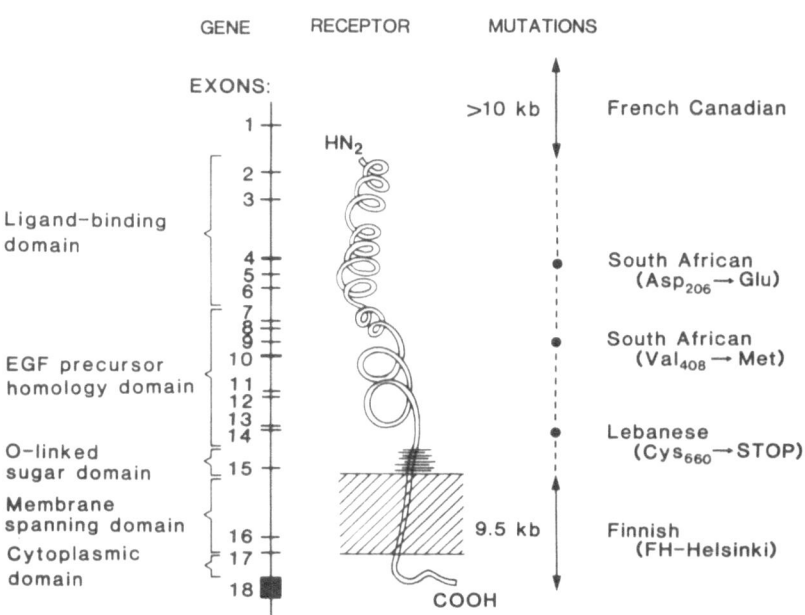

Fig. 1. Structure of the normal LDL receptor gene and the predicted architecture of the LDL receptor. Also shown are receptor mutations enriched in four different populations.

for valine (Fig. 1). Amino acid 408 is located in the domain homologous to the epidermal growth factor precursor which is known to be essential for the normal dissociation of the receptor from its ligand at the acidic pH in an endosome[6]. With disturbance of this dissociation the half-life of this mutant receptor is shorter than that of the normal one.

MOLECULAR BIOLOGY OF THE LDL RECEPTOR: THEORETICAL AND PRACTICAL IMPLICATIONS

Detailed characterization of the structure of the LDL receptor gene has greatly increased our understanding, not only on fundamentals of lipoprotein metabolism in particular, but also on the functioning of cell membrane receptors in general. On the basis of the pioneering studies of Brown, Goldstein and collaborators it came as no surprise that FH proved to be heterogeneous at the DNA level. Patients with FH seem to comprise an almost endless repertoir of living laboratories: abnormal structures combined with defects in function often permit strong reasoning of the essentials of the normal structure. Today, DNA techniques offer definite diagnostic tools for FH only in a few selected populations discussed above. It may be anticipated, however, that the rapidly progressing technology based on DNA amplification by polymerase chain reaction (PCR) as well as automated DNA sequencing will eventually provide means for molecular diagnosis of most, if not all, patients with FH.

REFERENCES

1. C. Müller, Xanthomata, hypercholesterolemia, angina pectoris. Acta Med. Scand. (Suppl.) 89:75, (1938).
2. J. L. Goldstein, and M. S. Brown, Familial hypercholesterolemia. in: "The Metabolic Basis of Inherited Diseases," J. B. Stanbury, J. B. Wyngaarden, D. S. Fredrickson, J. L. Goldstein, and M. S. Brown, eds. McGraw-Hill Book Co, New York (1983).
3. M. S. Brown, and J. L. Goldstein, A receptor mediated pathway for cholesterol homeostasis. Science 232: 34 (1986).
4. T. C. Sudhof, J. L. Goldstein, M. S. Brown, and D. W. Russell, The LDL receptor gene: a mosaic of exons shared with different proteins. Science 228: 815 (1985).
5. T. Yamamoto, C. G. Davis, M. S. Brown, W. J. Schneider, M. L. Casey, J. L. Goldstein, and D. W. Russell, The human LDL receptor: a cysteine-rich protein with multiple Alu sequences in its mRNA. Cell 39: 27 (1984).
6. C. G. Davis, J. L. Goldstein, T. C. Sudhof, R. G. W. Anderson, D. W. Russell, and M. S. Brown, Acid-dependent ligand dissociation and recycling of LDL receptor mediated by growth factor homology region. Nature 326: 760 (1987).
7. K. Kajinami, H. Mabuchi, H. Itoh, I. Michishita, M. Takeda, T. Wakasugi, J. Koizumi, and R. Takeda, New variant of low density lipoprotein receptor gene: FH-Tonami. Arteriosclerosis 8: 187 (1988).
8. B. Horsthemke, A. Dunning, and S. Humphries, Identification of deletions in the human low density lipoprotein receptor gene. J. Med. Genet 24: 144 (1987).
9. S. Langlois, J. J. P. Kastelein, and M. R. Hayden, Characterization of six partial deletions in the low-density-lipoprotein (LDL) receptor gene causing familial hypercholesterolemia (FH). Am. J. Hum. Genet. 43: 60 (1988).
10. H. H. Hobbs, E. Leitersdorf, J. L. Goldstein, M. S. Brown, and D. W. Russell, Multiple crm⁻ mutations in familial hypercholesterolemia: evidence for 13 alleles including four deletions. J. Clin. Invest. 81: 909 (1988).
11. E. Leitersdorf, A. Chakravarti, and H. H. Hobbs, Polymorphic DNA haplotypes at the LDL receptor locus. Am. J. Hum. Genet. 44: 409 (1989).
12. R. Norio, H. R. Nevanlinna, and J. Perheentupa, Hereditary diseases in Finland; rare flora in rare soil. Ann. Clin. Res. 5: 109 (1973).
13. K. Aalto-Setälä, H. Gylling, T. Miettinen, and K. Kontula, Identification of a deletion in the LDL receptor gene: a Finnish type of mutation. FEBS Lett. 230:31 (1988).

14. K. Aalto-Setälä, E. Helve, P. T. Kovanen, and K. Kontula, The Finnish type of LDL receptor gene mutation (FH-Helsinki) deletes exons encoding the carboxy-terminal part of the receptor and creates an internalization-defective phenotype. J. Clin. Invest. 84:499 (1989).

15. K. Aalto-Setala, The Finnish type of the LDL receptor gene mutation: molecular characterization of the deleted gene and the corresponding mRNA. FEBS Lett. 234: 411 (1988).

16. M. A. Lehrman, W. J. Schneider, M. S. Brown, C. G. Davis, A. Elhammer, D. W. Russell, and J. L. Goldstein, The Lebanese allele at the low density lipoprotein receptor locus: nonsense mutation produces truncated receptor that is retained in endoplasmic reticulum. J. Biol. Chem. 262:401 (1987).

17. H. H. Hobbs, M. S. Brown, D. W. Russell, J. Davignon, and J. L. Goldstein, Deletion in the gene for the low-density-lipoprotein receptor in a majority of French Canadians with familial hypercholesterolemia. N. Engl. J. Med. 317: 734 (1987).

18. E. Leitersdorf, E. J. Tobin, J. Davignon, and H. H. Hobbs, Common low-density lipoprotein receptor mutations in the French Canadian population. J. Clin. Invest. 85: 1014 (1990).

19. E. Leitersdorf, D. R. Van Der Westhuyzen, C. A. Coetzee, and H. H. Hobbs, Two common low density lipoprotein receptor gene mutations cause familial hypercholesterolemia in Africaners. J. Clin. Invest. 84: 954 (1989).

20. T. Yamamoto, R. W. Bishop, M. S. Brown, J. L. Goldstein, and D. W. Russell, Deletion in cysteine-rich region of LDL receptor impedes transport to cell surface in WHHL rabbit. Science 232: 1230 (1986).

21. V. Esser, and D. W. Russell, Transport-deficient mutations in the low density lipoprotein receptor: alterations in the cysteine-rich and cysteine-poor regions of the protein block intracellular transport. J. Biol. Chem. 263: 13276 (1988).

22. E. Leitersdorf, H. H. Hobbs, A. M. Fourie, M. Jacobs, D. R. Van Der Westhuyzen, and G. A. Coetzee, Deletion in the first cysteine-rich repeat of low density lipoprotein receptor impairs its transport but not lipoprotein binding in fibroblasts from a subject with familial hypercholesterolemia. Proc. Natl. Acad. Sci USA 85: 7912 (1988).

INVESTIGATION OF STRUCTURAL DOMAINS IN HUMAN SERUM LOW

DENSITY LIPOPROTEIN APOLIPOPROTEIN B100

Mojgan Djavaheri and Lawrence P. Aggerbeck

Centre de Génétique Moléculaire
Centre National de la Recherche Scientifique
91190 Gif-sur-Yvette, France

INTRODUCTION

Human serum low density lipoprotein (LDL) apolipoprotein B100 (apo-B100) plays a major role in lipid transport and cholesterol metabolism by mediating lipoprotein binding to specific cell surface receptors (the LDL receptor).[1] Amino acid mutations in either the LDL receptor or in the apo-B100 molecule can disrupt efficient binding of LDL to the receptor and result in dyslipidemia.[2,3] To understand fully how this protein functions at the molecular level, detailed knowledge of its structure must be obtained. The primary structure of apo-B100 has been established indicating that the protein is a single polypeptide chain composed of 4536 amino acid residues.[4,5,6] There is, however, little detailed information regarding the secondary and tertiary structure of the protein. The modulation of these different levels of protein structure must have important consequences for both lipoprotein structure and function.

Proteins which contain more than 200 residues are frequently organized into structural domains.[7,8] Although the word domain is frequently used to signify quite different characteristics of a protein (for example, lipid binding or receptor binding domains), the definition of a structural domain can be rather precise. As defined by Wetlaufer[9], structural domains are contiguous stretches of sequences which are also contiguous in three dimensions (implying compactness). Domains may have physical stability as separate structures isolated from the rest of the protein[10] and they may have important functional roles.[11] The identification of structural domains within a protein is essential for understanding the fashion in which a protein folds. Structural domains may represent intermediates along the folding pathway. Domains may fold independently and then be further assembled to give a fully folded protein. Finally, the fine-structure of a domain may be further modulated by tertiary or quaternary type interactions.[12]

Although structural domains have been described for a wide variety of proteins, they have not been extensively investigated in the case of serum apolipoproteins. Recently, two structural domains were described for apo-E3,[13,14] an amino-terminal and a

Hypercholesterolemia, Hypocholesterolemia, Hypertriglyceridemia
Edited by C.L. Malmendier *et al.*, Plenum Press, New York, 1990

carboxyl-terminal domain. The amino-terminal domain, which has been crystallized,[15] contains the information necessary for interaction with the LDL receptor whereas the carboxyl-terminal domain may function more effectively in interacting with lipids. In the case of apo-B100, previous X-ray[16] and electron microscopic[17] studies have suggested that the protein may contain

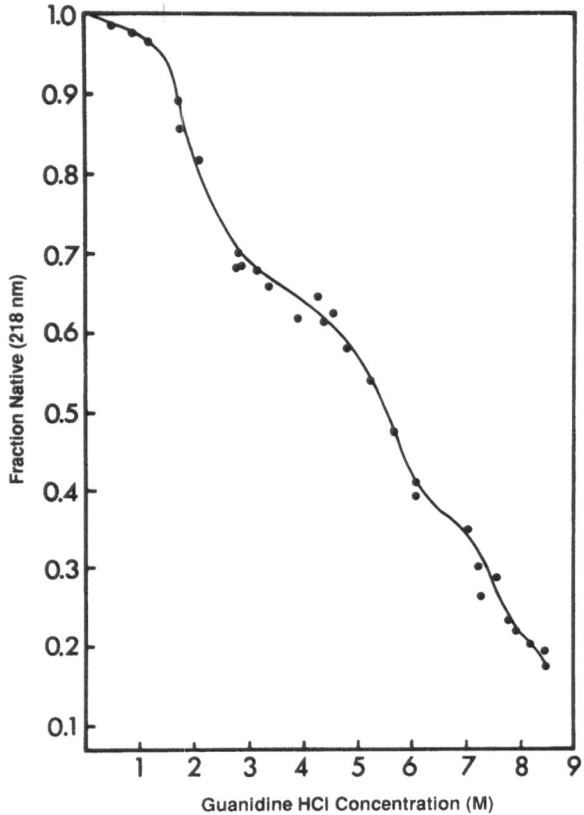

Figure 1. Denaturation curve of apo-B100 in human serum low density lipoprotein. The fraction of apo-B100 in the native state is plotted as a function of the concentration of guanidine hydrochloride.

a number of structural domains. Indeed, the large size of apo-B100 suggests that the protein could contain several structural domains. Digestion with several enzymes has shown that apo-B100 in LDL contains three large regions resistant to limited proteolysis separated by two protease susceptible regions.[18]

One of the objectives in our studies of low density lipoprotein is to look for, to isolate and to study the organization and folding of structural domains in apo-B100. Several techniques are potentially useful for demonstrating the presence of structural domains in proteins. One approach that is particularly powerful is the investigation of the stability of the protein to denaturation.

DENATURATION OF APOLIPOPROTEIN B100

The denaturation of apo-B100 with guanidine hydrochloride was assessed by far ultraviolet circular dichroism. The denaturation of the protein is quantitated by following the loss of secondary structure at a given wavelength, for example at 218 nanometers. The fraction of the amount of ordered secondary structure at 218 nm with respect to the native protein is plotted in figure 1 as a function of the guanidine hydrochloride concentration. As the concentration of guanidine hydrochloride in solution is increased, there is a progressive loss of ordered structure as indicated by the decrease in the fraction of native structure. It is clear that the curve is asymmetric with a number of inflection points. These results suggest that the denaturation of the protein is a weakly cooperative process which proceeds in an independent fashion for different parts of the protein. This type of behavior is consistent with the presence of several structural domains.

The denaturation of apo-B100 dissociated from its lipid moiety was also studied. Apo-B100 was prepared by high performance liquid chromatography in the presence of the non-ionic detergent $C_{12}E_8$ (n-octaethylene glycol monoether). The denaturation by guanidine hydrochloride of the apo-B100-detergent complex was also found to be asymmetric with a number of inflection points and this behavior is consistent with the presence of structural domains. However, as compared to the lipid bound form, the denaturation occurs at lower guanidine hydrochloride concentrations.

LIMITED PROTEOLYSIS OF APOLIPOPROTEIN-B100

In order to obtain further evidence for the existence of structural domains in apo-B100 and to tentatively identify which parts of the apo-B100 sequence are involved in domain structures, limited proteolysis was used as an independent technique. The rate of enzymatic hydrolysis of a peptide bond depends not only upon the chemical environment of the bond but also upon the physical stability of the bond which is determined by the tertiary structure of the protein.[7,8,19,20] Exposed or loosely folded polypeptide segments are rapidly hydrolyzed.[21] Structural domains appear relatively resistant to enzymatic hydrolysis presumably because most potentially hydrolyzable peptide bonds are not exposed to the enzyme or are stabilized by other forces--hydrophobic interactions, hydrogen bonds. Further, domains are relatively stable to unfolding which would otherwise make them increasingly susceptible to hydrolysis. Thus, the goals of the limited proteolysis experiments are, on the one hand, to identify stable intermediates during the course of proteolysis and, on the other hand, to locate regions highly susceptible to proteolysis.

The experimental approach used was to determine the time course of hydrolysis by incubating the lipoprotein and the proteolytic enzyme at room temperature. At different time points, aliquots of the reaction mixture were removed and the appropriate protease inhibitor was added to terminate the reaction. Peptide fragments were identified in terms of their molecular weights and immunoreactivities toward monoclonal antibodies having known epitopes. Certain fragments were further characterized with respect to their biophysical characteristics, such as their Stokes radii and were studied by electron microscopy to appreciate their structural characteristics.

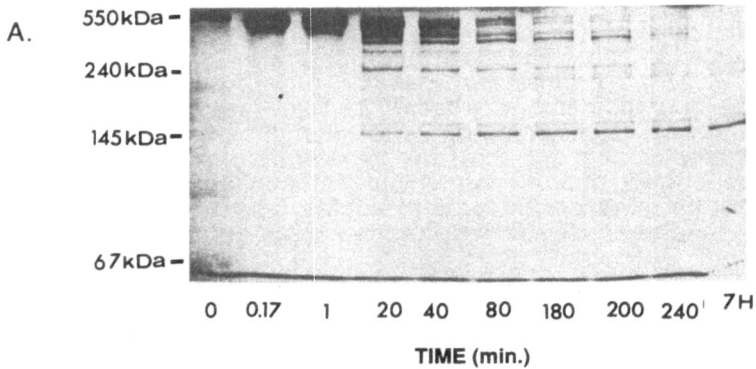

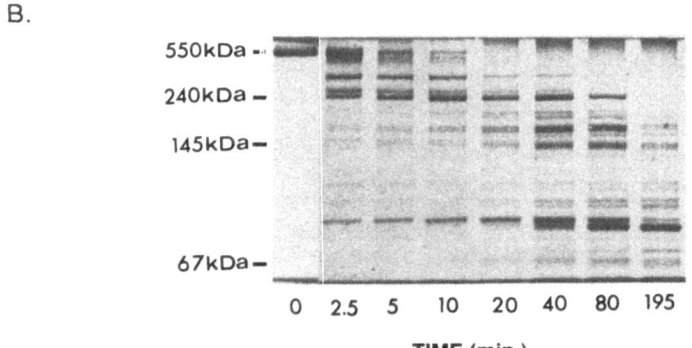

Figure 2. Time course of hydrolysis of human serum low density lipoprotein apo-B100 by various proteolytic enzymes as assessed by sodium dodecyl sulfate-polyacrylamide gel electrophoresis. LDL (5 mg/ml) in 10 mM Tris, 1 mM EDTA, pH 7.4 was incubated at room temperature with the indicated enzyme. At various times, 7 ug of protein was removed, the enzymatic reaction was terminated with phenylmethanesulfonyl fluoride and the reaction products analyzed by sodium dodecylsulfate-polyacrylamide gel electrophoresis. The mobility of peptides of known molecular weight is shown at the left. A. chymotrypsin digestion (E:S=1:100, w/w). B. elastase digestion (E:S=1:50, w/w).

Several enzymes having different specificities were used in the limited proteolysis experiments. Two examples serve to illustrate the important points derived from the limited proteolysis experiments. The time course of hydrolysis of LDL apo-B100 by chymotrypsin is shown in figure 2A. It is clear that in the initial states of proteolysis there are several classes of stable fragments of large molecular weight. It is important to emphasize the large size of these stable intermediates which are of the order of 145,000 and 240,000. Similar results were obtained with several enzymes. Stable classes of fragments of molecular weights around 145,000 and 240,000 were consistently noted. In addition to these large stable intermediates other classes of stable intermediates of somewhat smaller molecular weight have been observed as illustrated in figure 2B by the time course of subtilisin hydrolysis of LDL. In addition to the previously mentioned classes, there is another class of fragments of about 100,000 daltons mass.

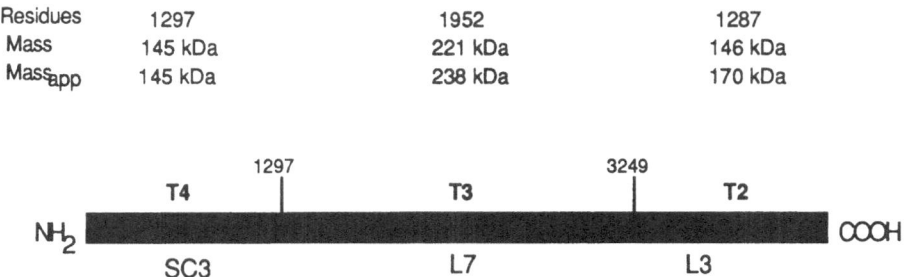

Residues	1297	1952	1287
Mass	145 kDa	221 kDa	146 kDa
Mass$_{app}$	145 kDa	238 kDa	170 kDa

Figure 3. Schematic diagram of the apo-B100 sequence. The diagram indicates the number of amino acid residues, the mass and the apparent mass on sodium dodecylsulfate-polyacrylamide gel electrophoresis for the thrombolytic fragments of apo-B100. The fragments recognized by the monoclonal antibodies SC3, L7 and L3 are also shown.

In order to assess the location of these stable fragments within the apo-B sequence, their immunoreactivity was assessed by Western blotting with three monoclonal antibodies (figure 3). Antibody L3 recognizes an epitope in the vicinity of residue 4355 situated at the carboxyl-terminal region of apo-B.[22] Antibody L7 recognizes an epitope in the vicinity of residue 2331 which is located in the central part of apo-B.[22] Finally, antibodly SC3 recognizes an epitope situated at the amino-terminal of apo-B (unpublished results). Using these antibodies it was determined that stable intermediates produced during the course of limited proteolysis are situated in all three regions of the apo-B100 molecule. This is consistent with the existence of several structural domains in the protein.

Since there are stable intermediates derived from several parts of the apo-B100 sequence, model fragments from each of the regions were investigated in more detail. One group of stable fragments of 145,000 daltons mass has its origin in the amino-terminal portion of the molecule. Therefore, the thrombolytic fragment T4 of apo-B was used as a model for a structural domain situated in this region of the molecule. Thrombin cleaves apo-B100 at residues 1297 and 3249 yielding three peptide fragments: T4 of about 140 kilodaltons mass, T3 of about 240 kilodaltons and T2 of 170 kilodaltons (figure 3). T3 and T2 are joined by a disulfide bond[4]. The fragment, T4, was isolated in the presence of a non-ionic, non-denaturing detergent to preserve its structural characteristics. LDL was hydrolyzed by thrombin and the apo-B100 fragments were prepared by high performance liquid chromatography in the presence of the detergent $C_{12}E_8$. As shown in figure 4, two protein fractions labelled I and II were obtained. Each fraction was further purified, after concentration, by rechromatography.

The identity of the two protein fractions obtained by high performance liquid chromatography was verified by polyacrylamide gel electrophoresis in the presence of SDS as shown in figure 5. Fraction I was shown to contain mainly the fragments T3-molecular weight 240000-and T2-molecular weight 170000-with a small amount of unhydrolyzed apo-B100 and T4. The fragments T3 and T2 are distinguishable since the gel is run under reducing conditions. Fraction II contains the fragment T4 with little contamination.

The purification of these fragments by high performance liquid chromatography provides, as well, a measure of their Stokes radii, and thus their form[23,24,25]. From the partition coefficients of T4, T32 and apo-B100, their Stokes radii could be estimated to be 52 A for T4, 110 A for T32 and 132 A for apo-B100. These values suggest a compact structure for T4 whereas T32 and apo-B100 are highly asymmetric.

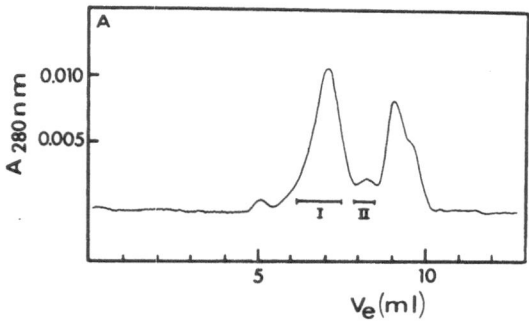

Figure 4. Preparation of thrombolytic fragments of apo-B100 by high performance liquid chromatography. LDL was proteolyzed with thrombin, incubated with an excess of $C_{12}E_8$ and 500 ug of LDL was injected onto a TSK 5000 column equilibrated at 0.5 ml/min. with 20 mM TAPS, 300 mM NaCl, 1 mM $C_{12}E_8$, pH 10 at 20°C.

Electron microscopy was also used to assess the structure of the purified fragments of apo-B100. The images of T4 (fraction II) obtained by low angle rotary shadowing of the molecules dried from glycerol containing solutions[26,27] is shown in figure 6. The fragment T4 is compact and globular. The size of the fragment, as assessed by electron microscopy, is in good agreement with the Stokes radius measured by high performance liquid chromatography.

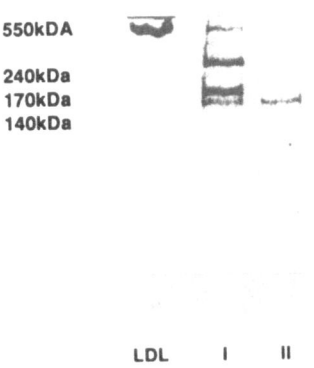

550kDA

240kDa
170kDa
140kDa

LDL   I   II

Figure 5. Sodium dodecylsulfate-polyacrylamide gel electrophoresis of the thrombolytic fragments of apo-B100 separated by high performance liquid chromatography. Peaks I and II, obtained by chromatography of LDL digested with thrombin as described in figure 4, were analyzed by electrophoresis in 5% polyacrylamide gels containing SDS. The mobility of peptides of known molecular weight is indicated on the left.

Fr-II

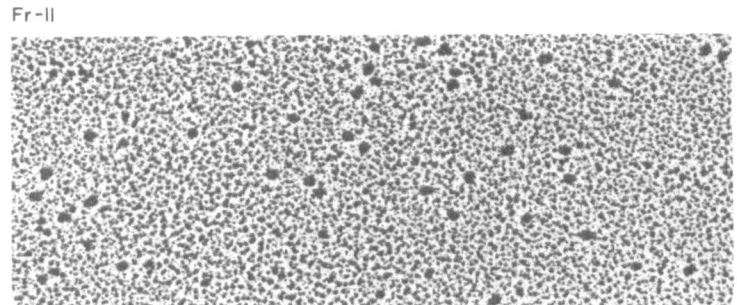

Figure 6. Electron microscopy of the fragment T4 of apo-B100. The fragment T4 produced by thrombin digestion of LDL and isolated by high performance liquid chromatography was examined by low angle rotary shadowing of samples dried from glycerol containing solutions. The fragment appears compact but slightly elongated.

The form of T4 was also studied by freeze-drying electron microscopy. In this technique, the sample is rapidly frozen to preserve its structure and the ice is then sublimed away gradually prior to shadowing with a heavy metal. The form of the particle indicates a compact globular form. The size agrees well with that obtained by rotary shadowing electron microscopy and high performance liquid chromatography.

CONCLUSIONS

In conclusion, our experiments involving denaturation of apo-B by guanidine hydrochloride and limited proteolysis suggest the presence of structural domains in apo-B100. To further study the characteristics of one of these domains we used the thrombolytic fragment T4 of apo-B as a model amino-terminal domain. After isolation in a non-ionic, non-denaturing detergent, T4 appeared to be compact and globular. Apo-B100 and the fragment T32 were highly asymmetric and flexible. Further studies of structural domains in apo-B100 such as these will help to define the physical organization of the protein and the relationship between its st ructure and its function.

ACKNOWLEDGEMENTS

The authors thank Dr. J. Yon-Kahn, University of Paris XI, Orsay, for use of the circular dichroism instrument and J. C. Dedieu for graphic arts. This work was supported in part by Research Grant HL 18577-11 from the National Institutes of Health.

REFERENCES

1. M. S. Brown and J. L. Goldstein, A receptor-mediated pathway for cholesterol homeostasis, Science 232: 34-47 (1986).
2. J. L. Goldstein, M. S. Brown, R. G. W. Anderson, D. W. Russell and W. J. Schneider, Receptor-mediated endocytosis: Concepts emerging from the LDL receptor system, Ann. Rev. Cell Biol. 1: 1-39 (1985).
3. L. F. Soria, E. H. Ludwig, H. R. G. Clarke, G. L. Vega, S. M. Grundy and B. J. McCarthy, Association between a specific apolipoprotein B mutation and familial defective apolipoprotein B-100, Proc. Natl. Acad. Sci. U.S.A. 86: 587-591 (1989).
4. T. J. Knott, R. J. Pease, L. M. Powell, S. C. Wallis, S. C. Rall, Jr., T. L. Innerarity, B. Blackhart, W. H. Taylor, Y. Marcel, R. Milne, D. Johnson, M. Fuller, A. J. Lusis, B. J. McCarthy, R. W. Mahley, B. Levy-Wilson and J. Scott, Complete protein sequence and identification of structural domains of human apolipoprotein B, Nature 323: 734-738 (1986).
5. C.-Y. Yang, S.-H. Chen, S. H. Gianturco, W. A. Bradley, J. T. Sparrow, M. Tanimura, W.-H. Li, D. A. Sparrow, H. DeLoof, M. Rosseneu, F.-S. Lee, A. M. Gotto and L. Chan, Sequence, structure receptor-binding domains and internal repeats of human apolipoprotein B-100, Nature 323: 738-742.
6. S. W. Law, S. M. Grant, K. Higuchi, A. Hospattankar, K. Lackner, N. Lee and H. B. Brewer, Jr., Human liver apolipoprotein B-100 cDNA: Complete nucleic acid and derived amino acid sequence, Proc. Natl. Acad. Sci. U.S.A. 83: 8142-8146 (1986).

7. M. G. Rossmann and P. Argos, Protein folding, <u>Annu. Rev. Biochem.</u> 50: 497-532 (1981).

8. P. S. Kim and R. L. Baldwin, Specific intermediates in the folding reactions of small proteins and the mechanism of protein folding, <u>Annu. Rev. Biochem.</u> 51: 459-489 (1982).

9. D. B. Wetlaufer, Nucleation, rapid folding and globular intrachain regions in proteins, <u>Proc. Natl. Acad. Sci. U.S.A.</u> 70: 697-701 (1973).

10. P. L. Privalov, Stability of proteins. Small globular proteins, <u>Adv. Protein Chem.</u> 33: 167-241 (1979).

11. J. Janin and S. J. Wodak, Structural domains in proteins and their role in the dynamics of protein function, <u>Prog. Biophys. Mol. Biol.</u> 42: 21-78 (1983).

12. J. S. Richardson, The anatomy and taxonomy of protein structure, <u>Adv. Protein Chem.</u> 34: 167-339 (1981).

13. J. R. Wetterau, L. P. Aggerbeck, S. C. Rall, Jr., and K. H. Weisgraber, Human apolipoprotein E3 in aqueous solution. I. Evidence for two structural domains, <u>J. Biol. Chem.</u> 263: 6240-6248 (1988).

14. L. P. Aggerbeck, J. R. Wetterau, K. H. Weisgraber, C.-S. C. Wu and F. T. Lindgren, Human apolipoprotein E3 in aqueous solution. II. Properties of the amino- and carboxyl-terminal domains, <u>J. Biol. Chem.</u> 263: 6249-6258 (1988).

15. L. P. Aggerbeck, J. R. Wetterau, K. H. Weisgraber, R. W. Mahley and D. A. Agard, Crystallization and preliminary X-ray diffraction studies on the amino-terminal (receptor binding) domain of human apolipoprotein E3 from serum very low density lipoproteins, <u>J. Mol. Biol.</u> 202: 179-181 (1988).

16. V. Luzzati, A. Tardieu and L. P. Aggerbeck, Structure of serum low density lipoprotein. I. A solution X-ray scattering study of a hyperlipidemic monkey low density lipoprotein, <u>J. Mol. Biol.</u> 131: 435-473 (1979).

17. T. Gulik-Krzywicki, M. Yates and L. P. Aggerbeck, Structure of serum low density lipoprotein. II. A freeze-etching electron microscopic study, <u>J. Mol. Biol.</u> 131: 475-484 (1979).

18. G. C. Chen, S. Zhu, D. A. Hardman, J. W. Schilling, K. Lau and J. P. Kane, Structural domains of human apolipoprotein B-100. Differential accessibility to limited proteolysis of B-100 in low density and very low density lipoproteins, <u>J. Biol. Chem.</u> 264: 14369-14375 (1989).

19. E. Mihalyi, "Application of Proteolytic Enzymes to Protein Structure Studies", 2nd Ed., Vol. 1, CRC Press, Inc., Boca Raton, FL. (1978).

20. E. Mihalyi, "Application of Proteolytic Enzymes to Protein Structure Studies", 2nd Ed., Vol. 2, CRC Press, Inc., Boca Raton, FL. (1978).

21. T. E. Creighton, "Proteins. Structures and Molecular Principles", Freeman Publications, San Francisco (1983).

22. R. J. Pease, R. W. Milne, W. K. Jessup, A. Law, P. Provost, J.-C. Fruchart, R. T. Dean, Y. L. Marcel and J. Scott, Use of bacterial expression cloning to localize the epitopes for a series of monoclonal antibodies against apolipoprotein B100, <u>J. Biol. Chem.</u> 265: 553-568 (1990).

23. M. leMaire, E. Rivas and J. V. Moller, Use of gel chromatography for the determination of size and molecular weight of proteins: Further caution, <u>Anal. Biochem.</u> 106: 12-21 (1980).

24. M. leMaire, L. P. Aggerbeck, C. Monteilhet, J. P. Andersen and J. V. Moller, The use of high performance liquid chromatography for the determination of size and molecular weight of proteins: A caution and a list of membrane proteins suitable as standards, <u>Anal. Biochem.</u> 154: 525-535 (1986).

25. M. leMaire, A. Ghazi, J. V. Moller and L. P. Aggerbeck, The use of gel chromatography for the determination of sizes and relative molecular masses of proteins: Interpretation of calibration curves in terms of gel-pore-size distribution, Biochem. J. 243: 399-404 (1987).

26. A. Gulik, L. P. Aggerbeck, J. C. Dedieu and T. Gulik-Krzywicki, Freeze-fracture electron microscopic analysis of solutions of biological molecules, J. Microscopy 125: 207-213 (1982).

27. J. E. B. Fox, L. P. Aggerbeck and M. C. Berndt, Structure of glycoprotein Ib-IX complex from platelet membranes, J. Biol. Chem. 263: 4882-4890 (1988).

IMPLICATIONS OF THIOLESTER LINKED FATTY ACIDS IN APOLIPOPROTEIN B

Diana M. Lee

Lipoprotein and Atherosclerosis Research Program
Oklahoma Medical Research Foundation
825 N.E. 13th Street, Oklahoma City, OK 73104 USA

INTRODUCTION

Apolipoprotein B (ApoB) is obligatory for triglyceride transport. Why ApoB is singled out among all apolipoproteins is not well understood. ApoB is the most hydrophobic apolipoprotein among all known apolipoproteins. It has the highest tendency to undergo aggregation. If one examines the amino acid composition of ApoB it resembles any other common water-soluble protein. Why, then, is ApoB so hydrophobic? We believe that the thiolester bound fatty acids recently found in ApoB[1,2] may play an important role in contributing to the hydrophobic nature of ApoB.

HYDROPATHY OF ApoB

One can determine the hydrophobicity of ApoB by calculating the average hydropathy value for ApoB (i.e., determination of hydrophobicity or hydrophilicity) using the hydropathy index of individual amino acids[3,4]. The high negative value represents high hydrophilicity and the high positive value represents high hydrophobicity. These values were assigned based primarily on the free energies of transfer for the side-chains of the amino acid from water phase into organic phase, and the numbers were tested with proteins with known characteristics[3]. We tabulated the amino acid composition of ApoB from one of the sequence data[5] (Table 1). The average hydropathy for ApoB was calculated to be -0.31 (Table 1). This negative value reflects the hydrophilic nature of ApoB based solely on the protein moiety. Comparing this value to that of the water-soluble serum albumin, which is calculated to be -0.39 (Table 1), the hydrophilicity of ApoB is not too far from that of albumin. For a protein with such a hydrophilic property, it is unreasonable to expect it to be insoluble in aqueous buffers.

Hypercholesterolemia, Hypocholesterolemia, Hypertriglyceridemia
Edited by C.L. Malmendier *et al.*, Plenum Press, New York, 1990

Table 1. Average hydropathy of ApoB and albumin calculated from amino acid composition.

Side chain	Hydropathy Index[a]	ApoB mols/mol[b]	Albumin mols/mol[c]
Lys	-3.9	356	59
His	-3.2	115	16
Arg	-4.5	148	24
Asp	-3.5	234	36
Asn	-3.5	244	17
Thr	-0.7	298	28
Ser	-0.8	392	24
Glu	-3.5	299	61
Gln	-3.5	230	21
Pro	-1.6	168	24
Gly	-0.4	207	12
Ala	1.8	267	62
Cys	2.5	25	35
Val	4.2	251	41
Met	1.9	78	6
Ile	4.5	288	8
Leu	3.8	524	61
Tyr	-1.3	152	18
Phe	2.8	223	31
Trp	-0.9	37	1
Total		4536	585
Average Hydropathy		-0.31	-0.39

[a]Kyte and Doolittle[3].

[b]Amino acid composition was tabulated from the sequence data of Chen et al.[5]

[c]Amino acid composition was tabulated from sequence data of Meloun et al.[6]

The hydropathy values of ApoB and albumin may be compared to those of other known proteins as reported by Kyte and Doolittle[3]. As shown in Fig. 1, nearly all the points for hydrophobic membrane proteins are above the zero line, whereas most of the points for hydrophilic soluble proteins are below the zero line. ApoB and albumin would fall near the median of the hydrophilic points even though the chain length of ApoB would place it on the fifth frame toward the right. These data show that, based on the amino acid composition, ApoB more closely resembles a water-soluble globular protein than an insoluble hydrophobic membrane protein.

Alternatively, one can calculate the hydropathy profile along the sequence to locate the hydrophobic regions, as many of the investigators have done with ApoB[5,7,8]. Olofsson et al.[7] found 39 sequences with high hydrophobicity, having an average hydropathy index of 2.6 ± 0.5, similar to membrane-spanning sequences. The average length of the sequences is only 7 ± 2 amino acid residues. Knott et al.[8] found 11 sequences lying at the lipid/ aqueous interface with strong hydrophobicity. For an apolipoprotein with amino acid residues over 4500, necessary to encompass an enormous quantity of neutral lipid core, this small number of hydrophobic regions appears to be extremely meager, even with the inclusion of a number of regions containing amphipathic β-sheet structure which has also been suggested to participate in lipid binding[7,8,9].

Thus, it is not surprising that bound fatty acids are present in ApoB to enhance its hydrophobicity.

LINKAGES BETWEEN FATTY ACIDS AND ApoB—THIOLESTER BONDS

Methylamine (MA), a reagent specifically reactive with thiolester and which forms a covalently modified product[10], was used to study the presence of thiolester bonds in ApoB[1,2]. Freshly isolated human plasma ApoB was first reduced and carboxymethylated to eliminate further participation of sulfhydryls and disulfide linkages in the carboxymethylation reaction of the newly generated sulfhydryls. The reduced and carboxymethylated (RCM-)ApoB

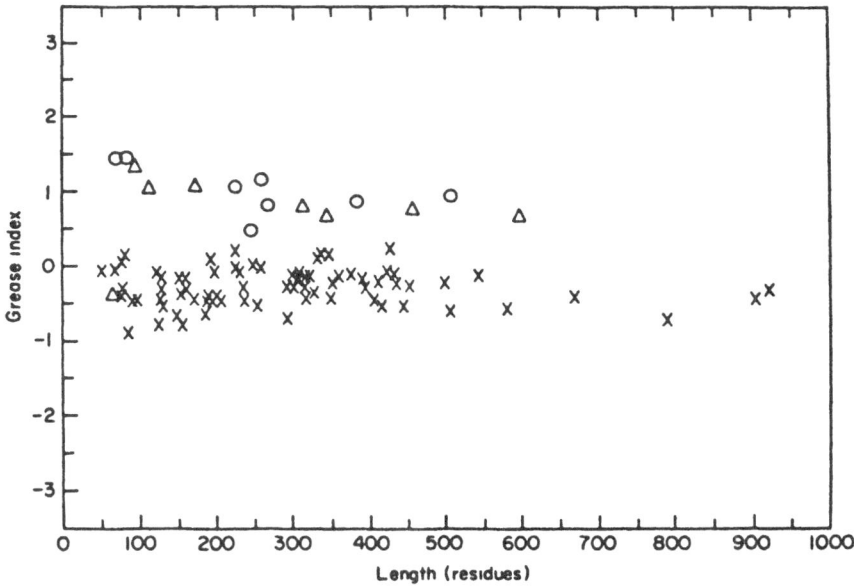

Fig. 1. Plot of average hydropathy (grease index) of various proteins against their chain length: (X) 84 soluble enzymes with known sequences; (O) 8 membrane-embedded proteins with known sequences; (Δ) 8 putative proteins of human mitochondria. Figure reproduced from Kyte and Doolittle[3] with permission.

Table 2. Covalent Incorporation of [14C]Methylamine and [14C]Iodoacetate in RCM-ApoB[1]

Incubation time (h)	(A) Incorporation of [14C]methylamine (mols/mol of ApoB)	(B) Incorporation of [14C]iodoacetate (mols/mol of ApoB)	(B)-(A) (mol/mol) of ApoB
0.5	3.3	3.2	-0.1
2.5	2.9	8.2	5.3
6.0	2.9	10.0	7.1
24.0	2.9	12.5	9.6

RCM-ApoB was incubated with (A) 50 mM [14C]methylamine (sp. act. 16.66 mCi/mmol) and (B) 50 mM methylamine (unlabeled) + 10 mM [14C]iodoacetate (sp. act. 13.33 mCi/mmol) at pH 8.5 and 30°C. Experimental detail, see Huang et al.[1]

in 6 M urea/Tris containing preservatives was incubated with [14C]MA at pH 8.5 and 30°C[1]. Covalent incorporation of [14C]MA was observed in RCM-ApoB with concomitant generation of new sulfhydryl groups, which could be blocked with [3H]- or [14C]iodoacetate. The molar ratio of S-[3H]CM/N-[14C]MA incorporation was near one at one-half hr incubation, suggesting the possible presence of intramolecular thiolester linkage(s) (see Table 2)[1,11]. As the incubation time increased, the incorporation of S-[3H]CM increased in RCM-ApoB without the concomitant increase in N-[14C]MA incorporation, suggesting the possible presence also of intermolecular thiolester linkages, because the ligand containing N-[14C]MA was separated from the protein. After 24 hr incubation, up to 10 mols of extra sulfhydryl groups were generated (Table 2). These N-[14C]MA-containing ligands were soluble in organic solvent. Following purification by thin-layer chromatography with two solvent systems (solvent system 1: hexane/diethyl ether/glacial acetic acid 113:35:3 v/v; solvent system 2: hexane/ethyl acetate/methanol/glacial acetic acid 90:20:20:2 v/v) the [14C]MA-derivatives were transesterified in methanolic HCl. Gas-liquid chromatography identified the methyl ester as palmitate and stearate at 1:1 molar ratio for human ApoB. Therefore, the palmitic and stearic acids are covalently linked to ApoB via intermolecular thiolester linkages to the side chain of cysteine residues. So what has happened is this (Fig. 2): thiolester linked fatty acids on ApoB are sensitive to alkaline attack. In the presence of [14C]MA, covalent products [14C]-methyl hexadecanoyl amide and [14C]-methyl octadecanoyl amide are formed. After transesterification in methanolic HCl these fatty acyl amides are converted to methyl fatty acyl ester and identified by GLC as fatty acids. In ApoB, new -SH groups are generated.

52

The same method was applied to rat plasma RCM-ApoB, thiolester linked fatty acids were also observed in rat ApoB[2]. However, in addition to the bound palmitate and stearate as found in human ApoB, a small amount of bound oleate was also observed in rat ApoB[2] (Table 3). The significance of this difference is not yet known.

Bound palmitic and stearic acids were also found by Fisher after hydrolysis of chymotryptic peptides of totally delipidized ApoB[12,13]. He estimated at least 8 mols of fatty acids were present in ApoB[12]. The presence of covalently bound palmitate was also observed recently in ApoB secreted by HepG2 cells[14].

Thus far, the only other apolipoprotein known to be acylated with long-chain fatty acid is apolipoprotein A-I, by ester linkage[15]. However, the fatty acid was reportedly removed before apolipoprotein A-I was secreted into the circulation[15]. Thus, ApoB is the only acylated apolipoprotein present in plasma and the only secreted apolipoprotein which is acylated via thiolester.

SIGNIFICANCE OF INTERMOLECULAR THIOLESTER BOUND FATTY ACIDS

Thiolester is a high energy labile linkage. It is readily broken _in vitro_. Under alkaline conditions, free fatty acids and sulfhydryls are generated. The free fatty acids may cause ApoB to aggregate in aqueous solu-

[^{14}C]METHYLAMINE TREATMENT OF RCM-ApoB

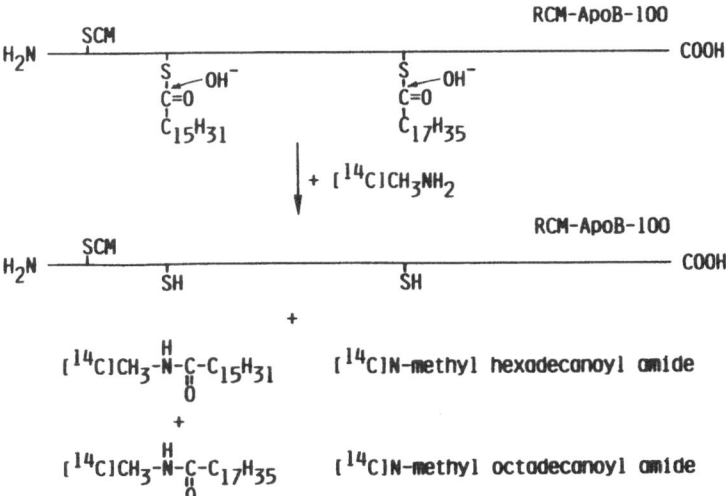

Fig. 2. Production of fatty acyl amides by the reaction of reduced and carboxymethylated ApoB with [^{14}C]methylamine.

Table 3. Thiolester Linked Fatty Acids in ApoB

Species	Fatty acids (mols/mol)
Human	Palmitate : Stearate (1:1)
Rat	Palmitate : Stearate : Oleate (4:4:1)

tions. If the newly generated sulfhydryls are exposed to the surface of ApoB, they are susceptible to autoxidation. This autoxidation process is via the free radical pathway[16]. Inter- or intramolecular -S-S- bridges may be produced by autoxidation of -SH groups and H_2O_2 is generated as a by-product[16]. In the presence of low density lipoproteins (LDL), H_2O_2 may oxidize lipids and causes lipid peroxidation of LDL. The presence of labile thiolester linked fatty acids in ApoB explains why ApoB has a strong tendency towards aggregation, why ApoB is sensitive to autoxidation[17], and why LDL is susceptible to lipid peroxidation.

PROTEINS ACYLATED WITH PALMITATE

Based on the literature information, a large number of proteins in eukaryotic cells are known to be acylated with palmitate (Table 4): the transforming protein of Harvey sarcoma virus, p21 *ras*[18-21], the surface glycoproteins of several enveloped viruses[22-25], the histocompatibility complex antigens[26], the mammalian transferrin receptor[27], the erythrocyte membrane skeleton protein ankyrin[28], the membrane glycoprotein rhodopsin[29], and the neuronal growth cone protein, GAP-43[30]. Palmitic acid is most often found to be linked through a thiolester bond to the cysteine residue. This has been established conclusively in the case of the G glycoprotein of vesicular stomatitis virus[25], p21 *ras*[19], the GAP-43[30], the transferrin receptor[31], and the histocompatibility complex antigens[26]. Unlike myristic acid, which is only found attached through an amide bond to amino-terminal residues (most frequently, glycine), palmitic acid is found to modify cysteines present within the sequence of a protein. These residues are generally found near membrane-binding domains of a protein and usually on the cytoplasmic face of a membrane[34]. Myristylation is a co-translational modification process while palmitylation is a post-translational process[34]. The presence of thiolester linked palmitate at Cys-186 in Harvey murine sarcoma (HMS) p21 *ras* protein was reported to be essential for the biological activities of *ras* proteins: for transformation of NIH/3T3 cells, for membrane localization, and for lipid binding[21]. The absence of Cys-186 or the replacement of Cys-186 with Ser-186 by point mutation causes the total loss of lipid binding capacity and the biological activities of HMS p21 *ras* proteins. In

Table 4. Palmitylated Proteins

Proteins	Linkages	References
p21 *ras*	Thiolester	18,19,20,21
Vesicular stomatitis virus G glycoprotein	Thiolester	22,23,24,25
Histocompatibility complex antigens	Thiolester	26
Mammalian transferrin receptor	Thiolester	27,31
Erythrocyte membrane skeleton ankyrin	(Ester)[a]	28
Rhodopsin	(Ester)[a]	29
Neuronal growth cone protein, GAP-43	Thiolester	30
Parainfluenza virus F glycoprotein		32
Influenza virus HA glycoprotein		32
SV 40 Virus large T antigen		33
Sindbis virus E2 glycoprotein		24
Apolipoprotein A-I	Ester	15
Apolipoprotein B	Thiolester	1,2,12,13,14

[a]The methodology employing alkaline and hydroxyamine cleavage for identification of ester bond does not exclude the possibility of the bond being thiolester.

addition, the presence of Cys at any other location in the sequence of *ras* proteins could not be palmitylated to acquire the biological activities. Interestingly, when the nonpalmitylated p21 *ras* derivatives were myristylated at their amino termini, the p21 *ras* became activated and restored the membrane association and full transforming activity. Furthermore, myristylated forms of normal cellular *ras* also became transforming[35]. Thus, the normal function of cellular *ras* is diverted to transformation by myristate and this suggests that the normal *ras* must be regulated ordinarily by some unique property of palmitate that myristate does not mimic[35]. These data point out the importance of post-translational events. They suggest that fatty acid acylation is 1) for a <u>functional</u> purpose, 2) <u>site specific</u> and 3) <u>fatty acid specific</u>.

Acylation of p21 N-*ras* is reportedly a dynamic event with a rapid acylation-deacylation cycle, (half life about 20 min) even in the absence of protein synthesis[36]. Acylation of p21 N-*ras*[36] and of ankyrin[24] is resistant to inhibitors of protein synthesis while most other cellular acyl proteins are sensitive to these agents. Most protein palmitylation occurs in the early Golgi complex[23,37], suggesting that the enzymatic machinery for acylation of p21 N-*ras* and ankyrin may differ from that of the bulk of cellular palmitylated proteins[36]. Alternatively, there may be two pools of acylation substrates, one freely accessible to the acylation machinery (e.g., p21 N-*ras*) and the other inaccessible due to intracellular localization[36].

The neuronal growth cone protein, GAP-43 is synthesized initially as a soluble protein that becomes attached to membranes posttranslationally after early fatty acylation at the only two cysteine residues, Cys-3 and Cys-4 and becomes an insoluble protein[30].

Taking this information all together, we may predict that fatty acid acylation of ApoB may also be reversible, which offers the potential for dynamic regulation of ApoB and the assembly of its lipoproteins. It is possible that ApoB may be synthesized initially as a soluble protein, like GAP-43[30], which becomes insoluble after acyltation with long chain fatty acids.

EFFECT OF FATTY ACYL CHAIN LENGTH

Using acylated synthetic peptides with varying acyl chain length, Ponsin[38] demonstrated that the hydrophobicity of the lipid-associating peptide increased with the acyl chain length and the binding constant to a phospholipid matrix or a high density lipoprotein model increased by 3 orders of magnitude as the length of acyl chain increased from 0 to 16 carbon units. It was concluded that for a given helical potential, the binding of a lipid-associating peptide to lipoprotein is governed by its hydrophobicity. Applying this knowledge to the acylation of ApoB, it suggests that the hydrophobicity of the loci with sterylation is higher than those with palmitylation, and far higher than the loci without acylation.

CONCLUSIONS

The literature information implies that we can predict that fatty acid acylation of ApoB is a means of increasing the hydrophobicity of the apolipoprotein for membrane attachment and lipid binding. Although the specific function of acylation of ApoB is still unknown, the selectivity suggests its possible role in assembly, secretion, lipid binding or transport. The acylation sites are most likely specific. The chain length of fatty acid and the possible reversibility of acylation may serve to modulate the function of ApoB. Probably it is this regulated fatty acid acylation that singles out ApoB and makes it obligatory for triglyceride transport.

REFERENCES

1. G. Huang, D. M. Lee and S. Singh, Identification of the thiol ester linked lipids in apolipoprotein B, _Biochemistry_ 27:1395 (1988).

2. V. S. Kamanna and D. M. Lee, Presence of covalently attached fatty acids in rat apolipoprotein B via thiolester linkages, Biochem. Biophys. Res. Commun. 162:1508 (1989).

3. J. Kyte and R. F. Doolittle, A simple method for displaying the hydropathic character of a protein, J. Mol. Biol. 157:105 (1982).

4. T. P. Hopp and K. R. Woods, Prediction of protein antigenic determinants from amino acid sequences, Proc. Natl. Acad. Sci. 78:3824 (1981).

5. S.-H. Chen, C.-Y. Yang, P. F. Chen, D. Setzer, M. Tanimura, W.-H. Li, A. M. Gotto and L. Chan, The complete cDNA and amino acid sequence of human apolipoprotein B-100, J. Biol. Chem. 261:12918 (1986).

6. B. Meloun, L. Morávek and V. Kostka, Complete amino acid sequence of human serum albumin, FEBS Lett. 58:134 (1975).

7. S.-O. Olofsson, G. Bjursell, K. Boström, P. Carlsson, J. Elovson, A. A. Protter, M. A. Reuben and G. Bondjers, Apolipoprotein B: structure, biosynthesis and role in the lipoprotein assembly process, Atherosclerosis 68:1 (1987).

8. T. J. Knott, R. J. Pease, L. M. Powell, S. C. Wallis, S. C. Rall, Jr., T. L. Innerarity, B. Blackhart, W. H. Taylor, Y. Marcel, R. Milne, D. Johnson, M. Fuller, A. J. Lusis, B. J. McCarthy, R. W. Mahley, B. Levy-Wilson and J. Scott, Complete protein sequence and identification of structural domains of human apolipoprotein B, Nature 323:734 (1986).

9. C.-Y. Yang, S.-H. Chen, S. H. Gianturco, W. A. Bradley, W.A., J. T. Sparrow, M. Tanimura, W.-H. Li, D. W. Sparrow, H. DeLoof, M. Rosseneu, F.-S. Lee, Z.-W. Gu, A.M. Gotto, Jr. and L. Chan, Sequence, receptor-binding domains and internal repeats of human apolipoprotein B-100, Nature 323:738 (1986).

10. B.F. Tack, R.A. Harrison, J. Janatova, M.L. Thomas and J.W. Prahl, Evidence for presence of an internal thiolester bond in third component of human complement, Proc. Natl. Acad. Sci. USA 77:5764 (1980).

11. D.M. Lee and S. Singh, Presence and localization of two intramolecular thiolester linkages in apolipoprotein B, Circulation 76II:286 (1988).

12. W.R. Fisher, The structure of lipoproteins: Covalently bound fatty acids, Ph.D. Dissertation, University of Pennsylvania, University Microfilms, Inc., Ann Arbor, MI (1964).

13. W.R. Fisher and S. Gurin, Structure of lipoproteins. Covalently bound fatty acids. Science 143:362 (1964).

14. J.M. Hoeg, M.S. Meng, R. Ronan, S.J. Demosky, Jr., T. Fairwell and H.B. Brewer, Jr., Apolipoprotein B synthesized by Hep G2 cells undergoes fatty acid acylation, J. Lipid Res. 29:1215 (1988).

15. J.M. Hoeg, M.S. Meng, R. Ronan, T. Fairwell and H.B. Brewer, Jr., Human apolipoprotein A-I. Post-translational modification by fatty acid acylation, J. Biol. Chem. 261:3911 (1986).

16. D.M. Lee, A.J. Valente, W.H. Kuo and H. Maeda, Properties of apolipoprotein B in urea and in aqueous buffers. The use of glutathione and nitrogen in its solubilization, Biochim. Biophys. Acta 666:133 (1981).

17. Y.M. Torchinsky, The chemical properties of SH groups. Sulfhydryl reagents, in: "Sulfur in Proteins", Y.M. Torchinsky, ed., Pergamon Press, New York, NY (1981).

18. J.E. Buss and B.M. Sefton, Direct identification of palmitic acid as the lipid attached to p21^{ras}, Mol. Cell. Biol. 6:116 (1986).

19. Z.-Q. Chen, L.S. Ulsh, G. DuBois and T.Y. Shih, Posttranslational processing of p21 ras proteins involves palmitylation of the C-terminal tetrapeptide containing Cysteine-186, J. Virology 56:607 (1985).

20. B.M. Sefton, I.S. Trowbridge, J.A., Cooper and E.M. Scolnick, The transforming proteins of Rous sarcoma virus, Harvey sarcoma virus and Abelson virus contain tightly bound lipid, Cell 31:465 (1982).

21. B.M. Willumsen, K. Norris, A.G. Papageorge, N.L. Hubbert and D.R. Lowy, Harvey murine sarcoma virus p21 ras protein: Biological significance

of the cysteine nearest the carboxy terminus, EMBO J. 3:2581 (1984).

22. M.F.G. Schmidt and M.J. Schlesinger, Fatty acid binding to vesicular stomatitis virus glycoprotein: a new type of posttranslational modification of the viral glycoprotein, Cell 17:813 (1979).

23. M.F.G. Schmidt and M.J. Schlesinger, Relation of fatty acid attachment to the translation and maturation of vesicular stomatitis and Sindbis virus membrane glycoproteins, J. Biol. Chem. 255:3334 (1980).

24. M.F.G. Schmidt, M. Bracha and M.J. Schlesinger, Evidence for covalent attachment of fatty acids to Sindbis virus glycoproteins, Proc. Natl. Acad. Sci. USA 76:1687 (1979).

25. J.K. Rose, G.A. Adams and C.J. Gallione, The presence of cysteine in the cytoplasmic domain of the vesicular stomatitis virus glycoprotein is required for palmitate addition, Proc. Natl. Acad. Sci. USA 81:2050 (1984).

26. J.F. Kaufman, M.S. Krangel and J.L. Strominger, Cysteines in the transmembrane region of major histocompatibility complex antigens are fatty acylated via thioester bonds, J. Biol. Chem. 259:7230 (1984).

27. M.B. Omary and I.S. Trowbridge, Covalent binding of fatty acid to the transferrin receptor in cultured human cells, J. Biol. Chem. 256:4713 (1981).

28. M. Staufenbiel and E. Lazaride, Ankyrin is fatty acid acylated in erythrocytes, Proc. Natl. Acad. Sci. USA 83:318 (1986).

29. P.J. O'Brien and M. Zatz, Acylation of bovine rhodopsin by [^{3}H]palmitic acid, J. Biol. Chem. 259:5054 (1984).

30. J.H.P. Skene and I. Virág, Posttranslational membrane attachment and dynamic fatty acylation of a neuronal growth cone protein, GAP-43, J. Cell Biol. 108:613 (1989).

31. S. Jing and I.S. Trowbridge, Identification of the intermolecular disulfide bonds of the human transferrin receptor and its lipid-attachment site, EMBO J. 6:327 (1987).

32. M.F.G. Schmidt, Acylation of viral spike glycoproteins: A feature of enveloped RNA viruses, Virology 116:327 (1982).

33. U. Klockmann and W. Deppert, Acylated simian virus 40 large T-antigen: a new subclass associated with a detergent-resistant lamina of the plasma membrane, EMBO J. 2:1151 (1983).

34. B.M. Sefton and J.E. Buss, The covalent modification of eukaryotic proteins with lipid, J. Cell Biol. 104:1449 (1987).

35. J.E. Buss, P.A. Solski, J.P. Schaeffer, M.J. MacDonald, C.J. Der, Activation of the cellular proto-oncogene produce p21 *ras* by addition of a myristylation signal, Science 243:1600 (1989).

36. A.I. Magee, L. Gutierrez, I.A. McKay, C.J., Marshall and A. Hall, Dynamic fatty acylation of p21^{N-ras}, EMBO J. 6:3353 (1987).

37. W.G. Dunphy, E. Fries, L.J. Urbani and J.E. Rothman, Early and late functions associated with the Golgi apparatus reside in distinct compartments, Proc. Natl. Acad. Sci. USA 78:7453 (1981).

38. G. Ponsin, Relationship between structure and metabolism of HDL apolipoproteins: Study with synthetic peptides, Adv. Exp. Med. Biol. 243:139 (1988).

FACTORS REGULATING THE DISTRIBUTION OF CHOLESTEROL BETWEEN LDL AND HDL

P.J. Barter, L.B.F. Chang, and O.V. Rajaram

Baker Medical Research Institute
Melbourne, Australia

The importance of understanding factors which influence the partitioning of cholesterol between different plasma lipoprotein fractions is highlighted by the observation that the risk of developing coronary heart disease correlates positively with the concentration of cholesterol in low density lipoproteins (LDL)[1] and negatively with that in high density lipoproteins (HDL)[2]. Most of the cholesterol in plasma exists as cholesteryl esters which reside with triacylglycerol in the hydrophobic core of lipoproteins. In human plasma, cholesteryl esters exchange between all lipoprotein fractions in a process of equilibration catalysed by the cholesteryl ester transfer protein (CETP)[3,4]. Since the rate of the CETP-mediated exchange between LDL and HDL in human plasma is rapid relative to the rate of catabolism of each lipoprotein fraction[5] (Fig.1), the cholesteryl esters in these two lipoproteins must be close to equilibrium *in vivo*. Thus, in terms of regulating the partitioning of cholesteryl esters between LDL and HDL, it is probable that the level of activity of CETP is not normally rate limiting. We have recently reported, however, that the CETP-mediated equilibrium between LDL and HDL can be disrupted by Na oleate which, as a consequence, promotes a shift in the partitioning of cholesteryl esters from the non-atherogenic HDL to the atherogenic LDL[6].

Samples of the plasma fraction of density 1.019-1.21 g/ml containing a mixture of HDL and LDL were incubated at 37°C for 24 hours in the presence of various additions[6]. Incubations performed in the presence of buffer alone produced no change in the distribution of cholesteryl esters between the two lipoproteins. Incubation in the presence of CETP (purified 5000-fold relative to lipoprotein-free plasma and added at a concentration of 2.4 units/ml), on the other hand, resulted in a net mass transfer of 40% of the HDL cholesteryl esters to the LDL fraction. When, however, identical mixtures of HDL, LDL and CETP were supplemented with fatty acid-poor human serum albumin at a protein concentration of 40 mg/ml, there was a 50% inhibition of the net mass transfer, with only about 20% of the HDL cholesteryl esters being redistributed to LDL in these incubations. By contrast, addition of fatty acid-rich human serum albumin did not inhibit the CETP-mediated net mass transfer of cholesteryl esters from HDL to LDL. The implication that non-esterified fatty acids may, therefore, have been involved in promoting the net mass

Hypercholesterolemia, Hypocholesterolemia, Hypertriglyceridemia
Edited by C.L. Malmendier *et al.*, Plenum Press, New York, 1990

59

transfers was supported by finding that when the mixtures of HDL, LDL and CETP were supplemented with Na oleate at a concentration of 0.12 mM, the CETP-mediated net mass transfers were markedly increased, with 66% of the HDL cholesteryl ester mass now being transferred to LDL during the 24 hours of incubation[6]. In other experiemnts it was found that the enhancement of the CETP-mediated net mass transfer by Na oleate was concentration dependent over the range 0.06 mM to 0.24 mM. In the absence of CETP, on the other hand, addition of Na oleate at concentrations of up to 0.24 mM had no effect on the distribution of cholesteryl esters between HDL and LDL.

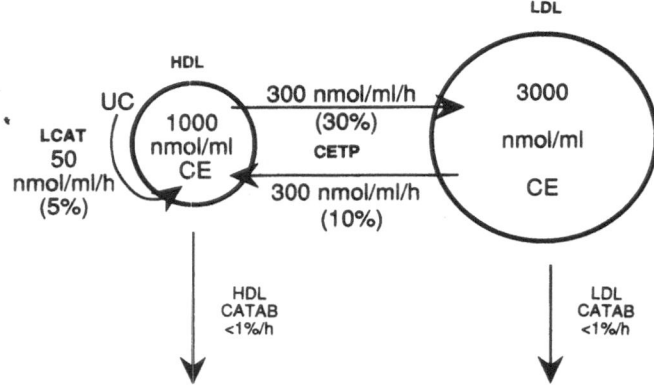

Fig. 1. Equilibration of cholesteryl esters between HDL and LDL. The rates shown are those which occur under physiological conditions.

Given that CETP promotes bidirectional transfers and thus an equilibration of cholesteryl esters between HDL and LDL (Fig.1), the observation of a net mass transfer from HDL to LDL in the presence of CETP *in vitro* indicates either that equilibration is still incomplete at the commencement of the incubation or, if complete, that the equilibrium is disrupted during the incubation. To differentiate between these two possibilities, time course studies were performed with mixtures of HDL and LDL supplemented with a tracer amount of HDL labelled isotopically in the cholesteryl ester moiety. In the presence of CETP, the isotopically-labelled cholesteryl esters equilibrated between HDL and LDL within three hours of incubation[6]. The net mass transfers, by contrast, were progressive for up to 24 hours. Furthermore, despite the effects of fatty acid-poor albumin and Na oleate in respectively decreasing and increasing the magnitude of the net mass transfers of cholesteryl esters from HDL to LDL, neither albumin nor Na oleate had any discernable effect on the rate at which CETP promoted the equilibration of the isotopically-labelled cholesteryl esters between the two fractions. Thus, the redistribution of cholesteryl ester mass was quite distinct from the isotopic transfers and could therefore not be explained simply as the completion of a process of equilibration. Rather, the fact of a net mass transfer between already equilibrated pools indicated that the equilibrium was being disrupted. We postulate that this disruption is caused by non-esterified fatty acids which, in the presence of CETP, promote a redistribution of cholesteryl esters from HDL to LDL.

The mechanism by which CETP and Na oleate interact to promote a net mass transfer of cholesteryl esters from HDL to LDL is not known. However, some insights may be obtained by reviewing the postulated mechanism of action of CETP. Two general models have been proposed: (i) a shuttle model[7] and (ii) a ternary collision complex model[8]. According to the shuttle model, CETP picks up molecules of cholesteryl ester and triacylglycerol and circulates as a CETP-lipid complex[9]. It has been postulated that this complex binds transiently to lipoprotein particles during which there is an exchange of lipids between CETP and the lipoprotein. The CETP-lipid complex then dissociates from the lipoprotein and circulates in plasma until again it binds to a lipoprotein particle and again exchanges lipids. In this way, CETP acts as a shuttle which promotes an exchange and thus an equilibration of cholesteryl esters and triacylglycerol between lipoprotein particles; this includes an exchange between particles within a given lipoprotein class as well as between particles in different classes[7]. The net effect of the process is an exchange of lipid molecules between lipoprotein particles, whether cholesteryl ester for cholesteryl ester, cholesteryl ester for triacylglycerol or triacylglycerol for triacylglycerol; it does not, however, result in a net change in the total core lipid content of lipoprotein particles. Thus, while the shuttle model can explain the heteroexchange of cholesteryl ester for triacylglycerol promoted by CETP in incubations of HDL and VLDL[10], it cannot account for a net mass transfer of cholesteryl esters from HDL to LDL[6].

CETP has also been suggested to act by mediating a ternary collision complex with HDL and LDL during which lipid constituents redistribute between the lipoprotein particles[8]. It is possible that the formation of a ternary complex of HDL, CETP and LDL may lead to a remodelling of the lipoprotein particles which results in a net movement of constituents from HDL to LDL. In these terms, one consequence of the formation of a ternary complex may be the conversion of HDL into the small, lipid-poor, protein-rich particles as observed recently in incubations of HDL, LDL, CETP and Na oleate[11]. Furthermore, since it is known that non-esterified fatty acids enhance the binding of CETP to lipoproteins[12], it is not unreasonable to speculate that non-esterified fatty acids may also enhance the formation of ternary complexes.

We postulate that both the shuttle and the ternary complex mechanisms operate and that CETP may act both to promote the shuttling of lipids between HDL and LDL and the formation of collision complexes between the lipoprotein particles. In the absence of non-esterified fatty acids on the lipoprotein surface, we postulate that the shuttle mechanism predominates and that the major lipid transfer process is one of exchange. As the non-esterified fatty acid concentration on the surface

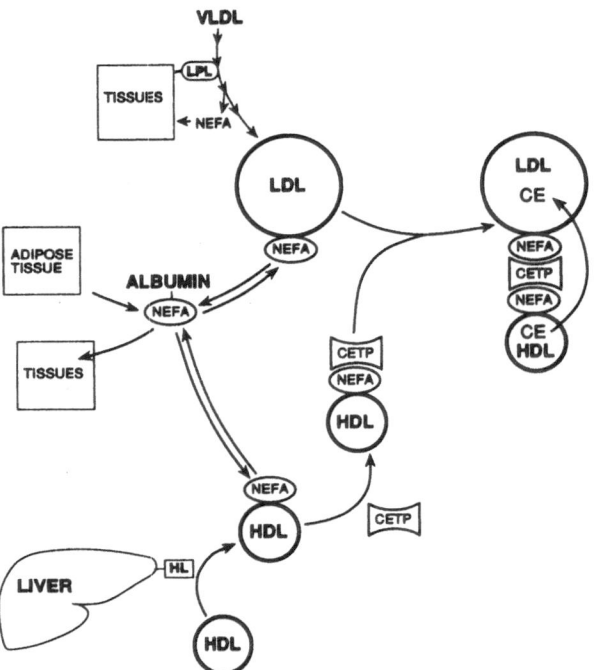

Fig. 2. Proposed mechanism of the interaction between CETP and non-esterified fatty acids (NEFA) in promoting a redistribution of cholesteryl esters from HDL to LDL.

of lipoproteins is increased and there is a consequent enhancement in the binding of CETP to the particles, we postulate that the formation of ternary complexes of HDL-CETP-LDL is favoured, resulting in a net mass transfer of core lipids from HDL to LDL (Fig.2).

There is strong circumstantial evidence from other studies that non-esterified fatty acids on the surface of plasma lipoproteins interact with and modify the function of CETP. For example, when a mixture of HDL and CETP is supplemented by the addition of very low density lipoproteins (VLDL) which have been pretreated with lipoprotein lipase, there is an enhanced formation of small HDL particles[13]. It has also been reported that lipolytic products, specifically non-esterified fatty acids, result in both an increase in the binding of CETP to lipoproteins and an increase in the rate of cholesteryl ester transfer from HDL to VLDL[12]. The synergistic effects of hepatic lipase and CETP in promoting the reduction in size of HDL[13] may also reflect an involvement of non-esterified fatty acids.

It has been observed that CETP promotes a net mass transfer of cholesteryl esters from HDL to LDL even in incubations which have not been supplemented with exogenous Na oleate[6]. However, the fact that this CETP-mediated net mass transfer was markedly inhibited by fatty acid-poor but not by fatty acid-rich albumin[6] suggested that albumin and CETP may compete for the endogenous non-esterified fatty acids which exist as normal components of the surface of plasma lipoproteins[12]. It will be of interest to investigate whether the capacity of albumin to inhibit the CETP-mediated net mass transfer of cholesteryl esters from HDL to LDL correlates with its content of non-esterified fatty acids over the range of concentrations encountered physiologically.

It is still uncertain whether an interaction of CETP and non-esterified fatty acids is of physiological importance in modulating the partitioning of cholesterol between HDL and LDL. Nevertheless, it is tempting to speculate that variations in plasma non-esterified fatty acid metabolism may play a major role in regulating the distribution of cholesterol between plasma lipoproteins. For example, the low concentrations of HDL cholesterol in obese subjects and in smokers may reflect increased concentrations of plasma non-esterified fatty acids in such subjects. There are also obvious implications for subjects in whom the non-esterified fatty acid concentration is increased as a consequence of exposure to chronic stress. Clearly, therefore, an involvement of non-esterified fatty acids in modulating plasma cholesterol transport is a phenomenon of great potential importance. Much more investigation is required to define the mechanism and both the physiological and pathological implications of the role played by non-esterified fatty acids in determining the partitioning of cholesterol between the non-atherogenic HDL and the atherogenic LDL fractions.

REFERENCES

1. W. B. Kannel, W. P. Castelli, T. Gordon and P. H. McNamara, Serum cholesterol, lipoproteins and the risk of coronary heart disease, Ann. Int. Med. 74:1 (1971).
2. G. J. Miller and N. E. Miller, Plasma-high-density-lipoprotein concentration and the development of ischaemic heart disease, Lancet i:16 (1975).

3. N. M. Pattnaik, A. Montes, L. B. Hughes and D. B. Zilversmit, Cholesteryl ester exchange protein in human plasma: isolation and characterisation, Biochim. Biophys. Acta 530:428 (1978).

4. P. J. Barter and J. I. Lally, In vitro exchanges of esterified cholesterol between serum lipoprotein fractions: studies of humans and rabbits, Metabolism 28:230 (1979).

5. P. J. Barter and M. E. Jones, Rate of exchange of esterified cholesterol between human low and high density lipoproteins, Atherosclerosis 34:67 (1979).

6. P. J. Barter, L. B. F. Chang, H. H. Newnham, K. A. Rye and O. V. Rajaram, Roles of lecithin: cholesterol acyltransferase, lipid transfer proteins and nonesterified fatty acids in plasma cholesterol transport, J. Drug Devel. (in press, 1990).

7. P. J. Barter and M. E. Jones, Kinetic studies of the transfer of esterified cholesterol between human plasma low and high density lipoproteins, J. Lipid Res. 21:238 (1980).

8. J. Ihm, D. M. Quinn, S. J. Busch, B. Chataing and J. A. K. Harmony, Kinetics of plasma protein-catalyzed exchange of phosphatidylcholine and cholesteryl ester between plasma lipoproteins, J. Lipid Res. 23:1328 (1982).

9. T. L. Swenson, R. W. Brocia and A. R. Tall, Plasma cholesteryl ester transfer protein has binding sites for neutral lipids and phospholipids, J. Biol. Chem. 263:5150 (1988).

10. G. J. Hopkins, L. B. F. Chang and P. J. Barter, Role of lipid transfers in the formation of a subpopulation of small high density lipoproteins, J. Lipid Res. 26:218 (1985).

11. P. J. Barter, L. B. F. Chang, H. H. Newnham, K. A. Rye and O. V. Rajaram, The interaction of cholesteryl ester transfer protein and unesterified fatty acids promotes a reduction in the particle size of high density lipoproteins, Biochim. Biophys. Acta (in press, 1990).

12. D. Sammett and A. R. Tall, Mechanisms of enhancement of cholesteryl ester transfer activity by lipolysis, J. Biol. Chem. 260:6687 (1985).

13. J. L. Ellsworth, M. L. Kashyap, R. L. Jackson and J. A. K. Harmony, Human plasma lipid transfer protein catalyses the speciation of high density lipoproteins, Biochim. Biophys. Acta 918:260 (1987).

14. G. J. Hopkins and P. J. Barter, Role of triglyceride-rich lipoproteins and hepatic lipase in determining the particle size and compositions of high density lipoproteins, J. Lipid Res. 27"1265 (1986).

REGULATION OF HEPATIC LIPASE EXPRESSION BY AN INTERMEDIATE

OF THE CELLULAR CHOLESTEROL BIOSYNTHETIC PATHWAY

Steven J. Busch, Gary A. Martin, Roger L. Barnhart
Margaret A. Flanagan and Richard L. Jackson

Merrell Dow Research Institute
2110 E. Galbraith Road
Cincinnati, OH 45215

INTRODUCTION

Two lipolytic enzymes, hepatic triglyceride lipase (H-TGL) and lipoprotein lipase (LpL), are responsible for the catabolism of lipoproteins in the circulation (1). H-TGL is bound to endothelial cells lining the sinusoidal cavities of the liver and has been identified on the adrenals and ovaries as well. These steroidogenic tissues utilize cholesterol derived from receptor-mediated and receptor-independent pathways (2-7). Recent studies suggest that only the hepatocyte is capable of synthesizing H-TGL, and that extrahepatic H-TGL is derived directly from hepatic secretion (8). The level of hepatic secretion of H-TGL has been shown to be partially responsive to specific hormonal concentrations in the plasma and several factors in situ. However, no evidence has yet linked H-TGL expression to regulation of the cholesterol biosynthetic pathway. In this study we demonstrate that the expression of H-TGL in the transformed hepatic cell-line, HepG2, is induced under conditions in which the cholesterol biosynthetic pathway is inhibited. We demonstrate that by depriving the cell of a biosynthetic intermediate prior to cholesterol, H-TGL expression is substantially induced.

MATERIALS AND METHODS

Materials. [2-^{14}C]Acetate, sodium salt (53 mCi/mmol), tri[1-^{14}C]oleoylglycerol (54.3 mCi/mmol), [α-^{32}P]dATP (400 Ci/mmol), and [α-^{32}P]UTP (400 Ci/mmol) were purchased from Amersham Corp. (Arlington Heights, IL). Mevinolin was obtained from Merck Sharpe and Dohme Research Laboratories (Rahway, NJ). LDL and HDL were isolated from normal human plasma by ultracentrifugation in KBr between densities 1.019-1.063 and 1.063-1.210 g/ml, respectively.

Hepatoma cells. HepG2 cells were obtained from American Type Culture Collection (Rockville, MD). They were maintained in Eagle's MEM supplemented with 10% fetal bovine serum (FBS) and 2 mM glutamine (maintenance medium) at 37°C in a humidified air atmosphere of 5% CO_2. Lipoprotein-deficient serum (LPDS) was prepared from FBS by ultracentrifugation in KBr at density 1.21 g/ml. After centrifugation for 48 h at 214,000 x g the lipoprotein layer was removed by aspiration. The bottom fraction was dialyzed against phosphate buffered saline (PBS, 10 mM

Hypercholesterolemia, Hypocholesterolemia, Hypertriglyceridemia
Edited by C.L. Malmendier *et al.*, Plenum Press, New York, 1990

potassium phosphate, pH 7.4, containing 120 mM NaCl and 2.7 mM KCl) and sterile filtered through a Corning 0.22 micron cellulose acetate membrane (Corning Glass Works, Corning, NY) prior to use.

Measurement of cellular cholesterol biosynthesis. Cholesterol biosynthesis was determined in cells by measuring the incorporation of [14C]acetate into total cholesterol. Confluent monolayers of HepG2 cells were established in 36 mm well plates (Falcon) and were fed with maintenance media. Prior to each experiment cells were incubated with 10% LPDS-MEM media for 4 h. Then mevinolin was added to the cell cultures and incubated for 2 h. Cells were pulsed with 1 µCi of [14C]acetate per well for 4 h. The medium was removed, and cell mono-layers were washed 2x with PBS. Lipids were extracted by adding 2 ml of hexane:isopropanol (3:2, v/v) to each well and allowing the extraction to proceed for 2 h at room temperature, taking care to avoid evaporation of the solvent. After removing the solvent, each well as washed 2x with 1 ml of the same solvent. The combined lipid extracts were dried under a stream of nitrogen, and each pellet was then resolubilized in 500 µl of the same solvent. Ten µl, containing 10 µg of unlabeled cholesterol and 5 µg of unlabeled cholesterol oleate added as carrier, were spotted on silica gel TLC strips and developed in heptane:diethyl ether:acetic acid (85:14:1, v/v/v). TLC strips were air dried and stained with iodine vapor to locate cholesterol and cholesterol oleate; spots corresponding to these standards were cut out and radioactivity determined. Following lipid extraction, cell protein was solubilized in 1 ml of 1 N NaOH and quantitated by the method of Lowry et al. (9) using bovine serum albumin (BSA) as a standard.

Assay for H-TGL activity. H-TGL activity in the HepG2 cell culture media was quantitated by measuring the lipolytic activity towards a substrate of tri[1-14C]oleoylglycerol emulsified with Triton N101 as described previously (10, 11).

Cell total cholesterol determination. HepG2 cells (approximately 2 x 10^8) were trypsinized and transferred to conical 50 ml centrifuge tubes. Cells were washed with two sequential washes of 15 ml Hank's Buffer and were extracted with 5 ml of hexane:isopropanol (3:2, v/v) overnight at 4°C. After centrifugation, the cells were re-extracted with 2 ml of solvent for 30 min at room temperature. The two extractions were combined, and total cholesterol was determined by gas chromatography after saponification and extraction with hexane; free cholesterol was determined by omitting the saponification step.

Measurement of H-TGL mRNA transcript levels. Clone pST668, used for solution hybridization to quantitate H-TGL mRNA transcripts, is a 668 bp Sst1-Sst1 insert (from nucleotides 115 to 783) of H-TGL pHL220 (10), ligated into the polylinker Sst1 site of the vector pSP6/T7-19 (BRL, Bethesda, MD). Antisense mRNA (cRNA) probes were synthesized on a linear Pvu11-EcoR1 isolated fragment from the pST668 plasmid, containing the entire 668 bp internal H-TGL cDNA insert and the T7 polymerase binding site. These labeled transcripts were synthesized by the manufacturers recommendation. Tracer excess solution hybridization was carried out essentially as described by Lee et al. (12). Total cellular RNA was isolated from HepG2 cells by the guanidinium isothiocyanate procedure followed by cesium chloride centrifugation (13). The total cellular RNA pellet was resolubilized in water and passed through a RNase free G-50 Sephadex column prior to determining the nucleic acid concentration by UV absorbance at 260 nm. H-TGL transcripts were calculated per HepG2 cell based on the mass of total RNA per cell of 6.9 pg (14).

Measurement of H-TGL and HMG-CoA reductase mRNA by Northern-blot

<u>analysis</u>. Northern-blot analysis for H-TGL was performed as described previously (10) using hybridization conditions of Busch et al. (11). For the Northern-blot analysis for HMG-CoA reductase a 2.5 kb Bgl 11 fragment containing the majority of the human HMG-CoA reductase cDNA (15) was cut from pHRED-102 (ATCC# 57042) and isolated by gel electrophoresis. This fragment was labeled with $[\alpha-^{32}P]ATP$ and $[\alpha-^{32}P]CTP$ to a specific activity of $1-2 \times 10^9$ cpm/µg. Prehybridization and hybridization were performed at 42°C in 0.2% BSA, 0.2% polyvinylchloride, 0.2% ficoll, 50 mM Tris-HCl, pH 7.5, 0.1% sodium pyrophosphate, 1.0% SDS, and 50% formamide; prehybridization was performed for 4 h and hybridization for 48-72 h. The final wash of the blots was at 65°C for 45 min in 0.1 x SSC (1 x SSC = 0.15 M NaCl, 0.015 M sodium citrate, pH 7.0), 1% SDS.

RESULTS

The human hepatoma cell line, HepG2, when treated with lipoprotein deficient serum, was inducible for cholesterol biosynthesis. Incorporation of $[^{14}C]$acetate into cholesterol increased approximately 2-fold from 16,654 to 35,553 cpm/mg of cell protein. This treatment did not result in altered levels of H-TGL secretion. Simultaneous addition of a cholesterol biosynthetic inhibitor, mevinolin, blocked 85% of the cholesterol biosynthesis and induced the level of secreted H-TGL activity by approximately 5-fold. H-TGL-specific mRNA levels were increased 2-fold by mevinolin treatment, whereas HMG-CoA reductase mRNA levels were induced 14-fold. The addition of LDL or HDL, in the absence of mevinolin treatment, had no significant effect on H-TGL expression. In contrast, simultaneous addition of LDL and mevinolin induced H-TGL expression 2-fold over mevinolin alone. This also resulted in a strong decrease in HMG-CoA reductase expression. These results demonstrate that a decrease in the flow of intermediates into the cholesterol biosynthetic pathway resulted in an increase in both H-TGL and HMG-CoA reductase whereas LDL feeding strongly suppressed HMG-CoA reductase. In contrast, this treatment did not suppress but instead further enhanced H-TGL expression induced by mevinolin.

Accumulation of cellular cholesterol was stimulated by mevalonic acid feeding of cells. In Table 1, the addition of mevalonic acid to cells alone caused an elevation in cellular cholesterol relative to cell protein and a decrease in the basal levels of H-TGL secretions. When mevalonic acid was added with mevinolin, cholesterol levels were also elevated and H-TGL secretion again decreased to below controls. These results show a differential response of H-TGL regulation to mevinolin + LDL compared to mevinolin + mevalonic acid and suggest that an intermediate in the cholesterol biosynthetic pathway other than cholesterol regulates H-TGL expression.

The results in which LDL feeding in the presence of mevinolin resulted in HMG-CoA reductase repression suggested that oxidized cholesterol was produced during the metabolism of LDL and that it caused this repression. In contrast, H-TGL expression was enhanced by this treatment. A similar effect was observed in cells treated simultaneously with mevinolin and 25-hydroxycholesterol. HMG-CoA reductase mRNA levels were reduced to below detectable levels as a result of treatment with 25-hydroxycholesterol while H-TGL mRNA levels were increased 4-fold with a concomitant increase in H-TGL secretion. We suggest that H-TGL expression is regulated by an intermediate of the cholesterol biosynthetic pathway and that maximum induction due to mevinolin and 25-hydroxycholesterol treatment results from complete blockage of this cholesterol biosynthetic pathway.

.

Table 1. Mevinolin and mevalonic acid induce changes in cell cholesterol content and in H-TGL expression.

Treatment	Total Cell Protein	Cell Cholesterol µg/mg Cell Protein	Secreted H-TGL Activity
LPDS only	10.8	36.1	42.4 ± 1.8
Mevalonic acid	13.2	45.3	20.2 ± 6.0
Mevinolin	10.2	27.4	169.0 ± 29.0
Mevinolin + Mevalonic acid	17.2	51.4	15.3 ± 1.1

HepG2 cell cultures were fed LPDS-MEM and subsequently refed with the indicated additions at the following concentrations; mevalonic acid, 1 mM; mevinolin, 37 µM. Medium was replaced every 24 h and the second change was recovered and quantitated for H-TGL activity as described in Materials and Methods. Cell protein and cholesterol were also quantitated for each culture plate as described. H-TGL activity is expressed as nmol oleic acid released/h/15 ml culture medium ± SD (n=3).

DISCUSSION

In this study the hepatic mesenchymal cell-line, HepG2, was utilized to demonstrate that H-TGL expression can be induced by reducing cellular concentrations of metabolites in the cholesterol biosynthetic pathway. Addition of mevinolin to HepG2 cells clearly reduced cellular cholesterol content, induced H-TGL mRNA levels and induced the level of H-TGL secreted activity. This suggests that H-TGL expression may be regulated by the level of cellular cholesterol or intermediates in the cholesterol biosynthetic pathway. Since mevalonic acid treatment reduced H-TGL expression, we suggest that H-TGL expression is regulated by a product at or beyond the level of mevalonic acid. Both LDL-cholesterol feeding and 25-hydroxycholesterol feeding in the presence of mevinolin enhanced the mevinolin-induction of both H-TGL mRNA and secreted lipolytic activity and simultaneously reduced HMG-CoA reductase expression. This suggests a differential regulatory point between HMG-CoA reductase and H-TGL expression. The mechanism of H-TGL regulation is not at present clear. Regulation is apparent at both the level of H-TGL-specific mRNA and H-TGL protein. Post-transcriptional regulation may also be suggested by the relatively modest increases in H-TGL secretion accompanying the 4-fold increase in H-TGL-specific mRNA levels due to mevinolin and 25-hydroxy-cholesterol treatment. Perhaps synthesis of an effector which influences either H-TGL mRNA turnover or transcription is blocked by lack of mevalonic acid. Transcriptional induction of H-TGL might occur as a consequence of decreased maturation of a repressor protein. In this regard, Beck et al. (1988) have described the requirement of mevalonic acid as a precursor for the formation of isoprenylated nuclear proteins. This isoprenyl-protein form is itself an obligate intermediate necessary for proper maturation of the nuclear laminins. Mevalonic acid depletion due to strong inhibition of HMG-CoA reductase by a combined treatment of mevinolin + 25-hydroxycholesterol would decrease the availability of mature repressor protein. Mevalonic acid feeding would therefore prevent induction by mevinolin (Table 1) and even lower basal expression when added in excess of normal cellular levels. Identification of the pathway intermediate(s) which is responsible for H-TGL repression will help to clarify the mechanism involved in H-TGL regulation.

ACKNOWLEDGEMENTS

The authors would like to thank Ms. Anjalee Jaganathen and Ms. Mary Lynn Points for help in preparation of this manuscript.

REFERENCES

1. R. L. Jackson, Lipoprotein lipase and hepatic lipase, The Enzymes 16:141 (1983).

2. H. Jansen and W. J. de Greef, L-type lipase activity in ovaries of superovulated rats. Relationship to cholesterol homeostasis, Mol. Cell. Endocrinol. 57:7 (1988).

3. M. S. Brown and J. L. Goldstein, A receptor-mediated pathway for cholesterol homeostasis, Science 232:34 (1986).

4. E. Reaven, Y.-D. I. Chen, M. Spicher, S.-F. Hwang, C. E. Mondon and S. Azhar, Uptake of low density lipoproteins by rat tissues: Special emphasis on the luteinized ovary, J. Clin. Invest. 77:1971 (1986).

5. J. T. Gwynn and J. F. Strauss III, The role of lipoproteins in steroidogenesis and cholesterol metabolism in steroidogenic glands, Endocr. Rev. 3:299 (1982).

6. H. Jansen, Hepatic triglyceride lipase and high density lipoproteins, Arztl. Lab. 34:29 (1988).

7. H. Jansen and W. C. Hülsmann, Enzymology and physiological role of hepatic lipase, Biochem. Soc. Trans. 13:24 (1985).

8. M. H. Doolittle, H. Wong, R. C. Davis and M. C. Schotz, Synthesis of hepatic lipase in liver and extrahepatic tissues, J. Lipid Res. 28:1326 (1987).

9. O. H. Lowry, N. J. Rosebrough, A. L. Farr and R. J. Randall, Protein measurement with the Folin phenol reagent, J. Biol. Chem. 193:265 (1951).

10. G. A. Martin, S. J. Busch, G. D. Meredith, A. D. Cardin, D. T. Blanken- ship, S. J. T. Mao, A. E. Rechtin, C. W. Woods, M. M. Racke, M. P. Schafer, M. C. Fitzgerald, D. M. Burke, M. A. Flanagan and R. L. Jackson, Isolation and cDNA sequence of human postheparin plasma hepatic triglyceride lipase, J. Biol. Chem. 263:10907 (1988).

11. S. J. Busch, G. A. Martin, R. L. Barnhart and R. L. Jackson, Heparin induces the expression of hepatic triglyceride lipase in a human hepatoma (HepG2) cell-line, J. Biol. Chem. 264:9527 (1989).

12. J. J. Lee, F. J. Calzone, R. J. Britten, R. C. Angerer and E. H. Davidson, Activation of sea urchin actin genes during embryogenesis: Measurement of transcript accumulation from five different genes in Strongylocentrus purpuratus, J. Mol. Biol. 188:173 (1988).

13. T. Maniatis, E. F. Fritsch and J. Sambrook, 1982, in: Molecular Cloning: A Laboratory Manual, Cold Spring Harbor Laboratories, Cold Spring Harbor, New York.

14. C. Sadhu and L. Gedamu, Regulation of human metallothionein (MT) genes. Differential expression of MT1-F, MT1-G, and MT11-A genes in the hepatoblastoma cell line (HepG2), J. Biol. Chem. 263:2679 (1988).

15. K. L. Luskey and B. Stevens, Human 3-hydroxy-3-methylglutaryl coenzyme A reductase. Conserved domains responsible for catalytic activity and sterol-regulated degradation, J. Biol. Chem. 260:10271 (1985).

16. L. A. Beck, T. J. Hosick and M. Sinensky, Incorporation of a product of mevalonic acid metabolism into proteins of Chinese hamster ovary cell nuclei, J. Cell. Biol. 107:1307 (1988).

Human Plasma Lecithin:Cholesterol Acyltransferase (LCAT)
On the role of essential carboxyl groups in catalysis

Matti Jauhiainen and Peter J. Dolphin[*]

Department of Biochemistry, National Public Health Institute, Helsinki, Finland and [*]Department of Biochemistry, Dalhousie University, Halifax, Nova Scotia, Canada

LCAT is one of the three major enzymes of plasma lipoprotein metabolism and catalyses the transacylation of the fatty acid at the sn-2 position of lecithin to the 3-hydroxyl group of cholesterol forming lysolecithin and cholesteryl ester within the plasma compartment. In a series of reports (1-4) we elucidated the chemical catalytic mechanism of this important enzyme which now appears to bear some similarities to that of the recently discovered lecithin:retinol acyltransferase (LRAT) (5). Incorporating the previous observation that LCAT can esterify lysolecithin present in LDL (6) and the recent observations of Sorci-Thomas et al. (7) the major features of the LCAT catalytic mechanism can be summarized as shown in Figure 1. The sn-2 carbonyl carbon of lecithin, present in plasma HDL, is initially subjected to nucleophilic attack by the oxygen atom of Serine-181 with the resultant formation of a transient tetrahedral adduct which decays with bond cleavage to yield a fatty acylated serine residue. The fatty acid is then internally transacylated to one of two vicinal cysteine residues (Cys-31 and Cys-184) by a mechanism that appears irreversible. Either cysteine residue can then donate its fatty acid moiety to cholesterol forming cholesteryl ester. The lysolecithin acyltransferase or LAT reaction is proposed to involve only Ser-181 which becomes acylated upon lecithin cleavage and donates its fatty acid back to another lysolecithin molecule when the concentration of the lysolipid is sufficiently high. Cholesteryl ester can be both formed and cleaved by LCAT (7). During cleavage the fatty acid is retained by the enzyme and transferred to another cholesterol molecule. Fatty acid originating as cholesteryl ester does not appear to be transferred to lysolecithin (7) and we thus propose that either of the vicinal cysteine residues mediates the cleavage of cholesteryl ester with the formation of a fatty acylated residue. This is designated as the CAT or cholesterol acyltransferase reaction in Fig. 1. The enzymatic cleavage of ester and amide bonds by serine esterases, proteases and triacylglycerol lipases is thought to involve a "catalytic triad" of three residues, serine, histidine and aspartic acid (8-10). These residues participate in a proton relay, the function of which is to retain the catalytic serine hydroxyl proton and thereby assist the nucleophilic attack of the serine oxygen atom upon the carbonyl carbon of the ester or amide bond.

Hypercholesterolemia, Hypocholesterolemia, Hypertriglyceridemia
Edited by C.L. Malmendier *et al.*, Plenum Press, New York, 1990

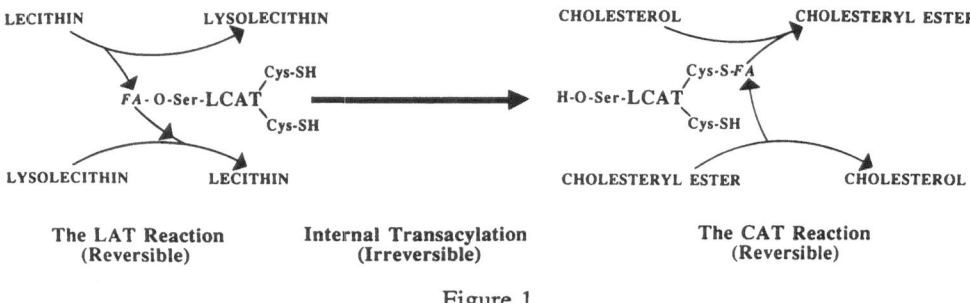

The LAT Reaction
(Reversible)

Internal Transacylation
(Irreversible)

The CAT Reaction
(Reversible)

Figure 1

Our previous studies utilizing specific chemical modification of highly purified LCAT
(1) and the inhibitor, phenylboronic acid (2) had clearly demonstrated the involvement of a
single serine and histidine residue in the first part of the LCAT reaction, lecithin cleavage. In
an attempt to determine if a proton relay system involving aspartic acid is present within the
catalytic site of LCAT we have recently evaluated the requirement for functional carboxyl
groups in lecithin cleavage by LCAT. Our proposed mechanism for the first portion of the
LCAT reaction is as shown in Fig. 2.

Our approach was to expose highly purified human plasma LCAT to 25 mM 1-Ethyl-
3-(3'-dimethylaminopropyl)-carbodiimide (EDAC), a specific carboxyl group reagent, in the
presence of 1-[^{14}C] Glycine methyl ester which replaces the carbodiimide and radiolabels the
activated carboxyl group. The results of these experiments are shown in Fig. 3 which clearly
demonstrates a progressive, time dependent inactivation of LCAT by carbodiimide treatment
with a concomitant increment in the number of carboxyl groups modified. Stoichiometric
analyses of these data in the reciprocal plot shows that a maximum of three carboxyl groups
are modified upon complete inactivation of the enzyme. Residues present within the catalytic
site of an enzyme are normally afforded protection against chemical modification when
incubated in the presence of their substrate(s). Contrary to the expected result, exposure of

MECHANISM OF LECITHIN CLEAVAGE BY LCAT

TRANSITION STATE

PROTON RELAY

SERINE OXYESTER

PROTON RELAY

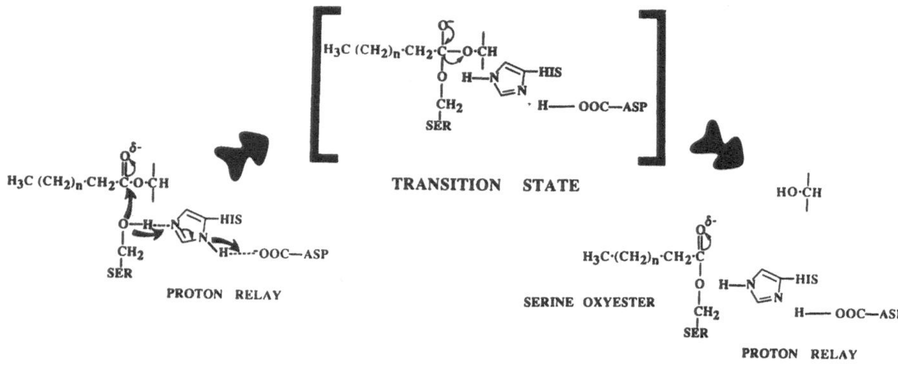

Figure 2

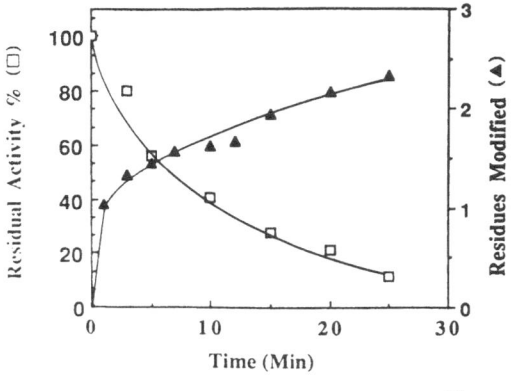
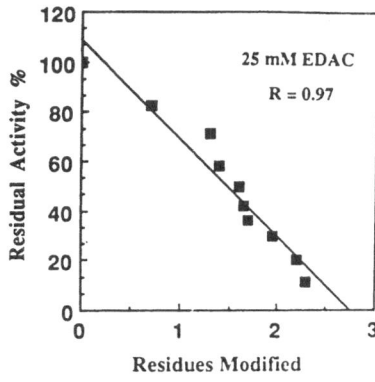

Figure 3

LCAT to 50 mM EDAC in the presence of an artificial proteoliposome substrate containing Lecithin:Cholesterol:apo A-I (3.2:250:0.8 mol:mol:mol) completely failed to protect LCAT against inactivation.

However, increasing the cholesterol content of the proteoliposome relative to a fixed mass of lecithin and apo A-I resulted in a progressive increment in the ability of these substrates to protect the enzyme against carbodiimide induced inactivation as shown in Fig. 4. Furthermore, proteoliposome with increased cholesterol content were better substrates for LCAT mediated cholesteryl ester formation than those with lower cholesterol content as shown in Table 1. Proteoliposomes with apo A-I content btween 0.2 and 2.4 mol relative to 250 mol of lecithin and 12.5 mol of cholesterol were all equally ineffective in protecting LCAT against EDAC inactivation. HDL_3, in contrast, effectively protected the enzyme. We therefore conclude that as the cholesterol to lecithin ratio of the artificial substrates approach that of HDL_3, a natural substrate of LCAT, their catalytic efficiency as reflected by their ability to fit the catalytic site is increased. There is one further consideration with respect to this and our previous (1) substrate protection experiments in which we had shown that the proteoliposome containing the least amount of cholesterol (Lecithin:Cholesterol:apo A-I (3.2:250:0.8 mol:mol:mol)) adequately protected the catalytic serine, histidine and cysteine residues against inactivation by chemical modification. In this instance it should be noted that the serine and cysteine residues would be fatty acylated during incubation with the substrate and thus unavailable to interact with the chemical modifiers. The histidine residue proximal to the catalytic serine residue may well be subject to steric hindrance following fatty acylation of the serine and thus protected against chemical modification. The carboxyl group of aspartic acid present within a proton relay system although "buried" within the tertiary structure of the protein is generally quite removed from the catalytic serine residue and does not participate directly in the catalytic mechanism via the formation of a transient tetrahedral adduct or via the formation of a relatively stable acylated species. Thus protection of the carboxyl groups of aspartic acid residues participating within a catalytic triad against chemical modification would

VARIABLE PROTECTION OF LCAT AGAINST INACTIVATION BY 50 mM EDAC WITH PROTEOLIPOSOMES OF INCREASING CHOLESTEROL / LECITHIN RATIO

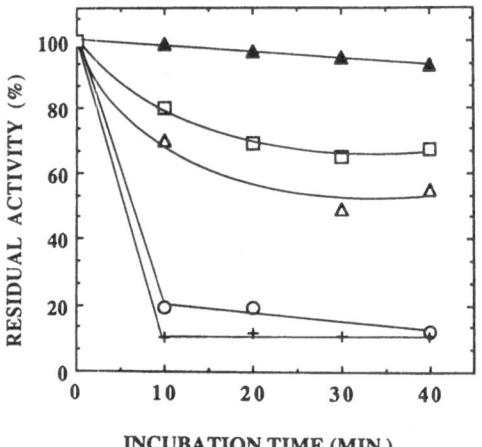

SYMBOL	CHOLESTEROL (mol)	LECITHIN (mol)	APO A-1 (mol)
▲	CONTROL		
+	3 2	250	0 8
○	6 4	250	0 8
△	22 4	250	0 8
□	28 8	250	0 8

Figure 4

be more dependent upon enzyme substrate affinity and the extent to which the substrate conformed to the geometry of the catalytic site. This possibility appears to be supported by the data in Fig.4 and Table 1.

In conclusion, these data clearly show that a maximum of three functional carboxyl groups are required for the LCAT mediated formation of cholesteryl ester from lecithin and cholesterol. The requirement for functional carboxyl groups indicates, but does not prove,

Table 1

THE EFFECTS OF INCREASING CHOLESTEROL TO PHOSPHOLIPID MOLAR RATIO UPON SUBSTRATE CATALYTIC EFFICIENCY

APO A-1	CHOLESTEROL (nmol)	LECITHIN (nmol)	CHOL / LECITHIN (mol / mol)	nMol CE Formed / h / ml.
0.8	3.2	250	0.0128	6 4
0.8	6.4	250	0.0256	12 9
0 8	22.4	250	0.0896	26.9
0 8	28.8	250	0.1152	34 5

HDL_3 has a molar cholesterol to phospholipid ratio of 0.288.

that a Ser - His - Asp catalytic triad similar to that present in serine-histidine esterases, proteases and triacylglycerol lipases may be operative in human plasma LCAT.

REFERENCES

1) Jauhiainen,M., and Dolphin,P.J. (1986) J. Biol. Chem. 261, 7032-7043.
2) Jauhiainen,M., Ridgway,N.D., and Dolphin,P.J. (1987) Biochim. Biophys. Acta. 918, 175-188.
3) Jauhiainen,M., Stevenson,K.J., and Dolphin,P.J. (1988) J. Biol. Chem. 263, 6525-6533.
4) Jauhiainen,M., Yuan,W., Gelb,M.H., and Dolphin,P.J. (1989) J. Biol. Chem. 264, 1963-1967.
5) MacDonald,P.N., and Ong,D.E. (1988) J. Biol. Chem. 263, 12478-12482.
6) Subbaiah,P.V., Albers,J.J., Chen,C-H., and Bagdade,J.D. (1980) J. Biol. Chem. 255, 9275-9280.
7) Sorci-Thomas,M., Babiak,J., and Rudel,L.L. (1990) J. Biol. Chem. 265, 2665-2670.
8) Kraut,J. (1977) Ann. Rev. Biochem. 46, 331-358.
9) Brady,L., Brzozowski,A.M., Derewenda,Z.S., Dodson,E., Dodson,G., Tolley,S., Turkenberg,J.P., Christiansen,L., Hugh-Jensen,B., Norskov,L., Thim,L., and Menge,U. (1990) Nature 343, 767-770.
10) Winkler,F.K., D'Arcy,A., and Hunziker,W. (1990) Nature 343, 771-774.

PLASMA LIPOPROTEIN PHENOTYPE IN RESPONSE TO CHOLESTERYL

ESTER TRANSFER PROTEIN LEVELS IN DYSLIPOPROTEINEMIA

Yves L. Marcel, Alan R. Tall, Mireille Hogue, Ross W. Milne
and Ruth McPherson

Laboratory of Lipoprotein Metabolism, IRCM, Montreal
Quebec; Department of Medicine, Columbia University College
of Physicians and Surgeons, New York, N.Y.; Lipid Research
Laboratory, Royal Victoria Hospital, McGill University
Montreal, Quebec

There are two mamallian systems for reverse cholesterol transport, one of which is dependent on the presence of cholesteryl ester transfer protein (CETP) activity[1,2]. The rat is typical of species lacking CETP activity and under these conditions, triglyceride-rich lipoproteins and their remnants transport only that cholesterol which was initially associated with the nascent particles. Rat HDL on the other hand, is rich in apoE and transports cholesterol derived from both splanchnic and peripheral organs, and esterified by LCAT and eventually returns it to the liver, via an apoE receptor. In contrast in the human system, the high CETP activity promotes the transfer of 2/3 or more of cholesteryl esters in HDL to the triglyceride-rich lipoproteins and their remnants. These recipient particles are then actively removed by hepatic apoE and apoB/E receptors[3]. In individuals with active receptors, this represents the major pathway for reverse cholesterol transport while a minority of HDL cholesterol is cleared directly by an apoE-mediated pathway. Direct selective uptake of HDL cholesteryl esters by the liver has also been suggested, but recent evidence indicated that this route is also apoE-mediated[4].

We do not know which of the two above pathways is the most efficient in returning cholesterol to the liver and most protective against atherosclerosis, it is thus of interest to consider the different lipoprotein phenotypes that occur under conditions of either low or high CETP activity. In genetic CETP deficiency, the classical phenotype is that of hyperalphalipoproteinemia with accumulation of large HDL particles[5,6]. ApoE levels have been found to be increased from about 1.5 to 6 fold above normal in different CETP deficient patients[6]. The lipoprotein profile observed is thus similar to that of the rat. The accumulation of cholesteryl esters in HDL suggests that in absence of CETP, maximal transport of cholesteryl esters cannot be maintained. However this apparently inefficient pathway represents a safe form for intravascular storage of cholesterol since HDL cholesterol is not directly taken up by any peripheral receptor.This pathway is apparently of adequate capacity in species with low LDL transport rates.

Hypercholesterolemia, Hypocholesterolemia, Hypertriglyceridemia
Edited by C.L. Malmendier *et al.*, Plenum Press, New York, 1990

77

In contrast, in situations where CETP activity is high, more cholesteryl esters are efficiently transfered to triglyceride-rich lipoproteins and their remnants, which provided that these particles contain sufficient apoE$_3$ or apoE$_4$ are actively cleared by hepatic apoE receptors. Thus the reverse cholesterol transport pathway operating with high CETP activity is also dependent upon apoE for clearance. When the available apoE is not an effective ligand for its receptor, accumulation of β-VLDL results in the expression of the type III phenotype.

However no mutation causing an increase in CETP levels and activity has been reported and therefore no example of the lipoprotein profile that would prevail under such conditions is available. Increased lipid transfer activity has been measured in certain human dyslipoproteinemia[8], notably in familial hypercholesterolemia and in dysbetalipoproteinemia, although decreased lipid transfer activity has also been reported[9]. Some of these apparent contradictions reflect the composite CETP activity that is measured in plasma. The assay of CETP activity is influenced by several factors, notably by the composition and ratio of donor and acceptor lipoproteins and by the presence of putative inhibitors of CETP. It is therefore important to measure the plasma levels of CETP in different types of dyslipoproteinemia where abnormal lipoproteins accumulate for a better understanding of the role of lipid transfer in the development of these lipoprotein phenotypes. Using a radioimmunoassay recently developped[10], we report here the preliminary data on plasma CETP levels in hyperlipoproteinemia.

RESULTS AND DISCUSSION

We have studied hyperlipoproteinemic patients recruted from the Lipid Research Clinic of the Royal Victoria Hospital. This included 17 patients with familial hypercholesterolemia, 18 with familial combined hyperlipoproteinemia, 10 with dysbetalipoproteinemia (type III), 12 with hypertriglyceridemia (typeIV), and 5 with chylomicronemia (type V). The diagnosis was made on the basis of appropriate family history, clinical and biochemical data. Total cholesterol, triglycerides, HDL cholesterol were measured by standard techniques[11] and the levels of apoA-I, A-II, B, and E by radioimmunoassays as described earlier[10]. The control population (79 subjects) was taken from the previously reported groups of normolipemic subjects which were studied under the same conditions and in whom plasma levels of CETP had also been measured[10].

Plasma levels of CETP were significantly increased in patients with hypercholesterolemia (+26%),and familial combined hyperlipidemia (+25%). However CETP levels were unchanged in the patients with moderate hypertriglyceridemia without chylomicronemia (type IV).The highest levels of CETP were observed in patients with type III hyperlipoprteinemia (+68%) and in patients with severe hyperchylomicronemia (+85%). The present results corroborate the earlier observations of Tall et al[8], who showed that CETP activity was increased in type III hyperlipoproteinemia. In hypercholesterolemia as in familial combined hyperlipidemia, we speculate that the modest increase in CETP may be related to increased delivery of cholesterol to the periphery which stimulates the synthesis of CETP by macrophages[12]. The higher levels of CETP which are observed in type V and type III may be derived from additional macrophagic intestinal[13] and hepatic[14,15] synthesis in response to the increased chylomicron and chylomicron remnant levels. Alternatively, assuming that CETP is cleared with these lipoproteins, the high level of CETP seen in types III and V may be related to the decreased catabolism of chylomicrons and their remnants. However the later hypothesis seems improbable since no CETP was found associated with chylomicrons in the plasma of fat-fed subjects[10].

Calculation of Pearson correlation coefficients between CETP levels and lipoprotein parameters for the total group of hyperlipoproteinemic patients demonstrated the existence of significant relationships between CETP and total cholesterol (r=0.52), CETP and VLDL cholesterol (r=0.63), CETP and triglycerides (r=0.53), and CETP and apoE (r=0.40). The correlation with triglycerides was only observed when hyperchylomicronemic subjects were included. In contrast to earlier observations in normolipemic subjects, CETP levels in hyperlipidemic patients were not correlated with apoA-I. The present observations confirm previous evidence that CETP levels vary in relation with cholesterol. In the cholesterol fed rabbit, the levels of hepatic mRNA are increased together with the plasma levels of CETP[16]. In response to the diet, apoE plasma levels are also increased[16] but hepatic mRNA levels are unchanged[17], demonstrating that although the genes for these two proteins are regulated by dietary cholesterol, there are important differences in tissue specificity for the control of their expression. This is also consistant with observations in human that both CETP and apoE are increased with probucol treatment[18] and with dietary cholesterol (McPherson et al, unpublished results).

In conclusion CETP levels are increased in conditions where cholesterol transport is increased, perhaps especially when peripheral delivery of cholesterol is increased. There is also evidence that CETP and apoE are coregulated by cholesterol although with different tissue specificities. Finally, elevated CETP is associated with a lipoprotein phenotype of decreased HDL cholesterol and increased cholesteryl esters and apoE in chylomicrons and in their remnants. This suggests that increase CETP activity may be essential for the expression of type III hyperlipoproteinemia in E2 homozygosity.

REFERENCES

1. A.R. Tall, Plasma lipid transfer proteins, J. Lipid Res. 27:361 (1986).

2. Y.C. Ha, and P.J. Barter, Differences in plasma cholesteryl ester transfer activity in sixteen vertebrate species, Comp. Biochem. Physiol. 71B:265 (1982).

3. R.J. Havel, Functional activities of hepatic lipoprotein receptors, Ann. Rev. Physiol. 48:119 (1986).

4. F. Rinniger, and R.C. Pitman, Mechanism of cholesteryl ester transfer protein-mediated uptake of high density lipoprotein cholesteryl esters by HepG2 cells, J. Biol. Chem. 264:6118 (1989).

5. J. Koizumi, H. Mabuchi, A. Yoshimura, I. Michishita, M. Takeda, H. Itoh, Y. Sakai, T. Sakai, K. Ueda, and R. Takeda, Deficiency of serum cholesteryl ester transfer activity in patients with familial hyperalphalipoproteinemia.

6. S. Yamashita, Y. Matsuzawa, M. Okazaki, H. Kako, T. Yasugi, H. Akioka, K, Hircho, and S. Tarui, human apolipoprotein E-rich high density lipoproteins (HDL$_c$) in subjects with deficiency of cholesteryl ester transfer , Atherosclerosis 70:7 (1988).

7. R. W. Mahley, Apolipoprotein E: cholesterol transport protein with expanding role in cell biology, Science 240:622 (1988).

8. A.R. Tall, E. Granot, R. Brocia, J. Tabas, C. Hessler, K. Williams, and M. Denke, Accelerated transfer of cholesteryl esters in dyslipidemic plasma, J. Clin. Invest. 79:1217 (1987).

9. P.E Fielding, C.J. Fielding, R.J. Havel, J. P. Kane, and P. Tun, Cholesterol net transport, esterification and transfer in human hyperlipidemic plasma, J. Clin. Invest. 71:449 (1983).

10. Y.L. Marcel, R. McPherson, M. Hogue, H. Czarnecka, Z. Zawadski, P.K. Weech, M.E. Whitlock, A.R. Tall, and R.W. Milne, Distribution and concentration of cholesteryl ester transfer protein in plasma of normolipemic subjects, J. Clin. Invest. 85:10 (1990).

11. Lipid Research Clinics Program, in "Manual of Laboratory operations". U.S. Government printing office, Bethesda Md, p. 1-81 (1974).

12. J.H. Tollefson, R. Faust, J.J. Albers, and A. Chait, Secretion of a lipid transfer proteinn by human moncyte-derived macrophages, J. Biol. Chem. 260:5887 (1985).

13. R.A. Faust, and J.J. Albers, Regulated vectorial secretion of cholesteryl ester transfer proteinn (LTP-I) by the Calo2 model of human enterocyte epithelium, J. Biol. Chem. 263:8786 (1988).

14. T.L. Swenson, J.S. Simmons, C.B. Hesler, C. Bisgaier, and A.R. Tall, Cholesteryl ester transfer protein is secreted by HepG2 cells and contains asparagine-linked carbohydrates and sialic acid, J. Biol. Chem. 262:16271 (1987).

15. R.A. Faust, and J.J. Albers, Synthesis and secretion of plasma cholesteryl ester transfer protein by human hepatocarcinoma cell line, HepG2, Arteriosclerosis 7:267 (1987).

16. E.M. Quinet, L.B. Agellon, P.A., Kroon, Y.L. Marcel, Y.C. Lee, M.E. Whitlock, and A.R. Tall, Atherogenic diet increased cholesteryl ester transfer protein mRNA levels in rabbit liver, J. Clin. Invest. 85:357 (19??).

17. Y.S. Chao, T.T. Yamin, G.M. Thompson, and P.A. Kroon, Tissue-specific expression of genes encoding apolipoprotein E in rabbits. J. Biol. Chem. 259:5306 (1984).

18. R. McPherson, M. Hogue, R.W. Milne, A.R. Tall, and Y.L. Marcel, Increase in plasma cholesteryl ester transfer protein during Probucol treatment: relationship to changes in HDL compostion, submitted for publication.

HDL RECEPTOR-MEDIATED CHOLESTEROL EFFLUX FROM CELLS AND ITS REGULATTION

E.L. Bierman[*], J. Oram, and A. Mendez

[*]Division of Metabolism, Endocrinology
   and Nutrition RG-26
University of Washington
Seattle, WA 98195

HDL_3 particles can nonspecifically pick up cholesterol from the plasma membrane by desorption, but specific high affinity interaction of particles containing intact apo AI with an HDL receptor protein (110 kDA) on the cell surface is required to signal cholesterol translocation from intracellular compartments to the cell surface, thereby enhancing cholesterol efflux (1). Once at the cell surface, cholesterol is removed from cells by appropriate acceptor particles including, but not limited to, HDL. Treatment of HDL particles with trypsin or tetranitromethane will destroy the ability of these particles to signal cholesterol translocation; however, the ability of trypsinized HDL to accept cholesterol from plasma membrane sterol is preserved (2). Thus, cells excrete intracellular cholesterol by a pathway that involves two major steps: the translocation of cholesterol from intracellular pools to the plasma membrane, and the removal of membrane-associated cholesterol from cells by exogenous acceptors. It is only the translocation step of this pathway that requires receptor binding of HDL and this is the step that is regulated.

One of the major signals promoting intracellular translocation of sterol to the plasma membrane appears to involve protein kinase-C (3). Treatment of cells with protein kinase activators (phorbol ester, diacylglycerol, calcium ionophore) will mimic the action of HDL, while use of protein kinase-C inhibitors (e.g., sphingosine) will block this effect. Direct measurements of protein kinase-C have shown that HDL binding to cells will activate the enzyme. The nature of the proteins phosphorylated by protein kinase-C and the mechanisms of sterol translocation from intracellular sites remain to be elucidated.

The number of HDL receptors on the cell surface appears to be regulated by cell cholesterol content and their proliferative state. HDL binding and HDL receptor-mediated sterol efflux from cells (fibroblasts, smooth muscle cells, endothelial cells, monocyte-derived macrophages) are up-regulated by an increase in cellular unesterified cholesterol, whether added from exogenous sources (solubilized cholesterol, LDL cholesterol or acetyl LDL cholesterol to cells with scavenger receptors) or endogenous sources (inhibition of esterification with an ACAT inhibitor) (4) (5) (6) Modification of HDL apoproteins with tetranitromethane or treatment with proteases will block HDL binding and subsequent sterol translocation (1) (7).

Hypercholesterolemia, Hypocholesterolemia, Hypertriglyceridemia
Edited by C.L. Malmendier *et al.*, Plenum Press, New York, 1990

Distinct cellular pools of cholesterol can be differentially radio-labeled. The intracellular pool of cholesterol can be radiolabeled by incubating cholesterol-loaded cells with tritiated mevalonic acid at 15°C. With this technique most of the label remains with newly synthesized intracellular sterol (1). The plasma membrane compartment can be labelled by brief incubation at 37°C with tritiated cholesterol. The specific translocation of intracellular cholesterol to the plasma membrane can be measured by washing and fixing cells and then incubating them with cholesterol oxidase, an enzyme that reacts only with sterol in accessible domains in the plasma membrane. Thus, cholesterol oxidase-sensitive sterol is presumed to be at the plasma membrane while cholesterol oxidase-resistant sterol is presumed to be in an intracellular compartment.

Since the growth state of cells influences cell cholesterol homeostasis, the question of whether alteration of cell proliferation would regulate HDL receptor activity was tested. Cholesterol-loaded fibroblasts were incubated 24 to 48 hours with growth factors including platelet-derived growth factor, insulin and IGF-1. All mitogens tested reduced HDL binding in a dose related manner in association with increased DNA synthesis (8) (9). HDL$_3$-mediated efflux of intracellular sterols was diminished by growth factors in parallel with decreased HDL binding. Decreased HDL receptor activity was associated with reciprocally increased LDL receptor activity. Thus, both mechanisms responsive to growth factors appear to be designed to make more cell cholesterol available for new membrane synthesis.

Conversely, growth inhibition increased HDL receptor activity. Both HDL binding and HDL mediated efflux of intracellular sterols from cells were increased in the presence of the growth inhibitor gamma interferon (10). Receptor up-regulation was associated with increased HDL receptor number (by kinetic analysis) and increased binding (by ligand blot) to the 110 kilodalton HDL receptor protein. Thus, by altering intracellular cholesterol levels in regulatory pools, changes in the proliferative state of cells can modulate HDL receptor activity to finely regulate cell cholesterol homeostasis (Table 1). Mitogens, by down-regulating HDL receptor-mediated cholesterol efflux from cells, could contribute to arterial cell cholesterol accumulation.

Table 1

Factors affecting HDL Receptor-Mediated Cholesterol Efflux

Increase	Decrease
cell cholesterol excess	cell cholesterol depletion
inhibition of ACAT	growth factors (mitogens)
inhibition of cell proliferation	insulin
activation of protein kinase-C	inhibition of protein kinase-C

The coupling of receptor binding of HDL to the cellular translocation pathway provides an efficient means for cells to protect themselves from accumulation of unesterified cholesterol. In the absence of HDL particles, excess cholesterol that accumulates in cells by any of a variety of delivery pathways is esterified and stored as esterified cholesterol in lipid droplets. When HDL receptors become occupied with lipoprotein particles, cells receive a signal that sterol acceptors are present in the extracellular fluid, and excess intracellular cholesterol is diverted from the microsomal esterifying enzyme to the plasma membrane for removal. By this process, HDL depletes cells of cholesterol and prevents the formation of cholesterol ester-rich foam cells. This receptor-mediated sterol translocation pathway may be the biochemical mechanism that underlies the apparent anti-atherogenic effect of HDL.

References

1)   Slotte JP, Oram JF, and Bierman EL, 1987, Binding of high density lipo-
     proteins to cell receptors promotes translocation of cholesterol from
     intra-cellular membranes to the cell surface. J. Biol. Chem. 262:12904-
     12907.

2)   Oram JF, Johnson CJ, Brown TA, 1987, Interaction of high density
     lipoprotein with its receptor on cultured fibroblasts in macrophages:
     Evidence for reversible binding at the cell surface without
     internalization. J. Biol. Chem. 262:2405-2410.

3)   Mendez, AJ, Oram JF, and Bierman EL, 1989, Sphingosine inhibits HDL-
     mediated efflux of intracellular sterols. Arteriosclerosis 9:720a.

4)   Oram JF, Brinton EA, Bierman EL, 1983, Regulation of high-density
     lipoprotein receptor activity in cultured human skin fibroblasts and
     human arterial smooth muscle cells. J. Clin. Invest. 72:1611-1621.

5)   Brinton EA, Kenagy R, Oram JF, Bierman EL, 1985, Regulation of high
     density lipoprotein binding activity of aortic endothelial cells by
     treatment with acetylated low density lipoprotein. Arteriosclerosis
     5:329-335.

6)   Aviram M, Bierman EL, Oram JF, 1989, High density lipoprotein
     stimulates sterol translocation between intracellular and plasma
     membrane pools in human monocyte-derived macrophages. J. Lipid Res.
     30:65-76.

7)   Brinton EA, Oram JF, Chen C-H, Albers JJ, Bierman EL, 1986, Binding of
     high-density lipoprotein to cultured fibroblasts after chemical
     alteration of apoprotein amino acid residues. J. Biol. Chem. 261:491-
     503.

8)   Oppenheimer MJ, Oram JF, Bierman EL, 1987, Downregulation of high
     density lipoprotein receptor activity of cultured fibroblasts by
     platelet-derived growth factor. Arteriosclerosis 7:325-332.

9)   Oppenheimer MJ, Sundquist K, Bierman EL, 1989, Downregulation of high
     density lipoprotein receptor activity in human fibroblasts by insulin
     and IGF-I. Diabetes 38:117-122.

10) Oppenheimer MJ, Oram JF, Bierman EL, 1988, Up-regulation of high
     density lipoprotein receptor activity by τ-Interferon associated with
     inhibition of cell proliferation. J. Biol. Chem. 263:19318-19323.

BINDING OF APOLIPOPROTEINS A TO ADIPOSE CELLS : ROLE OF RECEPTOR SITES IN CHOLESTEROL EFFLUX AND PURIFICATION OF BINDING PROTEIN(S)

Ronald Barbaras[1], Pascal Puchois[2], Anne Pradines-Figuères[1], Armin Steinmetz[2], Véronique Clavey[2] Nordine Ghalim[2], Jean-Charles Fruchart[2], and Gérard Ailhaud[1]

[1] Centre de Biochimie, Parc Valrose, 06034 Nice France
[2] Institut Pasteur/SERLIA, 59019 Lille, France

Epidemiological studies have shown a relationship between low concentrations of high density lipoprotein (**HDL**) cholesterol and the incidence risk of cardiovascular diseases[1,2]. Recent pharmacological studies[3] have clearly demonstrated the protective role of HDL and their involvment in reverse cholesterol transport *in vivo*[4,5]. In that respect apo E-free HDL has been long known to bind to a variety of cells and to promote cholesterol efflux[6]. Among peripheral tissues, adipose tissue is recognized both in man and rodents for its ability to accumulate, store and, when needed, mobilize a large pool of unesterified cholesterol[7,8]. Thus adipose cells represent a cell type suitable to study the first step in reverse cholesterol transport, i.e. cholesterol efflux. Unfortunately adipocytes isolated from adipose tissue loose their viability within a few hours, preventing the analysis of middle-term and long-term responses. During the last decade have been established in our laboratory preadipocyte cell lines from adipose tissue of genetically-obese ob/ob mice[9] and their lean counterpart[10]. The validity of these cellular models is supported by i) the biochemical properties of differentiated cells which are similar, if not identical, to those of adipocytes isolated from fat tissue and ii) the ability of undifferentiated cells to differentiate *in vivo* within a few weeks into fully mature fat cells after their injection into athymic mice, under conditions where these cells could be unambiguously demonstrated not to be fat cells originating from the host animal[11]. Most of the studies, if not otherwise stated, were performed with Ob1771 cells, a subclone of Ob17 cells established from ob/ob mice[12,13].

Hypercholesterolemia, Hypocholesterolemia, Hypertriglyceridemia
Edited by C.L. Malmendier *et al.*, Plenum Press, New York, 1990

85

CHARACTERIZATION OF LDL AND HDL BINDING SITES AND CHOLESTEROL FLUX/EFFLUX IN OB1771 CELLS

The binding of human apo AI, apo AII and apo AIV to mouse adipose cells and the study of their functional properties were made feasible owing to extensive homologies existing between rat, mouse and human apolipoproteins[14-16]. In addition important homologies do exist between rat (and likely mouse) and human apo B, including the consensus region of apo B and apo E which should be involved in the binding to the apo B,E receptor[17-20]. The binding of ^{125}I-LDL was competitively inhibited by LDL > VLDL > total HDL ; human LDL and mouse LDL were equipotent in competition assays. Methylated LDL and apo E-free HDL were not competitors. In contrast, the binding of ^{125}I-apo E-free HDL was competitively inhibited by apo E-free HDL > total HDL and that of ^{125}I-HDL$_3$ by mouse HDL. Thus mouse adipose cells possess distinct apo B,E and apo E-free HDL binding sites which can recognize heterologous or homologous lipoproteins. Further studies of apo E-free HDL binding sites revealed that the binding of ^{125}I-HDL$_3$ was competitively inhibited by apo AI/dimyristoylphosphatidyl-choline complexes > mouse HDL > HDL$_3$. To explore the possibility that apo AI, apo AII and apo AIV bind to the same sites, competition experiments were performed in which binding of either of the three apolipoproteins was performed in the presence of the two other unlabeled apolipoproteins. The results suggest strongly that apo AI, apo AII and apo AIV bind to common receptor sites[21]. This hypothesis is supported by the finding that a highly purified protein from Ob1771 cells remains able to bind the three apolipoproteins (ref.22 and *vide infra*). The observation that ~ 1 mol of apo AII is bound per 2 mol of apo AI or apo AIV (Table I) could be explained if one assumes that the receptor site is a dimeric structure (see Fig.1) which recognizes each monomer of dimeric apo AII in the same way that it recognizes two molecules of monomeric apo AIV or apo AI. In any event, the stoichiometry of apo AI (or apo AIV) *versus* apo AII binding has been consistently observed in intact cells and in homogenates after detergent solubilization as well as after extensive purification of binding proteins of 80 and 92 kDa (ref.22 and *vide infra*).

During the course of these studies, it was observed that the endogenous cholesterol synthesis was nil[12] but the most striking observation was the fact that long-term exposure of adipose cells to LDL and HDL$_3$ did not affect the number of apo B,E receptor sites and that of apo E-free HDL receptor sites. In other words, the "buffering" capacity of adipose cells seems limited with respect to the regulation of cholesterol content. This lack of cholesterol homeostasis would explain the rather unique ability of adipose tissue *in vivo* to accumulate and mobilize a large pool of unesterified cholesterol[7,8]. Since differentiated Ob1771 cells were able to find, internalize and degrade LDL[12], it appeared that adipose cells did not show an efficient cholesterol homeostasis *in vitro* and thus, as a first prediction, should accumulate cholesterol. The second prediction was that cholesterol-preloaded cells should mobilize cholesterol when exposed to appropriate lipoprotein particles. Both predictions were fullfilled. As shown in Table I, it is of interest to note

Table 1. Parameters of binding and cholesterol efflux of various apolipoprotein complexes

Ligand	Binding		Cholesterol efflux	
	K_d μM	B_{max} (sites/cell)	EC_{50} μM	$[C]_{max}$ μM
Apo AI[a]	1	193,000	1.5	3.2
Apo AII[a]	0.2	100,000	0	0
Apo AIV	0.32	223,000	0.3	1.1

[a] Apo AI, apo AII and apo AIV bind but apo AI and apo AIV only promote cholesterol efflux[13]. In addition both LpAI, LpAI:AII and likely LpAIV bind but LpAI and LpAIV only promote cholesterol efflux[21-23].

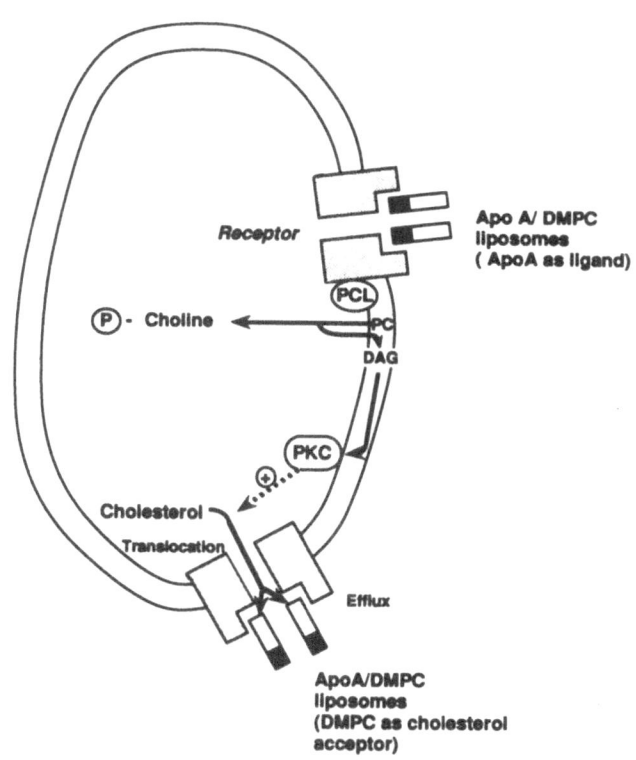

Fig. 1 Hypothetical model for cholesterol efflux from mouse adipose cells.

that comparisons between apparent K_d values for binding of apo AI and apo AIV and the EC_{50} values for cholesterol efflux are within the same range of concentrations[21-24]. These results suggest that specific binding to these distinct sites was a pre-requisite to cholesterol accumulation and subsequently to cholesterol mobilization. It is also of interest to note that cholesterol accumulation was taking place in the presence of LDL under the form of unesterified cholesterol only, in agreement with the fact that, at least in rat fat tissue, the majority (75-95%) of adipocyte cholesterol is unesterified and associated with central oil (triacylglycerol) droplet[7,8].

RELATIONSHIPS IN ADIPOSE CELLS BETWEEN THE PRESENCE OF RECEPTOR SITES FOR APO AI, APO AII AND APO AIV AND THE PROMOTION OF CHOLESTEROL EFFLUX

In order to establish whether receptor sites for HDL were indeed required for the promotion of cholesterol efflux, use was made of Ob17 cells in which have been induced genetically defined alterations of the growth control mechanism by transferring cloned oncogenes[25]. Ob17PY cells were obtained after transfer of the complete early region of polyoma virus whereas Ob17MT cells were obtained after transfer of a modified genome encoding only the middle T protein. The broad range of phenotypes thus generated has also offered us unique opportunities to study cholesterol efflux in adipose cells as cells of the Ob17MT18 subclone had a 3-fold higher number of HDL receptor sites than cells of the parental Ob17 clone whereas *growing* Ob17PY cells did not have any detectable sites (see below).

As a pre-requisite to study the critical role, if any, of HDL receptor sites and to undertake their purification, conditions for their visualisation were searched and found using bivalent cross-linking reagent discuccimidyl suberate at 4°C in the presence of apo AI-containing liposomes and intact Ob1771 cells or derived crude membranes[26]. The existence of two specific cell-surface protein components of M_r 100,000 and 130,000 was demonstrated. It is possible that two different proteins of M_r ~70,000 and ~100,000 able to bind one molecule of proteins of apo AI of M_r 28,000 are indeed present in adipose cells. Alternatively the possibility of either a single glycoprotein able to bind one molecule of apo AI but having different degrees of glycosylation, or a single glycoprotein able to cross-link one or two molecules of apo AI, could be envisionned. The key observation in our study on the role of HDL receptor sites in the promotion of cholesterol efflux was that *no binding* of HDL3, apo AI, apo AII or apo AIV was observed in growing Ob17PY cells and derived crude membranes, in contrast to growing or growth-arrested Ob1771 cells (see Table I) or Ob17MT18 cells (not shown). After thymidine block, growth-arrested Ob17PY cells became able to recover in parallel binding activities for HDL3, apo AI, apo AII and apo AIV. The possibility that this recovery was an event common to various cell surface receptors is not very likely since apo B,E and transferrin receptor sites were both present in growing and growth-arrested Ob17PY

cells as well as in Ob1771 cells. The recovery of HDL
receptor sites in growth-arrested Ob17PY cells was rapid
(16 h) and prevented in actinomycin D- or cycloheximide-
treated cells, adding further support to the conclusion that
these sites are protein component(s). When experiments of
cholesterol efflux were performed, the results showed that,
after cholesterol accumulation taking place in the presence of
LDL cholesterol, subsequent exposure to HDL3 or apo AI (but
again not apo AII) led to cholesterol efflux from Ob1771 cells
and growth- arrested Ob17PY cells but not from growing Ob17PY
cells[26]. Thus it appears that the presence of high-affinity
receptor sites for HDL in intact adipose cells is required for
the promotion of cholesterol efflux. The existence of cell
surface binding sites which recognize apolipoproteins A is
supported by recent experiments showing that the binding of
apo AI/DMPC complexes to intact Ob1771 cells was followed
within 1-2 minutes by the formation of diacylglycerol from
phosphatidylcholine as substrate ; it is of interest that
apo AII/DMPC complexes were inactive in that respect,
supporting the view that apo AII was playing the role of an
antagonist[21,24,27,28]. Altogether, these observations led us
to attempt in purifying apo A binding proteins by using
Ob17MT18 cells, a transformed cell line enriched 3-fold in
apo A binding sites as compared to the parental Ob17 cells.

The purification scheme is shown in Figure 2. An 1,400-
fold purification over the starting crude homogenate was
achieved[22]. The purified material contained two proteins that
were both able to bind apolipoproteins AI, AII and AIV but not
LDL. Glycopeptidase F treatment showed the existence of a
single protein bearing either N-linked high-mannose or complex
oligosaccharide chains. The purified material showed an
apparent molecular mass of 80 ± 9 kDa by high-pressure liquid
chromatography on TSKG 3000 SW column. Rabbit polyclonal
antibodies directed against the purified material revealed two
protein bands of 80 and 92 kDa after sodium dodecyl sulfate
polyacrylamide gel electrophoresis under reducing conditions
and immunoblotting. These bands were undetectable in growing
Ob17PY cells previously shown not to bind the various apo As
or not to undergo cholesterol efflux, whereas they were
conspicuous in growth-arrested Ob17PY cells which recovered
these properties. It is of utmost importance to recall that
these binding sites are present at the cell surface of intact
cells but more than 90% of apo A and apo B,E (LDL) binding
sites were shown to be present intracellularly[12]. This
situation is similar to that observed in skin fibroblasts
where a large proportion of LDL binding activity is also
present within the cells. Therefore both cell surface and
intracellular binding sites were purified in the present
study, but it must be recalled that the affinities of these
binding sites for their ligands were very similar in intact
Ob17 cells and derived crude membranes[12] and that the binding
parameters were found to be very similar for intact Ob17MT18
cells and the fraction purified from these cells by DEAE-
Trisacryl chromatography (Fig.2). Thus it is assumed that
cell surface binding sites and intracellular binding sites are
identical and that a receptor recognizing apolipoprotein A
has been purified. Figure 1 summarizes our working
hypothesis : it is possible that the functional apolipoprotein
A receptor, required for cholesterol efflux but not for

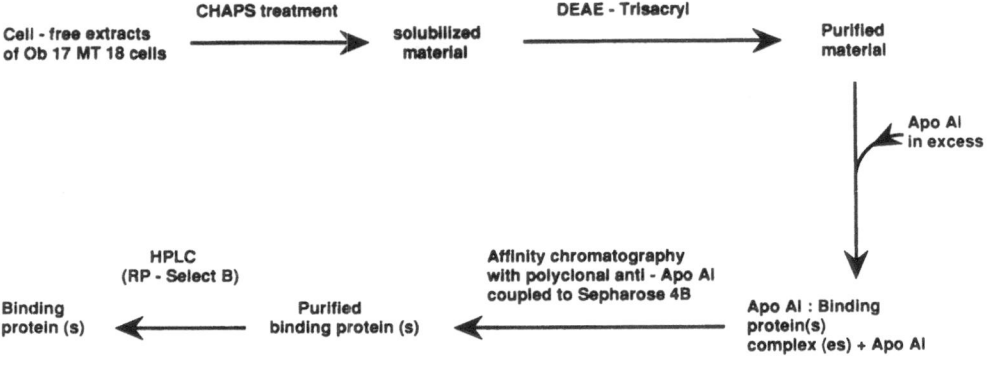

Fig. 2 Purification of apolipoprotein A binding
protein(s) from mouse adipose cells.

binding activity (Fig.1), is a dimer of two single polypeptide chains. This dimeric structure would be able to recognize either one mole of the dimeric apo AII or two moles of the monomeric apo AII or AIV. If so, we envision that the binding of one molecule of apo AI (or apo AIV), but not that of apo AII, might induce a conformational change allowing the binding of a second molecule of apo AI (or AIV) and the formation of an activated receptor. It is suggested that, within the apo AI (or apo AIV)/DMPC complexes, the apolipoprotein plays the role of a ligand triggering the PKC pathway (cholesterol translocation to the cell surface) whereas the liposomal structure *per se* plays the role of a cholesterol acceptor (cholesterol efflux from the cell surface). Recent experiments indicate that it is indeed the case and that a distinction between both events can be made experimentaly (N. Théret-Bidoui *et al.*, unpublished work).

ACKNOWLEDGEMENTS

The authors wish to thank Miss V. Boivin and Mrs. B. Barhanin for expert technical help, Dr. J. Barhanin for helpful advice in cross-link experiments and Mrs.G. Oillaux for expert secretarial assistance. This work was supported by the "Centre National de la Recherche Scientifique" (CNRS UPR 7300), by the "Fondation pour la Recherche Médicale Française" (Nice) and by "Institut Pasteur" (Lille).

REFERENCES

1. T. Gordon, W.P. Castelli, M. C. Hjortland, W. B. Kannel and T. R. Dawber, High density lipoprotein as a protective factor agains coronary heart disease: the Frammingham Study, Am. J. Med. 62:707 (1977).

2. N. E. Miller, O. H. Forde, D. S. Thelle and O. D. Mjos, The Tromso Heart Study: high-density lipoprotein and coronary heart disease: a prospective cas control study, Lancet 1:965 (1977).

3. M. Heikki Frick et al., Helsinki heart study: Primary-prevention trial with gemfibrozil in middle-aged men with dyslipidiema: safety or treatment, changes in risk factors, and incidence of coronary heart disease, N. Engl. J Med. 317:1237 (1987).

4. N. E. Miller, A. La Ville and D. Crook, Direct evidence that reverse cholesterol transport is mediated by high-density lipoprotein in rabbit, Nature 314:109 (1985).

5. Th. J.C. van Berkel, H. F. Bakkeren, F. Kuipers and R. J. Vonk in: Proceedings of Xth International Symposium on Drugs Affecting Metabolism, p.47 (1989).

6. A. R. Tall and D. M. Small, Body cholesterol removal: role of plasma high-density proteins, Adv. Lipid Res. 17:1 (1980).

7. R. K. Krause and A. D. Hartman, Adipose tissue and cholesterol metabolism, J. Lipid Res. 25:97 (1984).

8. A. Angel and B. Fong, Lipoprotein interactions and cholesterol metabolism in human fat cells, in: "The Adipocyte and Obesity: Cellular and Molecular Mechanisms", A. Angel, C. H. Hollenberg and D. A. K. Roncari, eds., Raven Press, New York (1983).

9. R. Négrel, P. Grimaldi and G. Ailhaud, Establishment of preadipocyte clonal line from epididymal fat pad ob/ob mouse that responds to insulin and to lipolytic hormones, Proc. Natl. Acad. Sci. USA 75:6054 (1978).

10. C. Forest, A. Doglio, L. Casteilla, D. Ricquier and G. Ailhaud, Expression of the mitochondrial uncoupling protein in brown adipocytes. Absence in brown preadipocytes and BFC-1 cells. Modulation by isoproterenol in adipocytes, Exp. Cell Res. 168:233 (1987).

11. D. Gaillard, P. Poli and R. Négrel, Characterization of ouabain-resistant mutants of the preadipocyte Ob17 clonal line. Adipose conversion in vitro and in vivo, Exp. Cell Res. 156:513 (1985).

12. R. Barbaras, P. Grimaldi, R. Négrel, and G. Ailhaud, Binding of lipoproteins and regulation of cholesterol synthesis in cultured mouse adipose cells, Biochim. Biophys. Acta 845:492 (1985).

13. R. Barbaras, P. Grimaldi, R. Négrel, and G. Ailhaud, Characterization of high-density lipoprotein binding and cholesterol efflux in cultured mouse adipose cells, Biochim. Biophys. Acta 888:143 (1986).

14. P. Forgez, M. J. Chapman, S. C. Rall Jr. and M. C. Camus, The lipid transport system in the mouse, Mus musculus: isolation and characterization of apolipoproteins B, AI, AII, and CIII, J. Lipid Res. 25:954 (1984).

15. C. G. Miller, T. D. Lee, R. C. LeBoeuf and J. E. Shively, Primary structure of apolipoprotein AII from inbred mouse strain BALB/c, J. Lipid Res. 28:311 (1987).

16. S. C. Williams, S. M. Bruckheimer, A. J. Lusis, R. C. LeBoeuf and A. J. Kinniburgh, Mouse apolipoprotein AIV gene: nucleotide sequence and induction by a high-lipid diet, Mol. Cell. Biol. 6:3807 (1986).

17. T. L. Innerarity, E. J. Friedlander, S. C. Rall, K. H. Weisgarber and R. W. Mahley, The receptor-binding domain of human apolipoprotein E, _J. Biol. Chem._ 258:12341 (1983).

18. T. J. Knott, R. J. Pease, L. M. Powell, S. C. Wallis, S. C. Rall Jr., T. L. Innerarity, B. Blankhart, W. H. Taylor, Y. Marcel, R. W. Mahley, B. Levy-Wilson and J. Scott, Complete protein sequence and identification of structural domains of human apolipoprotein B, _Nature_ 323:734 (1986).

19. C. Y. Yang, S. H. Chen, S. H. Gianturco, W. A. Bradley, J. T. Sparrow, M. Tanimura, W. K. Li, D. A. Sparrow, H. DeLoof, M. Rosseneu, F. S. Lee, Z. W. Gu, A. M. Gotto Jr. and L. Chan, Sequence, structure, receptor binding domains and internal repeats of human apolipoprotein B-100, _Nature_ 323:738 (1986).

20. A. J. Lusis, R. West, M. Mehrabian, M. A. Reuben, R. C. LeBoeuf, J. S. Kaptein, D. F. Johnson, V. N. Schumaker, M. P. Yuhasz, M. C. Schotz and J. Elovson, Cloning and expression of apolipoprotein B, the major protein of low and very low density lipoproteins, _Proc. Natl. Acad. Sci. USA_ 82:4597 (1985).

21. A. Steinmetz, R. Barbaras, N. Ghalim, V. Clavey, J.C. Fruchart, and G. Ailhaud, Human apolipoprotein AIV binds to apolipoprotein AI/AII receptor sites and promotes cholesterol efflux from adipose cells, _J. Biol. Chem._ 265:7859 (1990).

22. R. Barbaras, P. Puchois, J.C. Fruchart, A. Pradines-Figuères and G. Ailhaud, Purification of the apolipoprotein A receptor from mouse adipose cells, _Biochem. J.,_ in press.

23. R. Barbaras, P. Puchois, J.C. Fruchart and G. Ailhaud, Cholesterol efflux from cultured adipose cells is mediated by LpAI particles and not by LpaI:AII particles, _Biochem. Biophys. Res. Commun._ 142:63 (1987).

24. R. Barbaras, P. Puchois, P. Grimaldi, A. Barkia, J.C. Fruchart and G. Ailhaud, _in_: "Eicosanoids, Apolipoproteins, Lipoprotein Particles, and Atherosclerosis", C. L. Malmendier and P. Alaupovic, eds., Plenum Press, New York (1988).

25. P. Grimaldi, D. Czerucka, M. Rassoulzadegan, F. Cuzin and G. Ailhaud, Ob17 cells transformed by middle-T-only gene of polyoma virus differentiate in vitro and in vivo into adipose cells, _Proc. Natl. Acad. Sci. USA_ 81:5440 (1984).

26. R. Barbaras, P. Puchois, P. Grimaldi, A. Barkia, J.C. Fruchart and G. Ailhaud, Relationship in adipose cells between the presence of receptor sites for high density lipoproteins and the promotion of reverse cholesterol transport, _Biochem. Biophys. Res. Commun._ 149:545 (1987).

27. C. Delbart, N. Théret, G. Ailhaud, J.C. Fruchart, Phosphatidylcholine breakdown during receptor binding of HDL$_3$, _Circulation_ 80:487 (1989).

28. A. Barkia, P. Puchois, N. Ghalim, R. Barbaras, G. Ailhaud and J.C. Fruchart, Differential role of apolipoprotein AI-containing particles in cholesterol efflux from adipose cells, submitted.

LIPOPROTEIN A-I CONTAINING PARTICLES

Nicolas Duverger*, Nordine Ghalim*, Nathalie Theret*
Philippe Duchateau*, Gustave Aguie*, Gérard Ailhaud[+]
Graciela Castro*, Jean-Charles Fruchart*

* SERLIA et U. INSERM 325, Institut Pasteur, 1, rue du
Professeur Calmette, 59019 Lille Cédex, France
+ Centre de Biochimie, Parc Valrose, 06034 Nice, France

INTRODUCTION

Many epidemiological studies have indicated that the plasma level of high density lipoproteins (HDL) is inversely correlated with the risk for coronary artery disease[1]. It has been hypothesized that HDL exerts this protective effect by the "reverse" transport of excess cholesterol from peripheral tissues to the liver[2].

Nevertheless, in spite of its role in anti-atherogenesis, the true mechanisms of HDL uptake of peripheral cholesterol and subsequent delivery to the liver are still under investigation.

Conventionally HDL are isolated by ultracentrifugation in the density range of 1.063 to 1.21 g/ml. HDL represents a heterogenous population of particles which differ in size, in lipid and protein composition and overall in their metabolic functions.

Although nearly all the apoproteins that have been characterized thus far can be found in HDL in variable proportions, apo A-I and apo A-II together comprise 85-90% of the total HDL protein.

Ultracentrifugation has been an invaluable tool for subfractionation of lipoprotein particles but it has been shown that this procedure alters the structure and composition of the particles[3].

We have been using immunological procedures to isolate the different particles of the HDL fraction. Particles containing apo A-I and apo A-II (LpA-I:A-II) and particles that do not contain apo A-II (LpA-I).

Studies published by our group in collaboration with Ailhaud's group have shown that on long-term exposure to LpA-I and LpA-I:A-II particles ; only the LpA-I particles were able to promote cholesterol efflux from cholesterol preloaded, differentiated OB 1771 adipose cells[4].

More recently[5] it has been shown that particles containing apo A-IV are equally effective in producing cholesterol efflux.

Taking into account these results we decided to further study the different HDL particles ; to do so we prepared directly from whole plasma LpA-I and LpA-I:A-II particles and simultaneously we isolated from the same plasma sample, the particles containing apo A-IV (LpA-IV) and the particles containing apo A-I, A-IV and A-II (LpA-I:A-IV:A-II).

Hypercholesterolemia, Hypocholesterolemia, Hypertriglyceridemia
Edited by C.L. Malmendier *et al.*, Plenum Press, New York, 1990

The present studies were undertaken with two major goals :
- to define the lipid and protein composition of the different particles ;
- to correlate their composition to their ability to promote the cholesterol efflux from adipocytes in culture.

In order to understand the intracellular mechanism of cholesterol efflux and to appreciate the different roles of apo A-I and apo A-II, we decided to test the PKC involvement in this phenomenon.

It was recently described[6] that binding of HDL$_3$ to ^{3}H-phosphatidyl-choline (PC) labelled platelets stimulated a transient biphasic increase in diacylglycerol (DAG). We have analysed this stimulatory effect using HDL$_3$ and proteoliposomes containing apo A-I or A-II on incubation with the adipose cells OB 1771.

MATERIAL AND METHODS

The study was carried out on the plasma of five normolipidemic male subjects ; the blood was drawn into tubes containing EDTA and a mixture of protease inhibitors. The plasma was promptly separated by low-speed centrifugation at 4°C and immediately used for isolation of the parti-cles. All manipulations were performed at 4°C.

The particles were prepared by immunoaffinity chromatography as outlined before[1,2]. On each fraction, we determined : the lipid composi-tion by enzymatic methods ; the apoprotein composition by ELISA.

We measured the concentration of the two major phospholipids : phosphatidylcholine and shingomyelin isolated by thin layer chromato-graphy[7]. We estimated the proportion of the molecular species of fatty acids on the lipid extract purified by thin layer chromatography. The fatty acids hydrolyzed and methylated were measured on gas liquid chromatography[8].

The activity of the lecithin cholesterol acyl transferase (LCAT) was measured by the proteoliposome method of Chen and Albers[9]. Cholesterol efflux was determined on differentiated OB 1771 cells preloaded with ^{3}H-cholesteryl ester LDL prepared by the method of Craig et al[10].

^{3}H-diacylglycerol was measured after separation on thin layer chromatography. The apo A-I and apo A-II containing liposomes were prepared by the cholate dialysis procedure[9], molar ratio DMPC to protein 150:1.

Quantitation of apo A containing lipoprotein particles

In order to quantify LpA-I and LpA-I:A-II, two tests have been developed. To directly determine LpA-I:A-II, we have used an enzyme-linked differential antibody immunosorbent assay[11]. To directly determine LpA-I we have developed a simpler procedure using differential electro-immunoassay[12]. By using a large excess of anti A-II, LpA-I:A-II particles are retained in the first peak and LpA-I migrates as a second peak. A monoclonal anti A-I labelled with peroxidase revealed the two peaks while a monoclonal anti A-II revealed only one peak.

RESULTS

Figure 1 shows the representation of the composition by weight percentage of the different particles.

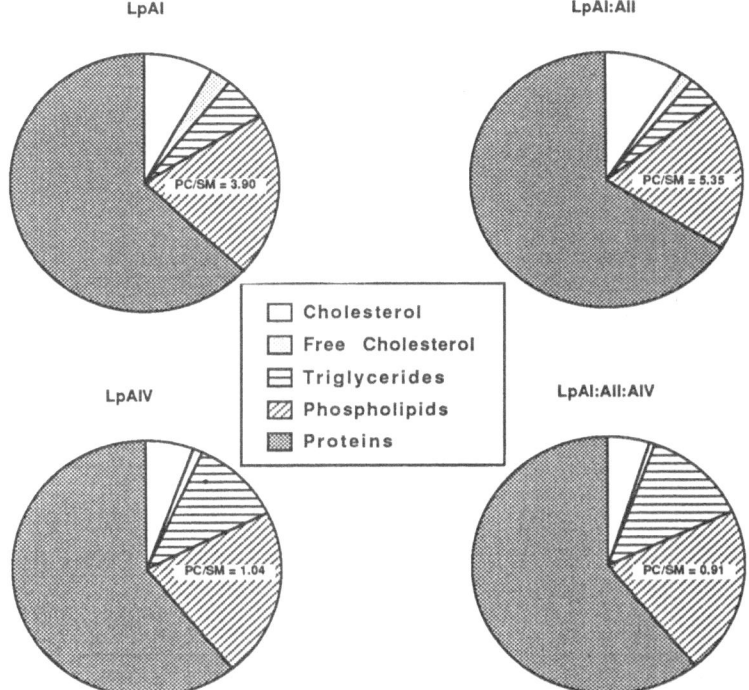

Fig. 1. Composition of the particles (wt %)

The four types of particles have about the same proportion of protein, but they show significant differences in cholesterol and trigly-ceride content. LpA-I and LpA-I:A-II particles contained more cholesterol (11% of the total mass) than LpA-IV and LpA-I:A-IV:A-II which contain 6%, conversely the particles containing apo A-IV have more triglycerides (12%).

Apolipoprotein analysis showed that LpA-I and LpA-IV contain a single apolipoprotein, 97 and 98.6% respectively, LpA-I:A-II contain 53% of apo A-I and 45% of apo A-II. The LpA-I:A-IV:A-II contain 65% of apo A-I, 18% of apo A-IV and 14% of apo A-II.

We determined the proportion of the different molecular species of fatty acids, the most striking results were found in the phospholipid fractions. The LpA-IV particles contain a high proportion of saturated fatty acids (76%) ; significatively different from LpA-I (58%) and LpA-I:A-II (47%).

Figure 2 shows the results of the determination of the activity of LCAT. Taking the value for the LpA-I particle as one hundred percent activity, we found that the LpA-IV particles have the most activity, followed by LpA-IV:A-I:A-II and the LpA-I ; the LpA-I:A-II particles have very little LCAT activity.

The phospholipid analysis revealed differences among the particles in the type of phospholipid constituents. The phosphatidylcholine/sphin-gomyelin ratios were 3.9, 5.3 for LpA-I, LpA-I:A-II respectively and around 1 for the particles containing apo A-IV.

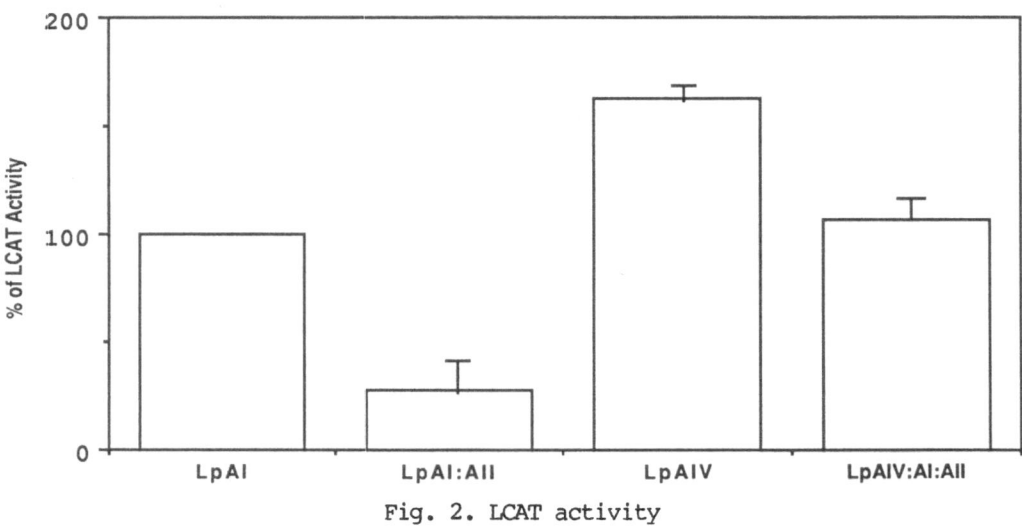

Fig. 2. LCAT activity

The characterized particles were used to test their capacity to promote the efflux of ^{3}H-cholesterol from cholesterol-preloaded OB 1771 cells. A rapid decrease in the cellular cholesterol content occurred during the incubation of the cells with LpA-I, LpA-IV:A-I:A-II and LpA-IV. This decrease was about 30 to 35% of the initial cellular cholesterol after 3 hours of incubation (Figure 3).

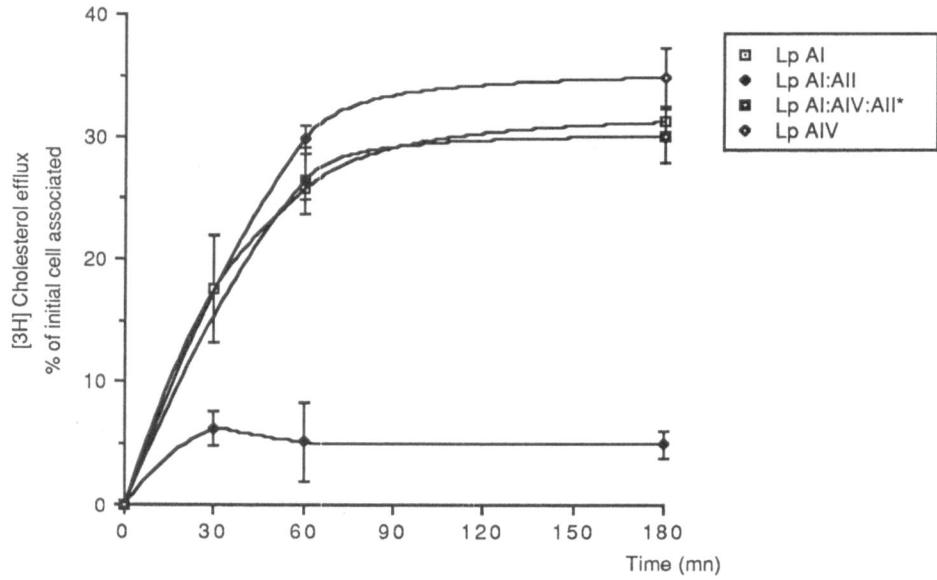

Fig. 3. Cholesterol efflux

Incubation of ^{3}H-PC prelabelled adipose cells in the presence of HDL$_3$ or liposome containing apo A-I results in ^{3}H-DAG production with a maximum at 5 min. On the other hand, tetranitromethane modified HDL$_3$ (TNM-HDL$_3$) which is not recognized by the HDL receptor, did not induce ^{3}H-PC breakdown and DAG release. Moreover despite their ability to be effective competitors for apo A-I binding sites, apo A-II containing liposomes were ineffective in stimulating DAG production by phospholipase activation. It seems likewise that only the binding of apo A-I to cell surface receptors is able to promote DAG generation (Figure 4).

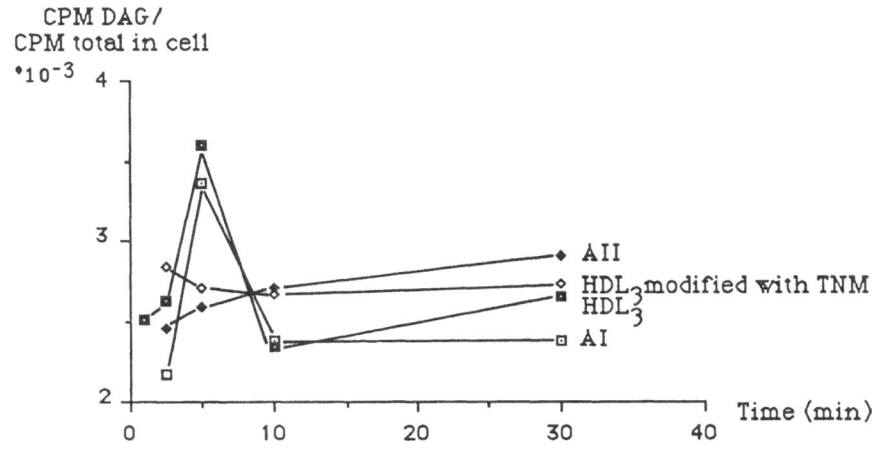

Fig. 4. 1-2 DAG formation after stimulation of ^{3}H-phosphatidylcholine prelabelled adipocytes with HDL$_3$, apo A-I or apo A-II containing proteoliposomes

The presence of phorbol esters in the incubation medium enhanced the cholesterol efflux efficiency in LDL-cholesterol loaded OB17 adipose cells (Figure 5). This data strongly suggests that PC-breakdown, DAG production and protein kinase C activation are involved in the cholesterol efflux.

DISCUSSION

Our results show that factors other than the lipid composition of the particles studied are the major determinant of the ability to promote cholesterol efflux.

The compositional data of LpA-I and the two apo A-IV containing particles reveals that these lipoproteins are very different, but on incubation with adipocytes they are equally effective in facilitating the efflux of cholesterol. On the other hand LpA-I:A-II are very similar in composition to LpA-I but are unable to promote cholesterol efflux. These results confirm the antagonist role proposed for apo A-II.

This antagonist role of apo A-II is demonstrated in our studies on the production of DAG and correlates very well with the results obtained in epidemiological studies.

Chol efflux
% of initial cell associated

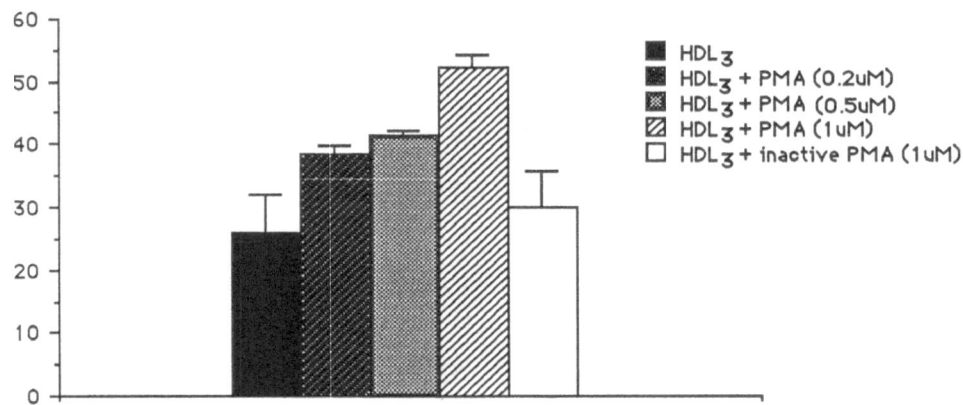

Fig. 5. Efflux of cholesterol induced by HDL₃ and Phorbol esters (PMA) on
OB 1771

We have shown recently that the lower apo A-I levels for patients
with significant coronary artery disease were reflecting, in fact, a
decrease in LpA-I particles[13]. Other data obtained in octogenarians[14]
supports, also, the view that LpA-I might represent the "anti-atherogenic" fraction of HDL.

Apo A-I in females and apo A-II in males were lower in octogenarians
while apo A-I in males and apo A-II in females were similar in
octogenarian and control subjects. However, LpA-I was significantly
elevated in octogenarian males and females by comparison with younger
control subjects.

Recently it has been observed that the level of LpA-I in children
whose patients suffer from premature coronary heart disease (CHD) was
lower than that of a control group without any familial history of CHD[15].

The clinical interest in the quantification of LpA-I and LpA-I:A-II
is illustrated by the effect of moderate alcohol consumption on HDL[16]. We
have measured HDL cholesterol, apo A-I, apo A-II, LpA-I and LpA-I:A-II in
plasma from three hundred and fifty male subjects matched for age and
clinical data and divided into five groups according to their alcohol
consumption. Results confirm that alcohol consumption increases
LpA-I:A-II and decreases LpA-I. These opposite variations are dose
dependent and the differences are highly significant. Our findings
indicate that an increase in HDL cholesterol can reflect an increase in
LpA-I:A-II and a decrease in LpA-I. Moreover, assuming that LpA-I is the
"anti-atherogenic" subfraction, alcohol would not have any anti-atherogenic effect through the increase in HDL.

REFERENCES

1. T. Gordon, W.P. Castelli, M.C. Hjortland, W.B. Kannel, T.R. Dawber,
   High density lipoprotein as protective factor against coronary
   heart disease, Am. J. Med. 62:707 (1977).

2. J. A. Glomset, The plasma lecithin-cholesterol acyl transferase
   reaction, J. Lipid Res. 9:155 (1968).

3. G. R. Castro, and C.F. Fielding, Evidence for the distribution of
   apolipoprotein E between lipoprotein classes in human normocholesterolemic plasma and for the origin of unassociated apolipoprotein E (LpE), J. Lipid Res. 25:58 (1984).

4. R. Barbaras, P. Puchois, J.C. Fruchart, and G. Ailhaud, Cholesterol efflux from cultured adipose cells is mediated by LpA-I particles but not by LpA-I:A-II particles, <u>Biochem. Biophys. Res. Commun.</u> 142:63 (1987).

5. A. Steinmetz, R. Barbaras, N. Ghalim, V. Clavey, J.C. Fruchart, and G. Ailhaud, Human apolipoprotein A-IV binds to apolipoprotein A-I/A-II receptor sites and promotes cholesterol efflux from adipose cells, <u>J. Biol. Chem.</u> 265:7859 (1990).

6. C. Delbart, N. Theret, G. Ailhaud, J.C. Fruchart, Phosphatidylcholine breakdown during receptor binding of HDL_3, 62nd Scientific Sessions of American Heart Association, New-Orleans, 13-16 novembre 1989, Circulation, 80/4, supplement II, II-487, abstract 1938 (1989).

7. A. Nouvelot, G. Sezille, P. Dewailly, J.C. Fruchart, Chromatographie monodimensionnelle des lipides polaires sur couche mince de gel de silice en gradient discontinu d'humidité, <u>Revue Française des Corps Gras</u> 7:251 (1977).

8. L. Hagenfelt, A gas chromatography method for the determination on individual free fatty acids in plasma, <u>Clin. Chim. Acta</u> 13:266 (1966).

9. C. H. Chen and J.J. Albers, Characterization of proteoliposomes containing apolipoprotein A-I : a new substrate for the measurement of lecithin cholesterol acyl transferase activity, <u>J. Lipid Res.</u> 23:680 (1982).

10. J. F. Craig, D.P. Via, B.C. Sharril, L.A. Labar, W.W. Mantulin, A.M. Gotto, and L. Smith, Incorporation of defined cholesteryl esters into lipoproteins using cholesterol ester-rich microemulsions, <u>J. Biol. Chem.</u> 257:330 (1982).

11. E. Koren, P. Puchois, P. Alaupovic, J. Fesmire, A. Kandoussi, J.C. Fruchart, Quantification of two different types of apolipoprotein A-I containing lipoprotein particles in plasma by enzyme linked differential antibody immunosorbent assay, <u>Clin. Chem.</u> 33:38 (1987).

12. H. J. Parra, H. Mezdour, N. Ghalim, J.M. Bard, J.C. Fruchart, Differential electroimmunoassay on ready-to-use plates for human LpA-I lipoprotein particles, <u>Clin. Chem.</u> in press (1990).

13. P. Puchois, A. Kandoussi, P. Fievet, J.L. Fourrier, M. Bertrand, E. Koren, J.C. Fruchart, Apolipoprotein A-I containing lipoproteins in coronary artery disease, <u>Atherosclerosis</u> 68:35 (1987).

14. G. Luc, J.M. Bard, S. Lussier-Cacan, H.J. Parra, J.C. Fruchart, J. Davignon, High density lipoprotein particles in octogenarians, <u>Metabolism</u> in press (1990).

15. P. Amouyel, H.J. Parra, D. Boute, J.M. Bard, D. Isorez, D. Barbier, J.C. Fruchart, G. Zylberberg, Inlfuence des antécédents cardio-vasculaires familiaux sur le profil lipoprotéique d'une population issue de la métropole lilloise, XVème Réunion de l'ADELF, Bordeaux, 12-15 septembre 1989.

16. N. Ghalim, P. Puchois, G. Zylberberg, P. Fievet, C. Demarquilly, J.C. Fruchart, Effect of alcohol intake on human apolipoproteins A-I containing lipoproteins subfractions, <u>Arch. Int. Med.</u> in press (1990).

DISTRIBUTION OF CHOLESTEROL WITHIN HIGH DENSITY LIPOPROTEINS FRACTIONATED BY IMMUNOAFFINITY CHROMATOGRAPHY

Daniel Pometta and Richard W. James

Division de Diabétologie, Département de Médecine
Hôpital cantonal universitaire, Geneva

INTRODUCTION

Cholesterol enjoys notoriety as a pro or anti-atherogenic lipid by virtue of the lipoprotein with which it is associated. This simplistic interpretation is most evident in the positive and negative correlations with the atherosclerotic process manifested by cholesterol associated with the most abundant lipoprotein species, respectively low density (LDL) and high density (HDL) lipoproteins [1,2]. The statement requires qualification, however, when lipoprotein sub-populations are contemplated. Thus, within the HDL density spectrum, the protective influence is largely believed to reside within the ultracentrifugally-defined lower density HDL-2 subclass [3]. HDL-3 enjoys much less support as an anti-atherogenic lipoprotein particle, a somewhat unsatisfactory state of affairs when considering mechanistic explanations of the function of HDL. Notably, particles of the size of HDL-3 are the principal acceptors of cellular cholesterol, the initiating step in reverse cholesterol transport [4]. In addition, they can act as a repository for lipids shed from triglyceride-rich lipoproteins, greatly facilitating the catabolic removal of these potentially atherogenic lipoprotein particles [5]. Further subfractionation of lipoproteins into, hopefully, metabolically homogenous sub-populations is one approach which should yield more satisfying explanations. This is the rationale behind the studies described in the present report. HDL subclasses 2 and 3, as defined by the physico-chemical criteria of ultracentrifugation, have been further fractionated using an immunoaffinity approach [6,7] targeting the predominant HDL apolipoprotein (apo) components, apos A-I and A-II. We have examined the cholesterol distribution of such immunoaffinity-defined fractions within HDL-2 and 3 isolated from

Hypercholesterolemia, Hypocholesterolemia, Hypertriglyceridemia
Edited by C.L. Malmendier *et al.*, Plenum Press, New York, 1990

healthy male and female populations. Furthermore, the distribution has been analysed in sub-groups corresponding to the lowest and highest total HDL-cholesterol quartiles of the same populations.

MATERIALS AND METHODS

Study populations

Healthy male (n=35) and female (n=33) subjects were recruited from the university hospital and medical research centre in Geneva. Basic clinical characteristics of these groups have been described previously [8]. Average HDL-cholesterol values were 1.22±0.19mm/l for men and 1.55±0.33mm/l for women.

The populations were also segregated into quartiles based on values of total HDL-cholesterol. For the male population, quartiles 1 and IV averaged HDL-cholesterol levels of 0.99±0.09mm/l (range 0.80-1.07mm/l) and 1.49±0.13mm/l (range 1.29-1.60mm/l). Corresponding values for female quartiles I and IV were 1.19±0.10mm/l (range 0.98-1.28mm/l) and 2.03±0.26mm/l (range 1.76-2.46mm/l).

Lipoprotein fractionation

Serial fractionation of high density lipoproteins from fasting plasma by ultracentrifugation and immunoaffinity chromatography was achieved as described [8]. The immunoaffinity procedure gave rise to two types of lipoprotein particle, described according to the presence of apos A-I and A-II. Thus LpAI,AII contains both apos, whereas LpAI contains apo A-I but no apo A-II. The particles are further defined by their subclass source ie HDL-2 or HDL-3.

Other analyses

Lipid and protein measurements and statistical analyses were performed as described previously [8,9,10].

RESULTS

Fig. 1 shows the distribution of cholesterol within the subfractions in both the male and female populations. Cholesterol was principally associated with HDL3-LpAI,AII and concentrations were not significantly different between the males and females. In contrast,

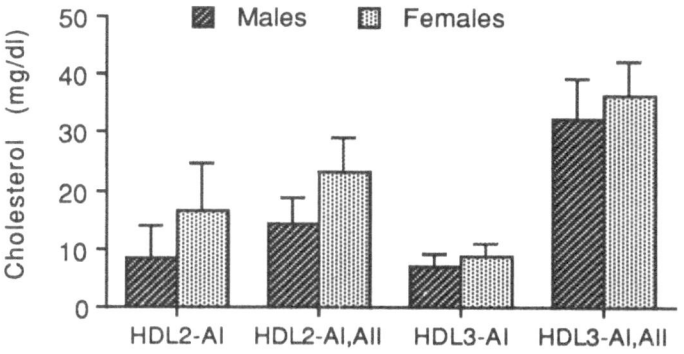

Fig. 1. Cholesterol concentrations of HDL subfractions

highly significant differences (p<0.0001 for HDL-2 derived subfractions; p=0.0001 for HDL3-LpAI) between populations were observed for the remaining subfractions. These differences were particularly marked for HDL2-LpAI and HDL2-LpAI,AII, being respectively 100% and 60% higher in the female group. Of the cholesterol associated with (AI,AII), 31% was within the HDL-2 density range in men, compared to 39% for women. For (AI), the male group had 54% of associated cholesterol within HDL-2, whereas the female group had 65% within the lower density subclass. Total HDL-cholesterol (measured after phosphotungstate precipitation) correlated strongly with HDL2-LpAI and HDL2-LpAI,AII cholesterol, with coefficients of +0.66 and +0.67 (men) and +0.83 and +0.82 (women) respectively. These contrast with

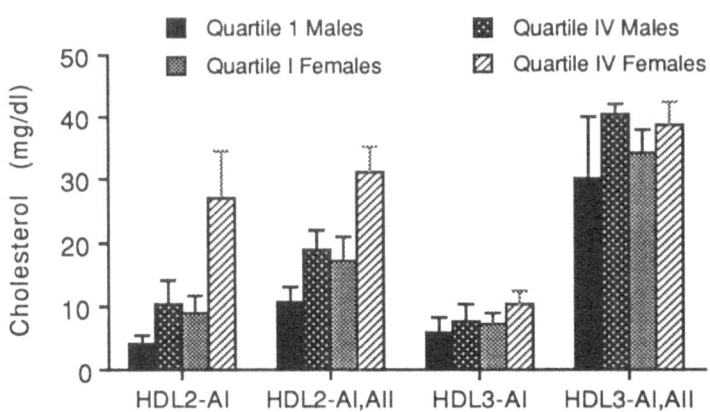

Fig. 2. Cholesterol in HDL subfractions of population quartiles

the coefficients observed for the quantitatively major fraction, HDL3-LpAI,AII cholesterol, with values of +0.53 (men) and +0.46 (women).

Cholesterol of both HDL-2 derived fractions showed negative correlations with plasma triglyceride levels. For HDL2-LpAI these were -0.53 for men and -0.36 for women: corresponding correlations for HDL2-LpAI,AII were -0.39 and -0.30 respectively.

The lipoprotein association of cholesterol was also examined in the first and fourth quartiles of both populations. For the female population, significantly lower cholesterol levels were found in quartile I in both HDL-2 and 3 density ranges (25.9 ± 5.0 v 58.1 ± 9.4 mg/dl ($p < 0.0001$) and 41.5 ± 4.9 v 50.7 ± 6.5 mg/dl ($p = 0.015$)). Further subfractionation (Fig. 2) showed that differences at the HDL-3 level were due to the (AI) lipoproteins ($p = 0.0005$), there being similar concentrations of HDL3-LpAI,AII ($p = 0.09$). Both immunoaffinity defined fractions from HDL-2 were highly significantly increased in quartile IV (LpAI, $p = 0.0002$; LpAI,AII, $p = 0.0001$).

When cholesterol associated with the subfractions was correlated with plasma triglyceride levels (Table 1) some interesting differences between the two quartiles emerged. Thus, for HDL-2, negative correlations were observed for LpAI for quartile I, but with LpAI,AII for quartile IV. Contrasting results were also evident for HDL-3 derived immunoaffinity-derived particles (Table 1): triglycerides were correlated in a positive manner with HDL3-LpAI,AII from quartile I, but negatively with the same fraction from quartile IV.

With respect to quartiles established for the male population, qualitatively similar conclusions could be drawn from an analysis of the lipoprotein association of cholesterol. Namely, quartile I had significantly lower cholesterol concentrations within subclasses HDL-2 (14.8 ± 3.1 v 29.4 ± 5.7 mg/dl; $p = 0.0007$) and HDL-3 (37.7 ± 6.0 v 48.1 ± 2.6 mg/dl; $p = 0.005$). Likewise, HDL2-LpAI ($p = 0.002$) and HDL2-LpAI,AII ($p = 0.0004$) were also significantly lower in quartile 1 (Fig. 2). Within HDL-3, LpAI,AII cholesterol was significantly higher ($p = 0.004$) in quartile IV, but there were no significant differences ($p = 0.12$) in cholesterol concentrations of LpAI (Fig. 2).

As with the female population, the male quartiles differed somewhat when examined in terms of triglyceridaemia. Whereas negative correlations were observed with fractions originating from HDL-2 of both quartiles, divergent correlations were found for HDL-3 derived fractions (Table 1). Both fractions of quartile I were positively correlated with triglyceride levels, in contrast to negative correlations exhibited for the same fractions from quartile IV.

Table 1. Correlations of plasma triglycerides with cholesterol concentrations of immunoaffinity-defined subfractions.

	HDL 2		HDL 3	
	AI	AI,AII	AI	AI,AII
Males				
Quartile I	-0.60	-0.61	+0.24	+0.73
Quartile IV	-0.68	-0.70	-0.49	-0.31
Females				
Quartile I	-0.50	+0.15	+0.14	+0.77
Quartile IV	+0.15	-0.72	+0.33	-0.45

Values are those for the correlation coefficients

In a final analysis, the correlations between HDL-cholesterol and cholesterol associated with the immunoaffinity-defined fractions were established. As shown in Table 2, there were distinct differences between quartiles I and IV for both males and females. Notable, strong correlations were observed only for HDL2-LpAI of quartile I, but with three fractions, HDL2-LpAI, HDL2-LpAI,AII and HDL3-LpAI, from quartile IV.

Table 2. Correlations of total HDL-cholesterol with subfraction concentrations of cholesterol.

	HDL 2		HDL 3	
	AI	AI,AII	AI	AI,AII
Males				
Quartile I	+0.50	+0.19	+0.13	+0.37
Quartile IV	+0.57	+0.72	+0.66	+0.37
Females				
Quartile I	+0.67	+0.30	+0.07	+0.01
Quartile IV	+0.77	+0.78	+0.74	+0.13

Values are those for the correlation coefficients

DISCUSSION

The dual fractionation procedure adopted herein provides a more precise definition of the cholesterol distribution within the high density lipoprotein spectrum. The results demonstrate the importance of both HDL2-LpAI and HDL2-LpAI,AII in determining total plasma levels of HDL-cholesterol. This is evident from the strength of the correlation coefficients when comparing total HDL-cholesterol with cholesterol in each subfraction. Moreover, male-female differences in HDL levels largely reside in the concentrations of HDL2-LpAI and HDL2-LpAI,AII (Fig. 1). Finally, within the same population, it is these two subfractions that essentially differentiate subjects in the first and fourth quartiles of HDL plasma concentrations. It suggests that physiological events giving rise to both fractions are important in determining HDL-cholesterol levels.

Although the major proportion of HDL-2 cholesterol is present in LpAI,AII in both males and females, it would appear that HDL2-LpAI is the more sensitive indicator of HDL-cholesterol levels. Thus, concentrations of HDL2-LpAI cholesterol are 2.5 to 3 fold higher in quartile IV as compared to quartile I. In contrast, levels of HDL2-LpAI,AII cholesterol in quartile IV are less than double those in quartile I. Further, in quartile I of both populations, only HDL2-LpAI cholesterol shows a strong correlation with total HDL-cholesterol. Interestingly, subjects in quartile IV also showed strong correlations between total HDL-cholesterol and HDL2-LpAI,AII cholesterol, again in both sexes. The latter is one observation that differentiates the two quartiles. Another is the correlation between triglyceridaemia and cholesterol levels. Notably, HDL3-LpAI,AII cholesterol showed a positive correlation with triglycerides in quartile I, but a negative correlation in quartile IV. Other differences were also evident (Table 2), although there was less of a parallel in the response of the male and female quartiles.

Overall, the results suggest that the combination of distinct fractionation procedures can be helpful in further defining the association of cholesterol with high density lipoproteins. It should provide information useful in determining the relative importance of the different subfractions to the anti-atherogenic effect of HDL.

ACKNOWLEDGEMENTS

The work reported herein was supported by grants 3.999-0.86 and 32.9484-88 from the Swiss National Research Fund.

REFERENCES

1. NIH Consensus Development Conference, Lowering blood cholesterol to prevent heart disease, JAMA., 253: 2080 (1985)
2. Study group, European Atherosclerosis Society, The recognition and management of hyperlipidaemia in adults: A policy statement of the European Atherosclerosis Society, Europ. Heart J., 9: 571 (1988)
3. N.E. Miller, Association of high density lipoprotein subclasses with ischaemic heart disease and coronary atherosclerosis, Am. Heart J., 113: 589 (1987)
4. J.F. Oram, Effects of high density lipoprotein subfractions on cholesterol homeostasis in human fibroblasts and arterial smooth muscle cells. Arteriosclerosis, 3: 420 (1983)
5. S. Eisenberg, High density lipoprotein metabolism, J. Lipid Res., 25: 1017 (1984)
6. P. Alaupovic, The physicochemical and immunological heterogeneity of human plasma high density lipoproteins, in: 'Clinical and metabolic aspects of high density lipoproteins,' N.E. Miller and G.J. Miller, eds., Elsevier, Amsterdam (1984)
7. M.C. Cheung and J.J. Albers, Characterisation of lipoprotein particles isolated by immunoaffinity chromatography. Particles containing A-I and A-II and particles containing A-I but no A-II, J. Biol. Chem., 259: 12201 (1984)
8. R.W. James and D. Pometta, Immunofractionation of high density lipoprotein subclasses 2 and 3. Similarities and differences of fractions isolated from male and female populations, Atherosclerosis, In Press
9. R.W. James, A. Proudfoot and D. Pometta, Immunoaffinity fractionation of high density lipoprotein subclasses 2 and 3 using anti-apolipoprotein A-I and A-II immunosorbent gels, Biochim. Biophys. Acta, 1002: 292 (1989)
10. R.W. James and D. Pometta, Differences in lipoprotein subfraction composition and distribution between diabetic patients and controls. A study in male, type I (insulin-dependent) diabetes, Diabetes, In Press

METABOLIC ROLE OF HUMAN APOPROTEIN A-IV

L. LAGROST, P. GAMBERT, A. ATHIAS, and C. LALLEMANT

Laboratoire de Biochimie des Lipoproteines,
Hôpital du Bocage, 2 bd Maréchal de Lattre de Tassigny
21034 Dijon Cedex, France

INTRODUCTION

Compared with the other human apoproteins, physiological variations and metabolic functions of apo A-IV remain obscure. The apo A-IV has been discovered in rat HDL and has been later found in human plasma, mesenteric lymph and interstitial fluid. Apo A-IV is synthetized by intestine and liver and is catabolized by liver and kidneys. As apo A-IV has been suspected to play a role in triglyceride transport, we undertook the study of its plasma concentration in a population of subjects with various plasma triglyceride levels. Moreover, recent studies brought some evidence for an implication of apo A-IV in the reverse cholesterol transport. Because recent reports suggested that HDL was the major lipoprotein class involved in this metabolic pathway, we investigated the structural and metabolic relation between apo A-IV and HDL.

ROLE OF APOPROTEIN A-IV IN TRIGLYCERIDE TRANSPORT

By using an apo A-IV competitive enzyme immunoassay we found a significant positive correlation between apo A-IV and triglyceride concentrations in human sera (1) (Figure1). These data were in good agreement with previous reports suggesting a relation between apo A-IV and triglyceride secretion. An increase of plasma apo A-IV concentration has been found after lipid feeding (2,3) and in circumstances associated with accumulation of remnants of triglyceride-rich lipoproteins, such as chronic renal failure treated by peritoneal dialysis or hemodialysis (4,5). Moreover, a relation between the apo A-IV mRNA synthesis and the triglyceride secretion has been observed in intestine and liver (6,7). As this correlation between apo A-IV triglycerides was found in non-chylomicronemic sera from fasting subjects, we can assume that triglycerides were contained mainly in VLDL from hepatic origin and that the liver, in fasted subjects, could contribute significantly to the plasma apo A-IV pool.

Hypercholesterolemia, Hypocholesterolemia, Hypertriglyceridemia
Edited by C.L. Malmendier *et al.*, Plenum Press, New York, 1990

109

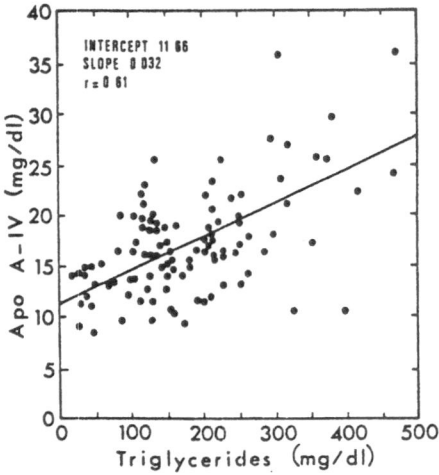

Fig. 1. Correlation of apo A-IV and triglycerides in 105 native sera

ROLE OF APOPROTEIN A-IV IN REVERSE CHOLESTEROL TRANSPORT

Distribution of apoprotein A-IV in normolipidemic human serum

As apo A-IV has been suspected to play a role in reverse cholesterol
transport, its structural relation with HDL was studied. Whereas it is
known that in rat most of apo A-IV associates with HDL, in human very
little apo A-IV has been found on lipoprotein particles isolated by
traditional ultracentrifugation methods. However, the distribution of apo
A-IV between lipoprotein free and lipoprotein fractions varies widely and
appears to be dependent on the techniques used to fractionate sera. About
20 to 35 % of apo A-IV were found associated with HDL after agarose gel
permeation chromatography of total plasma (2,3,8), while more than 90 %
of human apo A-IV localized in the lipoprotein-free fraction after ultracentrifugation of plasma at density 1.21 (2,9).

By using a high performance Superose 12-HR column (Pharmacia), we
were able not only to confirm the presence of apo A IV in the HDL
fraction but to demonstrate that, in fasting human sera, a majority of
apo A-IV eluted with the HDL fraction, mainly within the HDL2 size range
(10). Moreover, the gel filtration procedure allowed us to evidence the
potential disruptive effect of ultracentrifugation on the apo A-IV
distribution in total serum (Figure 2). This distribution of apo A-IV
could be attributable to the combined effects of high ionic strength and
high sheering forces of ultracentrifugation procedure.

Concurrently, the coprecipitation of the most part of apo A-IV after
incubation of total serum with anti-apo A-I antibodies indicated that
about 70 to 80 % of serum apo A-IV were carried by apo A-I containing HDL
(Figure 3). This proportion of apo A-IV coprecipitated with HDL is

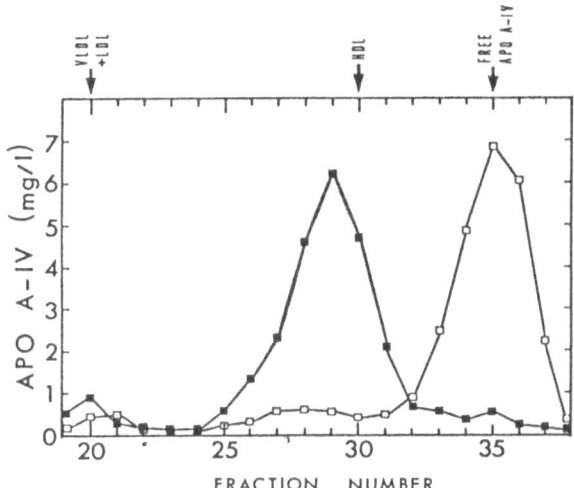

Fig. 2. Effect of ultracentrifugation on the distribution of serum
apo A-IV.(■): Gel filtration elution profile of A-IV in
native serum. (□): Distribution of apo A-IV after treatment
of the serum by sequential ultracentrifugation at d: 1.25

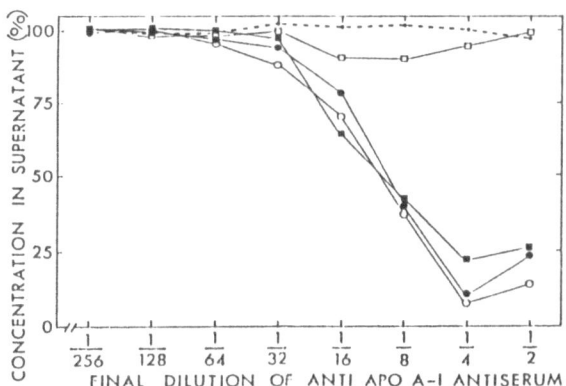

Fig. 3. Effect of HDL immunoprecipitation on serum apo A-IV: influence
of detergent. Total serum, supplemented (open symbols) or not
(closed symbols) with 0.1% Tween 20, was incubated with
various dilutions of anti-apo A-1-immunoglobulins.
Concentrations of apo A-1 (● and ○) and apo A-IV (■ and □)
were measured in the supernatant and expressed as percent
of those in control mixture in which antiserum was replaced by
dilution buffer. As a control experiment, immunoprecipitation
was conducted in the absence of Tween on a solution of pure
apo A-IV (11.0 mg/dl) (discontinuous line)

considerably lowered (less than 10 %) when HDL particles were dissociated by addition of Tween to the serum dilutions prior to the precipatation step.

These results provided evidence that, in fasted normal human sera, apo A-IV preferably associated with HDL and that this apo A-IV-HDL association is a weak one, easily disrupted. The remodeling of lipoprotein surfaces can modify the equilibrium between free and bound apo A-IV. Particularly, LCAT activity can induce a displacement of apo A-IV from lipoprotein-free fraction to the HDL particles.

Apoprotein A-IV and HDL conversion

Beyond its structural relation with HDL, apo A-IV can also play a dynamic role in the intravascular metabolism of HDL particles, especially in the size redistribution or conversion induced by the Cholesterol Ester Transfer Protein (CETP) (11). By using gradient gel electrophoresis, it was shown that the incubation of total lipoproteins in the presence of CETP at physiological levels induced a general displacement of HDL towards large size particles, a decrease in the HDL3 subpopulation and the appearance of small conversion products with mean diameters of 7.8 and 7.4 nm (Figure 4).

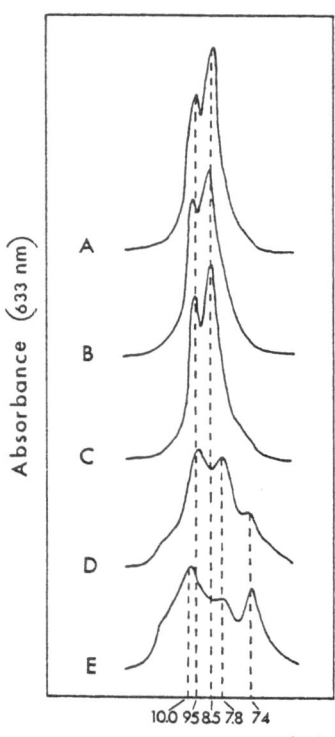

Fig. 4. Effect of CETP and apo A-IV on human HDL distribution profile.
Total lipoproteins were separated by gradient gel
electrophoresis on Pharmacia 4/30 slab gels.
A: Total lipoproteins, 4ºC
B: Total lipoproteins, 37ºC, 24 h
C: Total lipoproteins + apo A-IV, 37ºC, 24 h
D: Total lipoproteins + CETP, 37ºC, 24 h
E: Total lipoproteins + CETP + apo A-IV, 37ºC, 24 h

112

This phenomenon can be significantly altered by apo A-IV. Compared with incubation with CETP alone, incubation with CETP and apo A-IV increased the size redistribution of HDL and particularly favored the formation of the very small sized lipoprotein particles (mean diameter 7.4 nm). Incubation of total lipoproteins and apo A-IV, in the absence of CETP, do not modify the HDL distribution profiles.

DESCRIPTION OF THE MULTIPOTENTIAL INVOLVEMENT OF APO A-IV IN THE REVERSE CHOLESTEROL TRANSPORT

The results presented above, together with previously reported data, show that apo A-IV is involved at different levels of HDL metabolism and suggest that this apoprotein could play a major role in cholesterol transport from peripheral tissues to the liver. More precisely, the localization of apo A-IV within HDL, its implication in the formation of very small sized lipoprotein particles, its activating role in plasma cholesterol esterification by LCAT and its potential involvement in the cholesterol movements between intra- and extra-cellular media suggest that this apoprotein could play a major role in the reverse cholesterol transport (Figure 5).

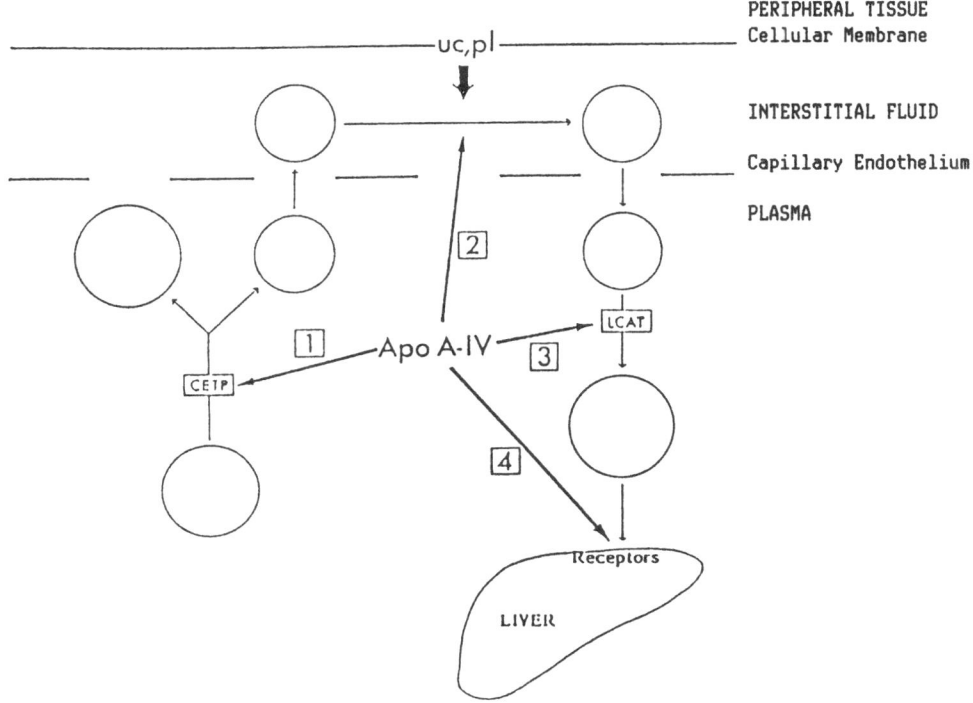

Fig. 5. Role of apoprotein A-IV in reverse cholesterol transport.
  1: Formation of very small sized lipoprotein particles.
  2: Cellular cholesterol efflux.
  3: Cholesterol esterification.
  4: Hepatic uptake of cholesteryl esters.

By its combined effects with CETP, apo A-IV can promote the formation of very small sized HDL particles. Such particles are susceptible to migrate easily in the interstitial space where they could participate to the cell cholesterol efflux (11). Moreover, as apo A-IV is relatively stable in aqueous solution, the free apoprotein itself could also participate directly to the peripheral cholesterol uptake. The particles issued from interstitial space, enriched with free cholesterol and phospholipids can then interact with plasma LCAT. The small HDL, enriched with unesterified cholesterol (UC) and phospholipids (PL), are known to be good substrates for this enzyme (12) which can be furthermore activated by apo A-IV (13, 14). These uptake and esterification of cholesterol, which constitute a key step in the Reverse Cholesterol Transport, lead to the formation of large sized HDL2-like particles enriched in esterified cholesterol. In vitro studies on cell cultures have indicated that apo A-IV could specifically interact with rat hepatocytes (15,16). As we observed that apo A-IV in human serum localized mainly in large sized HDL particles, it can be postulated that apo A-IV could facilitate specific uptake of HDL by the liver.

In conclusion, by promoting efflux of peripheral cholesterol, by activating cholesterol esterification in plasma, and by facilitating cholesterol uptake by the liver, apo A-IV could play a central role in the Reverse Cholesterol Transport.

REFERENCES

1. L. Lagrost, P. Gambert, S. Meunier, P.Morgado, J. Degres, P. d'Athis, and C. Lallemant, Correlation between apolipoprotein A-IV and triglyceride concentrations in human sera, J. Lipid Res., 30:701 (1989).
2. P. H. Green, R.P. Glickman, J.W. Riley, and E. Quinet, Human apolipoprotein A-IV : intestinal origin and distribution in plasma, J. Clin. Invest., 65:911 (1980).
3. C. L. Bisgaier, O. P. Sachdev, I. L. Megna, and R. M. Glickman, Distribution of apolipoprotein A-IV in human plasma, J. Lipid Res., 26:11 (1985).
4. P. J. Nestel, N. H. Fidge, and M. H. Tan, Increased lipoprotein-remnant formation in chronic renal failure, N. Engl. J. Med., 307:329 (1982).
5. M. Seishira, and Y. Muto, An increased apo A-IV serum concentration of patients with chronic renal failure on hemodialysis, Clin. Chim. Acta, 167:303 (1987).
6. N. A. Elshourbagy, M. S. Boguski, W. S. L. Liao, L. S. Jefferson, J. I. Gordon, and J. M. Taylor, Expression of rat apolipoprotein A-IV and A-I genes : mRNA induction during development and in response to glucocorticoids and insulin, Proc. Natl. Acad. Sci. USA, 82:8242 (1985).
7. M. Pessah, C. Salvat, S. R. Wang, and R. Infante, In vitro synthesis of apo A-IV and apo C by liver and intestinal mRNAs from lean and obese Zucker rats. Biochem. Biophys. Res. Commun, 142:78 (1987).
8. M. Rosseneu, G. Michiels, W. De Keersgieter, J. Bury, J. P. De Slypere, H. Dieplinger, and G. Utermann, Quantification of human apolipoprotein A-IV by"sandwich"-type enzyme-linked immunosorbent assay, Clin. Chem., 34:739 (1988).
9. G. Utermann, and U. Beisiegel, Apolipoprotein A-IV : a protein occurring in human mesenteric lymph chylomicron and free in plasma. Isolation and quantification, Eur. J. Biochem., 99:333 (1979).

10. L. Lagrost, P. Gambert, M. Boquillon, and C. Lallemant, Evidence for high density lipoprotein as the major apolipoprotein A-IV containing fraction in normal human serum, <u>J. Lipid Res.</u>, 30:1525 (1989).
11. P. Gambert, L. Lagrost, A. Athias, S. Bastiras, and C. Lallemant, Role of apolipoprotein A-IV in the interconversion of HDL subclasses, <u>Advances in Experimental Medicine and Biology</u>, 243:263 (1988).
12. P. J. Barter, G. J. Hopkins, and L. Gorjatschko, Lipoprotein substrates for plasma cholesterol esterification. Influence of particle size and composition of the high-density lipoprotein subfraction 3, <u>Atherosclerosis</u>, 58:97 (1985).
13. A. Steinmetz, and G. Utermann, Activation of lecithin : cholesterol acyltransferase by human apolipoprotein A-IV, <u>J. Biol. Chem.</u>, 260:2258 (1985).
14. C. H. Chen, and J. J. Albers, Activation of lecithin : cholesterol acyltransferase by apolipoprotein E-2, E-3, and A-IV isolated from plasma, <u>Biochim. Biophys. Acta</u>, 836:279 (1985)
15. E. Dvorin, N. L. Gorder, D. M. Benson, and A. M. Gotto, Jr., Apolipoprotein A-IV. A determinant for binding and uptake of high density lipoproteins by rat hepatocytes, <u>J. Biol. Chem.</u>, 261:15714 (1986).
16. Y. B. Mitchel, V. A. Rifici, and H. A. Eder, Characterization of the specific binding of rat apolipoprotein E-deficiency HDL to rat hepatic plasma membranes, <u>Biochim. Biophys. Acta</u>, 917:324 (1987).

REGRESSION OF ATHEROMA AND PUTATIVE ROLE OF CETP

IN CHOLESTERYL ESTER REMOVAL

Yechezkiel Stein and Olga Stein

Lipid Research Laboratory, Department of Medicine B
Hadassah University Hospital, and
Department of Experimental Medicine and Cancer Research
Hebrew University-Hadassah Medical School
Jerusalem, Israel

Evidence for regression of atherosclerosis induced by cholesterol feeding has been provided by several investigators [1-3]. However, since the extent of atherosclerotic involvement is quite variable, quantitative evaluation of regression is difficult. We have used ^{3}H-cholesteryl linoleyl ether (^{3}H-CLE), a nonhydrolyzable analog of cholesteryl easter as a stable marker for the quantitation of atherosclerotic involvement [4] and evaluated the potential usefulness of ^{3}H-CLE in the evaluation of regression of atheromatosis [5]. To that end, 20 rabbits were kept on a purina diet enriched with 1% cholesterol for 1 month and then on alternate weeks for an additional 2 months. The animals were randomized into two groups according to their plasma cholesterol levels and injected with autologous plasma labeled with ^{3}H-CLE [5]. The baseline group was killed 10-12 days after injection, while the regression group was fed purina fortified with 3% cholestyramine and killed 8-11 months after injection of the ^{3}H-CLE. We investigated the following: 1. Will the ^{3}H-CLE remain in the aorta during the 11-month period of regression? 2. If ^{3}H-CLE is retained, then the specific activity expressed as ^{3}H-CLE/CE mass should rise with CE loss during regression; 3. Is the loss of CE during regression similar from the different parts of the aorta?

At the end of the cholesterol feeding period, the mean plasma cholesterol was 1298 mg/dl. The amount of labeled ^{3}H-CLE in the aorta varied markedly among the individual rabbits, but was highly correlated (r = 0.875) with the amount of aortic cholesteryl ester determined in the base-

Hypercholesterolemia, Hypocholesterolemia, Hypertriglyceridemia
Edited by C.L. Malmendier *et al.*, Plenum Press, New York, 1990

117

line group (Fig. 1). The results presented in Fig. 2 compare the total and esterified cholesterol in the entire aorta of the baseline (10-12 days) and regression groups (11 months). In the baseline group, the mean total cholesterol was 13.2 ± 2.1 mg/aorta and the esterified cholesterol was 7.6 ± 1.3 mg/aorta. After 11 months of the regression regimen, the total cholesterol was 9.5 ± 1.9 mg/aorta, while cholesteryl ester decreased to

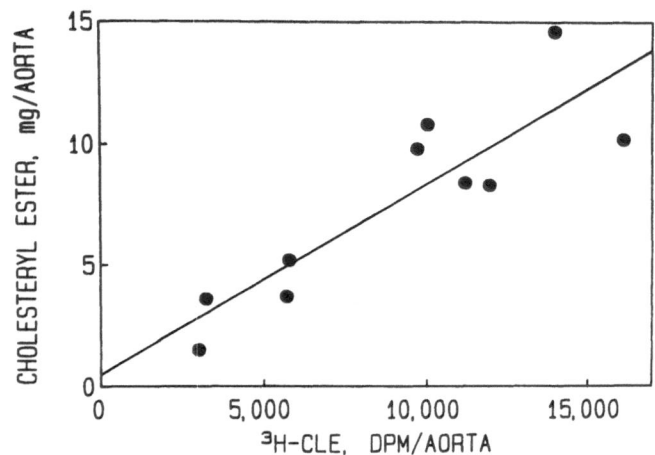

Fig. 1.   Linear regression analysis of correlation between aortic cholesteryl ester and ^{3}H-cholesteryl linoleyl ether (^{3}H-CLE) of the baseline group (r = 0.875).

3.1 ± 0.7 mg/aorta. The loss of cholesteryl ester was significant (p < 0.01). On the other hand, the amount of ^{3}H-CLE in the regression group was not different from that seen in the baseline group (Fig. 2). The mean specific activity of ^{3}H-CLE/CE was compared in the arch, thoracic and abdominal aorta of the baseline and regression groups (Fig. 3). As can be seen in the baseline group, the specific activity in the three regions examined was quite similar. A much higher specific activity of ^{3}H-CLE/CE was found in the regression group, the highest being seen in the region of the aortic arch.

These results permitted us to conclude that ^{3}H-CLE injected into
cholesterol fed rabbits was retained in the aorta for up to 11 months of
the regression period, while cholesteryl ester content decreased. The
retention of ^{3}H-CLE in the rabbit aorta, in face of high plasma CETP, would
not favour the role of CETP in CE removal from the aorta. This could have
been due to several possibilities, among them that the ^{3}H-CLE was not
accessible to the transfer protein. Indeed, in a model system in culture
[6], we have shown that while ^{3}H-CLE present in lipoproteins and bound to

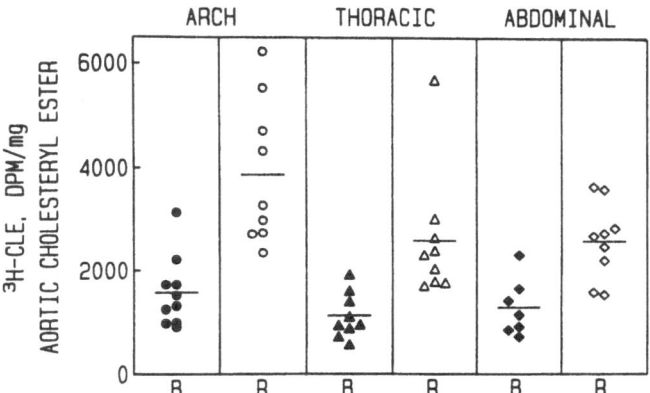

Fig. 2.   Specific activity of ^{3}H-cholesteryl linoleyl ether (^{3}H-CLE) in
           different regions of atherosclerotic rabbit aorta. B = baseline;
           R = regression. From Stein et al. [5]. Reproduced with permission
           of the Editor of Arteriosclerosis.

extracellular matrix was accessible to CETP and could be released into the
culture medium; once the lipoprotein had become ingested by a cell such as
a macrophage it became inaccessible to CETP [6]. Recently, Morton [7] has
presented evidence that CETP is able to remove cholesteryl ester from
intact macrophages. We proposed to test the putative role of CETP in CE
egress from reticuloendothelial cells in vivo in an animal model in which
one can modulate plasma CETP levels by dietary means. Son and Zilversmit

[8] have shown that cholesterol feeding in rabbits is accompanied by an increase in plasma CETP. We looked, therefore, for a smaller animal which would respond to cholesterol feeding with a rise of plasma CETP in analogy to the rabbit. In view of the studies of Dietschy et al. [9, 10], the hamster appeared to be a suitable model, but there were no data in the literature with respect to plasma CETP in the hamster. Therefore, we have examined hamsters for CETP activity and found measurable activity under

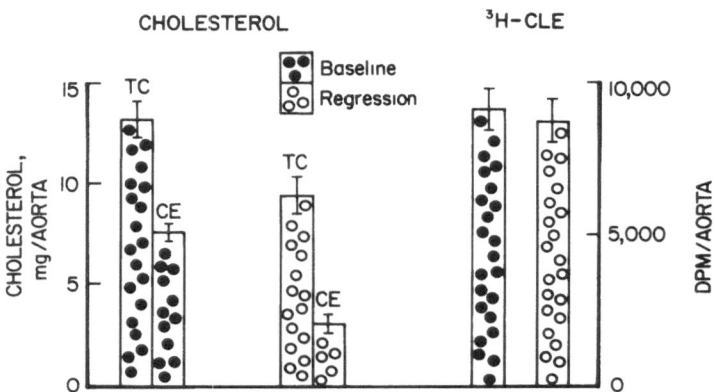

Fig. 3.   Total and esterified cholesterol and ^{3}H-cholesteryl linoleyl ether (^{3}H-CLE) at the end of induction and of regression period. TC = total cholesterol; CE = cholesteryl ester. Reproduced by permission of Elsevier Science Publishers, B.V.

control dietary conditions. We were able to modulate this activity by feeding diets enriched in cholesterol and fat [11]. As seen in Table 1, the hamsters responded to a high fat-high cholesterol diet with a significant increase in CETP activity. Therefore, we decided to use hamsters to evaluate the role of CETP in cholesteryl ester efflux from cells in vivo. The approach was based on our previous findings [12] that when acetylated LDL is labeled with ^{3}H-CLE and injected into rats, it disappears from the liver at a very slow rate. Since the rat does not have measurable CETP

Table 1. Effect of Cholesterol in Diet on CETA in Hamster Plasma

Supplement to Diet	Weeks on Diet	Plasma Lipids, mg/dl		CETA, units
		Chol-esterol	TG	
None		78	190	5.4 ± 0.7
Cholesterol	2	231	301	11.6 ± 2.7
	4	257	464	13.0 ± 0.6
Cholesterol + Margarine	2	232	303	16.1 ± 1.4
	4	394	395	19.4 ± 3.5
Cholesterol + Butter	2	557	1595	15.8 ± 3.0
	4	813	905	22.1 ± 2.2
	6	751	1445	27.8 ± 1.5

Adapted from Y.Stein et al. [11].

under normal dietary conditions or even after feeding of high fat and cholesterol [11], a comparison of loss of ^{3}H-CLE labeled acetylated LDL from rat and hamster liver could provide some information with respect to the role of CETP in cholesteryl ester removal from cellular elements in vivo. These experiments are now in progress and preliminary results suggest that under these experimental conditions, loss of ^{3}H-CLE from the liver is not increased by CETP.

REFERENCES

1. D.Vesselinovitch, R.W.Wissler, K.Fisher-Dzoga, R.Hughes, and L.Dubien, Regression of atherosclerosis in rabbits. Part 1. Treatment with low-fat diet, hyperoxia and hypolipidemic agents, Atherosclerosis 19:259 (1974).
2. R.W.St.Clair, Atherosclerosis regression in animal models: Current concepts of cellular and biochemical mechanisms, Prog.Cardiovasc. Dis. 26: 109 (1983).

3. M.R.Malinow, Experimental models of atherosclerosis regression. Atherosclerosis 48: 105 (1983).

4. Y.Stein, O.Stein, and G.Halperin, Use of ^{3}H-cholesteryl linoleyl ether for the quantitation of plasma cholesteryl ester influx into the aortic wall in hypercholesterolemic rabbits, Arteriosclerosis 2:281 (1982).

5. O.Stein, G.Hollander, Y.Dabach, G.Halperin, and Y.Stein, Use of ^{3}H-cholesteryl linoleyl ether as a quantitative marker for loss of cholesteryl ester during regression of cholesterol-induced aortic atheromas in rabbits, Arteriosclerosis 9:247 (1989).

6. O.Stein, G.Halperin, and Y.Stein, Cholesteryl ester efflux from extra-cellular and cellular elements of the arterial wall. Model systems in culture with cholesteryl linoleyl ether. Arteriosclerosis 6:70 (1986).

7. R.E.Morton, Interaction of plasma-derived lipid transfer protein with macrophages in culture, J.Lipid Res. 29:1367 (1988).

8. Y.-S.C.Son and D.B.Zilversmit, Increased Lipid Transfer Activities in hyperlipidemic rabbit plasma, Arteriosclerosis, 6:345 (1986).

9. D.K.Spady, and J.M.Dietschy, Dietary saturated triacylglycerols suppress hepatic low density lipoprotein receptor activity in the hamster, Proc.Natl.Acad.Sci.USA, 82: 4526 (1985).

10. D.K.Spady, and J.M.Dietschy, Interaction of dietary cholesterol and triglycerides in the regulation of hepatic low density lipoprotein transport in the hamster, J.Clin.Invest. 81:300 (1988).

11. Y.Stein, Y.Dabach, G.Hollander, and O.Stein, Cholesteryl ester transfer activity in hamster plasma: increase by fat and cholesterol rich diets, Biochim.Biophys.Acta 1042: 138 (1990).

12. Y.Stein, Y.Kleinman, G.Halperin, and O.Stein, Hepatic retention and elimination of cholesteryl linoleyl ether after injection of labeled acetylated LDL or chylomicrons, Biochim.Biophys.Acta 750:300 (1983).

SYMMETRY OF THE SURFACE, AND STRUCTURE OF THE CENTRAL CORE OF HUMAN LDL PARTICLES, ANALYZED BY X-RAY SMALL ANGLE SCATTERING

M.W. Baumstark[1,2], W. Kreutz[2], A. Berg[1], J.Keul[1]

[1]Med. Universitätsklinik, [2]Inst. f. Biophysik, Universität Freiburg
D-7800 Freiburg, F.R.G.

Although there exists a widely accepted model of the general structure of the LDL particle, there are still some important open questions, the most interesting being the three dimensional structure of the apoB molecule. There is very little knowledge of the exact conformation of this large polypeptide chain, containing many hydrophobic residues. Despite this hydrophobicity it was shown by several techniques (MABs, NMR, FTIR) that many domains of this protein are localized at the particle surface or are at least accessible to the solvent. An elaborate evaluation of X-ray scattering data from three LDL subfractions shows that

a)   The apoB molecule has to cover *at least* 47% of the surface of small LDL particles, and 37% of the surface of large LDL.

b)   Most of the mass of the apoB molecule is located within a small shell of 2.2nm width directly at the particle surface.

c)   There is no evidence for symmetries other than spherical, which means, that there are no marked 'spikes' of protein at the surface.

An other uncertain point is the organization of the central cholesterol ester (CE) core below the phase transition temperature. Our X-ray data confirms those models in which the cholesterol moieties of the CE-molecules are located at two concentric shells of 3.2 and 6.4 nm radius. In contrast to previous models we propose, based on detailed space filling calculations, that the CE molecules are arranged in an alternating orientation in such a way that about half of the acyl chains of the CE-molecules in each shell point towards the center of the particle, the other half pointing to the surface. This model facilitates an interdigitation of acyl chains of cholesterol esters with each other and with surface phospholipids. The interdigitation of acyl chains of core and surface lipids has also been proposed in the case of HDL and protein free models of LDL, and seems to be a general feature of lipoprotein structure.

INTRODUCTION

Our current knowledge on the molecular structure of low density lipoproteins (LDL) originates to a large extent from X-ray and neutron small-angle studies that have been published by several groups. For references see[1,2,3,4]. In all of these studies evaluation theories assuming monodisperse, radially symmetric particles were used. Only Luzzati et al.[5] considered deviations from spherical symmetry. While still assuming monodispersity, deviations from spherical symmetry turned out to be essential in Luzzati's interpretation of the data. Our evaluation theory[6,7] predicts the exact scattering intensity of a "polydisperse ensemble of quasi radially symmetric particles", and consequently needs neither to assume monodispersity nor perfect radial symmetry. It has to be emphasized that assuming monodisperse particle populations is in obvious contrast to biochemical and metabolic features of lipoproteins.

In this paper we address mainly two questions: the structure of the apoB molecule and the

Hypercholesterolemia, Hypocholesterolemia, Hypertriglyceridemia
Edited by C.L. Malmendier *et al.*, Plenum Press, New York, 1990

structure of the central lipid core. Differences between LDL subfractions are discussed in detail in[4]. The aim of our study was to find a model of the LDL structure which reproduces the scattering intensity of the sample within the experimental error (noise band) of the X-ray scattering data. None of the models published up to now has been shown to be consistent with this requirement. Since both, polydispersity and deviations from radial symmetry are quantitatively treated in our model, our method is well suited for investigating the question of whether the assumption of a multipole component, as proposed by[5,8], is in fact necessary to explain the experimental data. In addition, our approach allows to give an estimate of the random errors of the resulting electron density profiles. New aspects arise from the precisely determined electron density profiles and from the fact that we determined the electron density profiles of three different LDL subfractions. Based on our molecular model and chemical analysis, we present a detailed comparison of volumes, areas, and radii given by the electron density profiles to the space requirement of the molecular components.

MATERIALS AND METHODS

Samples

Serum was obtained from freshly drawn blood (50 ml without anticoagulant) of male, clinically healthy donors. LDL (d = 1.006-1.063 g/ml) was isolated by standard methods[9]. LDL subfractions were prepared as described previously[4]. After centrifugation the material of subfractions LDL-1 (d = 1.006-1.031 g/ml), LDL-3 (d = 1.034-1.037 g/ml), and LDL-6 (d = 1.044-1.063 g/ml) was dialyzed against a buffer containing 0.196 mol/kg NaCl, 0.5 g/l NaN$_3$ and 0.1 g/l EDTA. An immersible CX-30000 ultrafilter (Millipore) was used to concentrate the samples to total cholesterol concentrations of up to 50 mg/ml.

Chemical analysis

In all subfractions total cholesterol, free cholesterol, triacylglycerol (all Boehringer, Mannheim), and phospholipids (bioMérieux, Nürtingen) were determined. CE concentrations were calculated as (total cholesterol - free cholesterol) x 1.68. All tests were standardized according to the manufacturers' instructions. ApoB concentrations were determined by kinetic rate nephelometry using an automated Beckman ICS Analyzer II. This test was carefully standardized using amino acid analysis (for details see[4]). The given protein concentrations therefore represent amino acid concentrations and do not include the carbohydrate bound to apoB. For that reason in all subsequent calculations a molecular weight of 513,000 based on the amino acid sequence[10,11] is used. The results of the chemical analysis are given in Tab. 1.

X-ray small-angle scattering

X-ray small-angle diagrams were recorded as described in detail in a previous publication[12]. We used temperatures of 4°C and 37°C to record the scattering diagrams. These temperatures are below and above the phase transition temperature[13] of the cholesterol ester molecules inside the LDL particle. The actual phase transition temperatures T$_m$ of our LDL preparations (LDL-1: 21.1°C, LDL-3: 29.7°C, LDL-6: 29.9°C) were monitored by differential scanning calorimetry (DSC).

Table 1. Chemical composition of LDL subfractions

	LDL-1	LDL-3	LDL-6
% Apolipoprotein B	14.6±2.3	17.7±2.3	20.8±3.3
% Phospholipid	21.5±2.9	21.8±2.5	20.9±2.0
% Cholesterol	10.1±1.4	9.9±1.5	8.8±1.7
% Cholesterol ester	37.9±4.3	42.3±3.8	41.7±3.5
% Triacylglycerol	15.9±6.9	8.4±6.5	7.8±3.2

n=61, means ± SD

Evaluation method

The method for evaluating the X-ray small-angle scattering diagrams was described previously[4,6,7,12]. In brief, the particle structure is described by a set of parameters, which are adjusted by a non-linear least-square fit procedure to give an optimum fit to the experimentally determined X-ray scattering curve.

RESULTS AND DISCUSSION

Radial symmetry

To test the hypothesis of radial symmetry we tried to obtain fits of the scattering intensities by using the model "polydisperse ensemble of radially symmetric particles". No deviations from radial symmetry were allowed. It was possible to obtain satisfactory fits without any difficulty (Fig. 1a) in the case of scattering intensities recorded at 37°C. This holds for all types of LDL subfractions investigated in this study. Therefore there is no need to introduce symmetries other than radial to explain the observed scattering intensities of the LDL particles at 37°C.

Fits of the intensities recorded at 4°C were not as perfect as those of the high temperature form. In particular, the width of the scattering maximum at 1/3.7 nm^{-1} could not be reproduced completely satisfactorily by our model (Fig. 1b). A detailed analysis of this very small deviation turned out to be outside the precision of our scattering data. Although an asymmetry of the LDL particle at 4°C cannot be completely excluded on the basis of our data, there is no evidence to assume the existence of such deviations from radial symmetry. Other deviations from our model could exist and are even more probable. For example one could speculate that the ordered, liquid-crystal-like core is of the same size in all particles within the polydisperse ensemble. In fact such a "monodisperse" core would produce a sharper maximum at 1/3.7 nm^{-1} than predicted by our model.

We conclude that with the assumption of a polydisperse ensemble of <u>radially symmetric</u> particles it was possible to obtain fits of the scattering curves that were within the precision of the experimental data, and therefore our data does not support a shape of the LDL particle with marked 'spikes' of protein as proposed by[6,8]. It should be noted that this was not the case with HDL particles[12], where quadrupole-like deviations from radial symmetry had to be assumed to explain the measured scattering curves. The radial symmetry of the LDL particle together with the electron density profile implies a spread out conformation of the protein. Although apoB contains many hydrophobic domains[10,11], it must be localized predominantly inside the outer surface shell of high electron density, which has a width of about 2.2nm. This conformation is further supported by the fact that a large area of the particle surface has to be covered by protein since the surface of the LDL particle cannot be covered by phospholipid headgroups and free cholesterol alone[2,14]. An area of at least 470 nm^2 has to be covered by apoB, even if one assumes relatively

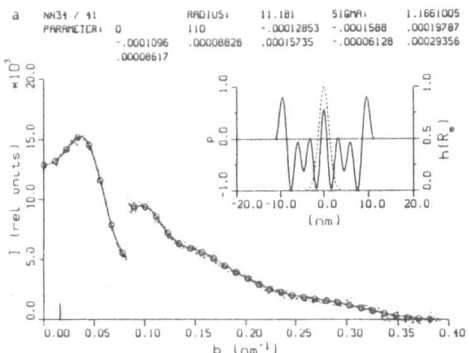
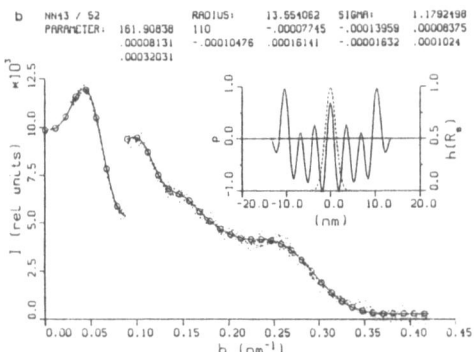

Fig. 1.   Fits (o-o-o) of the experimental scattering intensity (....). The outer part of the intensity curves was mutiplied by a constant weighting factor, which was also used for fitting. Inset: model electron density (——) and radius distribution function (----). (a) Fit of a scattering intensity recorded at 37°C. (b) Fit of a scattering intensity recorded at 4°C. (From[4] with permission).

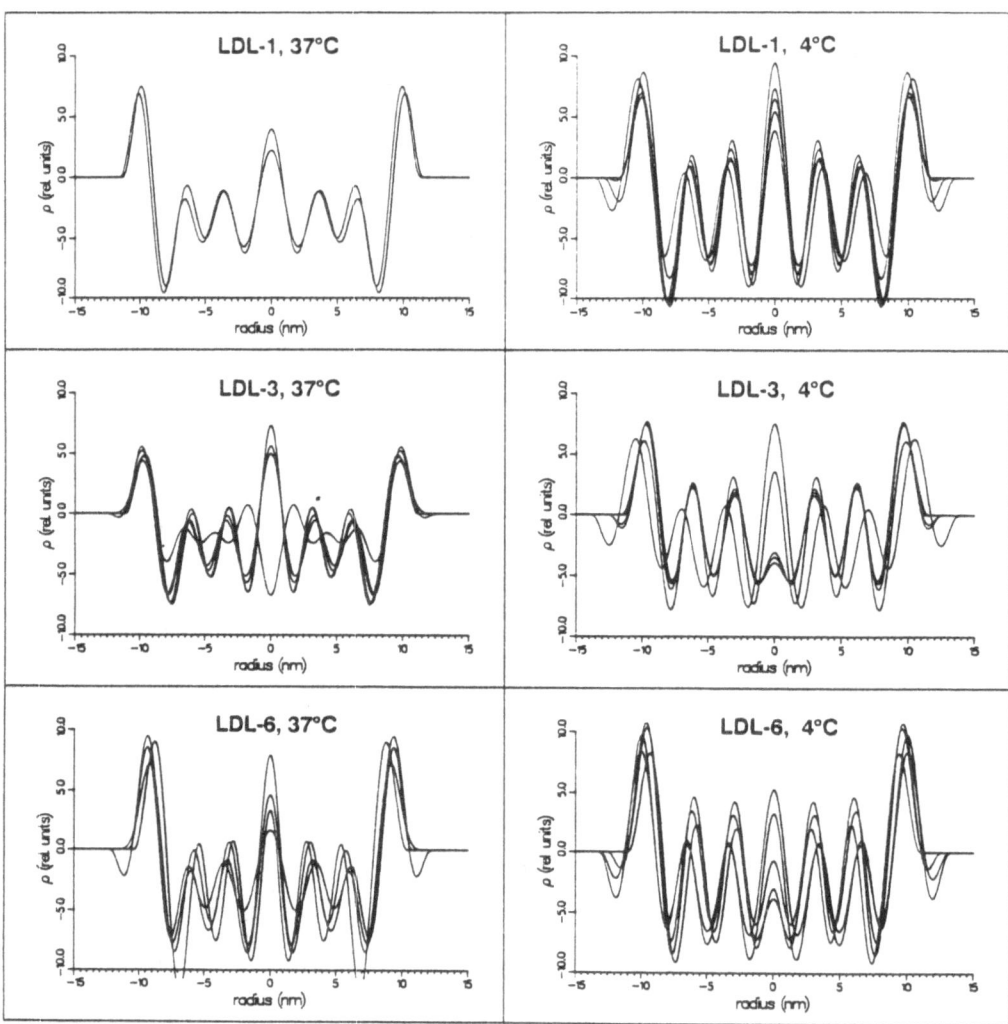

Fig. 2. Electron density profiles of LDL particles from different density classes recorded at 4 and 37°C, as indicated in the figure. (From[4] with permission).

large and additive areas of 0.7nm² per phospholipid and 0.41nm² per cholesterol molecule, which are areas observed at the air-water interface for single compounds[16]. Areas observed in mixed bilayers are usually smaller[16] and would suggest a larger apoB area covered by apoB. It is remarkable that the calculated area covered by apoB is nearly the same for all three LDL subfractions, which strongly supports our conclusion. The view that the apoB molecule covers a large amount of the particle surface is supported by many studies: it was shown by tryptic digestion[11] and by binding studies with monoclonal antibodies[17] (reviews: [18,19]) that domains in many regions of apoB are located at the particle surface. Using nuclear magnetic resonance[20,21] it could be demonstrated that about 20 % of the phospholipid headgroups are immobilized by interaction with apoB. Fast Fourier infrared spectroscopy[22] indicated that the α-helical, random coil and ß-turn structures may be situated on or near the surface of the LDL particle, while the ß-strands have no or only restricted contact with the external solution. It has been hypothesized that amphiphilic ß-strands[23] are the major lipid binding structure of apoB[10].

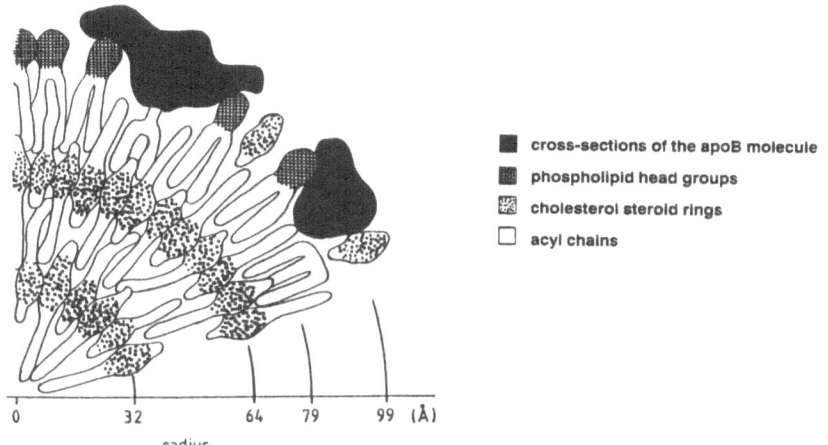

Fig. 3. Molecular model of the LDL particle at 4 °C as consistent with the measured electron density profiles, and considerations regarding the space filling of the chemical components. (From[4] with permission).

High and low temperature form of the LDL particle

We now want to compare the determined electron density profiles of the low temperature form with the high temperature form of the LDL particle: As shown in Fig. 2 a common feature of all profiles measured at 4 °C and at 37 °C is the outer double shell consisting of a shell of high electron density (HEDS) at the particle surface, and a shell of low electron density (LEDS) just below this shell. Both, HEDS and LEDS are similar in all particles. Great differences exist inside the core region. For the high temperature form of the core only comparably small oscillations were found which did not cross the zero level. In contrast there exists a pronounced and reproducible shell structure inside the core of the LDL particle below the phase transition temperature. We find two peaks of high electron density at average radii of 6.37 ± 0.29 nm and 3.23 ± 0.21 nm (average of LDL-1, LDL-3, and LDL-6, at 4 °C). In accordance with[24,25] these two peaks in the electron density profile can well be explained by a localization of the steroid moiety of the cholesterol esters on two concentric shells (Fig. 3) at radii of 6.37 and 3.23 nm, respectively. Such a model results in a cross-sectional area of 0.33 to 0.38 nm² per steroid system, a value close to that obtained for cholesterol esters in crystal packing, where values of 0.35 nm²[26] and 0.366 nm²[27] have been reported. These values are smaller than the one measured at the air-water interface (0.405 nm²)[15].

Details on the arrangement of the cholesterol ester acyl chains can be deduced from the electron density profile by a careful comparison of the measured dimensions with the volumes required to arrange the cholesterol esters in a certain way. The corresponding calculation shall be shown in detail for the case of an average LDL-3 particle which, in average, contains 1886 cholesterol ester molecules: We assume that the steroid shells are centered at radii of 3.2 and 6.4 nm, and that they have a width of 1.7 nm, corresponding to the length of a cholesterol molecule[27]. According to the available surface in each shell 377 cholesterol ester molecules are located at the inner steroid shell and 1509 molecules at the outer steroid shell. If a volume of 0.46 nm³ per acyl chain is assumed[28], 118 chains can be packed inside the inner shell (below 3.2 - 1.7/2 = 2.35 nm). Consequently the remaining 259 acyl chains have to be localized between the two steroid shells. This shell (4.05 to 5.55 nm) has room for further 693 acyl chains. The remaining 816 acyl chains from the second steroid shell are localized at a radius of > 7.25 nm, and are pointing towards the particle surface. The volumes required for the cholesterol moieties alone (238 and 951 nm³) correspond well with the volumes of the steroid shells given by the model, which are 224 nm³ (2.35 to 4.05 nm) and 880 nm³ (5.5 to 7.25 nm). Although the above calculations contain some simplifications, they undoubtedly show that some sort of alternating orientation of the acyl chains has to exist. One simple model is that about half of the acyl chains of each shell point radially towards the center of the particle, the other half pointing to the surface (Fig. 3). Such an arrangement is further supported by neutron small-angle scattering data since it perfectly reproduces the radii of gyration (R_g) measured by Laggner et al.[29]: Laggner published a mean radius of gyration of 6.0 ± 0.2 nm for the fully deuterated acyl chain, and a value of 7.0 ± 0.3 nm for the deuterated C-25 isopropyl groups[7]. From our model we calculate average values of 5.95 nm for the acyl chains and 7.03 nm for the C-25 isopropyl groups. Laggner's preferred model (all acyl chains pointing to the center of the particle) gives values of 5.0 and 6.9 nm, respectively.

In addition, our model facilitates an interdigitation of acyl chains of cholesterol esters and surface phospholipids. This interdigitation of acyl chains of core and surface lipids was proposed in the case of the particles[3,12,30], as well as protein free models of LDL[31], and seems to be an essential principle in lipoprotein structure.

At 37°C, above the phase transition, the arrangement of the cholesterol esters appears to be much less ordered. The oscillations of the electron density are of only half the amplitude compared to those at 4°C. Model calculations showed that they might originate from cut-off effects of the Fourier series used to represent $\rho_m(r)$. Furthermore, it should be mentioned that the electron density near the very center of the particle is poorly defined, since the errors in determining the electron density profile are proportional to $1/r$[12,32].

Comparing the size of LDL particles at 4 and 37°C, one finds slightly larger particle sizes for LDL below the phase transition point. This larger radius at 4°C is almost completely explained by an increase of the thickness of the outer surface shell ($HEDS_{width}$). Only in the case of the smallest LDL particles (LDL-6) a larger core is additionally measured. The small increase of $HEDS_{width}$, consistently found for all three subfractions, indicates that the phase transition of the core is accompanied by structural changes of the particle surface. Such changes are most probably related to a conformational change of the apoB molecule, as proposed by Laggner and Kostner[33], based on ESR spectroscopic data. Interestingly circular dichroism spectra of native LDL are identical at 4 and 50°C[34].

In conclusion we were able to demonstrate that our procedure of X-ray data evaluation is able to resolve differences between human LDL subfractions, and the corresponding high and low temperature forms of the particle core. Besides new information on structural differences between LDL subfractions and conformation of apoB a novel interpretation of the arrangement of the cholesterol esters in the ordered state is proposed.

This work was supported by the Deutsche Forschungsgemeinschaft, SFB 60, Teilprojekt D8.

REFERENCES

1. D. Atkinson, and D. M. Small, 1986, Recombinant lipoproteins: implications for structure and assembly of native lipoproteins, Ann. Rev. Biophys. Chem., 15:403.
2. D. Atkinson, D. M. Small, and G. G. Shipley, 1980, X-ray and neutron scattering studies of plasma lipoproteins, in: "Lipoprotein structure," A. M. Scanu, and F. R. Landsberger, eds., Annals of the New York Academy of Sciences, Vol. 348:284.
3. P. Laggner, and K. W. Müller, 1978, The structure of serum lipoproteins as analysed by X-ray small-angle scattering, Q. Rev. Biophys. 11:371.
4. M. W. Baumstark, W. Kreutz, A. Berg, I. Frey, and J. Keul, 1990, Structure of human low-density lipoprotein subfractions, determined by X-ray small-angle scattering, Biochim. Biophys. Acta 1037:48.
5. V. Luzzati, A. Tardieu, and L. P. Aggerbeck, 1979, Structure of serum low-density lipoprotein I. Solution X-ray scattering study of a hyperlipidemic monkey low-density lipoprotein, J. Mol. Biol. 131:435.
6. M. W. Baumstark, W. Welte, and W. Kreutz, 1982, A theory for the evaluation of small-angle scattering diagrams of quasi radially symmetric particles considering polydispersity and deviations from radial symmery, Acta Cryst. A38:835.
7. M. W. Baumstark, and W. Kreutz, 1982, Iterative Deconvolution Method for Evaluating X-Ray Small Angle Scattering Diagrams., in: "Deconvolution Reconvolution," M. Bouchy, ed., E.N.S.I.C. - I.N.P.L., Nancy.
8. T. G. Gulik-Krzywicki, M. Yates, and L. P. Aggerbeck, 1979, Structure of serum low-density lipoprotein. II. A freeze- etching electron microscopy study, J. Mol. Biol. 131:475.

9. F. T. Lindgren, Preparative ultracentrifugal laboratory procedures and suggestions for lipoprotein analysis, 1975, in: "Analysis of lipids and lipoproteins," E. G. Perkins, ed., American Oil Chemists' Society, Champaign, Ill.

10. T. J. Knott, R. J. Pease, L. M. Powell, S. C. Wallis, S. C. Rall jr, T. L. Innerarity, B. Blackhart, W. H. Taylor, Y. Marcel, R. Milne, D. Johnson, M. Fuller, A. J. Losis, B. J. McCarthy, R. W. Mahley, B. Levy-Wilson, and J. Scott, 1986, Complete protein sequence and identification of structural domains of human apolipoprotein B, Nature 323:734.

11. C. Y. Yang, S. H. Chen, S. H. Gianturco, W. A. Bradley, J. T. Sparrow, M. Tanimura, W. H. Li, D. A. Sparrow, H. DeLoof, M. Rosseneu, F. S. Lee, Z. W. Gu, A. M. Gotto jr, and L. Chan, 1986, Sequence, structure, receptor-binding domains and internal repeats of human apolipoprotein B-100, Nature 323:738.

12. M. W. Baumstark, W. Welte, and W. Kreutz, 1983, Electron-density determination of three high-density lipoprotein subfractions, considering polydispersity and deviations from radial symmetry, Biochim. Biophys. Acta 751:108.

13. R: J. Deckelbaum, G. G. Shipley, and D. M. Small, 1977, Structure and interactions of lipids in human plasma low density lipoproteins, J. Biol. Chem. 252:744.

14. P. Laggner, G. M. Kostner, U. Rakusch, and D. Worcester, 1981, Neutron small angle scattering on selectively deuterated human plasma low density lipoproteins, J. Biol. Chem. 256:11832.

15. G. L. Gaines jr, 1966, "Insoluble monolayers at lipid-gas interface," Interscience, New York.

16. H. Lecuyer, D. G. Dervichian, 1969 , J. Mol. Biol. 45:39.

17. Y. L. Marcel, T. L. Innerarity, C. Spilman, R. W. Mahley, A. A. Protter, and R. W. Milne, 1987, Mapping of human apolipoprotein B antigenic determinants, Arteriosclerosis 7:166.

18. W.- H. Li, M. Tanimura, C.- C. Luo, S. Datta, and L. Chan, 1988, The apolipoprotein multigene family: biosynthesis, structure, structure-function relationships, and evolution, J. Lipid Res. 29:245.

19. G. Schonfeld, and E. S. Krul, 1986, Immunologic approaches to lipoprotein structure, J. Lipid Res. 27:583.

20. E. G. Finer, R. Henry, R. B. Leslie, and R. N. Robertson, 1975, NMR studies of pig low- and high-density serum lipoproteins. Molecular motions and morphology, Biochim. Biophys. Acta 380:320.

21. P. L. Yeagle, R. G. Langdon, and R. B. Martin, 1977, Phospholipid-protein interactions in human low density lipoprotein detected by 31P nuclear magnetic resonance: Biochemistry 16:3487.

22. E. Herzyk, D. C. Lee, R. C. Dunn, K. R. Bruckdorfer, and D. Chapman, 1987, Changes in the secondary structure of apolipoprotein B-100 after Cu2+ -catalysed oxidation of human low-density lipoproteins monitored by Fourier transform infrared spectroscopy, Biochim. Biophys. Acta 922:145.

23. D. Osterman, R. Mora, F. J. Kézdy, E. T. Kaiser, and S. C. Meredith, 1984, A synthetic amphiphilic ß-strand tridecapeptide: a model for apolipoprotein B, J. Am. Chem. Soc. 106:6845.

24. D. Atkinson, R. J. Deckelbaum, D. M. Small, and G. G. Shipley, 1977, Structure of human plasma low-density lipoproteins: Molecular organisation of the central core, Proc. Natl. Acad. Sci. USA 74:1042.

25. P. Laggner, G. Degovics, K. W. Müller, O. Glatter, O. Kratky, G. Kostner, and A. Holasek, 1977, Molecular packing and fluidity of lipids in human serum low densiy lipoproteins, Hoppe Seyler's Z. Physiol. Chem. 358:771.

26. R. P. Rand, and V. Luzzati, 1968, X-ray diffraction study in water of lipids extracted from human erythrocytes, Biophys. J. 8:125.

27. B. M. Craven, and G. T. De Titta, 1976, Cholesteryl myristate: structure of the crystalline and mesophases, J. Chem. Soc., Perkin Trans. 27:814.

28. C. Tanford, 1980, "The hydrophobic effect: Formation of micelles & biological membranes," 2nd edition, John Wiley & Sons Inc., New York.

29. P. Laggner, G. M. Kostner, G. Degovics, and D. L. Worcester, 1984, Structure of the cholesteryl ester core of human plasma low density lipoproteins: Selective deuteration and neutron small-angle scattering, Proc. Natl. Acad. Sci. USA 81:4389.

30. W. Stoffel, O. Zierenberg, B. Tunggal, and E. Schreiber, 13C Nuclear magnetic resonance spectroscopic evidence for hydrophobic lipid-protein interactions in human high density lipoproteins, Proc. Nat. Acad. Sci. USA 71:3696.

31. G. S. Ginsburg, D. M. Small, and D. Atkinson, 1982, Microemulsions of phospholipids and cholesterol esters. Protein-free models of low density lipoprotein , J. Biol. Chem. 257:8216.

32. A. Tardieu, L. Mateu, C. Sardet, B. Weiss, V. Luzzati, L. Aggerbeck, and A. M. Scanu, 1976, Structure of human serum lipoproteins in solution. II. Small-angle X-ray scattering study of HDL3 and LDL , J. Mol. Biol. 101:129.

33. P. Laggner, and G. M. Kostner, 1978, Thermotropic changes in the surface structure of lipoprotein B from human-plasma low-density lipoproteins, Eur. J. Biochem. 84:227.

34. G. S. Ginsburg, M. T. Walsh, D. M. Small, and D. Atkinson, 1984, Reassembled plasma low density lipoproteins. Phospholipid-cholesterol ester-apoprotein B complexes, J. Biol. Chem. 259:6667.

ROLE OF AMPHIPATHIC HELIXES IN HDL STRUCTURE/FUNCTION

G. M. Anantharamaiah, C. G. Brouillette, J. A. Engler,
H. De Loof, Y. V. Venkatachalapathi, J. Boogaerts, and
J. P. Segrest

Departments of Medicine, Biochemistry and the Atherosclerosis
Unit, UAB Medical Center, Birmingham, Alabama 35294

ABSTRACT

   In a recent analysis we classified amphipathic helix domains into a
minimum of seven distinct classes. Four amphipathic helix classes are found
in lipid-associating proteins: apolipoproteins, certain polypeptide
hormones, polypeptide venoms and antibiotics, and certain complex transmembrane proteins. Three amphipathic helix classes are involved in both intra-
and intermolecular protein-protein interactions: calmodulin-regulated
protein kinases, coiled-coil containing proteins that include the so-called
leucine zipper, and globular helical proteins.

   Three central hypothesis have been developed in our studies of the
apolipoprotein class of amphipathic helixes: 1) The "Snorkel" hypothesis
proposes that when the amphipathic helix is associated with phospholipid,
amphipathic basic residues extend toward the polar face of the helix to
insert their charged residues into the aqueous milieu: thus the entirety of
the uncharged van der Waals'surface of the amphipathic helix is buried
within the lipid. 2) We have formulated a hypothesis that Glutamyl residues
located at positions 78 and 111 in apolipoprotein A-I on the nonpolar face
of two amphipathic helical domains are critical to LCAT activation. 3) The
hinged-domain hypothesis was proposed to explain the structural basis for
the quantization of HDL subspecies, protein-protein interactions in HDL,
and the HDL disc to sphere transformation.

INTRODUCTION

   The amphipathic helix is an often encountered secondary structural
motif in biologically active peptides and proteins. The amphipathic helix
was first described as a unique structure/function motif involved in lipid
interaction by Segrest *et al.* in 1974 (1). Amphipathic helical domains have
been described for other lipid associating proteins, including polypeptide
hormones such as endorphins, polypeptide venoms such as bombolitin, polypeptide antibiotics such as the meganins, and certain complex transmembrane
proteins, such as bacteriorhodopsin (2). Based upon a detailed analysis of
their physical-chemical and structural properties, we have grouped amphipathic helixes into seven classes (A, H, L, G, K, C, and M: Fig 1) (2).

Hypercholesterolemia, Hypocholesterolemia, Hypertriglyceridemia
Edited by C.L. Malmendier *et al.*, Plenum Press, New York, 1990

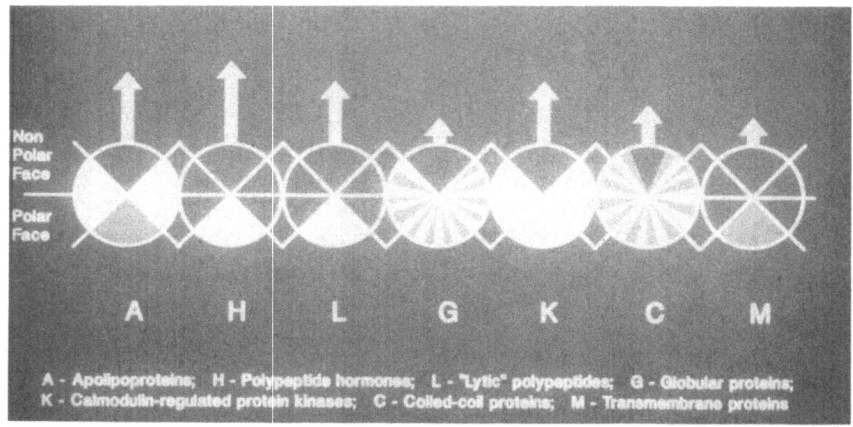

Fig. 1. Schematic representation of seven different classes of the
amphipathic helix. Different circles represent each class.
Arrows above each class indicate the relative mean hydrophobic moment.

The exchangeable apolipoproteins from plasma lipoproteins (apo A-I,
A-II. A-IV. C-I. C-II, C-III and E) belong to the class A amphipathic
helixes. They differ from the rest of the classes in that they possess
positively charged residues at the polar-nonpolar interface and negatively
charged residues at the center of the polar face. The most striking feature
of these exchangeable apolipoproteins is the presence of internal
22-residue-long repeats (3). More importantly, this repeating unit has the
periodicity of an amphipathic   helix. These observations form a major
basis for the investigations of the structure and functions of the apolipoproteins in our laboratory.

THE "SNORKEL HYPOTHESIS"

Our approach to understanding the factors responsible for the lipid
association of the apolipoprotein class of amphipathic helix was the
synthesis of model peptide analogs of this class. Our initial studies of
peptide analogs involved two sets of peptides (4-7). One set of model
peptides were designed to mimic the class A amphipathic helix; i.e.,
positively charged residues at the polar-nonpolar interface and negatively
charged residues at the center of the polar face. The second set of analogs
had a reverse charge distribution (Fig. 2). Examination of the lipid-
associating properties of these peptide analogs showed that the class A
mimicking peptide analogs interacted much more effectively to form stable
discoidal peptide:lipid complexes than the analogs with the reverse charge
distribution. These analogs for the class A amphipathic helix were also
able to competitively displace apo A-I from HDL compared to the analogs
with reverse charge distribution. From these studies we recognized that the
positively charged residues at the interface, because of their longer acyl
chainlength, may be playing an important role in increasing the lipid
affinity of the peptide analogs of class A amphipathic helix.

As part of a study to understand the molecular properties of apo A-I
we synthesized analogs for the 22mer consensus lipid-associating domain
with the sequence PVLDEFREKLNE x EEALKQKLK. As can be seen in Fig. 3, the
13th position (denoted by X in the sequence) is located at the nonpolar

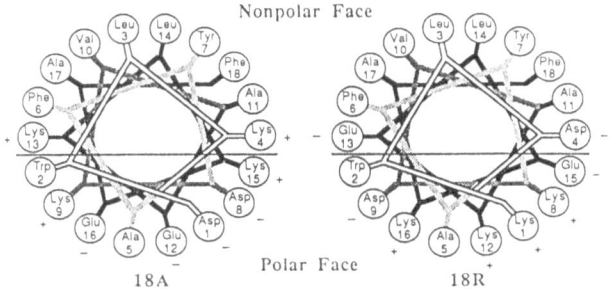

37aA = 18A - Ala - 18A

37pA = 18A - Pro - 18A

Fig. 2. Helical wheel representation of model 18A and 18R peptide analogs.

face, 40° from the polar-nonpolar interface of the amphipathic helix. Amino acid susbtitutions were therefore carried out in this position to see the effect of the acyl chain length on lipid affinity. Three peptide analogs were synthesized with X at the 13th position substituted by Glu, Lys and

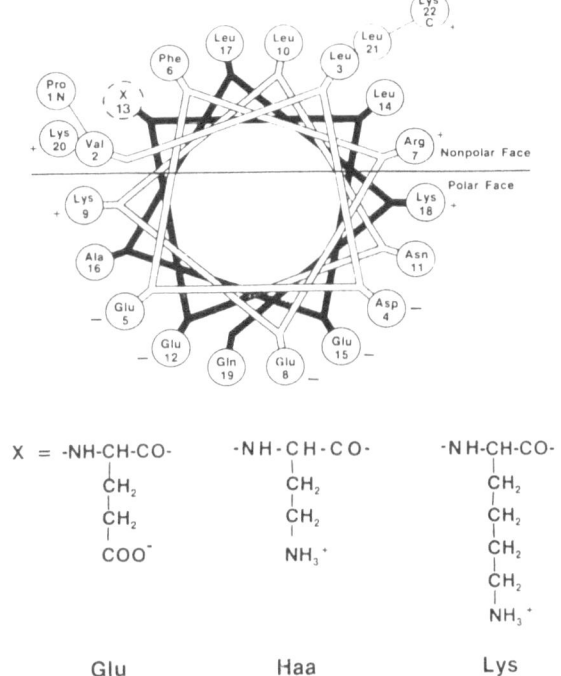

Fig. 3. Helical wheel representation viewed from the amino terminal end of the consensus 22mer peptide for the 22mer repeating units of apo A-I. X = Glu, Lys or Haa.

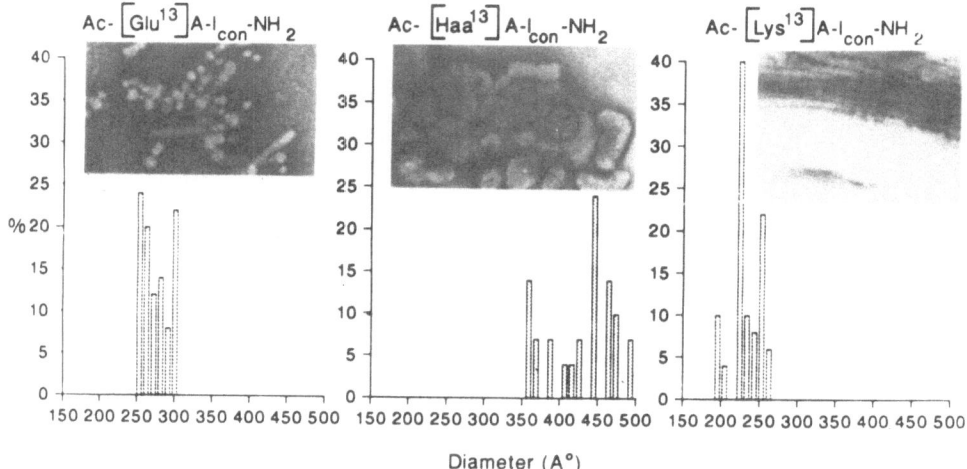

Fig. 4. (A) Negative stain electron microscopy of peptide DMPC complexes at
1:1 weight ratio. (B) Morphographic analysis of diameters of
discoidal complexes.

L-homoaminoalanine (Haa). The peptide containing Haa served as a control
peptide to study the effect of acyl chain length on lipid affinity because
this amino acid has the same acyl chainlength as Glu but a positively
charged side chain, similar to Lys. Thus the effect of the longer acyl
chain length of Lys on lipid affinity when it is close to the polar-
nonpolar interface was studied.

The peptides were compared for their ability to interact with multilamellar vesicles of dimyristoyl phosphatidylcholine (DMPC); the extent of
interaction was estimated by their ability to clarify turbid multilamellar
vesicles of DMPC and to form stable discoidal complexes. Turbidity clarifi-
cation studies showed that the Lys analog was the only peptide which clari-
fied the multilamellar vesicles of DMPC. The Lys analog formed the smallest
discoidal complex of the three peptides (Stokes diameters: $[Lys^{13}]A-I_{con}$,
110 ± 30 A; $[Glu^{13}]A-I_{con}$, 200 ± 50 A; and $[Haa^{13}]A-I_{con}$, 310 ± 30 A). We
have shown previously that the size of the discoidal complexes varies
inversely with the lipid affinity of the peptide. Thus, based on these
studies, the Lys analog has a greater lipid affinity than the other two.

Dye leakage experiments using fluorescence-entrapped egg PC liposomes
also suggested that the Lys analog possesses increased lipid affinity
compared to the other two analogs. The results of the fluorescence leakage
are shown in Fig. 5. The results show that $[Lys^{13}]A-I_{con}$ at 200 g equals
Triton X-100. The other two analogs at the same peptide concentration
released only about 50% of the entrapped dye.

Based on these results we propose the "Snorkel" model shown in Fig. 6
for the lipid association of apolipoprotein class amphipathic helix. The
bulk of the van der Waals surface areas of the positively charged residues
are hydrophobic. We propose that the amphipathic basic residues, when
associated with phospholipid, extend toward the polar face of the helix to
insert their charged moieties into the aqueous milieu for solvation. We
suggest that essentially all of the uncharged van der Waals surface of the
amphipathic helical domains of the apolipoproteins can be buried within the
interior of a phospholipid monolayer. Compared to the other classes of the
amphipathic helix, the arrangement of the charged residues found in the

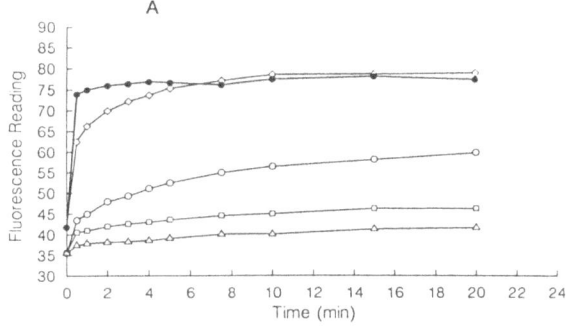

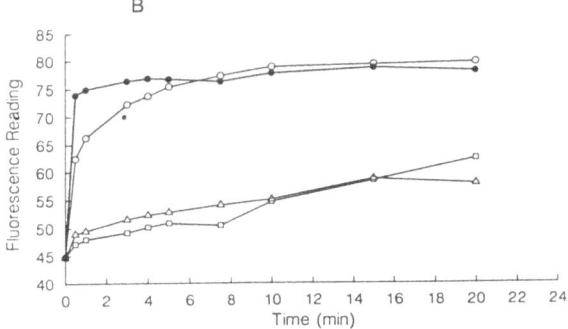

Fig. 5. Fluorescence dye leakage from egg PC entrapped vesicles: Carboxy fluorescein was trapped in small unilamellar egg PC vesicles. Perturbation by the protein of the lipid bilayer was monitored by the rate of leakage and dequenching of fluorescent dye into the media. A. Triton X-100 ● - ● and ◆ - ◆ 200 ɣg, ○-○ 100 ɣg, □ - □ 50 ɣg and ▲ - ▲ 25 ɣg of A-I$_{con}$Lys^{13}. B. ○ - ○ A-I$_{con}$Lys^{13}, □ - □ A-I$_{con}$Haa^{13} and ▲ - ▲ A-I$_{con}$Glu^{13}.

apolipoproteins provides for a deeper helix insertion into a monolayer and thus a greater lipid affinity.

MECHANISMS OF LCAT ACTIVATION BY APO A-I

Apolipoprotein apo A-I is the major co-factor for activation of the plasma enzyme lecithin:cholesterol acyltransferase (LCAT). LCAT is thought to mediate reverse cholesterol transport by trapping cholesterol in the form of cholesteryl ester in the HDL particle for removal by the liver. Many apolipoproteins also can serve as co-factors for the enzyme LCAT. However, the extent of activation for these proteins is similar to that of the nonspecific activators such as non-homologous synthetic peptides (5). Human apo A-I has therefore been studied extensively to localize the LCAT activating domain(s) in its sequence (6, 7). Based on studies of the CNBr fragments and the synthetic peptides corresponding to the CNBr fragmentation products the LCAT activating domain(s) could not be localized. We therefore examined the repeating 22mer units of apo A-I to attempt to understand the structure and function of apo A-I.

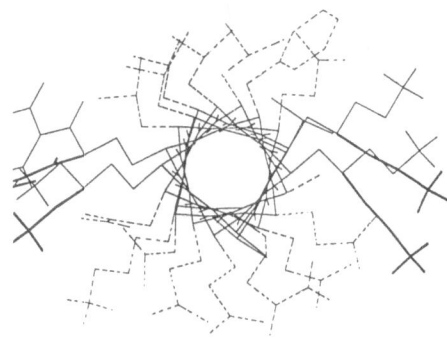

Fig. 6. Computer model of an amphipathic peptide indicating features of the
"Snorkel" model. The consensus class A apolipoprotein amphipathic
helical peptide (apo A-I$_{con}$) was modeled using the "SYBYL" program
package (TRIPOS Associates Inc.). An idealized   helix was built
with backbone dihedral angles Y = 58° and F = -47°. One side chain
dihedral angle of each of the four positively charged residues was
changed from the trans to gouche conformation so as to bring the
end of the side chains closer to the hydrophilic side of the helix.
Both the initial and the "snorkel" structure were subsequently
energy minimized with SYBYL using AMBER force-field parameters. A
superposition of the two minimized helixes is shown in the figure.
The energy difference between them was very small, showing the
"snorkel"-configuration is possible without bad intramolecular
steric constraints. Positively charged residues are shown in solid
light (initial) and dark (snorkel).

There are eight 22mer tandem repeats at the C-terminal end of apo A-I.
In every case but one, the eight tandem 22mer repeats are punctuated by a
single proline. Each tandem 22mer repeat in apo A-I is an amphipathic
helix. Because each tandem 22mer in apo A-I appears to represent a dupli-
cation of a primordial gene sequence (8), as part of our studies on the
molecular properties of amphipathic helixes, a consensus sequence approxi-
mation of the primordial 22mer was derived. The sequence, called A-I$_{con}$, is
as follows: Pro Val Leu Asp Glu Phe Arg Glu Lys Leu Asn Glu X Leu Glu Ala
Leu Lys Gln Lys Leu Lys (Fig. 3). This sequence represents the most preva-
lent amino acid residue at each position of the eight 22mers. Although the
13th residue is positioned at the nonpolar face of the consensus sequence,
six out of the eight tandem repeats of apo A-I have a charged residue at
this position. The anomalous nature of this residue suggested to us the
possibility that this position might play a role in the LCAT-activating
properties of apo A-I. To investigate this possibility we synthesized four
22mer peptide analogs differing only in the 13th position. Three homo and
one heterodimeric combinations were synthesized (Table 1).

Table 1.

```
E    PVLDE FREKL NEELE ALKQK LK
R    PVLDE FREKL NERLE ALKQK LK
K    PVLDE FREKL NEKLE ALKQK LK
H    PVLDE FREKL NEHLE ALKQK LK
EE   PVLDE FREKL NEELE ALKQK LK PVLDE FREKL NEELE ALKQK LK
RE   PVLDE FREKL NERLE ALKQK LK PVLDE FREKL NEELE ALKQK LK
RR   PVLDE FREKL NERLE ALKQK LK PVLDE FREKL NERLE ALKQK LK
HH   PVLDE FREKL NEHLE ALKQK LK PVLDE FREKL NEHLE ALKQK LK
```

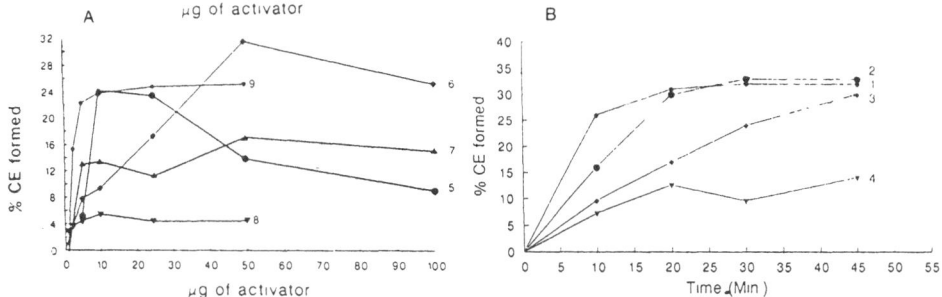

Fig. 7. Activation of LCAT by synthetic peptide analogs as measured by: (A)
egg lecithin small unilamelar vesicle procedure; (B) egg lecithin
discoidal particle prepared by the cholate dialysis procedure (10,
13).

These peptides were studied for their ability to activate the enzyme
LCAT using two different assay systems: i) the egg PC vesicular assay, and
ii) the discoidal assay method using the cholate dialysis procedure (10,
13). The results were compared with the ability of apo A-I to activate LCAT
in these two assay systems. The results of the LCAT activation using the
dimer peptide analogs are shown in Fig. 7. In both assay systems the
homodimer analog containing Glu at the 13th position equaled apo A-I in its
ability to activate LCAT.

Examination of the order of distribution of 13th amino acid in the
eight tandem repeats of apo A-I (Fig. 8) shows that residues 66-121 of apo
A-I are similar to the Glu-dimer analog. We therefore propose that the
major LCAT activating domain of apo A-I resides in 66-121 region of apo
A-I. These results thus explain the reason for failure in localizing the
LCAT activating domain of apo A-I. The 66-121 region of apo A-I contains
three Methionyl residues. CNBr cleavage would thus fragment this region of
apo A-I and the fragments would be devoid of 66-121 apo A-I sequence.
Recent studies by Jonas *et al.* (12) provide additional support for this
hypothesis.

These results led us to believe that glutamic acids at residues 78 and
111 of apo A-I are critical for the LCAT activating ability of apo A-I.
With a view to test this we have prepared recombinant apo A-I and a mutant
apo A-I in which the 78th Glu residue is changed to Ala. These two proteins
were compared with human apo A-I for their ability to activate the enzyme
LCAT following the assay systems used for synthetic peptide analogs.
Results from these preliminary$_{78}$LCAT studies are shown in Fig. 9. These
results indicate that the Glu^{78} -> Ala apo A-I is half as effective in

Apolipoprotein A-I

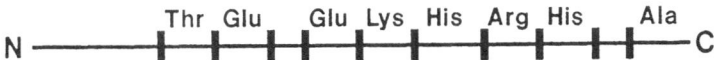

Fig. 8. Order of distribution of 13th amino acid in the eight tandem 22mer
repeats of apo A-I. Two vertical lines between the amino acid
residues represent 22mer. The peptide ([Glu 13]A-I$_{con}$)2 is similar
to the two glutamic acid-containing 22mers in having glutamic acid
residues at the same position as the two 22mers.

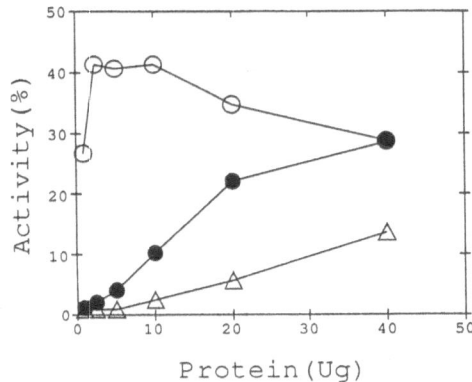

Fig. 9. Activation of LCAT by: (1) plasma apo A-I, (2) recombinant apo A-I
and (3) mutant (Glu[78] -->Ala) apo A-I as measured by the egg
lecithin small unilamellar vesicles assay method.

activating the enzyme LCAT as the recombinant apo A-I with the sequence
identical to that of plasma apo A-I. These results support the hypothesis
that sequence specific glutamic acid residues are important for the LCAT
activation by apo A-I.

THE "HINGED-DOMAIN" HYPOTHESIS

Analysis of the complexes formed between apo A-I and DMPC has shown
the formation of a highly controlled stoichiometric class of particles of
discrete size containing 2 apo A-I molecules (R-2 complexes) and three apo
A-I molecules (R-3 complexes) (11). Variation of the molar ratio of lipid
to protein produced discoidal complexes which varied in diameter by a uni-
formly spaced increment. These observations led to the proposal that the
size of the disc produced by this reaction is controlled by conformational
changes in apo A-I that results in the all-or-none binding of complete
helical (hinged) domains (11). Recently, this hypothesis has been supported
by the preliminary studies of apo A-I/POPC micellar subspecies (12).

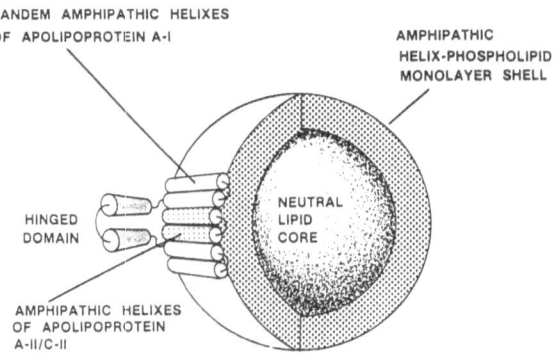

Fig. 10. Model of the "hinged-domain" hypothesis.

The mechanism of subspecies formation in HDL based on the "hinged domain" hypothesis is illustrated in Fig. 10.

Using affinity columns containing anti-apo A-I or anti-apo A-II, nine plasma HDL subpopulations were isolated and characterized using single vertical spin centrifugation (13, 14), electronmicroscopy and nondenaturing gradient gel electrophoresis (8). These were designated as HDL 1, 2, 3, 4, 5, 6, 7a,7b,7c, 8, and 9. The difference in surface area of HDL 4 compared to HDL 3, HDL 5 compared to HDL 4, and HDL 6 compared to HDL 5 averaged 26000 ± 300 A2, the differences in surface area of HDL 7b compared to HDL 7a, and HDL 7c compared to HDL 7b average 3400 ± 100 A^2, while the difference in surface area of HDL 7a compared to HDL 6 was 56000 A^2. This is reminiscent of the results of apo A-I:DMPC recombinants described earlier (11) in which incremental changes in discoidal diameter between several discrete R-2 and R-3 recombinant particles were observed. The quantized changes observed in the HDL population and apo A-I:DMPC discoidal structure could be best explained by the stoichiometric and conformational changes to apo A-I in the hinged-domain.

CONCLUSIONS

1. Class A amphipathic helixes, via the Lys/Arg "snorkel" process, mediate strong lipid affinity without damage or alteration to cell membranes. (Class H and L amphipathic helixes mediate strong lipid affinity but damage or alter cell membranes).

2. Amphipathic helical repeats 2 and 3 represent the major LCAT-activating domain in apo A-I. This domain activates LCAT in a 2-step process:

Step 1. The amphipathic helix produces a protein-lipid cleft recognized by the active site of LCAT, allowing the enzyme to get at the Sn-2 bond of phosphatidylcholine.

Step 2. Sequence- and structure-specific glutamyl residues in repeats 2 and 3 create and activated cholesterol intermediate.

3. There are one, or probably two, hinged-domains in apo A-I that have an amphipathic -loop- helical motif. These domains possess two reversible states (lipid-bound and lipid-unbound) that mediate the following:

a. A low free energy mechanism for reorganization of the amphipathic helical arrays on the edge of discoidal HDL to an array that covers the surface of spheroidal HDL.
b. Interactions of apo A-I with other apos (E.g., apo A-I).
c. HDL subspecies (including size quantization).

REFERENCES

1. J. P. Segrest, R. L. Jackson, J. D. Morrisett. and A. M. Gotto, Jr., FEBS Lett., 38:247 (1973).
2. J. P. Segrest, H. De Loof. J. G. Dohlman, C. G. Brouillette, and G. M. Anantharamaiah, Proteins, In press (1990).
3. C -C. Luo, W -H. Li, M. N. Moore, and L. Chan, J. Mol. Biol., 187:325 (1986)
4. G. M. Anantharamaiah, J. L. Jones. C. G. Brouillette, C. F. Schmidt. B. H. Chung, T. A. Hughes, A. S. Bhown, and J.P. Segrest, J.Biol. Chem., 260:10248 (1985).

5. G. M. Anantharamaiah, Synthetic peptide analogs of apolipoproteins, In:
    Methods of Enzymology, J. P. Segrest and J. J. Albers, eds.,
    Academic Press, New York, 128:627 (1986).
6. R. M. Epand, A. Gawish, M. Iqbal, K. B. Gupta, C. H. Chen, J. P.
    Segrest, and G. M. Anantharamaiah, J. Biol. Chem., 262:9389 (1987).
7. Y. V. Venkatachalapathi, K. B. Gupta, H. De Loof, J. P. Segrest, and
    G. M. Anantharamaiah, In: Peptides, ESCOM Press (J. Rivier Ed.) 672
    (1990)
8. C. J. Fielding, V. G. Shore, P. E. Fielding, Biophys. Biochem. Res.
    Commun, 46:1493 (1972).
9. J. J. Albers, J. T. Lin, G. P. Roberts, Artery, 5:61 (1979)
10. G. M. Anantharamaiah, Y. V. Venkatachalapathi, C. G. Brouillette, and
    J. P. Segrest, Arteriosclerosis, 10:95 (1990).
11. C. G. Brouillette, J. L. Jones, T. C. Ng, H. Kercert, B. H. Chung, and
    J. P. Segrest, Biochemistry, 23:359 (1984)
12. A. Jonas, K. E. Kezdy, and J. H. Wald, J. Biol. Chem, 264:4818 (1989).
13. M. C. Cheung, J. P. Segrest, J. J. Albers, J. T. Cone, C. G.
    Brouillette, B. H. Chung, M. Kashyap,, A. Glasscock, and G. M.
    Anantharamaiah, J. Lipid Res., 28:913 (1987).
14. B. H. Chung, J. P. Segrest M. J. Ray, J. D. Brunzell, J. E. Hokanson,
    R. M. Krauss, K. Beaudrie, and J. T. Cone, Single vertical spin
    density gradient ultracentrifugation, In: Methods of Enzymology,
    J. P. Segrest and J. J. Albers, eds, Academic Press, New York,
    128:181 (1986).

CONFORMATIONAL PROPERTIES OF APOLIPOPROTEINS

STUDIED BY COMPUTER GRAPHICS

J-L. De Coen, J-P. Kocher, C. Delcroix, J-F. Lontie
and C.L. Malmendier

Fondation de Recherche sur l'Atherosclerose
2, rue Evers, B-1000 Brussels

INTRODUCTION

The physical properties of apolipoproteins in solution and when
interacting with lipids have been extensively studied using various
experimental methods (1). In water, these molecules appear to be in
equilibrium between a statistical coil and some folded conformers
containing α -helices. When interacting with lipids, the α -helical
content increases to reach values around 60% for apolipoprotein A-II
for example. Taking these observations into account and looking at
the results obtained when rules to predict secondary structure are
applied to the sequence of apo A-II we proposed recently (2) that
the tertiary structure of this molecule could be similar to the one
of uteroglobin which is a little steroid binding protein of known
crystalline structure (3). In particular, we observed that the
regions of high helical content appear to be located at similar positions in the two molecules being in both cases delineated by proline
residues. Furthermore, the occurrence of these helical regions gives
rise for the two molecules to a clustering of hydrophobic residues.
In the crystal, uteroglobin is a dimer made of two identical monomer
units containing each four α -helices which are packed in space according to the scheme depicted on Fig. 1. The antiparallel arrangement

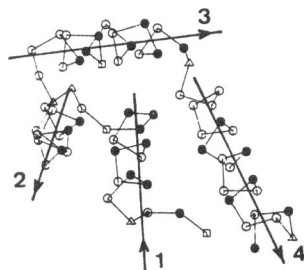

Fig. 1 Folding pattern of the monomer of uteroglobin in the
crystal. The course of the helices are identified by
arrows. Classes of residues are represented by symbols:
● hydrophobic, △ proline, □ glycine, ○ others.

Hypercholesterolemia, Hypocholesterolemia, Hypertriglyceridemia
Edited by C.L. Malmendier *et al.*, Plenum Press, New York, 1990

141

of helices 1, 2 and 4 is such that most of the hydrophobic residues
of these amphipatical helices are located on the same side of the
protein and define with the hydrophobic cluster of helix 3 a continuous hydrophobic surface which is buried from the solvent by the
formation of the dimer. In the present paper, we show that a similar
hydrophobic surface occurs when a model protein is built with the
amino acid sequence of apo A-II put on the backbone of the tertiary
structure of uteroglobin.

RESULTS

The modeling of the tertiary structure of apo A-II was carried
out in three steps using the computer graphics software BRUGEL (4).
First,the best alignment of the sequences of apo A-II and uteroglobin
was determined. The resulting alignment led to the introduction of
two insertions. The first one was the insertion of residue Lys-27 of
apo A-II inside the turn between helices 1 and 2 and the second one
was an extension of helix 3 by three residues Glu-58, Gln-59 and
Thr-60. These insertions were carried out without any difficulties
using a subroutine which is able to replace a fragment of a protein
by another one. The routine cuts out a fragment of a protein and
replaces it by a longer or a shorter one taken out the protein
crystallographic data bank. By this procedure the tertiary structure
is modified only at the local level while the overall folding
of the protein remains unchanged. Secondly, the side-chain confor-
mations of the model of apo A-II were fitted at best with those
occurring in uteroglobin when the substituted residues are similar
and were choosen to avoid van der Waals contacts with the backbone
when the residues are different. Thirdly,the conformational energy
of the model was calculated and minimized using a conjugated gradient
algorithm.

The stereoviews shown in Fig. 2a and b correspond respectively
to the tertiary stucture of the monomer of uteroglobin as it appears
in the crystal and to the model of apo A-II built by computer graphics. The plots of the amino-acid residues occurring at the hydrophobic surfaces of uteroglobin and the model of apo A-II are reported
in Fig. 3a and b.

DISCUSSION

The conformation of uteroglobin in the crystal shows several
interesting features concerning the intramolecular interactions provided by the side chains. Tyr-21 and Leu-25 seem to play a crucial
role for the stabilization of the observed folding by filling a gap
between helices 1 and 3 (see Fig. 2a). These two hydrophobic amino
acids are present at exactly the same positions in the sequence of
apo A-II and accordingly can be involved in a similar stabilizing
effect in the case of this protein (see Fig. 2b).

From a comparison of Fig. 3a and 3b it is clear that the hydrophobic surface observed in uteroglobin remains mainly hydrophobic
for apo A-II. However in the case of apo A-II, one observes that a
few polar residues are located among the hydrophobic ones. These
polar residues are mainly Lys, Gln and Ser. Lys 39, for example,
fill the gap above residues Tyr-21 and Leu-25 which occur in uteroglobin due to the presence of a Gly at position 38. Another Lys
side-chain located at position 46 in apo A-II intercalates between
Gln-13 and Tyr-14 giving rise to a polar spot which was completely
hydrophobic in the case of uteroglobin being made of Leu-13, Leu-14
and Leu-45. Ser-9 is the only other polar residue which appears sur-
rounded by hydrophobic residues in the model of apo A-II. The other
polar residues, namely Lys-3, Lys-28, Gln-35, Ser-45, Thr-72, Gln-73
and Gln-77 appear to be located mostly at the periphery of the

UTEROGLOBIN

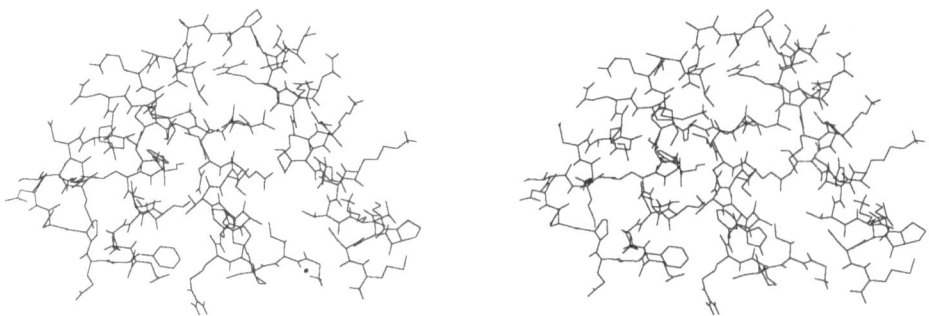

Fig. 2a Stereoview of the tertiary structure of the monomer of
uteroglobin as it is observed in the crystal.

APOLIPOPROTEIN A-II

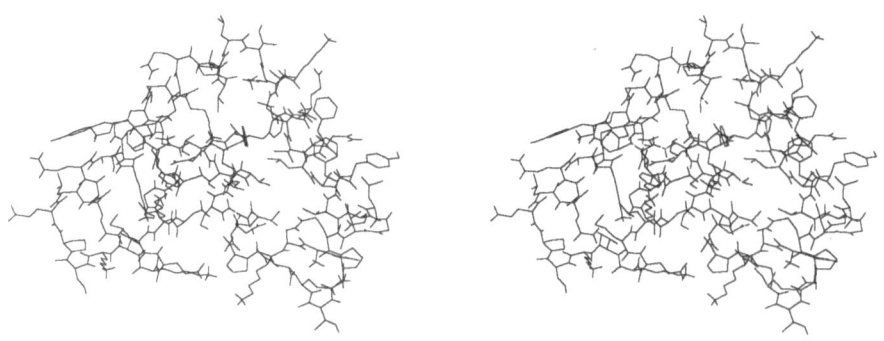

Fig. 2b Stereoview of the model of apo A-II built by computer graphics
using the backbone structure of uteroglobin as template.

UTEROGLOBIN

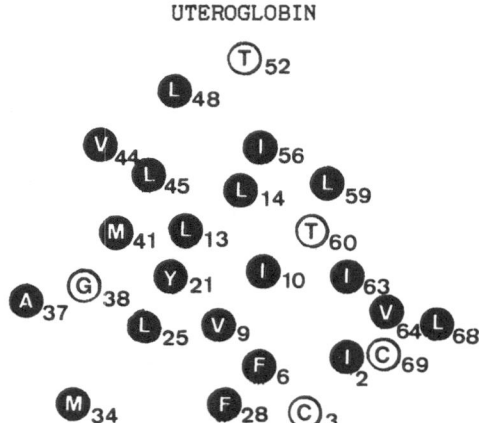

Fig. 3a Distribution of residues observed on the hydrophobic surface
of the monomer of uteroglobin in the crystal.
Hydrophobic residues are marked out in black.

APOLIPOPROTEIN A-II

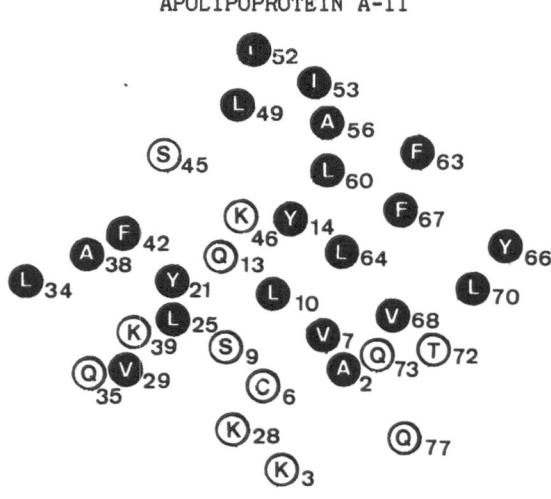

Fig. 3b Distribution of residues occuring on the hydrophobic surface
of the model of apo A-II builded by computer graphics.

hydrophobic surface.This is also the case for the Cys-6 residue
which in the dimer of apo A-II gives rise to the formation of a disulfide bridge.

The presence of polar residues on the hydrophobic surface of
the model of apo A-II could explain the experimental evidence that
electrostatic interactions play some role in the binding process of
this apolipoprotein with lipids (5). These polar residues
might been involved in some specific interactions: the amino groups
of the Lys could interact specifically with the phosphate moieties
of the phopholipids while the hydroxyl groups of Ser-9 and
Tyr-21 as well as the amide group of Gln-13 could allow the formation of specific hydrogen bonds.

In conclusion, one can say that it is possible to build
a sterically acceptable model of apo A-II on the folding pattern
observed in uteroglobin. The presence in this model of an hydrophobic
surface which is studded with some polar residues might be related
to some specific binding of apo A-II at the surface of lipoprotein
particles made of phospholipids,triglycerides and cholesterol.

REFERENCES

1. H.J. Pownall, Q. Pao, D. Hickson, J.T. Sparrow, A.M. Gotto Jr.
   Thermodynamics of lipid-protein association in human plasma
   lipoproteins, Biophys. J., 37:175 (1982)
2. J-L. De Coen, M. Deboeck, C. Delcroix, J-F. Lontie, C. L. Malmendier, A proposed folding pattern for apolipoprotein A-II based on
   a structural analogy with uteroglobin, Proc. Natl. Acad. Sci. USA,
   85:5669 (1988).
3. I. Morize, E. Surcouf, M.C. Vaney, Y. Epelboin, M. Fridlansky,
   F. Milgrom, J-P. Mornon, Refinement of the C2221 crystal form of
   oxidized uteroglobin at 1.34 A resolution, J. Mol. Biol. 194:725
   (1987)
4. P. Delhaise, M. Bardiaux, S. Wodak, Interactive computer animation
   of macromolecules, J. Mol. Graphics 2:103 (1984)
5. F. Soetewey, M-J. Lievens, R. Vercaemst, M. Rosseneu, H. Peeters,
   V. Brown, Ionization behaviour of native apolipoproteins and
   their complexes with dimyristoyl lecithin, Eur. J. Biochem.,
   79:259 (1977)

ALTERATIONS OF HIGH DENSITY LIPOPROTEINS INDUCED BY THYROID HORMONES

IN MAN AND RAT

Gabriel Ponsin, Catherine Vialle-Valentin, and F. Berthezene

INSERM U. 197, Laboratoire de Métabolisme des Lipides
Hôpital de l'Antiquaille
Lyon, France

INTRODUCTION

In humans, thyroid diseases are accompanied by alterations in the plasma lipoproteins. Hypothyroidism is associated with hypercholesterolemia and an increased risk for atherosclerosis (1). The concentrations of apolipoprotein B and LDL cholesterol are elevated (Table 1), and the cholesterol net transport between cultured fibroblasts and the plasma of hypothyroid patients is dramatically decreased (2,3). In hypothyroidism, HDL cholesterol has been reported to be normal or elevated. Apo A-I is very often increased while apo A-II remains in the normal range (2,4). Conversely, the plasma concentrations of HDL cholesterol and apo A-I are decreased in hyperthyroid patients (5,6). Therefore, thyroid diseases constitute paradoxical situations where HDL cholesterol increases with the atheromatous risk. It has been shown that both in hypo- and hyperthyroidism, the variations in the HDL2 subfraction account for the most part of the total HDL changes (4,5). These abnormal HDL profiles have been tentatively explained on the basis of two findings. Firstly, it has been shown that the HDL clearance rate is decreased in hypothyroidism (1). Secundly, thyroid hormones have been demonstrated to stimulate the hepatic lipase (HL) activity which is known to be involved in the transformation of HDL2 to HDL3 (6,7). Therefore, the increase or decrease of the HDL2/HDL3 ratio respectively observed in hypo- and hyperthyroidism might be due to the variations of HL activity resulting from the changes of thyroid hormone concentrations. However, other causes might be partly responsible for the alterations of HDL pattern. These include the remodelling of HDL particles resulting from the lipid exchanges between lipoproteins mediated by Lipid Transfer Proteins (LTP).

The aim of the present study was to investigate the effect of hyperthyroidism on the structural and metabolic properties of HDL in the rat, a species with no detectable lipid transfer activity (8).

PLASMA CONCENTRATIONS OF APO B CONTAINING LIPOPROTEINS

Female rats (Sprague-Dawley, 180-200 g) were made hyperthyroid by continuous infusion of triiodothyronine (10 µg/100 g body weight/day) for 8 days, through osmotic pumps. This treatment induced of 4 fold increase in the plasma concentration of T3 (264 ± 20 ng/ml vs 61 ± 4 ng/ml; $m\pm$SEM). After overnight fasting, control as well as treated animals were exsanguinated. VLDL, LDL and HDL were then isolated by sequential ultracentrifugations.

Hypercholesterolemia, Hypocholesterolemia, Hypertriglyceridemia
Edited by C.L. Malmendier *et al.*, Plenum Press, New York, 1990

Table 1. Schematic representation of the lipoprotein profile
according to thyroid status, in humans

	Hypothyroidism	Euthyroidism	Hyperthyroidism
Plasma TG	++/+++	++	++
Plasma chol.	+++	++	+
LDL chol.	+++	++	+
HDL chol.	++/+++	++	+
HDL2/HDL3	+++	++	+
Apo B	+++	++	+
Apo A-I	+++	++	+
Apo A-II	++	++	++

As shown in figure 1, hyperthyroidism induced a 40 to 50 % decrease
in all VLDL lipid components, including phospholipids, triglycerides and
free as well esterified cholesterol. These data suggest that VLDL particles
were quantitatively reduced rather than qualitatively altered, which is in
agreement with previous reports indicating that thyroid hormones inhibit
VLDL synthesis in rats (9). No significant change was observed in LDL
fraction.

STRUCTURAL ALTERATIONS OF HDL

In HDL, the most important consequence of hyperthyroidism was a 45 %
increase of phospholipids (figure 2). A much smaller enhancement of cholesterol esters was also noted. These modifications resulted in a significant
decrease of the core/surface lipid ratio. Since others have shown that
thyroid hormones were able to stimulate the synthesis of apo A-I in rats
(9), our data might result, at least in part, from an increase of HDL
synthesis.

Consistent with their composition, HDL isolated from hyperthyroid rats
exhibited a shift to higher densities,after ultracentration in a NaBr
gradient (figure 3). Analysis of HDL samples in non denaturing gradient gel
electrophoresis (4-30 %) indicated a decrease in the mean particle size of
hyperthyroid HDL (figure 4).

Collective consideration of our data show that in rats, hyperthyroidism induces the formation of smaller and denser HDL particles which might
correspond to the accumulation of premature HDL, as suggested by their
relatively low core/surface lipid ratio.

METABOLISM OF HDL

To study the metabolic fate of HDL, we used a series of synthetic
acylated Lipid Associating Peptides (acyl LAPs). These consist of one
single peptide bearing saturated fatty acyl chains of various lengths (0
to 16 carbon units). The acyl peptides have a hydrophobic content that
increase with the acyl chain length. In a previous report, we have shown
that the partition coefficient of the acyl LAPs between HDL and the
aqueous phase was directly related to their degree of hydrophobicity (10).

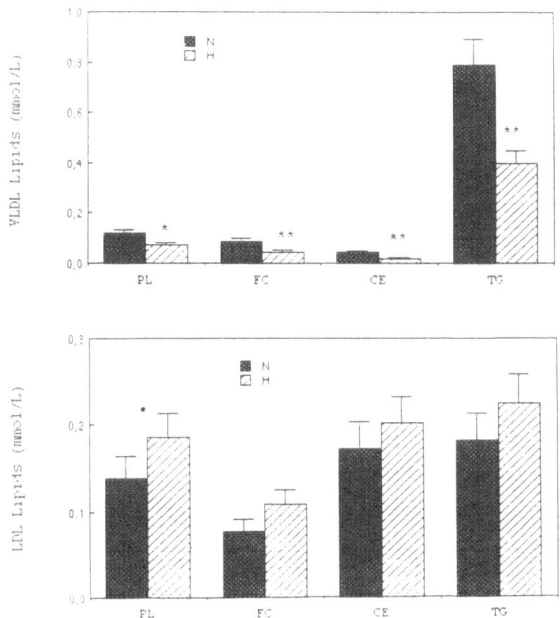

Fig. 1. Plasma concentrations of phospholipids (PL) free cholesterol
(FC), cholesterol esters (CE) and triglycerides (TG) in VLDL
or LDL of normal (N) and hyperthyroid (H) rats.

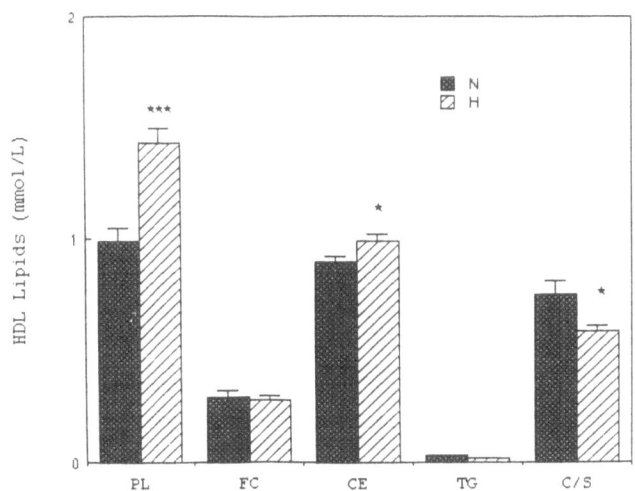

Fig. 2. Plasma concentrations of HDL lipids. The symbols are the same
as those described in fig. 1. The core/surface lipid ratio in
HDL (C/S) was calculated as (CE+TG)/(PL+FC).

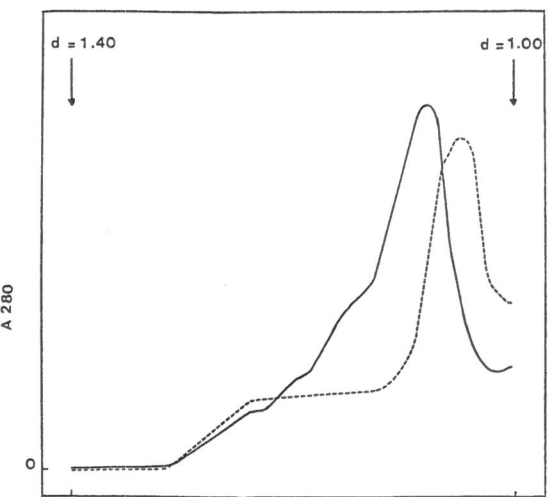

Fig. 3. Ultracentrifugation in NaBr gradient of HDL isolated from normal (---) or hyperthyroid (——) rats.

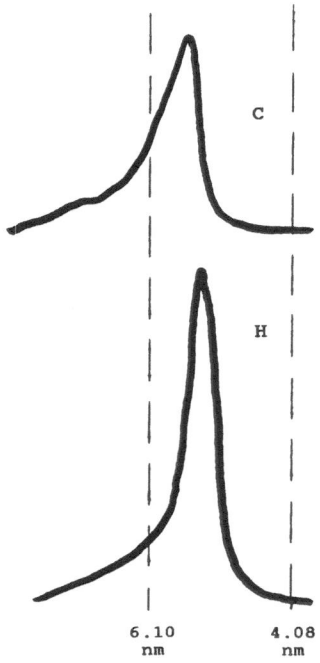

Fig. 4. Non denaturing gradient gel electrophoresis (4-30 %) of HDL from control (C) or hyperthyroid (H) rats. Following staining with Coomassie blue, the HDL peaks were determined by densitometry.

Then we studied the in vivo metabolism of these peptides in normal rats(11). The data were fitted to a model described in details elsewhere (12). According to this model, the partitioning of the LAPs between HDL and the aqueous phase of plasma can be calculated from their tissue distribution volumes.

In the present work, the metabolism of the LAPs were studied in hyperthyroid as well as normal rats, as previously described (11). Figure 5 shows that per gram of tissues, HDL-bound peptides distributed mainly in adrenals, liver and ovaries while they did not significantly associate with kidneys. No difference was found between normal and hyperthyroid rats.

The partition coefficient of each acyl LAP was calculated as the B/F ratio where B and F are the fraction of the peptide bound to HDL and the fraction of the peptide free in the aqueous phase, respectively. In normal as well as hyperthyroid rats, there was an identical Log-linear relationship between B/F and the acyl chain length of the LAPs (figure 6). This indicated that the binding capacity of HDL for the peptides was not modified by the hyperthyroid status. A similar conclusion could be drawn from the measurements of the tissue distribution volumes of apo A-I (figure 7). It has been shown that, in contrast to liver, adrenals and ovaries, the binding of apo A-I to kidneys was representative of free rather than HDL-bound apo A-I (11,13). Therefore, since the profile of distribution of apo A-I in tissues was not altered by hyperthyroidism, we concluded that the HDL-bound/free apo A-I ratio was the same as in control animals. This observation was of interest in view of our finding that the plasma concentration of HDL-phospholipids was increased by 45 % in hyperthyroid animals. The constant of equilibrium for the binding of apo A-I to HDL can be calculated from (11) :

$$Keq = \frac{B}{F} \text{ x } \frac{[W]}{[HDL \text{ phospholipids}]}$$

where [W] is the molar concentration of water (55.5 M). Since in hyperthyroid rats B/F remained unaltered in spite of the enhancement of the HDL-phospholipids concentration, one can calculate that Keq decreased by more than 30 %. This suggests that the remodelling of HDL structure in hyperthyroid rats resulted in a decrease in the affinity of HDL surface for apo A-I. To test this hypothesis, we measured the maximum wavelength (λmax) of tryptophan fluorescence of HDL apolipoproteins at various HDL dilutions (figure 8). The results clearly indicated that in hyperthyroid rats, the binding of apo A-I to HDL was more labile than in control animals.

CONCLUSIONS

In rats, hyperthyroidism induces structural alterations of HDL particles that are smaller and denser than in normal animals. This is likely due, at least in part, to an increase in the synthesis of HDL surface components. These structural alterations result in a decrease of the affinity of apo A-I for HDL. In humans, the changes in HDL induced by hyperthyroidism seem to be different in nature. They are mainly characterized by a decrease of HDL-cholesterol concentration and a decrease in the HDL2/HDL3 ratio occuring through the activation of hepatic lipase activity by thyroid hormones. Therefore, the rat does not appear to be a good model for humans to study the effects of thyroid status on lipoprotein metabolism.

ACKNOWLEDGMENTS

The authors are grateful to Drs H.J. Pownall and J.T. Sparrow for providing the acylated Lipid-Associating Peptides.

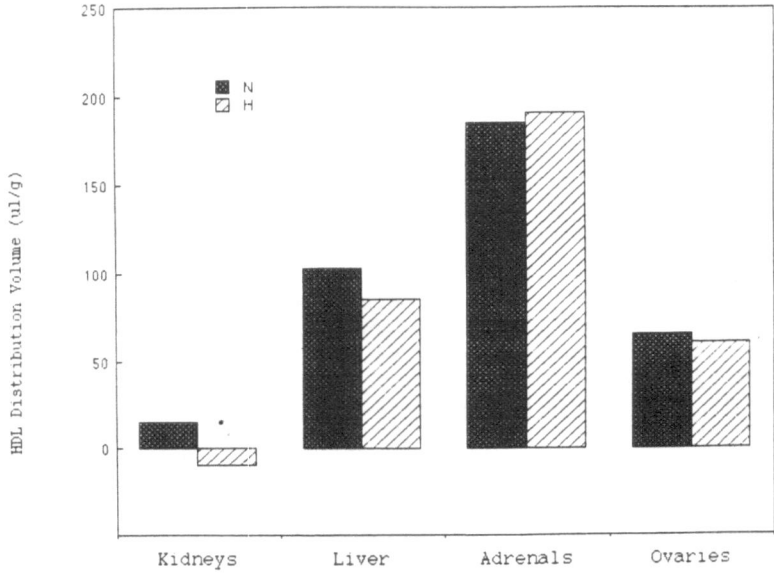

Fig. 5. The distribution volumes of HDL in normal (N) or hyperthyroid (H) rats were calculated after injection of 6 different radio-iodinated acyl LAPs having various hydrophobic contents.

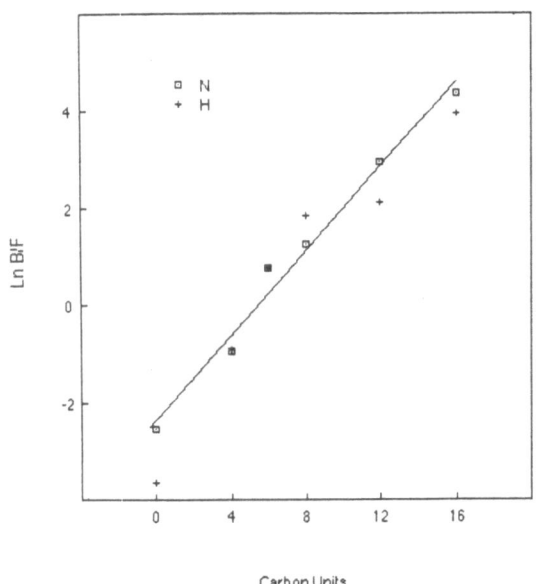

Fig. 6. Effect of the acyl chain length on the partition coefficients (B/F) of acylated LAPs between HDL and the aqueous phase, in normal (N) or hyperthyroid (H) rats.

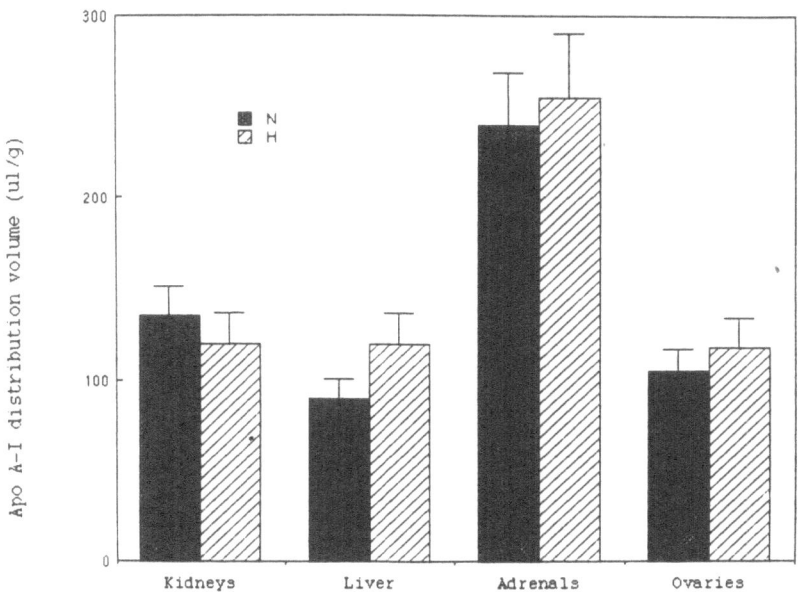

Fig. 7. Distribution volumes of radioiodinated apo A-I in various
tissues of normal (N) or hyperthyroid (H) rats.

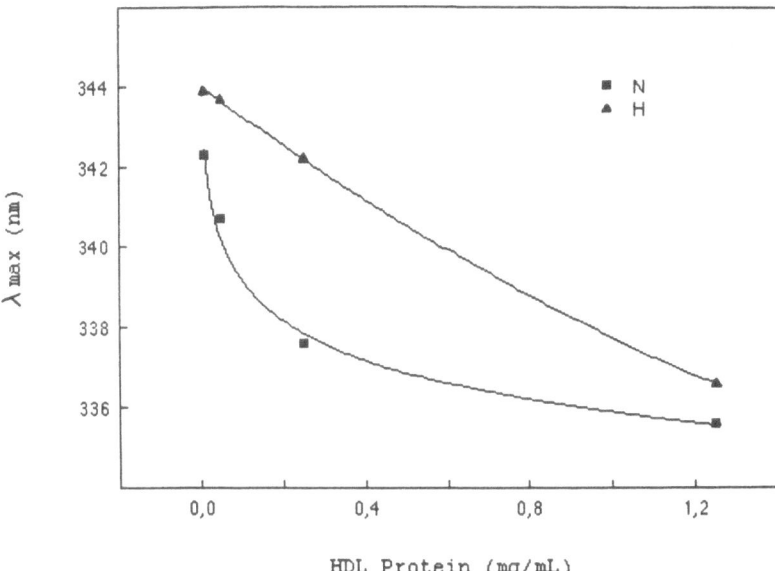

Fig. 8. Determination of the wavelengths of maximum tryptophan
fluorescence (λ max) of apolipoproteins, at various HDL
concentrations. The excitation wavelength was 280 nm.
Each point corresponds to the mean of 6 independent samples
of HDL isolated from normal (N) or hyperthyroid (H) rats.

REFERENCES

1. M. Heimberg, J.O. Olubadewo, and H.G. Wilcox, Plasma lipoproteins and regulation of hepatic metabolism of fatty acids in altered thyroid states, Endocrine Reviews 6:590 (1985).
2. E. Muls, M. Rosseneu, V. Blaton, E. Lesaffre, G. Lamberigts, and P. de Moor, Serum lipids and apolipoproteins A-I, A-II and B in primary hypothyroidism before and during treatment, Eur. J. Clin. Invest. 14:12 (1984).
3. L. de Parscau, P. Moulin, M.C. Bourdillon, L. Perrot, G. Ponsin, and F. Berthezène, Defect in the net transport of cholesterol from cultured fibroblasts to serum of hypothyroid patients, Horm. Metab. Res. 21:400 (1989).
4. C. Verdugo, L. Perrot, G. Ponsin, C. Valentin, and F. Berthezène, Time-course of alterations of high density lipoproteins (HDL) during thyroxine administration to hypothyroid women, Eur. J. Clin. Invest. 17:313 (1987).
5. E. Muls, M. Rosseneu, J. Bury, M. Stul, G. Lamberigts, and P. de Moor, Hyperthyroidism influences the distribution and apolipoprotein A composition of the high density lipoproteins in man, J. Clin. Endocrinol. Metab. 61:882 (1985).
6. C. Vialle-Valentin, L. Richard, L. Perrot, G. Ponsin, and F.Berthezène, Hyperthyroidism induces a change in the profile of high density lipoprotein subfractions related to that of plasma lipase activity, Exp. Clin. Endocrinol. (Life Sci. Adv.) 7:51 (1988).
7. J.R. Patsch, S. Prasad, A.M. Gotto, Jr., and G. Bengtsson-Olivecrona, Postprandial lipemia. A key for the conversion of high density lipoprotein 2 into high density lipoprotein 3 by hepatic lipase , J. Clin. Invest. 74:2017 (1984).
8. B.J. Meyer, Y.C. Ha, and P.J. Barter, Effects of experimental hypo-thyroidism on the distribution of lipids and lipoproteins in the plasma of rats, Biochim. Biophys. Acta 1004:73 (1989).
9. N.O. Davidson, R.C. Carlos, M.J. Drewek, and T.G. Parmer, Apolipo-protein gene expression in the rat is regulated in a tissue-specific manner by thyroid hormone, J. Lipid Res. 29:1511 (1988).
10. G. Ponsin, K. Strong, A.M. Gotto, Jr., J.T. Sparrow, and H.J.Pownall, In vitro binding of synthetic acylated lipid-associating peptides to high-density lipoproteins: Effect of hydrophobicity, Biochemistry 23:5337 (1984).
11. G. Ponsin, J.T. Sparrow, A.M. Gotto, Jr., and H.J. Pownall, In vivo interaction of synthetic acylated apopeptides with high density lipoproteins in rat, J. Clin. Invest. 77:559 (1986).
12. G. Ponsin, and H.J. Pownall, Equilibrium of apoproteins between high density lipoprotein and the aqueous phase: modelling of in vivo metabolism, J. Theor. Biol. 112:183 (1985).
13. C.K. Glass, R.C. Pittman, G.A. Keller, and D. Steinberg, Tissue sites of degradation of apoprotein A-I in the rat, J. Biol. Chem. 258: 7161 (1983).

THE HELP-SYSTEM IN THE TREATMENT OF SEVERE HYPERCHOLESTEROLAEMIA:

ACUTE AND LONG-TERM EXPERIENCE

D. Seidel

Institute for Clinical Chemistry
Klinikum Großhadern
Marchioninistraße 15
8000 München 70, FRG

SUMMARY

The **HELP** procedure provides a new means of treating high LDL concentrations in severe hypercholesterolemia with the unique additional effect of lowering Lp(a) and fibrinogen. In combination with HMG-CoA-reductase inhibitors a mean interval value of -75% for LDL as compared to the starting concentration may be achieved. The treatment has the advantage that the patient is not exposed to foreign proteins or compounds with attendant immunological problems. It displays a high degree of reproducibility and an almost unlimited capacity guaranteeing a constant therapy independent of the clinic performing the treatment. The first coronary angiographies after 2 years of HELP treatment in over 50 patients (to be reported elsewhere) give support to the hope that regression of coronary heart disease is possible in humans. Further studies and observations should eventually tell us at what level of LDL, Lp(a) and fibrinogen this may be expected.

We trust that the clinical benefit of this treatment regimen will be substantial for those patients who have problems in clearing LDL from their plasma pool and who are at the same time sensitive to elevated LDL-levels by the development of premature coronary sclerosis.

INTRODUCTION

Today a large body of evidence links coronary risk with elevated plasma levels of low density lipoproteins (LDL), Lp(a) and fibrinogen. In many patients, suffering from coronary heart disease (CHD), all three compounds are elevated and may potentiate the cardiovascular risk. Most forms of hyper-beta-lipoproteinemia result from a defect in the removal of LDL from plasma by the liver and the LDL receptor is now recognized as the crucial element in the control of LDL-cholesterol homeostasis. If the physiological clearing mechanism is insufficient, diet and drug therapy alone is often ineffective. This also holds true for Lp(a) and fibrinogen, which can hardly be lowered by diet or drugs.

In humans plasma LDL-C levels below 120 mg/dl seem to be necessary to inhibit the development of atherosclerosis or to induce regression of the vessel wall lesions. We have investigated the efficiency and safety of a newly introduced selective plasma therapy (Heparin mediated Extracorporal LDL:Fibrinogen Precipitation), the **HELP**-system alone and in combination with HMG-CoA reductase inhibitors in the treatment of severe hypercholesterolemia associated with coronary heart disease.

Hypercholesterolemia, Hypocholesterolemia, Hypertriglyceridemia
Edited by C.L. Malmendier *et al.,* Plenum Press, New York, 1990

155

THE HELP-APHERESIS SYSTEM*

The technique operates by an increase of the positiv charges on LDL- and Lp(a) particles at low pH, allowing them to specifically form a network with heparin and fibrinogen in the absence of divalent cations. Only a limited number of other heparin binding plasma proteins are coprecipitated by heparin at low pH (plasminogen C3 and C4 complement). Other protein such as apo A_1, apo A_2, albumin or immunglobulins do not bind to heparin and are not precipitated.

The HELP-system has unique features;
1. it removes LDL, Lp(a) and fibrinogen with high efficiency
2. it uses only disposable material and avoids regeneration of any of the used elements
3. it avoids the use of compounds with immunogenic or immunostimulatory activity
4. it is a technically safe and well standardized procedure.
5. short and long term treatment tolerance is excellent.

The major characteristics of the HELP system are illustrated in the flow sheet (Fig.1):

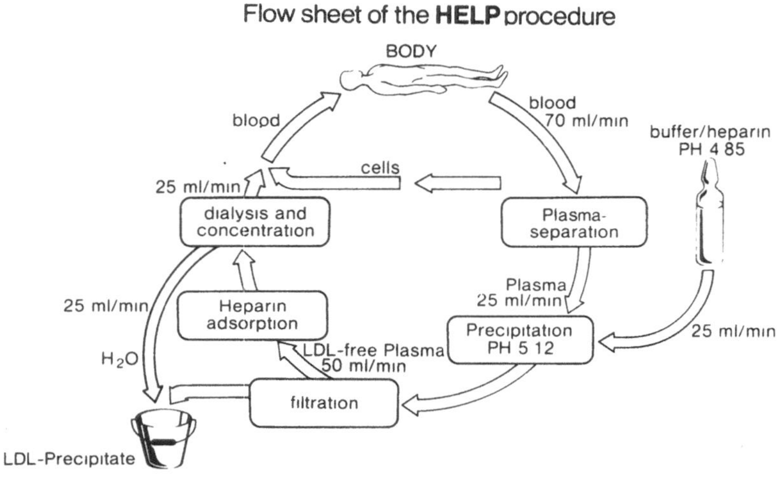

Flow sheet of the **HELP** procedure

Fig. 1
Flow sheet

In the first step plasma is obtained by filtration of whole blood through a plasma separator. This is then mixed continuously with a 0.3 M acetate buffer of pH 4.85 containing 100 IU heparin/ml. The sudden precipitation occurs at a pH of 5.12 and the suspension is circulated through a 0.4 M polycarbonate filter to remove the precipitated LDL, Lp(a) and fibrinogen. Excess heparin is absorbed by passage through an anion-exchange column which binds only heparin at the given pH. The plasma buffer mixture is finally subjected to a bicarbonate dialysis and ultra filtration to remove excess fluid and to restore the physiological pH, before the plasma is mixed with the blood cells and returned to the patient. All filter and tubings required for the treatment are sterile, disposable and are intended for single use only. This makes it easy and reliable to work with the system and guarantees a steady quality for each treatment, independent of a clinic performing the procedure. Safety is assured by a visual display and two microprocessors operating in parallel. Due to the excellent tolerance of the procedure the patients leave the hospital shortly after the end of the treatment.

The clinical experience with the HELP system goes back to 1985. Since then 220 patients were treated in over 12.000 single treatments. Up to now two patients are treated for more than four years, three patients are treated for more than 3 years; 69 for more than 2 years and 105 for more than 1 year and 41 for less than one year. Currently the system operates in 30 centers in Germany, Italy, the US, Austria, Israel and Ireland.

The efficiency of the system is 100% for the elimination of LDL, Lp(a) and fibrinogen. Per single treatment (lasting 1.5 to 2 hours) 2.8 to 3 liter of plasma are filtered, causing a reduction of 60 to 65 % for the three compounds.

The rates of return to pre-apheresis concentrations for LDL differ between normocholesterolemics, heterozygous and homozygous FH patients, while they are almost identical for Lp(a). Normocholesterolemics return rather fast towards the steady state pretreatment levels. Heterozygous FH patients display a rate of return intermediate between normocholesterolemics and a homozygous FH child, which was slowest in its rate of return to pretreatment LDL concentrations. The pretreatment values usually reached a new steady state after 4-8 treatments.

Long term effects of the HELP treatment based on interval values between two treatments (C after HELP + C before HELP : 2) and expressed as percentage of concentrations at start, revealed a mean reduction of

-51% for LDL, of
-45% for Lp(a), and of
-46% for apo B,

while
HDL was increased by +12%,
apo A_1 by +9%.
Fibrinogen was decreased by -46%.

The HELP treatment also significantly improves plasma viscosity (-15 %), erythrocyte aggregation (-50 %), and erythrocyte filtration (+15 %), which is followed by an acute (20-30%) increase of the oxygen tension in the muscle. The changes in plasma viscosity are primarily due to the reduction of LDL; the change in erythrocyte aggregation by the fibrinogen reduction. Changes in erythrocyte filtrability correlate with an improvement of the cholesterol/phospholipid ratio of the cell membrane. It is tempting to associate the rheological findings with the impressive relief from angina, the improvement in exercise EKG and in physical capacity that we observe in most (over 90%) of the patients shortly (2-3 months) after start of the therapy.

Experience with a combined HELP- and MHG-CoA reductase inhibitor therapy.

In cases with plasma cholesterol levels exceeding 300 mg/dl the use of diets and specified drugs may not be sufficient if LDL concentrations < 120 mg/dl and regression of CHD is approached as a means of secondary intervention.

We have therefore investigated the efficiency of a combined therapy, using MHG-CoA reductase inhibitors (Lovastatin[R], Simvastatin[R], Pravastatin[R]) together with the HELP apheresis and treated approximately 20 patients with severe FH on a long term basis.

These compounds significantly decrease the rate of return after HELP apheresis in both heterozygous and homozygous FH patients (20 - 30%). When the two treatments are combined, a reduction of the interval LDL-C level of 70 - 80% may be achieved (see Fig. 2 and 3) while Lp(a) and fibrinogen are not further affected. In the combined form therapy intervals between the HELP treatments may in many cases be stretched from 7-14 days, depending on the synthetic rates for LDL or the severity of CHD.

Maximaltherapie der Hypercholesterinämie bei KHK-Patienten

Pat. O.H.♂, 38 J. Basiswert: LDL-C 463 mg/dl ± 15

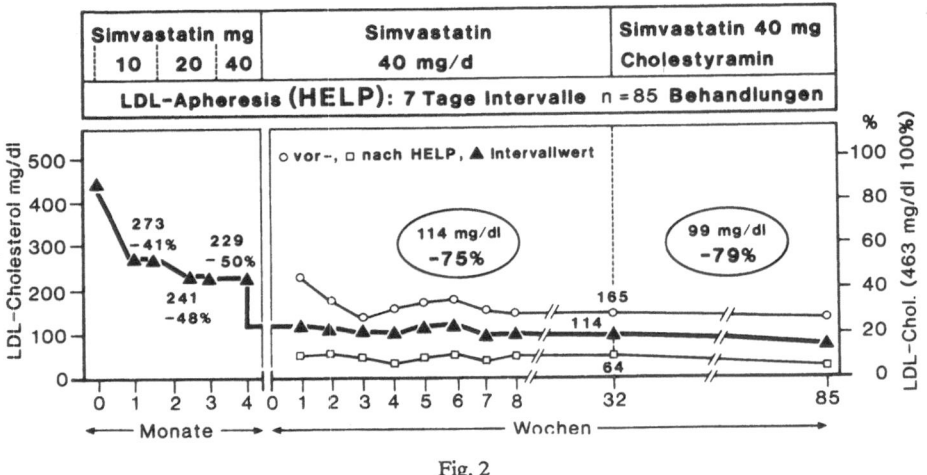

Fig. 2

Typical LDL follow-up kinetics under maximal therapy with the HELP-system and in combination with Simvastatin[R] and Cholestyramine[R].

An impressive reduction of LDL cholesterol is apparent in this form of familial heterozygous hypercholesterolemia and severe coronary heart disease. The patient was almost resistent to the lipid lowering therapy which was tried earlier.

Maximaltherapie der homozygoten familiären Hypercholesterinämie

Pat Ch J.♀, 7J homozyg FH Basiswert LDL-C 820 mg/dl

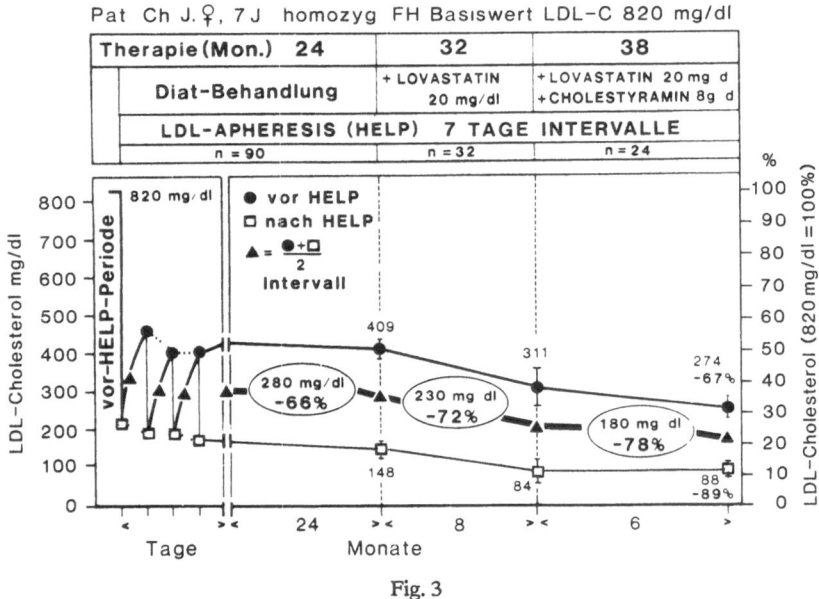

Fig. 3

LDL follow-up on long term treatment with the HELP-system alone and in combination with Lovastatin[R] and Cholestyramin[R] in a homozygous FH child. The therapy is excellently tolerated and

Overall treatment tolerance has been very good and no major complications have been observed after approximately 12.000 treatments in 220 patients. The effect maintained on long term treatment for over 4 years. As for proteins that are not precipitated by heparin at low pH, plasma concentrations at the end of the HELP therapy were generally in the range of 80 to 90% of the initial values and returned to their original level no later than 24 hours after the end of treatment. Substitution of any kind has not been necessary in 4 years of clinical experience. Special attention has been focussed to the effect of HELP on hemeostasis. All post treatment controls were typical for extracorporal procedures and no critical bleeding complications have been observed. Plasma electrolytes, hormones, vitamins, enzymes, and immunoglobulin concentrations as well as hematological parameters remained virtually unchanged at the end of each treatment and on long term application (see Tab. 1).

Tab. 1a and 1b

HELP-therapy in combination with HMG-CoA reductase inhibitors
Laboratory data (konventional units) after 1 year of combined treatment

HELP-therapy in combination with HMG-CoA R.I.
Laboratory data I (konv units)

Parameters	Baseline		24 months Simvastatin +HELP treatment	
	x̄	·SEM	x̄	·SEM
Substrates				
Sodium	140.0	0.7	141.0	0.3
Potassium	3.9	0.12	4.0	0.05
Calcium	9.2	0.11	8.9	0.1
Phosphate	3.7	0.16	3.3	0.03
Iron	88.2	9.3	95.5	3.9
Creatinine	0.85	0.04	0.9	0.02
BUN	15.2	1.7	14.5	0.4
Uric acid	5.3	0.4	5.3	0.4
Glucose	94.0	0.9	100.0	6.4
Tot Bilirubin	0.43	0.03	0.56	0.4
Tot Protein	7.0	0.1	6.9	0.1
Albumin%	61.6	1.73	61.4	0.42
α_1-Protein%	3.6	0.3	3.6	0.1
α-Protein%	8.0	0.42	8.3	0.14
β_2-Protein%	13.0	0.56	12.0	0.03
γ-Protein%	13.7	0.99	14.8	0.14
Enzymes				
ALAT (GOT)	10.0	0.4	13.5	0.4
ASAT (GPT)	11.0	2.0	19.0	1.0
γ-GT	21.0	5.8	25.0	2.1
CK	45.0	7.0	45.0	2.0
LDH	143.0	10.8	151.0	4.6
Amylase	16.0	2.7	16.0	0.3
CHS	5151.0	525.0	5455.0	530.0
ALP	101.0	6.6	110.0	2.8

HELP-therapy in combination with HMG-CoA R.I.
Laboratory data II (konv units)

Parameters	Baseline		24 months Simvastatin +HELP treatment	
	x̄	·SEM	x̄	·SEM
Hematological indices				
Hemoglobin	14.0	0.44	14.3	0.07
Hematokrit	41.8	1.1	42.0	0.73
Erythrocytes	4.4	0.14	4.6	0.1
Thrombocytes	226.0	10.2	220.0	9.5
Leucocytes	5.18	0.39	5.22	0.48
Lymphocytes	37.4	2.76	33.3	2.1
Monocytes	7.2	1.02	6.2	2.45
Neutrophils	51.3	3.16	57.4	3.12
Eosinophils	2.6	0.43	1.8	0.61
Basophils	0.7	0.18	0.7	0.1
Hemostasis				
Quick Test (PT)%	98.0	1.25	99.0	0.91
TT [sec]	14.0	0.12	14.0	0.21
Endocrinological indices.				
Cortisol	12.6	1.05	13.3	1.15
Testosterone	6.7	1.07	3.4	0.26
ACTH	40.3	3.78	40.4	6.18
LH*	15.9	8.36	11.1	5.91
FSH*	16.0	0.22	28.0	10.3
T3	133.5	7.35	123.5	12.4
T4	7.0	0.61	7.3	0.1
FT4	7.5	0.62	7.5	0.6
FT3	142.5	7.64	137.5	6.01

*(♂resp premenopausal♀)

the result achieved is maximal as compared with anything else reported in the literature for this type of disease. It is of interest to note (reported elsewhere) that neither Cholestyramine[R] nor Lovastatin[R] application alone - without the HELP therapy - showed a significant lipid lowering effect in this child.

FAMILIAL HYPOCHOLESTEROLEMIA AND HDL DEFICIENCY

Marie-France DUMON, Monique FRENEIX-CLERC
Marie-José MAVIEL and Michel CLERC

Laboratoire de Biochimie Médicale A, Université de Bordeaux
II. 146, Rue Léo Saignat, 33076 - Bordeaux-Cedex

INTRODUCTION

Familial hypocholesterolemia and HDL deficiency are often associated in a heterogeneous family of hereditary diseases.

These diseases though they are clinically, biochemically and genetically heterogeneous, may be divided in two groups and classified as primary and secondary. The former correspond to anomalies of the biosynthesis of HDL's major apoproteins. The later correspond to disorders of other origin (enzyme deficiency, absence or excess of lipoprotein enzyme cofactors, anomaly of lipid transfer protein, lipoprotein receptor disorders). Both may be classified as follows:

CLASSIFICATION OF FAMILIAL HYPOCHOLESTEROLEMIAS ASSOCIATED WITH HDL DEFICIENCY

PRIMARY

a - Familial hypocholesterolemia [1]

b - HDL deficiency with planar xanthomas [2]

c - Familial A-I and C-III deficiency [3]

d - Familial A-I, C-III and A-IV deficiency [4]

e - A-I absence [5]

f - A-1 variants (Milano, Marburg, Munster, Giessen, Norway) [6]

Hypercholesterolemia, Hypocholesterolemia, Hypertriglyceridemia
Edited by C.L. Malmendier *et al.*, Plenum Press, New York, 1990

SECONDARY

g - Tangier disease [7]

h - "Fish-eye" disease [8]

i-j - Hyperchylomicronemia (type IA or IB) [9]

k - LCAT deficiency [10].

As shown below these two groups may be opposed. In primary forms anomalies in chromosome 11 have been shown. They are localized in the cluster "AI, CIII, AIV" and are transmitted on the autosomal codominant mode (except hypoalphalipoproteinemia which is dominantly inherited). These forms are always associated with premature coronary-artery disease (except A-I Milano). In secondary forms other chromosomes than 11 are incriminated. They are transmitted on the autosomal recessive mode and premature coronary-artery disease is at least not the major clinical defect.

GENETICS AND ATHEROSCLEROSIS [11-16]

Primary Forms	Autosomal Inheritance	Gene	Chr	HDLc mg/l	Premature C.A.D.
a	Dominant	Multiple RFlPs in A-I, C-III, A-IV complex	11	<180	+
b	Codominant	?	11	<40	+
c	Codominant	A-I, C-III, DNA rearrangement	11	<50	+
d	Codominant	entire gene complex deletion	11	<50	+
e		deletion ?	11	<15	+
f	Codominant	mutations	11		+ (Milano)
Secondary Forms					
g	Recessive	A-I (no proof)	11 ?	<10	- (+)
h	"	AlphaLCAT ?	16 ?	<60	-
i	"	LPL-HL	18-15	<200	-
j	"	C-II	19	<200	-
k	"	Mutations ?	16		+

162

To explain the hypocholesterolemia associated with HDL deficiency numerous mechanisms have been demonstrated or invoked. In primary forms arteries are the major target as a consequence of a high decrease in cholesterol efflux correlated to A-I deficiency and of oxidized-LDL up-take by scavengers in the arterial wall correlated both with increased residence time of LDL and with apo A-IV deficiency (which induces vit.E deficiency). In secondary forms, on the contrary, arteries appear as a minor target. In this case the common mechanism looks like to a metabo-lic deviation of modified lipoproteins which are directed toward histiocytes (perhaps sub-species dependent on their anatomical locali-zation: tonsils, gut, liver, spleen, lymph modes, peripheral nerves, skin, cornea). The modifications implied in each case may be the consequence of enzymes, or receptors or lipid transfer proteins anoma-lies (LPL, HTGL, LCAT, CETP, HDL receptors ?) [17]. The distribution as well as the structure of apolipoproteins and/or lipids may be responsible of the modification (A-II enriched VLDL [18], triglycerides rich Tangier LDL2 [19], Triglycerides poor HDL in fish-eye disease [20]).

Two patients discovered in our laboratory (an homozygous Tangier patient and a case of Fish-eye disease) are subjected to on-going research. They seem able to give new insights in the pathophysiology of these secondary forms.

MATERIALS AND METHODS

CASES REPORT

The Tangier patient has been described in previous reports [7, 21]. Briefly, he is an homozygous patient 43 years old with total absence of normal HDL and not any sign of cardio-vascular disease. However massive lipid depositions of cholesteryl esters have been observed all along his digestive tract, the whole digestive mucosae being lined with small yellowish gray plaques. Furthermore he presents high cornea opacities at slit lamp examination.

The "Fish-Eye" disease (FED) patient, a man, born in 1936 in Constantine (Algeria) is from a Jewish family of Caucasian origin and he was referred to our hospital for general examination because he presented corneal opacities with a slight impairment of vision, parti-cularly in twilight. At 10-20 years he was recognised to have very pale

blue eyes and a gradual fading of the cornea. Vision slowly deteriora-
ted from 40 years of age and was corrected for astigmatism. Slit lamp
examination showed a lipidic degeneration of both corneae filtering
through the stroma and looking finely granular. Intraocular pressure
was normal. He had no other medical history. Tonsils and digestive
mucosa were normal, neither xanthelasma nor xanthoma were observed.
Blood pressure gradients (130/80 mm Hg), electrocardiogram at rest and
during exercise as echocardiography or supraaortic Doppler and echogra-
phy of abdominal aorta showed no anomaly. The digestive track, liver,
spleen and thyroid gland were normal in harmony with the good general
health of the patient.

METHODS

The electrophoretic and immunochemical methods of analysis have
been previously described [7, 19, 22, 23]. "A" Esterase activity was
measured as described by MACKNESS [24]. Heparin affinity chromatography
was used to separate HDL from lipoproteins containing apo B as reported
by DUMON [25]. Immunofixation tests were performed according to CAWLEY
[26] using sera anti-human apo AI, AII and B form IMMUNO AG. (Vienna).

RESULTS AND DISCUSSION

Lipid stained Agarose as polyacrylamide electrophoretic gels
(fig. 1) showed in both patients the absence of HDL and modifications
in LDL mobility. On the contrary immunoelectrophoretic and immunofixation patterns (which are revealed by protein stains) visualized HDL. In
Tangier serum apo A-I, A-II and B comigrated in VLDL (fast band in
agarose gel - fig. 2) and LDL (slow band), that explains why they are
electrophoretically absent in the HDL normal region. In FED serum apo
AI and AII were differently distributed and corresponded to HDL of
decreased mobility, furthermore small amounts of AI and AII were
revealed with apo B in VLDL and LDL. FED-HDL being abnormally poor in
triglycerides and consequently unstained by Sudan-Black (according
with the concept of alpha-LCAT deficiency of Holmquist [27] they did
not appear in electrophoretic gels.

To try to separate lipoproteins containing apo B from other
lipoproteins heparin-affinity-chromatography was used. As previously
described [25] this method provides an alpha fraction containing normal

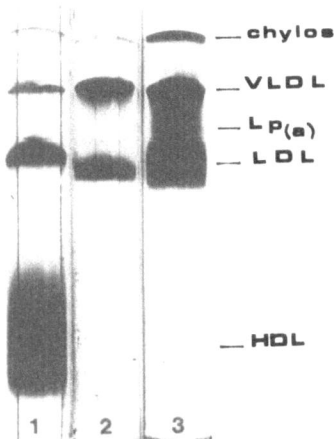

Fig. 1. Discontinuous polyacrylamide gel electrophoresis clearly shows the
absence of HDL in sera from FED (2) and Tangier (3) patients as
compared to control (1). This gel contains acrylamide: 2% cathodic
(from chylomicrons at the top to VLDL), 3,5% middle (between VLDL
and LDL), 3% anodic (from LDL to the bottom).

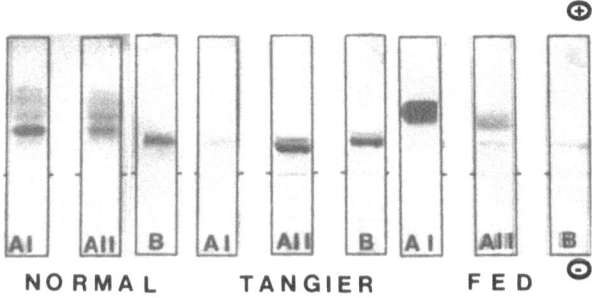

Fig. 2. Immunofixation tests were performed separating serum lipoproteins
by electrophoresis in agarose gel in a first time and incubating
each electrophoretic strip during 30 min. with serum anti apo AI,
AII or B in a second time.
The anti serum was thinly layered on each gel. After incubation
each gel was thoroughly washed and coloured with Amidoschwartz 10 B
in acetic acid 1%. In Tangier serum AI, AII and B were abnormally
distributed and comigrating with VLDL (fast-cathodic band) and LDL
(slow cathodic band). On the contrary in FED serum HDL of slow
mobility containing A-I and A-II but not B were visualized whereas
small amounts of A-I and A-II were present with apo B in VLDL and
LDL regions.

165

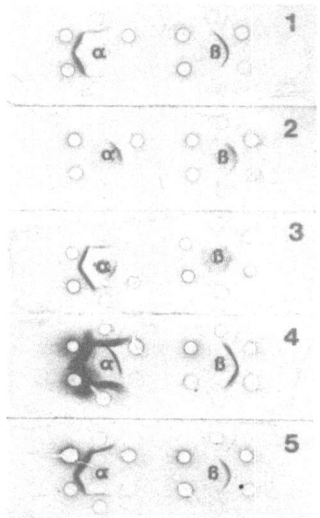

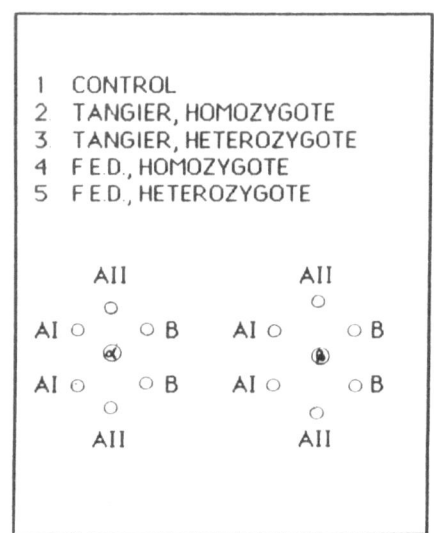

1 CONTROL
2 TANGIER, HOMOZYGOTE
3 TANGIER, HETEROZYGOTE
4 F.E.D., HOMOZYGOTE
5 F.E.D., HETEROZYGOTE

Fig. 3. The double immunodiffusion method of OUCHTERLONY has been applied
to the alpha and beta fractions eluted from heparin affinity chromatography columns. 10 µl of alpha or beta fractions were poured in
the central hole and serum anti AI, AII and B (10 µl of each in one
hole and 20 µl of each in another hole) around the central hole.
The alpha fraction from control serum (1) contained AI and AII but
not B. On the contrary B was present as AI and AII in the alpha
fractions from Tangier homozygote (2) Tangier heterozygote (3) FED
homozygote (4). FED heterozygote (5) was normal as compared to
control (1).

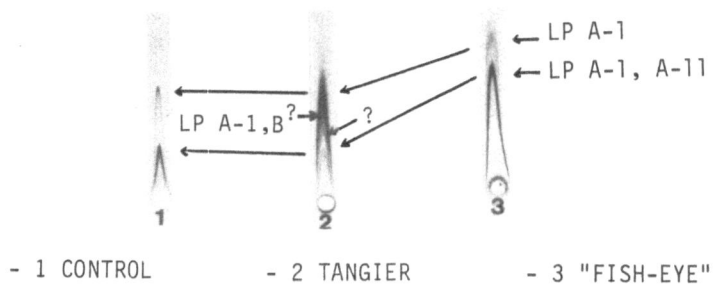

- 1 CONTROL   - 2 TANGIER   - 3 "FISH-EYE"

Fig. 4. AI and AI - AII lipoparticles from serum were studied using agarose
gel lipoparticles A-I plates from SEBIA (France). In each hole
different volumes were poured: 3 µl of normal serum (1), 30 µl of
Tangier serum (2), 10 µl of FED serum (3). (1) and (3) revealed the
same types of lipoparticles though in FED serum Lp AI - AII figured
in larger amount. In Tangier serum two supplementary rockets
between Lp AI and Lp AI - AII are visualized and supposed to be
Lp AI - B and Lp AI - AII -B.

HDL and a bêta fraction containing normal VLDL and LDL. As shown in fig. 3 the alpha fractions from Tangier and FED patients contained apo B which is absent in normal HDL. This anomaly could be due either to the enrichment of LDL and/or VLDL by triglycerides which could mask the apo B heparin binding sites or to a linkage between apo AI and the apo B heparin binding sites. First, LDL2 were ultracentrifugally isolated from both FED and Tangier 12 H fasting serum. FED-LDL2 were shown enriched in triglycerides as compared to the mean of five male controls of the same age (2.87 and 0.37 ± 0.09 mmole/L-1 respectively). Tangier-LDL2 were studied using the fluorescence quenching by iodide ions [19] and demonstrated enriched in triglycerides as compared to control (28.3 % and 9.3 %, in LDL2 dry weight, respectively). Secondly using Lp AI lipoparticle gel plates from SEBIA (France) we observed in F.E.D. serum the predominance of Lp AI - AII and the presence of Lp AI. Whereas in Tangier serum Lp AI and Lp AI - AII were shown associated with two other sorts of lipoparticles (Fig. 4) which could be Lp AI - B and Lp AI - AII - B as compared to immunofixation test data. Finally triglycerides enriched LDL2 which were at one and the same time observed in FED and Tangier serum could be proposed to be involved in corneae lipid deposition (the analysis of one cornea from our FED patient has been found enriched in cholesteryl-esters and triglycerides: unpublished data). Furthermore, Lp AI - B or Lp AI - AII - B particles could be proposed as modified lipoproteins directed toward digestive histiocytes which are the main target in Tangier disease. A putative linkage between AI and B in Tangier serum in the state of our investigations may be supported by the following arguments and observations:

- Ionic interactions between anionic residues of heparin and cationic residues of apo B binding sites have been postulated to explain the heparin affinity chromatography uptake of LDL and VLDL [28, 29].

- Different polyanions can interact with the heparin binding sites of apo B and inhibit the LDL and VLDL uptake by heparin affinity columns [29].

- Apo AI is excreted by hepatocytes as a phospho-apolipoprotein [30]. The absence of dephosphorylation related to a deficiency of the corresponding phosphatase could justify an interaction between phosphorylated apo AI and apo B heparin binding sites. This phosphatase deficiency could be the originary and hereditary defect in Tangier disease.

- Another way (same type as above) could implicate anionic organic redidues as a consequence of "A-esterase" deficiency [24] which could represent a candidate to explain the hereditary defect in Tangier disease.

Finally post-translational modifications of apo A-I which could implicate both apo AI and the heparin binding sites of apo B are the subject of our on-going research, all the more because no evidence has been found for mutations (protein sequence) and for major deletions or insertions in the apo AI gene [23, 31].

CONCLUSION

Tangier and Fish-Eye diseases represent secondary forms of familial HDL deficiency with hypocholesterolemia. Their origin, at the gene level, is unknown. In both diseases abnormal lipoproteins and/or lipoparticles have been observed that seem preferentially recognized by other target tissues than the arterial wall. Triglycerides enriched LDL2 have been found in both diseases and the cornea could be, the main target of modified LDL2: furthermore, because not any evidence of anomaly in the apo AI gene has been found, especially in Tangier disease, different hypothesis implicating a post-translational esterification of apo AI and a deficiency of the corresponding candidate hydrolyzing enzyme (Phosphatase, "A-esterase"?) have been postulated. Such hypothesis are supported by the phospho-apo AI physiologically excreted by human hepatocytes as demonstrated by Brewer and al. (Deficiency in the corresponding phosphatase need to be discovered) as well as by total deficiency in "A-esterase" activity in Tangier disease (the corresponding esterified apo AI need to be discovered). Both hypotheses could justify ionic interactions between apo AI and the apo B heparin binding sites and could explain the abnormal lipoparticles AI - B and/or AI - AII - B observed in Tangier patient's serum.

REFERENCES

1. C. Vergani and G. Bettale, Familial hypo-alpha-lipoproteinemia, Clin. Chim. Acta, 114, 45-52 (1981).

2. A. Gustafson, W.Nc. Conathy, P. Alaupovic, M.D. Curry and B. Persson, Identification of apoprotein families in a variant of human apolipoprotein A deficiency, Scand. J. Clin. Lab. Invest., 39, 377- 383 (1979).

3. R.A. Norum, J.B. Sakier, S. Goldstein, A. Angel, R.B. Goldberg, W.D. Block, D.K. Noffee, P.J. Dolphin, J. Edelglass, D.D. Borograd and P. Alaupovic, Familial deficiency of apolipoproteins A-I and C-III and precocius coronary-artery disease, N. Engl. J. Med., 306, 1513-1519 (1982).

4. C.S. Shoulders, M.J. Ball and F.E. Beralle, Variation in the apo A- I, C-III, A-IV gene complex : its association with hypertrigly-ceridemia, Atherosclerosis, 80, 111-118 (1989).

5. E.J. Schalffer, W.H. Heaton, M.G. Wetzel and H.B. Brewer, Plasma apolipoprotein A-I absence associated with a marked reduction of high density lipoproteins and premature coronary artery disease, Arteriosclerosis, 2, 16-26 (1982).

6. A.V. Eckardstein, H. Funke, A. Heuke, K. Atland, A. Benninghoven and G. Assmann, Apolipoprotein A-I variants, J. Clin. Invest., 84, 1722-1730, (1989).

7. M.F. Dumon, M. Clerc and M. Clerc, Apolipoprotein A-I deficiency in Tangier disease, in : "Eicosanoids, apolipoproteins, lipoprotein   particles, and atherosclerosis", C.L. Malmendier and P. Alaupovic (Eds), Plenum Press, New-York, 67-73 (1988).

8. L.A. Carlson and B. Philipson, "Fish-Eye" disease a new familial condition with massive corneal opacities and dyslipoproteinemia, Lancet, II, 921-923 (1979).

9. P.N. Herbert, G. Assamnn, A.M. Gotto and D.S. Fredrickson Familial lipoprotein deficiency, in : "Metabolic Basis of inherited disease", M.S. Brown (Ed), Mac Graw Hill, New-York, 589-621 (1983).

10. K.R. Norum and F. Djone, Familial plasma lecithin : cholesterol acyltransferase deficiency. Biochemical study of a new inborn error of metabolism, Scand. J. Clin. Lab. Invest., 20, 231-238 (1967).

11. S.K. Karathanasis, R.A. Norum, V.I. Zannis and J. Breslow, An inherited polymorphism in the human apolipoprotein A-I gene locus related to development of atherosclerosis, Nature, 301, 718-720 (1983).

12. A. Rees, J. Stocks, C. Schoulders, L.A. Carlson, F.E. Baralle and

D.J. Galton, Restriction enzyme analysis of the apolipoprotein A-I gene in Fish-eye disease and Tangier disease, Acta Med. Scand., 215, 235-237 (1984).

13. S.K. Karathanasis, Apolipoprotein multigen family : tandem organization of human apolipoprotein A-I, C-III and A-IV genes, Proc. Natl. Acad. Sci. USA, 82, 6374-6378 (1985).

14. H.N. Ginsberg, N.A. Le, I.J. Goldberg, J.C. Gibson, A. Rubinstein, P.W. Iverson, R. Norum and V. Brown, Apolipoprotein B metabolism in subjects with deficiency of apolipoproteins C-III and A-I. J. Clin. Invest., 78, 1287-1295 (1986).

15. A.M. Kessling, J. Rajput-Williams, D. Baiton, J. Scott, N.E. Miller, I. Baker and S.E. Humphries, DNA polymorphisms of the apoliprotein A-II and A-I, C-III, A-IV genes : A study in men selected for differences in high density - lipoprotein cholesterol concentration, Am. J. Hum. Genet., 42, 458-467 (1988).

16. J.L. Breslow, Apolipoprotein genetic variation and human disease, Physiol. Rev., 68, 85-132 (1988).

17. G. Schmitz, G. Assmann, H. Robenek and B. Brennhausen, Tangier disease : A disorder of intracellular membrane traffic, Proc. Natl. Acad. Sci. USA, 82, 6305-6309 (1985).

18. C.S. Wang, P. Alaupovic, R.E. Gregg and H.B. Brewer, Studies on the mechanism of hypertriglyceridemia in Tangier disease. Determination of plasma lipolytic activies, K1 values and apolipoprotein composition of the major lipoprotein density classes, Biochim. Biophys. Acta, 920, 9-19 (1987).

19. M.F. Dumon, R.Q. Dang, R. Salvayre, M. Clerc and L. Douste-Blazy, Modified lipid protein interactions in Tangier LDL2 demonstrated by fluorescence quenching, Chem. Phys. Lipids, 49, 153-160 (1988).

20. M.F. Dumon, M. Freneix-Clerc, E. Peuchant and M. Clerc, Apos A-I A-II et B dans deux situations d'hypocholesterolemie héréditaire. Symp. Behring, Paris, 20 oct., C.R. p. 313 (1989).

21. M.F. Dumon, L. Dubourg, T. Jasawant, Y. Auche et M. Clerc, Bilan biochimique et bioclinique d'un nouveau cas de maladie de Tangier, Ann. Biol. Clin., 4, 681 (1985).

22. M.F. Dumon, S. Visvikis, T. Manabe and M. Clerc, Immunochemical study of the plasma low and high density lipoproteins in Tangier disease, FEBS Letters, 201, 163-167 (1986).

23. S. Visvikis, M.F. Dumon, J. Steinmetz, T. Manabe, M.M. Galteau, M. Clerc and G. Siest, Plasma apolipoproteins in Tangier disease : a study by two dimensional electrophoresis, Clin. Chem., 33, 120- 122 (1987).

24. M.M. Mackness, E. Peuchant, M.F. Dumon, C.H. Walker and M. CLerc, Absence of "A"Esterase activity in the serum of a patient with Tangier disease, Clin. Biochem., 22, 475-478 (1989).

25. M.F. Dumon and M. Clerc, Combination of affinity chromatography and analytical polyacrylamide gel electrophoresis for rapid measurement of human HDL apolipoproteins, Anal. Biochem., 41, 25-32 (1984).

26. L.P. Cawley, Immunofixation electrophoretic technique applied to identification of proteins in serum and cerebrospinal fenid, Clin. Chem., 22, 1262-1268 (1976).

27. L. Holmquist and L.A. Carlson, Alpha Lecithin : cholesterol acyl-transferase deficiency-lack of both phospholipase A2 and acyl-transferase activities characteristic of high density lipoprotein LCAT in Fish-Eye disease, Acta. Med. Scand., 222, 23-26 (1987).

28. K.H. Weisgraber and S.C. Rall, Human apolipoprotein B-100 heparin binding sites, Journ. Biol. Chem., 262, 11007-11103 (1987).

29. Y.T. Pan, A.W. Kruski and A.D. Elbein, Binding of [3H] heparin to human plasma low density lipoprotein, Arch. Biochem. Biophys., 189, 231-240 (1978).

30. Z.H. Beg, J.A. Stonik, J.M. Hoeg, S.T. Demosky, T. Fairwell and H.B. Brewer, Human apolipoprotein A-I. Post-translational modification by covalent phosphorylation, Journ. Biol. Chem., 264, 6913-6921 (1989).

31. A. Ress, J. Stocks, C. Shoulders, L.A. Carlson, F.E. Baralle and D.J. Galton, Restriction enzyme analysis of apolipoprotein A-I gene in Fish-Eye disease and Tangier disease, Acta. Med. Scand., 215, 235-237 (1984).

MECHANISMS OF HYPOCHOLESTEROLEMIA

C. L. Malmendier, J -F. Lontie, and D. Y. Dubois

Fondation de Recherche sur l'Athérosclérose, Brussels
Belgium

Although clinically significant hypocholesterolemia occurs less frequently than hypercholesterolemia, it has usually important consequences that involve derangement of major metabolic roles of lipoproteins.

WHAT IS HYPOCHOLESTEROLEMIA?

To give a definition of hypocholesterolemia is ambiguous as "normal" values differ strongly according to influence from genetic factors (sex, ethnic group) and age, but also to dietary habits and environmental factors (exercise...). Arbitrary limits of plasma cholesterol have been given according to these influences.

Influence of sex

Total cholesterol is usually lower in males than in females between age 25 and 50 years but after 50, the reverse is true. The difference is generally due to a lower LDL-C and a higher HDL-C in females (1). However in some populations male and female subjects had similar HDL cholesterol levels (2).

Influence of age

Total cholesterol regularly increases with age. HDL-C increases from age 5 to 70 in females but in males its value is lower between ages 25 and 54 (1). At birth, cholesterol, LDL-C, apoproteins B and A-I are low but already at 6 months values reach the low adult normal range (3, 4).

Influence of ethnic group

Only US black males have cholesterol values lower than white caucasians between 25 and 50 years. Black females have higher values than white females (5). Pima Indians have lower total cholesterol and HDL-C values than caucasians 2). The same is true for Tarahumara Indians of Mexico (6).

Hypercholesterolemia, Hypocholesterolemia, Hypertriglyceridemia
Edited by C.L. Malmendier *et al.*, Plenum Press, New York, 1990

173

Influence of diet

Vegetarians consuming a diet with a low cholesterol, low fat, high P/S (1.9) content show all cholesterol (total, LDL and HDL) and apolipoproteins A-I and B values lower than normal (7). Intake of dairy products (more saturated) by the vegetarians is associated with an increase of total cholesterol and LDL-C but much less of HDL-C (8). High carbohydrate, low fat, low cholesterol diet causes a significant decrease in total, LDL-C, and HDL-C levels (9).

Hypocholesterolemia may be considered when plasma levels are under 100 mg/dl.

PRIMARY (GENETIC) DISORDERS

Among genetic diseases, total cholesterol is reported normal in fish-eye disease (10), apo A-I Milano (11), and LCAT deficiency (12).

There are 4 primary disorders displaying low to very low cholesterol values: Tangier disease (20-45 mg/dl) (13), apolipoprotein A-I absence (111 mg/dl) (14), abetalipoproteinemia (26 mg/dl) (15), and hypobetalipoproteinemia.

Hypobetalipoproteinemia

Hypobetalipoproteinemia is accompanied by low cholesterol values (16). However the level remains around 100 mg/dl as the drop in LDL-C is counterbalanced by an increase sometimes marked in HDL-C. The decrease in LDL-C (under 50 mg/dl) accompanied by a much reduced apolipoprotein B level results from a decreased fraction of VLDL-apo B converted into LDL-apo B in addition to an increased FCR of LDL by both receptor and receptor-independent pathways (17). The reduced conversion results from an increased VLDL turnover attested by a very fast catabolism of apo C-II and C-III (18) and the very low level of triglycerides. A normal or high HDL-C with an increase of HDL$_2$ is generally observed. Both the high level of HDL-C and the accumulation of cholesterol in HDL$_2$ may be explained by a defect of cholesterol ester transfer in presence of a lack of available acceptor particles secondary to the fast VLDL turnover (19).

SECONDARY CAUSES

Among secondary causes of hypocholesterolemia, are found malnutrition, parasitoses (malaria, trypanosomiasis, leishmaniasis), hematological disorders, lymphokine induction and severe liver disease.

Malnutrition

Prolonged malnutrition is accompanied by hypocholesterolemia, reversible on correction of the nutritional status (20). It is surprising that, although kwashiorkor is an unfortunately very common condition, recent lipid and apoprotein plasma values are rarely found in the literature (21).

Visceral Leishmaniasis (Kala-azar)

Hypocholesterolemia in Kala-azar children is associated with high levels of triglycerides. Drastic hypocholesterolemia occurs because even if VLDL-C is 4-5 fold increased, LDL-C is reduced to the half and HDL-C even more reduced (22). The determination of HDL-C may be subject to error in presence of high concentrations of VLDL and TG. Consequently the estimation of LDL-C using Friedewald's formula becomes more imprecise (Table 1).

TABLE 1. Lipid and apolipoprotein plasma levels
in severely ill Kala-azar children

mg/dl	Patients n = 8		Controls n = 13	
TG	330	± 73	83	± 19
TC	94	± 13	151	± 34
HDL-C	2.5	± 1.9	39	± 13
LDL-C	26	± 12	90	± 38
VLDL-C	65	± 14	16	± 4
A-I	24	± 8	113	± 15
A-II	7	± 2	44	± 6
B	49	± 14	70	± 29
C-II	3.8	± 1.9	2.4	± 0.7
C-III	7.5	± 4.2	6.2	± 0.8
S(AA)	1.9	± 1.4	< 0.4	

Putative origins of lipoprotein modifications in these children are:
1) a reduction of VLDL conversion to LDL is attested by a reduction of LDL peak in gel filtration profiles using Superose 6 column (FPLC), as well as by LDL-C and apo B reduction
2) normal levels of apoproteins C-II and C-III indicate a normal hepatic synthesis and apo S(AA) irregularly increased, an acute phase response
3) hypertriglyceridemia may have many explanations:
    a) the origin of VLDL increase may result from a defective uptake of triglyceride-rich particles by the liver
    b) a mobilization of energy (adipose and muscle) stores accompanying parasitic infections. Rabbits infected with trypanosoma brucei developed cachexia and suppression of lipoprotein lipase activity (23). The loss of LPL caused a defect in the clearance of triglycerides from the serum and a resultant hypertriglyceridemia. The mediator responsible for these responses is cachectin or TNF (24). In fact elevated levels of cachectin were detected in patients with chronic parasitic infections (Kala-azar and malaria) (25). However a deficiency in lipoprotein lipase activity was not mentioned in Kala-azar.

Besides decreased apo B levels, the major findings are the low to very low levels of apo A-I and A-II. An unusual profile is observed for lipoprotein particles. Two-dimensional electrophoretic pattern demonstrated the presence of LP A-II particles without apo A-I and the bulk of LP A-I particles was devoided of apo A-II (22). The presence of LP A-II particles has been shown in Tangier disease (26) and in some hypertriglyceridemic patients (27).

HDL reduction in Kala-azar may have different origins:
    a) a reduction of HDL hepatic synthesis: this seems unlikely since normal or subnormal levels of apoproteins B, C-II, C-III, and low levels of apo S(AA) were observed
    b) an increase in catabolism: HDL binds quickly to the membrane of trypanosoma brucei. There is an interaction between HDL (trypanocidal factor) and the trypanosome surface (28). We have no experiment on the catabolism of HDL in these parasitoses. However it has been shown in malaria that changes in HDL and VLDL are related to the metabolism of the plasmodium. This parasite takes cholesterol from its host. In leishmaniasis the link between HDL and the parasite may lead to a sequestration and/or degradation of HDL in the tissues (liver, spleen,

RTE) where the parasite accumulates, provoking a severe hepatosplenomegaly
c) another hypothesis for HDL sequestration and/or degradation is the
formation of immune complexes between HDL and autoantibodies and their
fast elimination. This is sustained by the hyperglobulinemia observed
in Kala-azar (29) and could lead to accelerated immune elimination of
the antigen-antibody complex.

Hematological disorders

Hypolipidemia was said to occur in anemic patients (30) and in a
patient with congenital microspherocytosis cholesterol level was partially
corrected by splenectomy which may indicate that splenomegaly is related to
hypocholesterolemia and that increased phagocytic activity of the RTE is
one mechanism responsible suggesting an autoimmune process. Hypocholesterolemia in myeloproliferative disease (myeloid metaplasia, chronic
granulocytic leukemia) has been described by Gilbert (31) and Bases (32).
It is known that proliferating cells have increased rates of LDL uptake and
degradation (33). Freshly isolated mononuclear cells from patients with
myelogenous leukemia show increased degradation of LDL through the
receptor-mediated pathway (34). Hypocholesterolemia observed in leukemia
was explained by the high LDL-receptor activity of leukemic cells. Moreover
chemotherapy results in a rise in cholesterol levels concomitantly with the
disappearance from the blood of leukemic cells (35).

Lymphokine induction (treatment of cancer)

The cholesterol level is usually not much lowered in cancer (36, 37)
and even advanced cancer per se is often accompanied by limited hypolipidemia and hypoapolipoproteinemia. Recently different cytokines have
been assayed in the treatment of human metastatic cancer.

Adoptive immunotherapy consists in the injection of cells with
antitumor reactivity. Treatment with IL-2 and LAK cells triggers off
quickly a toxic process causing dramatic decrease in cholesterol. Twelve
patients with melanoma, hypernephroma or colon cancer (38) were treated by
intravenous injection of recombinant IL-2 and LAK cells in vitro incubated
with IL-2 according to Rosenberg's schedule (39).

TABLE 2. Lipid and apolipoprotein plasma levels (mean ± SD) in mg/dl
for 12 advanced cancer patients treated with IL-2 and LAK cells.

	TC	HDL-C	LDL-C	A-I	A-II	B
Before	166±30	31±10	111±21	119±3 6	23.6±7.5	74±14
	(98-195)	(17-41)	(69-148)	(60-185)	(8.5-37	(44-92)
Day* 8-9	88±27	6.7±3.0	42±22	48±15	12.5±6.5	56±14
	(52-101)	(3-11)	(13-104)	(18-85)	(1.6-23)	(28-78)
Day** 13-16	59±18	5.9±4.6	24±1	39±21	11.0±7.7	38±14
	(20-115)	(1-19)	(6-46)	(18-70)	(0.7-27)	(9-61)
1-2 months after	175±38	27±9	118±27	106±32	22.5±8.6	81±18
	(131-266)	(11-37)	(81-179)	(66-166)	(11-38)	(63-115)

* 3 days after cessation of IL-2
** At the middle of IL-2 +LAK cells treatment
Extreme values are in parentheses.

HDL-C is reduced by a mean of 71% and apo A-I and A-II by 67 and 54% respectively (Table 2). This HDL decrease is due to:
-anorexia leading to a reduction of apo A-I intestinal production and
-inhibition of apoproteins A-I and A-II synthesis in the liver following an acute phase response (apo S increase) induced by IL-1. IL-2 and LAK cells produced also an increase in triglycerides, VLDL and VLDL-C and by inhibition of LPL activity by TNF and IL-1 a reduced conversion of VLDL to LDL. This explains the low LDL-C (78% reduction) and apo B (49%) at the peak of treatment. Low plasma LDL-C may also result from an increased cellular demand for cholesterol in relation with lymphocyte activation and proliferation (40, 41). The drastic changes caused by treatment were quickly and completely reversible (Table 2). Similar effects on lipids have been reported also by Wilson *et al.* (42). Thus the metabolic action of lymphokines on lipid and apolipoprotein metabolism is highly sensitive.

Treatment of solid tumors (carcinomas of lung, breast, colon, kidney, ovaries) by intravenous rH-TNF (discontinuous courses) (43) is also associated with decreased HDL as well as increase in TG and VLDL. The latter is explained by the fact that TNF suppresses the activity and synthesis of lipoprotein lipase at the level of mRNA (43). GMCSF administered in severe aplastic anemia caused also a decrease in cholesterol (44). On the contrary α, β, and γ interferons (IFN) did not induce true hypocholesterolemia but a limited reduction of total cholesterol (45, 46, 47). Interestingly β IFN induced a reduction of LDL-C and hypertriglyceridemia at high dosages (46).

Liver disease and transplantation

Some liver diseases (fulminans hepatitis, severe hepatic parenchymal disease) are known to lower cholesterol. In hepatic disorders accompanied by portal hypertension, severe hepatic insufficiency liver transplantation may become the only treatment. In transplanted patients total cholesterol dropped from about day 2 after the stop of perfusions to reach its lowest value from day 4 to day 8 despite normal re-feeding (48) (Figure 1). Cholesterol decrease concerned LDL-C as well as HDL-C but the drop is much more pronounced for the later. This drop parallels that of apo A-I and apo A-II. This apoprotein decreasase might be secondary to a defect in synthesis besides a persistent normal or even reduced catabolism. LDL-C is reduced partly in relation with the reduced conversion VLDL to LDL and partly by normal function of peripheral receptors. We do not have any information about the functional status of hepatic receptors after liver transplant. When triglycerides are increased along with apoprotein B but not apo C-III (days 4 to 8) it may be a defect in LPL activity secondary to cytokine production due to the induction of cellular immunity. It is why cyclosporin or other immunosuppressive agents and corticoids are administered quickly at high dosages. Later on TG increase is in relation with a reduced conversion VLDL to LDL and a defective hepatic uptake of VLDL remnants.

MECHANISMS LEADING TO HYPOCHOLESTEROLEMIA

The major mechanisms leading to hypocholesterolemia are the following:
- a deficiency of synthesis due to a genetic defect (abetalipoproteinemia) or a toxic effect at the hepatic level (lymphokine, parenchymal liver disease)
- a rapid and altered catabolism of HDL-apoproteins (Tangier), or an increased catabolism of large VLDL (hypobetalipoproteinemia) or of LDL (hypobetalipoproteinemia, myeloproliferative diseases)
- a defect of conversion causing hypertriglyceridemia (and high VLDL-C) and low LDL-C but not affecting HDL-C (hypobetalipoproteinemia, Kala-azar, lymphokine, liver disease)

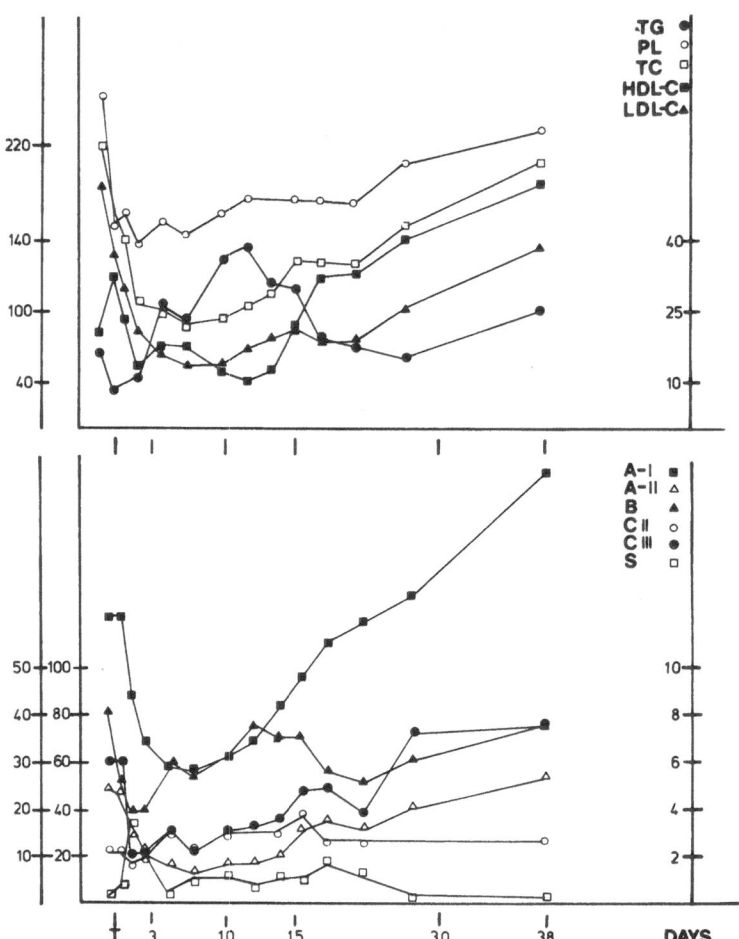

Figure 1. Typical lipid and apolipoprotein (mg/dl) changes following liver
transplantation (T = day of surgery) in a case showing good
recovery

- a sequestration leading to cholesterol accumulation in tissues (liver,
spleen, RTE) (Tangier, Kala-azar, myeloproliferative disorders) or an
increased uptake by the liver to enhance the excretion of cholesterol in
biliary acids.

Thus the etiology of hypocholesterolemia may have different origins
(isolated or combined) as summarized in Table 3.

CONCLUSIONS

1. Hypocholesterolemia might be considered as a negative factor of risk for
atherosclerosis. Thus a genetic disorder such as hypobetalipoproteinemia
with low LDL-C and high HDL-C, generally symptom-free, is favourable and
may be included in the longevity syndromes (49).
2. As a difference of risk exists depending on whether LDL-C (positively
correlated with atherosclerosis) or HDL-C (negatively correlated) is
diminished, an accurate HDL-C/LDL-C ratio may be useful in the prognosis

Table 3. Etiology of Hypercholesterolemia

	DEFICIENCY OF SYNTHESIS	INCREASED CATABOLISM	DEFECT OF CONVERSION	SEQUESTRATION
TANGIER	+		+	
APO A-I ABSENCE	+			
ABETALIPOPROTEINEMIA	+			
HYPOBETALIPOPROTEINEMIA		+	+	
MALNUTRITION	+	?		+
KALA-AZAR (LEISHMANIASIS)	?	+ ?	+	+
MYELOPROLIFERATIVE		+		+
IL-2 + LAK CELLS	+	+ ?	+	+
LIVER TRANSPLANTATION	+	?	+	

particularly when VLDL are largely increased.
3. As we have observed transient dramatically low cholesterol levels in drug-induced hypocholesterolemia (treatment of tumors by cytokines) and in liver transplanted patients, these levels are not life-threatening because, in absence of reject phenomenon, they are generally completely and more or less quickly reversible (from 4 to 30 days).
4. Along with total, LDL and HDL cholesterol levels, major informations about the origin and gravity of disorders are furnished by the determination of the different apolipoproteins.

PERSPECTIVES

Two important questions may be raised:
1) What is the optimal level of plasma cholesterol?
2) Is hypocholesterolemia despite a lower risk of atherosclerosis harmful? Neither question was definitely answered.

LDL-C of about 25 mg/dl (or more than 110 mg/dl of total cholesterol) seems to be innocuous since normal healthy infants showed such plasma levels (50). This level would be sufficient to provide enough cholesterol to body cells, but also fat-soluble vitamins A and E necessary fot the metabolism of myelin and development of brain and nervous tissue. Levels lower than those are found only in all primary and secondary disorders discussed above. Abetalipoproteinemia and homozygous hypobetalipoproteinemia with very low total and LDL cholesterol levels are generally accompanied by signs of impaired myelination leading to neurological symptoms, even to Friedreich's ataxia. The symptoms may be absent or less severe in presence of high HDL-C levels via apo E-rich HDL compensation (51). Thus the optimal level of total and LDL cholesterol appear to be comprised between 110-150 and 25-60 mg/dl respectively (50). Lower levels may be harmless if they are compensated by elevated HDL-C values as it is often the case in HßL. These patients are with hyperalphalipoproteinemia considered as longevity syndromes (49). Long lasting very low levels of cholesterol may impair normal membrane integrity and function with a resultant loss of resistance to cancerous change.

An acute, short-lived and dramatic reduction of cholesterol induced by cytokines may not provoke permanent damage because it is quickly and completely reversible upon cessation of administration. A similar reduction was observed in liver transplanted patients, the return to normal lasting longer as a function of complete hepatic recovery.

REFERENCES

1. The Lipid Research Clinics Program Epidemiology Committee: Plasma lipid distributions in selected north american populations: the lipid research clinics program prevalence study. Circulation, 60:427 (1979).
2. B. V. Howard, M. P. Davis, D. J. Pettitt, W. C. Knowler, and P. H. Bennett, Plasma and lipoprotein cholesterol and triglyceride concentrations in the Pima Indians: distributions differing from those of caucasians, Circulation, 68:714 (1983).
3. W. J. McConathy and D. M. Lane, Studies on the apolipoproteins and lipoproteins of cord serum, Pediat. Res., 14:757 (1980).
4. B. Christensen, C. Glueck, P. Kwiterovich, I. Degroot, G. Chase, G. Heiss, R. Mowery, I. Tamir, and B. Rifkind, Plasma cholesterol and triglyceride distributions in 13,665 children and adolescents: the Prevalence study of the lipid research clinics program, Pediat. Res.,14:194 (1980).
5. H. A. Tyroler, C. J. Glueck, B. Christensen, and P. O. Kwiterovich, Plasma high-density lipoprotein cholesterol comparisons in black and white populations. The Lipid research clinics program prevalence study, Circulation, 62, IV-99 (1980).
6. W. E. Connor, M. T. Cequeira, R. W. Connor, R. B. Wallace, R. Malinow, and H. R. Casdorph, The plasma lipids, lipoproteins and diet of the Tarahumara Indians of Mexico, Am. J. Clin. Nutr., 31-1131 (1978).
7. J. Burslem, G. Schonfeld, M. A. Howald, S. W. Weidman, and J. P. Miller, Plasma apoprotein and lipoprotein lipid levels in vegetarians, Metabolism, 27:711 (1978).
8. F. M. Sacks, D. Ornish, B. Rosner, S. McLanahan, W. P. Castelli, and E. H. Kass, Plasma lipoprotein levels in vegetarians. The effect of ingestion of fats from dairy products, J.A.M.A., 254:1337 (1985).
9. E. J. Schaefer, R. I. Levy, N. D. Ernst, F. D. Van Sant, and H. B. Brewer, Jr., The effects of low cholesterol, high polyunsaturated fat, and low fat diets on plasma lipid and lipoprotein cholesterol levels in normal and hypercholesterolemic subjects, Am. J. Clin. Nutr., 34:1758 (1981).
10. L. A. Carlson and B. Philipson, Fish-eye disease, a new familial condition with massive corneal opacities and dyslipoproteinemia, Lancet, ii:921 (1979).
11. G. Franceschini, C. Sirtori, A. Capurso, K. H. Weisgraber, and R. W. Mahley, A-I Milano apoprotein, J. Clin. Invest., 66:892 (1980).
12. E. Gjone, Familial LCAT deficiency, Acta. Med. Scand., 194:353 (1973).
13. P. Alaupovic, E.J. Schaefer, W. J. McConathy, J. D. Fesmire, and H. B. Brewer, Jr., Plasma apolipoprotein concentrations in familial apolipoprotein A-I and A-II deficiency (Tangier disease), Metabolism, 30:805 (1981).
14. E. J. Schaefer, W. H. Heaton, M. G. Wetzel, and H. B. Brewer, Jr., Plasma apolipoprotein A-I absence associated with a marked reduction of high density lipoproteins and premature coronary artery disease, Arteriosclerosis, 2:16 (1982).
15. D. R. Illingworth, W. E. Connor, and P. Alaupovic, High density lipoprotein metabolism in a patient with abetalipoproteinemia, Ann. Nutr. Metab., 25:1 (1981).
16. J -F. Lontie, C. L. Malmendier, C. Sérougne, D. Y. Dubois, C. Dachet, J. Férézou, and D. Mathé, Plasma lipids, lipoproteins and apolipoproteins in two kindreds of hypobetalipoproteinemia, Atherosclerosis, 1990. To be published.
17. C. L. Malmendier, J -F. Lontie, C. Delcroix, C. Sérougne, J. Férézou, and L. M. Veiga, Low density lipoprotein metabolism in a kindred with familial hypobetalipoproteinemia, Submitted to Metabolism, 1990.

18. C. L. Malmendier, C. Delcroix, J -F. Lontie. D. Y. Dubois, Apolipoprotein C-II and C-III metabolism in a kindred of familial hypobetalipoproteinemia, Submitted to Metabolism, 1990.

19. T. Magot, K. Ouguerram, J -F. Lontie, and C. L. Malmendier, In vivo study of lipoprotein cholesteryl ester movements between plasma lipoproteins in familial hypobetalipoproteinemia, Congrès du Gerli, Nice, April 1990.

20. A. L. Truswell, J. D. L. Hansen, C. E. Watson, and P. Wannenburg, Relation of serum lipids and lipoproteins to fatty liver in kwashiorkor, Am. J. Clin. Nutr., 22:568 (1969).

21. D. Lemonnier and Y. Ingenbleek, Les carences nutritionnelles dans les pays du Tiers Monde, GERM, Paris, Karthala-ACCT (1990)

22. D. Bekaert, R. Kallel, M -E. Bouma, J -F. Lontie, A. Mebazaa, C. L. Malmendier, and M. Ayrault-Jarrier, Plasma lipoproteins in infantile visceral leishmaniasis: deficiency of apolipoproteins A-I and A-II, Clin. Chim. Acta, 184:181 (1989).

23. C. A. Rouzer and A. Cerami, Hypertriglyceridemia associated with Trypanosoma brucei infection in rabbits: role of defective triglyceride removal, Mol. Biochem. Parasitol., 2:31 (1980).

24. B. Beutler and A. Cerami, Tumor necrosis, cachexia, shock, and inflammation: a common mediator, Ann. Rev. Biochem., 57:505 (1988).

25. P. Scuderi, K. E. Sterling, K. S. Lam, P. R. Finley, K. J. Ryan, C. G. Ray, E. Petersen, D. J. Slymen, and S. E. Salmon, Raised serum levels of tumor necrosis factor in parasitic infections, Lancet, ii:1364 (1986).

26. G. Assman, D. Fredrickson, P. Herbert, T. Forte, R. Heinen, An A-II lipoprotein particle in Tangier disease, Circulation, 50:259 (1974).

27. W. März and W. Gross, Immunochemical evidence for the presence in human plasma of lipoproteins with apolipoprotein A-II as the major protein constituant, Biochem. Biophys. Acta, 962:155 (1988).

28. M. R. Rifkin, Interaction of high-density lipoprotein with trypanosoma brucei: effect of membrane stabilizers, J. Cell. Biochem., 23:57 (1983).

29. B. Galvao-Castro, J. A. Sa Ferreira, K. F. Marzochi, M. C. Marzochi, S. G. Coutinho, and P. H. Lambert, Polyclonal B-cell activation, circulating immune complexes and autoimmunity in human visceral leishmaniasis, Clin. Exp. Immunol., 56:58 (1984).

30. B. M. Rifkind and M. Gale, Hypolipidaemia in anaemia. Lancet, ii:640 (1967).

31. H. S. Gilbert, H. Ginsberg, R. Fagerstrom, and W. V. Brown, Characterization of hypocholesterolemia in myeloproliferative disease. Relation to disease manifestations and activity, Am. J. Med., 71:595 (1981).

32. R. E. Bases and I. H. Krakoff, Studies of serum cholesterol levels in leukemia, J. Reticuloendoth. Soc., 2:8 (1965).

33. Y. K. Ho, R. G. Smith, M. S. Brown, and J. L. Goldstein, Low-density lipoprotein (LDL) receptor activity in human acute myelogenous leukemia cells, Blood, 52:1099 (1978).

34. H. Ginsberg, H. S. Gilbert, J. C. Gibson, N -A. Le, and W. V. Brown, Increased low-density-lipoprotein catabolism in myeloproliferative disorders, Ann. Intern. Med., 96:311 (1982).

35. S. Vitols, M. Björkholm, G. Gahrtob, and C. Peterson, Hypocholesterolaemia in malignancy due to elevated low-density-lipoprotein-receptor activity in tumor cells: evidence from studies in patients with leukaemia, Lancet, i:1150 (1989).

36. U. E. Nydegger and R. E. Butler, Serum lipoprotein levels in patients with cancer, Cancer Res., 32:1756 (1972).

37. J -F. Lontie and C. L. Malmendier, unpublished observations.

38. C. L. Malmendier, J -F. Lontie, J. P. Sculier, and D. Y. Dubois, Modifications of plasma lipids, lipoproteins and apolipoproteins in advanced cancer patients treated with recombinant interleukin-2 and autologous lymphokine-activated killer cells, Atherosclerosis, 73:173 (1988).

39. S. A. Rosenberg, M. T. Lotze, L. M. Muul, S. Leitman. A. E. Chang. S. E. Ettinghausen, Y. L. Matory, J. M. Skibber, E. Shiloni, J. T. Vetto, C. A. Seipp. C. Simpson, and C. M. Reichert, Observations on the systemic administration of autologous lymphokine-activated killer cells and recombinant interleukin-2 to patients with metastatic cancer, N. Engl. J. Med., 313:1485 (1985).

40. J. A. Cuthbert and P. E. Lipski. Promotion of human T lymphocyte activation and proliferation by fatty acids in low density and high density lipoproteins, Proc. Natl. Acad. Sci. USA, 81:4539 (1984).

41. M. T. Lotze. Y.L. Matory, S.E. Ettinghausen. A.A. Rainer, S. O. Sharrow, C. A. Y. Seipp, M. C. Custer, and S. A. Rosenberg, In vivo administration of purified human interleukin 2. II. Half life. immunologic effects, and expansion of peripheral lymphoid cells in vivo with recombinant IL2. J. Immunol., 135:2865 (1985).

42. D. E. Wilson, G. R. Birchfield, J. S. Hejazi, J. P. Ward, and W. E. Samlowski, Hypocholesterolemia in patients treated with recombinant interleukin-2: appearance of remnant-like lipoproteins, J. Clin. Oncol., 7:1573 (1989)

43. M. L. Sherman, D. R. Spriggs, K. A. Arthur, K. Imamura, E. Frei III, and D. W. Kufe, Recombinant human tumor necrosis factor administered as a five-day continuous infusion in cancer patients: Phase I toxicity and effects on lipid metabolism, J. Clin. Oncol., 6:344 (1988).

44. S. D. Nimer, R. E. Champlin. and D. W. Golde, Serum cholesterol-lowering activity of granulocyte-macrophage colony-stimulating factor, JAMA, 260:3297 (1988).

45. C. Ehnholm, K. Aho. J. K. Huttunen, E. Kostiainen, K. Mattila, J. Pikkarainen. and K. Cantell, Effect of interferon on plasma lipoproteins and on the activity of postheparin plasma lipases, Arteriosclerosis, 2:68 (1982).

46. I. B. Rosenzweig. D. A. Wiebe, E. C. Borden, B. Storer, and E. S. Shrago, Plasma lipoprotein changes in humans induced by ß-interferon, Atherosclerosis, 67:261 (1987).

47. R. Kurzrock, M. F. Rohde, J. R. Quesada, S. H. Gianturco, W. A. Bradley, S. A. Sherwin, and J. U. Gutterman, Recombinant γ interferon induces hypertriglyceridemia and inhibits post-heparin lipase activity in cancer patients, J. Exp. Med., 164:1093 (1986).

48. C. L. Malmendier, J -F. Lontie, D. Y. Dubois, and D. Mathé, Lipids and apolipoproteins in liver transplanted patients, In preparation.

49. C. J. Glueck, P. Gartside. R. W. Fallat. J. Sielski. and P.M. Steiner. Longevity syndromes: familial hypobeta and familial hyperalphalipo-proteinemia, J. Lab. Clin. Med.. 88:941 (1976).

50. M. S. Brown and J.L. Goldstein, A receptor-mediated pathway for cholesterol homeostasis, Science, 232:34 (1986).

51. C. B. Blum, R. J. Deckelbaum, L. D. Witte, A. R. Tall, and J. Cornicelli, Role of apolipoprotein E-containing lipoproteins in abetalipoproteinemia, J. Clin. Invest., 70:1157 (1982).

ORIGIN OF CHOLESTEROL AND BILE ACIDS IN THE DIVERTED BILE OF TWO PATIENTS

WITH TOTAL SMALL BOWEL RESECTION

J. Férézou, P. Beau[*], M. Parquet, T. Hajri, T. Magot,
C. Matuchansky[*] and C. Lutton

Physiologie de la Nutrition, Université Paris-Sud, 91400 Orsay
[*]Service de Gastroentérologie et Assistance Digestive, CHU, 86000
Poitiers

INTRODUCTION

An isotope study of cholesterol and bile acid metabolism was carried out in two patients who had undergone home cyclic parenteral nutrition for several months after total small bowel resection. This rare steady-state situation implies that cholesterol is eliminated from the body by an external biliary fistula which replaces the fecal output of cholesterol and bile acids that occurs in healthy subjects. This study gave evidence that newly synthesized hepatic cholesterol becomes the main source of cholesterol and bile acids eliminated into the bile in patients with nil digestive function and total interruption of the bile acid cycle.

METHODS

The isotope study started after at least one year of total parenteral nutrition, in two patients (A and B, respectively 62 and 57 years old) who differed by their surgical history. After a subtotal gut resection for mesenteric infarction following radiation enteropathy, patient A was fitted with a terminal duodenostomy which allowed complete diversion of the bile. After total small bowel resection due to mesenteric infarctus following a duodeno-pancreatectomy for nesidioblastoma, patient B was fitted with a gastrostomy and an external biliary fistula and his diabetes was perfectly equilibrated by means of a portable insulin pump. They were fed by cyclic parenteral nutrition, which provided during the night (14 hours) 30 Kcal/Kg/day, with 55% carbohydrates, 15% amino-acids and 30% lipids (as 250 ml/day Intralipid 20%).

The two patients received an intravenous pulse (2 hr infusion) of $1,2-{}^3$H-cholesterol (45×10^6dpm) dissolved in a sterile mixture of ethanol/Intralipid 20% (10/100, v,v). Blood samples (10-20 ml) were drawn at D1, D2, D3, D8, D14, D28, then every month for 6 months. Bile samples (20-30 ml) were collected every day during the first week, then every week for another three weeks, and the corresponding daily bile volumes were measured. The decay of plasma total cholesterol labeling was studied by input-output analysis[1] using a pluriexponential regression method[2] to measure the plasma cholesterol input from endogenous and exogenous origins. The mean daily output of cholesterol and bile acids and the lithogenic index of the bile[3] were assessed from chemical analyses of bile

Hypercholesterolemia, Hypocholesterolemia, Hypertriglyceridemia
Edited by C.L. Malmendier *et al.*, Plenum Press, New York, 1990

183

Table I. Cholesterol input-output analysis

| | Patients | | Controls[*] |
	A	B	(range)
Plasma inputs (mmol.day^{-1})	5.76	7.71	2.20-3.80
Endogenous (Internal Secretion)	5.61	7.55	1.80-3.00
Exogenous	0.15	0.15	0.30-0.80
R M P (mmol)	40.39	58.29	65.00-80.00
FCR (%.day^{-1})	14.3	13.2	3.0 - 5.0

Table II. Body cholesterol balance

| | Patients | | Controls[*] |
	A	B	(range)
Total output	(Bile)		(Feces)
(mmol. day^{-1})	14.20	18.19	2.40-3.60
Cholesterol from:	1.51±0.26	0.89±0.04	1.20-1.80
RMP	0.64±0.10	0.55±0.04	0.70-1.40
External secretion[**]	0.87±0.30	0.34±0.05	0.10-0.40
Non absorption of dietary cholesterol[**]			0.30-0.80
Bile acids from:	12.70±1.91	17.30±0.72	1.20-1.80
RMP	5.16±0.58	5.85±0.52	1.20-1.80
External secretion[**]	7.54±1.51	11.45±0.71	0
Lithogenic index	1.24±0.09	0.56±0.02	0.50-1.00
Total input (mmol. day^{-1})	14.02	19.35	2.40-3.60

[*] from ref. 4-6
[**] without prior exchange with RMP
RMP: rapidly miscible pool, FCR: fractional catabolic rate

samples. The ratio of the specific radioactivity (dpm/mmole) of biliary cholesterol or bile acids to that of plasma cholesterol measured at the same time (from the 3rd to the 28th day of the experiment) was used to calculate the fraction of cholesterol or bile acids which had the same labeling than plasma cholesterol. During this interval, the same decay was observed for free and esterified plasma cholesterol labeling. Therefore, the daily amounts of cholesterol and bile acids eliminated by the fistula were assessed according to their origin: rapidly exchangeable cholesterol for the labeled fraction, synthesized cholesterol directly poured into the bile for the unlabeled fraction.

RESULTS AND DISCUSSION

Results were compared to data already reported for non obese healthy subjects[4-6] (Table I). The total plasma cholesterol input was markedly higher in patients than in controls, especially the input of synthesized cholesterol, named internal secretion. In contrast, the exogenous flow provided by parenteral nutrition was lower than the absorption of dietary cholesterol in healthy subjects. The fraction of body cholesterol in permanent and rapid exchange with plasma cholesterol, or rapid miscible pool (RMP), includes cholesterol of blood, the liver and some other organs[4]. Probably because of small bowel resection, this pool was smaller in the two patients than in controls. The FCR (fractional catabolic rate) values indicated that the cholesterol turnover in this pool was accelerated in the two patients who displayed a permanent low plasma cholesterol level (1 mg/ml or less).

As shown in Table II, the daily output of cholesterol by the fistula was similar to the daily fecal elimination of neutral sterols in healthy subjects. In contrast, for bile acids, the output by the fistula largely exceeded the fecal elimination in controls. Because the specific radio-activity of cholesterol and bile acids was regularly lower than that of plasma cholesterol measured at the same experimental times, it can be concluded that only a fraction of biliary cholesterol and bile acids derived from cholesterol which initially entered the vascular pool before biliary elimination. A consistent fraction of cholesterol newly produced in the liver was directly poured into the bile (without prior exchange with the vascular pool), according to a process named external secretion. In patients, this cholesterol was largely converted into bile acids.

In terms of body cholesterol balance, the overall cholesterol production represents the sum of both internal and external secretions and the total cholesterol input the sum of exogenous and synthesized cholesterol flows. In healthy subjects, the external secretion occurs only from the gut and represents a minor source for fecal cholesterol[6]. In the two patients, the cholesterol production was strongly stimulated. It reached a mean value of 14.02 and 19.35 mmol.day^{-1} (5.40 and 7.47 g.day^{-1}) in patients A and B respectively, with a large fraction which did not pass via the vascular pool before its elimination. The production of bile acids (from cholesterol of both origins in the liver : RMP or external secretion) was also dramatically increased. In patient A, this conversion seemed to be relatively more efficient when cholesterol was derived from the vascular pool than from liver external secretion. The mean participation of newly synthesized cholesterol in the biliary cholesterol output was 58% in patient A and 38% in patient B. This participation was 59% and 66% respectively in bile acids eliminated by the fistula. Therefore the two patients differed by the mean lithogenic index measured in the bile samples. In patient A, the supersaturation of the bile with cholesterol was probably due to the limited capacity of the liver to convert into bile acids the newly synthesized cholesterol which is directly eliminated into the fistula without prior passage in the vascular pool.

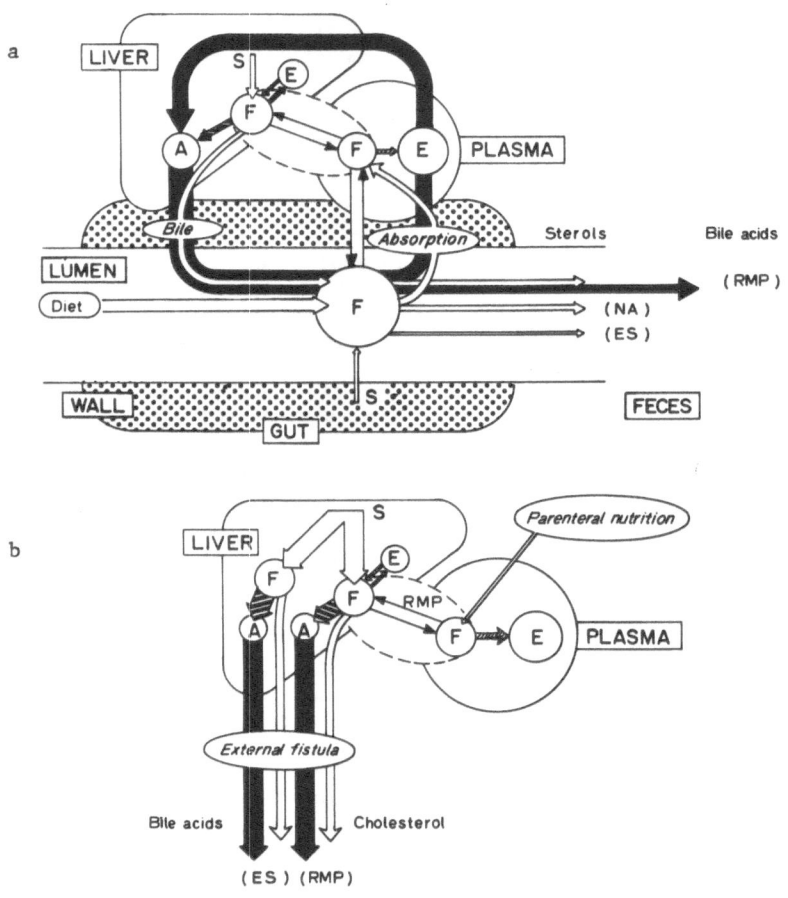

Figure 1. Schematic representation of cholesterol biodynamics in man
a: under normal dietary conditions
b: after total enterectomy and prolonged bile diversion

F	free cholesterol	⟶	free cholesterol (exchange)
E	esterified cholesterol	⟹	cholesterol (net flow)
A	bile acids	⟹	bile acids (net flow)
S	synthesis		conversion into bile acids
			esterification
			hydrolysis

Origin of biliary or fecal cholesterol and bile acids:

(RMP)	rapidly miscible pool
(ES)	external secretion
(NA)	non absorption

Our results can be compared to those previously obtained in six patients observed after several days (2 to 25) of complete bile diversion, but without gut resection[7]. Cholesterol production reached a mean value of 9.15 mmol.day^{-1} (3.5 g.day^{-1}) and the mean participation of newly synthesized cholesterol in the elimination of cholesterol and bile acids into the diverted bile was only 20% and 31% respectively. Probably because of total enterectomy, disturbances of the cholesterol turnover were even more marked in the two patients of the present study.

Under normal dietary conditions (Fig. a), there are two cholesterol inputs into the plasma (or RMP): the internal secretion of endogenous cholesterol (produced by the liver and gut) and the intestinal absorption of exogenous cholesterol, which is generally lower. The majority of esterified cholesterol in plasma appears by esterification in situ. Esterified cholesterol can be formed (and hydrolyzed) in the liver. Because of permanent and rapid movements of cholesterol (mainly in the free form) between plasma and liver, the cholesterol flow from the plasma into the liver largely exceeds that of the local cholesterol production. Therefore in the liver free cholesterol (from both exogenous and endogenous origins in the RMP) is largely converted into bile acids which enter the entero-hepatic cycle. Because of the efficient intestinal bile acid absorption, fecal bile acids are almost totally derived from RMP cholesterol, whereas fecal cholesterol has three distinct origins: the main is RMP cholesterol (including biliary cholesterol and cholesterol which comes from plasma through the intestinal wall), then non absorption of dietary cholesterol and external secretion from the gut[6,8]. These two last processes do not participate in the plasma cholesterol turnover.

In the two patients of the present study (Fig. b), the gut was omitted from this biodynamic pathway. Because the intravenous fat infusion represented a minor exogenous source for plasma cholesterol, the liver became almost the unique source of cholesterol in the body. The bile diversion dramatically enhanced the liver production of cholesterol (6-7 fold). Only part of newly synthesized cholesterol entered the plasma by internal secretion. The majority (60%) was directly eliminated by the fistula, largely converted into bile acids. Therefore, cholesterol and bile acids in the diverted bile were partly derived from RMP and partly from external secretion. This situation strongly contrasts with that observed in healthy subjects when the intact entero-hepatic circulation of bile acids continuously represses the cholesterol and bile acid productions.

It can be concluded from this study that after total enterectomy and prolonged bile diversion in man, cholesterol and bile acids eliminated into the bile mainly derive from newly synthesized hepatic cholesterol.

ACKNOWLEDGEMENTS

This work has been partly supported by a CRE grant awarded by INSERM.

REFERENCES

1. W. Perl and P. Samuel, Circulation Res. 25: 191-199 (1969).
2. K. Yamaoka, T. Tanigawara and T. Uno, J. Pharm. Dyn. 4: 879-885(1981)
3. P.J. Thomas and A.F. Hofmann, Gastroenterol. 65: 698-700 (1973)
4. P. Samuel and S. Lieberman, J. Lipid Res. 14: 189-196 (1973)
5. A.F. Hofmann, J. Clin. Gastroenterol. 10 (Sup. 2): S1-S11 (1988)
6. J. Férézou, T. Coste and F. Chevallier, Digestion 21: 232-243 (1981)
7. C.C. Schwartz, M. Berman, Z.R. Vlahcevic and L. Swell, J. Clin. Invest. 70, 863-876 (1982)
8. C. Lutton, Digestion 14: 342-356 (1976)

IN VIVO KINETICS

DR. MONES BERMAN

BERMAN'S SIMULATION ANALYSIS AND MODELING

Loren A. Zech[*†], Daniel J. Rader[†], and
Peter C. Greif[*]

[*]National Cancer Institute
[†]National Heart Lung and Blood Institute

Bethesda, Maryland USA 20892

Dr. Mones Berman, whose photograph appears at the beginning of this section, had an enduring interest in the theoretical aspects of lipid, lipoprotein and apolipoprotein metabolism, which stemmed from a more general interest in the systematic analysis of the kinetics and dynamics of metabolic molecules in the biological system. When tracer molecules are used to observe and measure the kinetics of a substance of interest, the observations are frequently related to a specific biological models for complete analysis. Possible models should be restricted to those that are compatible with other information about the system. Dr. Berman spent his life developing a formalism for the systematic analysis of tracer data taken from dynamical biological systems in both the steady and changing state.

The Berman View

Dr. Berman preferred to work in the domain of the theorist but was forced by the nature of the questions he asked to devote considerable time and resources to the computational domain. The collection of these latter efforts (1,2) have come to be known as SAAM[1] and CONSAM. SAAM and CONSAM in themselves were not an end result but a set of unfinished tools, which lagged behind the building and testing of theories using compartmental models. To understand Dr. Berman's objectives, one must examine and categorize his work as a scientist.

Since the beginning of time, science has been concerned with the observation. However, since the first application of science to a problem calling for a prediction, the practice of science has been divided into the experimental domain and the theoretical domain. With the development of the calculus in the 17th century, the mathematical model became a part of the theoretical

---

[1] SAAM for Simulation Analysis And Modeling
CONSAM for CONversational SAaM

Hypercholesterolemia, Hypocholesterolemia, Hypertriglyceridemia
Edited by C.L. Malmendier *et al.*, Plenum Press, New York, 1990

domain of science. Mones Berman began his scientific career at the moment of the birth of the electronic computer. The computer was a unique asset for the theorist, as it enabled him to simulate his models.

In addition to his contributions to the general aspect of compartmental modeling, Dr. Berman contributed to the design and construction of scientific instruments, the theory of ordinary differential equations, and the calculation of radiation dose. In addition, he built models used in the study of the kinetics of albumin, glucose, insulin, phosphorus, cholesterol, aldosterone, amino acid transport, magnesium, calcium, iodine, thyroid hormone, fatty acids, triglyceride, collagen synthesis, oxidative metabolism, lithium, glycerol, nucleotide regulation, methotrexate, pertechnetate, zinc, and adriamycin.

Many have said that Mones Berman was primarily a builder of compartmental models, but a brief examination of his contributions to the theory of the many systems listed above shows that he operated on a much higher plane. He was in fact a theoretical biologist. In the next few paragraphs we consider his approach, starting with his view of kinetic pools.

Whether observable or unobservable, a pool is some collection or amount of similar biological material which exhibits a range of behavior. Because each unit of material may not be identical or may not exhibit identical properties, each pool has average properties. When the range of these average properties is of no consequence to the experimental outcome, the pool is considered homogeneous. If a pool is not homogeneous by this criterion it is not a viable pool, but rather can be divided into several homogeneous pools.

In general a pool may be in one or more physiologic spaces and therefore pool is a more general term than a space or area. While a pool must be within some volume, as all material takes up space, the volume of distribution is not necessarily a direct property of the pool unless the material of a pool is separable based on the volume in which it is distributed. For example, a pool can be on a cell surface on in the plasma volume. The definition of a pool of biological material is thus a flexible physiologic concept. In contrast, a compartment in Dr. Berman's view is an amount of material that behaves kinetically as a distinct, homogeneous, well mixed quantity of material. A compartment is distinguished from a physiological space or a physiologic volume in that it has a theoretical property associated with movement of material into and away from the compartment: this theoretical property is kinetic behavior. It is this kinetic property of a compartment which makes the mathematical theory of compartmental models useful in the description of biological systems. Dr. Berman viewed a compartment or group of compartments as a theoretical construct which contains some amount of material. That quantity of material characterizes the state of the compartment or group of compartments at any point in time. The state is related to the amount of material within a designated physical space and/or physiological volume of a biological system. In other words, the compartmental model is an analogy of the biological system. Dr. Berman's compartmental models describe this relationship. Because compartmental models can be simulated using electronic computers, hypotheses about biologic systems can be tested and compared with observed data. It is this process of testing compartmental models and comparing the results with observed experimental results which Dr. Berman called model building. It was the give and take of this testing which intoxicated him. Dr. Berman continually pointed out that this testing was a process of model elimination. This process increased confidence in any model which had not been eliminated after a large number of such tests. It was also his contention that while the development process of model testing (analogy testing) did not lead to unique compartmental models for biological systems, this process did lead to the development of very useful tools for scientific investigation. Furthermore, it was with these tools that useful quantitative descriptions and estimates of the kinetic and dynamic properties of a biological system could be made.

For purposes of biological systems analysis, a mathematical description of a compartment results in a description of the state of a compartment at any time t compared to the state of a compartment at a reference time t_0. The physical analogy to the states are quantities such as concentration. In other words, the description of a compartment consists of the mathematical relationship between states or the mathematical description of a change in the state of a compartment.

Because the most common biological processes are first order[2] processes, Dr. Berman chose to build his compartmental models using compartments which could be described using first order[3] linear[4] differential equations. When this could not be accomplished, he built models as groups of modified first order compartments. According to Dr. Berman, the basic unit of a mathematical description of a compartment with state F_0 at time t=0 can be described by the following differential equation:

$$dF/dt = -LF$$

with the solution:

$$F(t) = F_0 e^{-Lt} .$$

This view should not be confused with the often heard statement that Dr. Berman was only interested in compartmental models made up of compartments described by first order linear differential equations. He believed that zeroth order, second and higher order, and other nonlinearities of interest such as Michaelis-Menten could be viewed as modifications of the basic linear differential equation. This belief was a consequence of the environment in which Dr. Berman practiced theoretical biology.

---

[2] The order of a basic process is the sum of the exponents of the states which describe the process.

[3] The order of a differential equation is calculated as the difference between the highest and the lowest order derivatives of the state. ie. $dF/dt = -Lt$ is first order.

[4] Linear indicates that the largest sum of the exponents of all states and their derivatives in any term of the differential equation is less than two. ie. $dF/dt = -Lt$ is linear.

It was also Dr. Berman's hypothesis that the system or environment was not changing rapidly compared to the kinetics of interest and that the material which flowed into the environment did not change that environment. In addition to flow from the simple compartment into the environment, Dr. Berman conceptualized flow from the environment into the basic compartment either as a constant or as changing in a prescribed manner not dependent on the state of the compartment. In addition, he espoused the view that this simple modification of the basic compartment could be described by the following modification to the above differential equation:

$$dF/dt = -LF + U.$$

This equation has the solution:

$$F(t) = (F_0 - U/L)e^{-Lt} + U/L$$

when

$$t = 0, \qquad\qquad F(t) = F_0,$$

when

$$L = 0, \text{ and } t > 0, \quad F(t) = F_0 + Ut,$$

when

$$F_0 = 0, \qquad\qquad F(t) = U/L(1 - e^{-Lt}),$$

and when

$$U = 0, \qquad\qquad F(t) = F_0 e^{-Lt}.$$

Modifying this concept as necessary, Dr. Berman constructed multicompartmental models, letting the output of one compartment be the input for another. In other words, Dr. Berman viewed the theoretical process of building a compartmental model of a system with its component parts of basic or modified compartments as making the connections between compartments, and making the connections to the environment. In his view, the model was a theoretical construct useful for hypothesis testing and making predictions about the biological system. Furthermore, his ability to build models for a number of biological systems allowed him to be an important investigator of experimental results and a useful practitioner of theoretical biology.

Tools for Building Models

In order to practice the art of model building, Dr. Berman was required to expend considerable effort in the computational domain of science. The basic tool of theorists is the simulator. Dr. Berman spent many hours modifying, building, and refining numerical tools. It is in this domain that Dr. Berman made his most lasting contribution by refining the numerical techniques used in building theoretical models and by teaching others how to use these techniques.

Dr. Berman was frequently asked: where does the first or starting model come from? While he often answered this question by describing the proposed starting model as a modification of some previously developed complicated model, close examination of his publications shows that he usually started with a very simple one or two compartmental model or a series of one or two compartmental models connected together. Dr. Berman would answer this question by suggesting that one should start with the simplest model possible, consisting of no more than a few compartments. In Dr. Berman's words (3): "If presumptive information about the model is not available, one may start with a mathematical model, specifically, with sums of exponentials, since this can eventually lead to the class of compartmental models. Transformations to other models can be made subsequently, if necessary." Even though he may have generated some confusion by starting his theories as mathematical models, it is important to note that he drew a distinction between theory and mathematical models. He also stated (4) that "Although no formalism has been developed as yet for the building of models, one may distinguish several stages in this process: a) choice is made for the type of model, b) degree of complexity of the model is defined, c) values of the parameters of the model are calculated, d) judgement is made whether the model is compatible with the data, and, e) the model is revised when it is inconsistent with the data." The number of stages was not fixed, and the theoretical and numerical tools to accomplish each stage continued to be developed even at the time of his death. In fact a pseudo-formalism has developed and continues to be practiced by a handful of his students today. A rough outline of this pseudo-formalism can be appreciated by the numerical tools in SAAM and CONSAM which have been developed and which are still under development.

There is no single characteristic which sets SAAM or CONSAM apart from all other simulators or sets of numerical tools for model building. On the other hand, a case can be made that nowhere else except in these software packages are all of specific set of characteristics present in the precise configuration in which they are found within SAAM/CONSAM. This point is both crucial and cardinal. An example may help illustrate this point. The flag of the United States of America has individual characteristics which it shares with the flags of other nations of the world. First, it is tri-colored. But so, too, are the flags of Australia, Belgium, the United Kingdom, France, Italy, and others. Second, its three colors are red, white, and blue. But so, too, are the flags of Czechoslovakia, France, the Netherlands, Panama, the United Kingdom, and, Yugoslavia. Third, it has stars in its basic design. But so, too, do the flags of Australia, the People's Republic of China, Honduras, and Venezuela. Despite these common characteristics, no flag in the world is identical to the flag of the United States. What is distinctive about the United States' flag is not any one of its several characteristics but the precise configuration of these characteristics. So, too, it is with SAAM/CONSAM in relation to all the other simulators and programs for building compartmental models.

As its very name suggests, SAAM (Simulation Analysis And Modeling) is characterized by an openness to all the procedures and numerical tools necessary for compartmental modeling. Some procedures in SAAM/CONSAM are well developed, many are in a lesser stage of development, and several are only planned for future development. SAAM is also amenable to the incorporation of a wide range of information about the system under investigation and to all observations on the system under investigation. SAAM/CONSAM is comprehensive and all-embracing with respect to the building of compartmental models in all the theoretical and computational diversity of the experience. With the building of compartmental models, we come to the point where

the distinctive Berman understanding and practical use of
SAAM/CONSAM most clearly emerges. These principles, at once
philosophical, theoretical, and practical, have and continue to
shape SAAM/CONSAM.

SAAM consists of three major sections: an input section, an
equation section, and a results section. These are connected by
four major sets of procedures used to solve the model and
process the data as they move from section to section. Three of
the four processes are designated by arrows around Figure 1. The
character interface contains a description of the model, observed data, and operating instructions in strictly formatted
lines of information. This interface was originally intended to
be used with punched cards, but it remains simple to use, and is
in universal ASCII characters. A compartmental modeling problem
can be described using a line editor and simple computer. While
Dr. Berman realized the broad utility of a character interface,
he planned for a graphical interface in which a visual representation of the model associated with format free listing of the
observed data and parameters values would replace the character
interface in CONSAM. He did not intend to replace SAAM with
CONSAM.

The first major operation in SAAM results when a series of
algorithms process the character based description of the model
and read the observed data creating an equation storage area and
a results storage area. These two sections of SAAM are collec-
tively called the KOMN section of SAAM. Dr. Berman had made
plans to modify the extent of KOMN so that the number of data
points, components, and compartments could be easily increased
or decreased. Much of this has been accomplished according to
his plans in later versions of SAAM, as shown in Table 1 below.

The second major operation in SAAM is the solution of the
equations and storing the calculated results for comparison to
the observed values. The third major process in SAAM is out-
putting the results in juxtaposition to the observed values in

SAAM

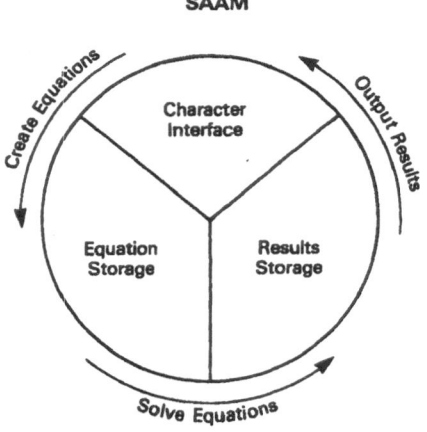

Fig 1.

CONSAM

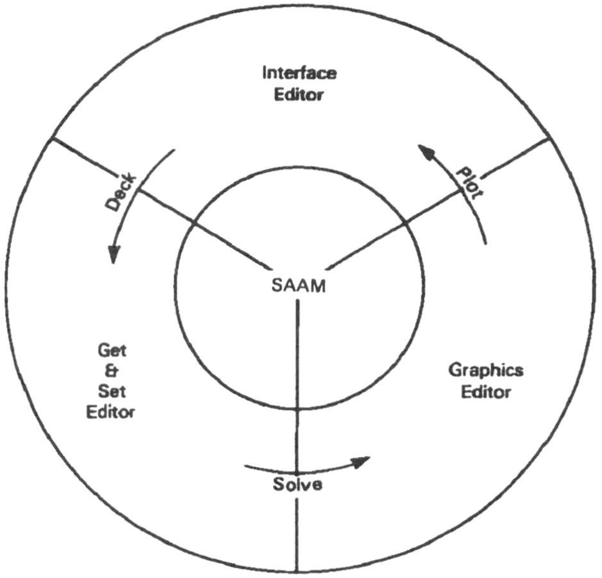

Fig 2

both character form and graphical format. The forth process, not
shown on the diagram, is the process of adjusting the parameters
of the model to decrease the difference between the observed and
calculated values, and resolving the equations with these
modified parameters. This procedure is carried out successively
until the difference becomes small and the final results are
printed and plotted.

Since Dr. Berman's death, several of his plans have been
implemented in newer versions of SAAM. First, the number of
data points was doubled, then tripled, and finally quadrupled.
Versions of SAAM with even larger numbers of data points are
planned. Second, the number of compartments and adjustable
parameters was increased by 20%, then doubled, and finally
tripled.

CONSAM is an acronym for CONversational SAaM and is an
interactive version of SAAM (Figure 2). This means that CONSAM,
above all else, is in essence an environment in which to use
SAAM. In its simplest form this software consists of three
editors and four groups of commands (DECK, SOLVE, ITERATE, and
PLOT) corresponding to the three sections SAAM and the four
procedures of SAAM. In addition, CONSAM also has a few additional procedures associated with the management of the editors
and processing of results. The character interface editor is a
simple line editor with display, insert, delete, and modify
commands which operate on the character interface in which the
model and data are given. The DECK command translates the
character description to equations and locations for the storage
of observed data and solutions.   The get and set editor enables
the user to modify the equation parameter values. It is this
editor which permits a change in the solution without a change

195

in the structure of the equations. The SOLVE command causes the equations to be solved and the results to be stored at the designated locations. The Graphics Editor provides clipping, scaling, sizing, and positioning of the graph or graphs. The PLOT command activates the output procedures producing a plot of the results. The fourth SAAM procedure is invoked from CONSAM by the ITERATE command.

This brief outline of SAAM and CONSAM is sufficient to discuss the plans which Mones Berman had for his software and to

Table 1. Recent saam versions

SAAM27	Dr. Berman's last version 25 compartments and 250 data points
SAAM28	250 data points increased to 500
SAAM29	500 data points increased to 750
SAAM30	25 compartments increased to 75 25 adjustable parameters increased to 75 750 data points increased to 1000

outline the progress has recently been made. Table 1 lists the last four versions of SAAM. Following plans made before 1980 for a larger version of SAAM resulted in SAAM28 in 1985, SAAM29 in 1986, and SAAM30 in 1989 with corresponding versions of CONSAM. This year a PC version of SAAM and CONSAM has been completed. The minimal configuration of computer hardware and software to run the PC version of SAAM30 and CONSAM30 is listed in Table 2.

Table 2. Minimal PC Hardware and Software for SAAM30 and CONSAM30

1. 16 Mhz. Intel 80386SX processor and 80387SX coprocessor.

2. 5 Megabytes of extended random access memory.

3. 10 Megabytes of hard disk space.

4. 3.5 or 5.25 inch floppy disk drive.

5. EGA video controller and 12 inch monochrome monitor.

6. DOS 4.01 operating system and GSS graphics driver software.

The new PC version of CONSAM also has on-line pages from the manual for most commands, expanded numbers of data types (i.e. error _vs_ observed values), and publication quality graphics in a format compatible with many word processors.

Dr. Berman had many other plans for the improvement of SAAM and CONSAM based on his insight and experience. Two of these improvements, still unfinished, are discussed below. First, motivated by his experience, Dr. Berman was keenly aware

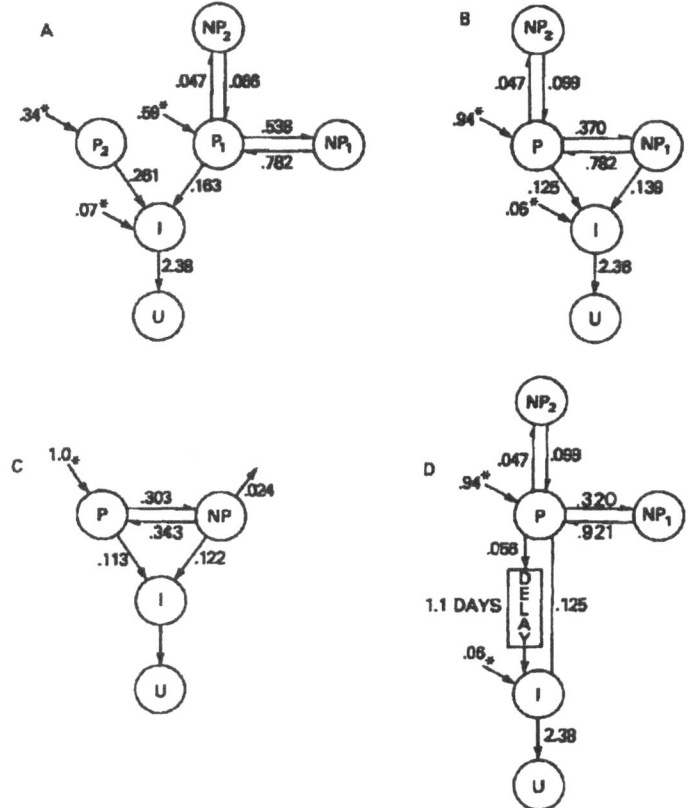

Fig 3.

that good models were built when several sets of parameter estimates for a model were combined to form a population estimate In his view, these population estimates of the adjustable model parameters and the corresponding population correlation matrix should be used as apriori constraints on the estimates of a new set of parameters. While he had included several procedures for estimating and incorporating the population parameters and their variances and covariances as constraints using SAAM, he did not completed this project. Recently these procedures have been improved and will soon be better integrated into SAAM.

Another, as yet unfinished project, was the development of a graphical user interface to be used in parallel with the character interface. It was Dr. Berman's plan to develop a translator which would translate a model diagram (graphical description) into a character description of the model and an inverse translator to convert a character description of a model into a diagram of a model. Slightly less than half of planned improvement of CONSAM was completed at the time of his death: for a restricted set of terminals and computers, a model diagram could be drawn from a character description. The remainder of this project has not been completed.

An Example of Dr. Berman's Modeling Contributions: The HDL Model

Dr. Berman made many contributions to the fields of lipid and lipoprotein modeling. The HDL compartmental model is an example of his involvement in lipoprotein modeling. A two compartment HDL model (5) appeared in 1973 as part of a model for apoC kinetics. The HDL compartmental model proposed in Blum, et al., 1977, ref. 6, and drawn in Figure 3C, was a modified and improved two compartment model with a new feature, irreversible loss from both compartments. A portion of this irreversible loss from the nonplasma pool did not move into the urine suggesting the existence of a second nonplasma pool. This paper (6) also discusses the possibility that a small portion of the plasma HDL did not exchange with the nonplasma compartment. This second feature was not included in the model diagram as this kinetic component was detectable in only a few subjects.

Several years later, (7) using data from simultaneous apoA-I and apoA-II kinetic turnover studies, separate compartmental models were developed for apoA-I and A-II, (Figures 3A and 3D). Comparing these models, apoA-I had a second, faster, plasma component not found in the apoA-II model. This was interpreted as a moiety of apoA-I which did not associate with apoA-II. This second apoA-I component was different in males than in females but did not correspond to overall apoA-I levels. We now understood the extra HDL plasma component which Dr. Berman predicted resulted from the contribution of apoA-I kinetics to total HDL kinetics, and was not a necessary part of the apoA-II model, Figure 3D. Figure 3B is an equivalent apoA-II model to that in Figure 3D and is included for purposes of comparison to Figure 3A and 3C.

Recent immunologic techniques have allowed the separation of HDL apoA-I into two lipoproteins, the LpA-I lipoprotein and the LpA-I,A-II lipoprotein. Following the simultaneous injection of radioionated ApoA-I and ApoA-II, plasma samples were collected and separated into apoA-I and apoA-I,A-II lipoproteins (8). The plasma decay curves for both the labeled apoA-I and the labeled apoA-II on the LpA-I,A-II lipoprotein and the labeled apoA-I on the LpA-I lipoprotein are shown in Figure 4. While the

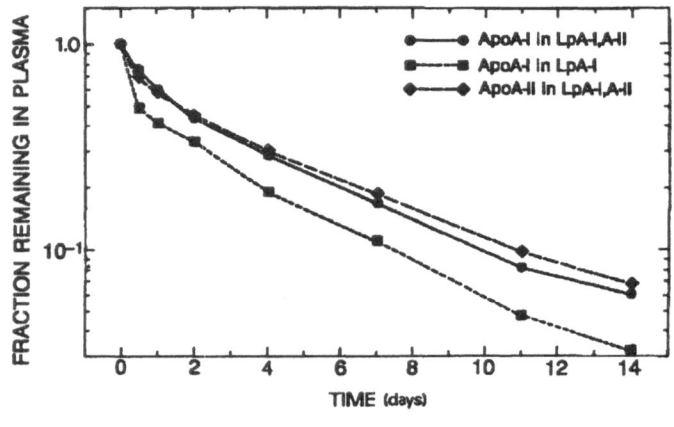

Fig 4.

apoA-I moiety on the LpA-I,A-II lipoprotein turnover is slightly
faster than that of apoA-II moiety on the same lipoprotein
fraction, the apoA-I moiety of the LpA-I lipoprotein decays
faster than both labels on the LpA-I,A-II lipoprotein. It is
this second difference which has been interpreted as the second
plasma component of apoA-I originally described by Dr. Berman.

In summary, Dr. Mones Berman contributed both quantitatively
and theoretically to the biosciences for more than three
decades. This exceptional scientist maintained the dream that
one day the theoretical and mathematical tools of modeling would
be routinely used to further the science of lipid and lipoprotein metabolism. Judging from the increased interest in lipoprotein and lipid kinetics, the large number of new users of SAAM
and CONSAM, and the recent acceptance of compartmental modeling
by many in the audience, Dr. Berman's boldest dream is about to
become reality.

REFERENCES

1. Berman,M. and Weiss, M.F., 1978 "SAAM Manual" US DHEW Pub.
   No. (NIH)75-180. pp196

2. Berman,M., Beltz,W.F., Greif,P.C., Chaby,R., and
   Boston,R.C. 1983 "CONSAM Users Guide" PHS Pub. No. 1983-
   421-132:3279. U.S. Gov. Printing Office, Washington, DC.

3. Berman,M. 1963 The formulation and testing of models.
   Ann. N. Y. Acad. Sci. 108:182-194

4. Berman,M., Shahn,E., and Weiss,M.F. 1962 The routine
   fitting of kinetic data to models: A mathematical formalism
   for digital computers. Biophys. J. 2:275-287

5. Hall,M., Bilheimer,D., Phair,R., Levy,R., and Berman,M.
   1974 A mathematical model for apoprotein kinetics in normal
   and hyperlipidemic patients. Circulation 50:III-114

6. Blum,C.B., Levy,R.I., Eisenberg,S., Hall,M.,Gobel,R.H., and
   Berman,M. 1977 High density lipoprotein metabolism in man.
   J. Clin. Invest. 60:795-804

7. Zech,L.A., Schaefer,E.J., Osborne,J.C.,Jr., Aamodt,R.L.,
   and Brewer,H.B.,Jr. 1984 inetics of Apolipoprotein A-I and
   A-II. in "Pathophysiology of Plasma Protein Metabolism"
   Giulino Mariana, ed., Plenum Publishing Corp., New York.

8. Rader,D.J., Castro,G.R., Kindt,M.R., Zech,L.A.,
   Fruchart, J.C. 1990 Differential in vivo metabolism of HDL
   subclasses LpA-I and LpA-I,A-II in man. Clin. Res. 38:240A.

CONSIDERATIONS IN DESIGNING AND ANALYZING

DATA FROM APO-B TURNOVER STUDIES

David M. Foster,
P. Hugh, and R. Barrett

Center for Bioengineering, FL-20
University of Washington
Seattle, Washington 98195

INTRODUCTION

The kinetics of apo-B transport through VLDL, IDL and LDL can be described by models developed from tracer data (1,2). From these models, one obtains estimates of metabolic parameters of interest such as (i) fractional catabolic rates, (ii) production rates, and (iii) transport from one lipoprotein class to another. The models, however, have assumptions built into them which affect the interpretation of the data and parameter estimates. With many new experimental procedures being developed, our knowledge of the structure and function of the plasma lipoproteins is rapidly increasing. Many of these new procedures lend themselves to turnover studies with the result that tracer data of increasing complexity are being generated. These include techniques to subfractionate the traditional lipoprotein populations (3,4) and the utilization of stable as well as radioactive isotopes (5,6). The interpretation of these data means that many of the previously developed models may have to be altered or changed altogether as the assumptions upon which they were based prove no longer to be valid. With tracer studies being used to elucidate altered metabolic pathways in pathophysiological conditions or in response to therapy, one runs a high risk of misinterpreting turnover data using models with underlying false assumptions.

The purpose of this work is to discuss some of these assumptions in light of the new experimental techniques. We will focus on some specific problems associated with interpreting the fractional catabolic rate and estimating the fraction of VLDL apo-B that is converted to LDL. The three questions to be addressed in particular are:

1. How can you estimate the fractional catabolic rate (FCR), and what affect does heterogeneity have on this estimate?
2. How can you estimate *de novo* production (PR)?
3. How do you quantify transport among lipoprotein moieties?

Some recommendations in terms of experimental design will be given.

VLDL TURNOVER STUDIES

The kinetics of VLDL apo-B have been studied using both exogenously and endogenously labeled VLDL apo-B. Both will be discussed below.

Hypercholesterolemia, Hypocholesterolemia, Hypertriglyceridemia
Edited by C.L. Malmendier *et al.,* Plenum Press, New York, 1990

Exogenous labeling of VLDL

Many studies have been performed in which VLDL is isolated and labeled with an isotope such as radioiodine. The labeled material is reinjected, serial plasma samples taken in which tracer levels are quantitated. Metabolic parameters such as the FCR and PR can be estimated from the resulting tracer concentration curve. Over the years, the plasma sampling schedule for VLDL apo-B has changed to include more frequent early samples, and the duration of the typical study has been extended from 12 to 15 hours to 48 to 72 hours. More frequent early samples resulted in the postulation of the "delipidation cascade" (discussed below) while the later samples arose when a slowly decaying "tail" was observed (1,2).

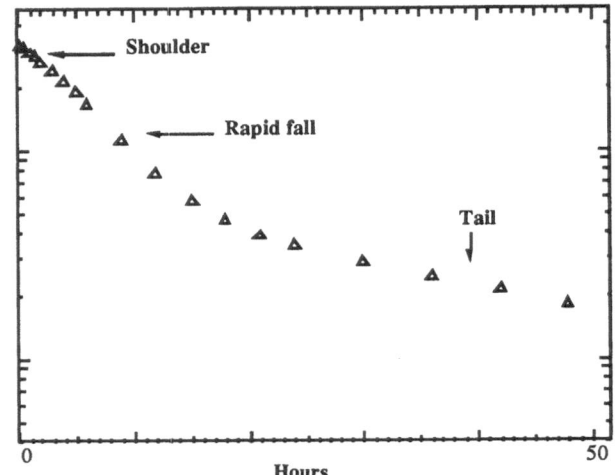

Figure 1. A typical set of VLDL apo-B turnover data; see text for explanation.

A typical set of data from a VLDL apo-B turnover study is shown in Figure 1. No units are shown on the ordinate to underline the fact that the kinetic information comes only from the shape of the curve; the units on the ordinate give some measure of tracer concentration such as dpm/ml, specific activity, or isotopic enrichment. The key features of this curve are the shoulder, the rapid fall and the slowly turning over tail.

Several models have been proposed to analyze such data; three are described in Figure 2 below (1,2).

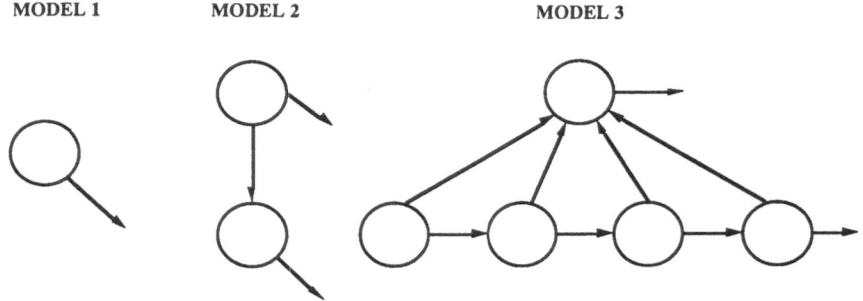

Figure 2. Three models used to analyze VLDL apo-B turnover data; see text for explanation.

Model 1 is a single pool model for VLDL apo-B which, although used early in the history of tracer VLDL apo-B studies when the typical study lasted only 12 - 18 hours, is now widely recognized as inappropriate since it cannot account for the tail portion of the curve. Despite this fact, this model is frequently used (as discussed below) in the analysis of VLDL apo-B tracer data when stable isotope-labeled amino acid precursors are infused! Model 2 is a typical two compartment model which displays many of the characteristics of the data shown in Figure 1; it cannot, however, describe the shoulder portion of the curve. As described in Berman (1), Model 3 is essentially the model proposed by Phair and Berman; it consists of the chain of compartments called the "delipidation cascade", and a slowly turning over compartment which is "fed" from each compartment in the cascade. New data resulting from subfractionating VLDL will force a critical look at this model especially in terms of physiological counterparts to the individual compartments in the model.

These models are used to estimate VLDL apo-B FCR; how do these estimates compare for both the tracer and tracee? The results are summarized below.

Table 1. Estimated FCR (hr^{-1})

	TRACER	TRACEE
Model 1	.12	.12
Model 2	.076	.14
Model 3	.079	.085

Notice the tracer and tracee FCR are different except for model 1. The difference between tracer and tracee FCR is a function of both the model used and the initial distribution of label in each compartment of the model. When VLDL apo-B is exogenously labeled, the slowly turning over particles, if initially labeled in proportion to mass, show a greater slow component in the plasma curve compared to their relative production rate. The difference in tracer and tracee FCR can therefore be attributed to labeling in proportion to mass and not production rate. In contrast, when endogenous labels are used the FCR of the tracer will always equal that of the tracee.

Besides estimating the FCR, one usually wants to estimate total apo-B transport, PR, through VLDL. This is the link between the tracer and tracee study. The formula to estimate the PR is: PR = FCR * apo-B mass. To estimate apo-B mass, one must measure plasma VLDL apo-B concentration AND the initial space in which the labeled lipoprotein is distributed. The equation to estimate the PR is then: PR = FCR * apo-B mass = FCR * [VLDL apo-B] * V_I, where V_I is the initial volume in which the labeled VLDL apo-B distributes. What does one use for V_I? Most researches use an estimate of plasma volume V_P as the initial distribution space. However, if the study is designed correctly, one should be able to use the isotope dilution technique to estimate a true V_I (Shipley and Clark (7)). This requires frequent early samples and a knowledge of the amount of tracer introduced into the system at time zero. In our experience, V_I measured in this way is slightly larger than V_P.

For the set of data given in Figure 1, models 1, 2, and 3 predict different values for V_I because of different extrapolation to time 0 from which different estimates of PR follow. Using these three models and assuming a VLDL apo-B plasma concentration of 0.075 mg/ml, the predicted V_I and PR is:

Table 2. Estimated V_I and PR

	Estimate of V_I	Estimate of PR
Model 1	2789 ml	25.0 mg/hr
Model 2	2728 ml	28.8 mg/hr
Model 3	3150 ml	20.1 mg/hr

In this table, the estimated FCR for the tracee was correctly used in the estimation of the PR. If one used the estimated FCR of the tracer in this calculation, which would be incorrect, one would obtain yet another set of estimates for PR.

Finally, if one used plasma volume V_P for this calculation, then with V_P equal to 2577 ml, the estimated PR is 16.4 mg/hr which is 82% of the estimate for model 3 given in Table 2. It is important to note that the choice of the "volume" to use will affect not only the estimated PR of VLDL, but, as will be seen below, the percent of LDL apo-B arising from VLDL.

What might account for a discrepancy between plasma volume and the initial volume of distribution of labeled VLDL apo-B? A possible answer comes from how VLDL are actually metabolized. At any one point in time a fraction of VLDL is bound to the endothelium where the triglycerides are being lypolyzed. With the tracer and tracee following the same metabolic pathways in the same proportion, the same will be true for the tracer. Hence at any point in time, a certain fraction of tracer will not be available for plasma sampling. Since the kinetics of the interaction between plasma VLDL and the binding sites cannot be resolved from the turnover study, this will result in an initial volume of distribution that is slightly larger than plasma volume.

Summary from exogenously labeled VLDL apo-B turnover studies

In summary the following points can be made:

1. The FCR of VLDL apo-B depends upon the model chosen, and
2. The PR of VLDL apo-B depends both upon the FCR and the volume chosen.

Endogenous labeling of VLDL

VLDL apo-B can also be labeled endogenously be either giving a bolus, a constant infusion, or a primed, constant infusion of a labeled precursor. The labeled precursor is normally an amino acid which can be labeled either with a radioactive or stable isotope. With the increased accuracy of measuring isotopic enrichment, many investigators are resorting to stable isotopes. Many of these investigators are resorting to simpler methods to estimate kinetic parameters which essentially disregards the work done over the years using radioactive isotopes; it is tantamount to saying that the kinetics of VLDL apo-B depends upon the isotope and not the protein.

In what follows, we will discuss the primed constant infusion since this protocol is frequently used in stable isotope studies. A typical set of data obtainable from such a study is shown in Figure 3. The precursor curve is shown together with VLDL apo-B data collected during the 15 hour infusion, and then following the washout to 50 hours. As with Figure 1, no units are shown to underline the fact that the kinetic information comes only from the shape of the tracer concentration curve.

The same models shown in Figure 2 can be used to analyze these data; the results are summarized in Figure 4.

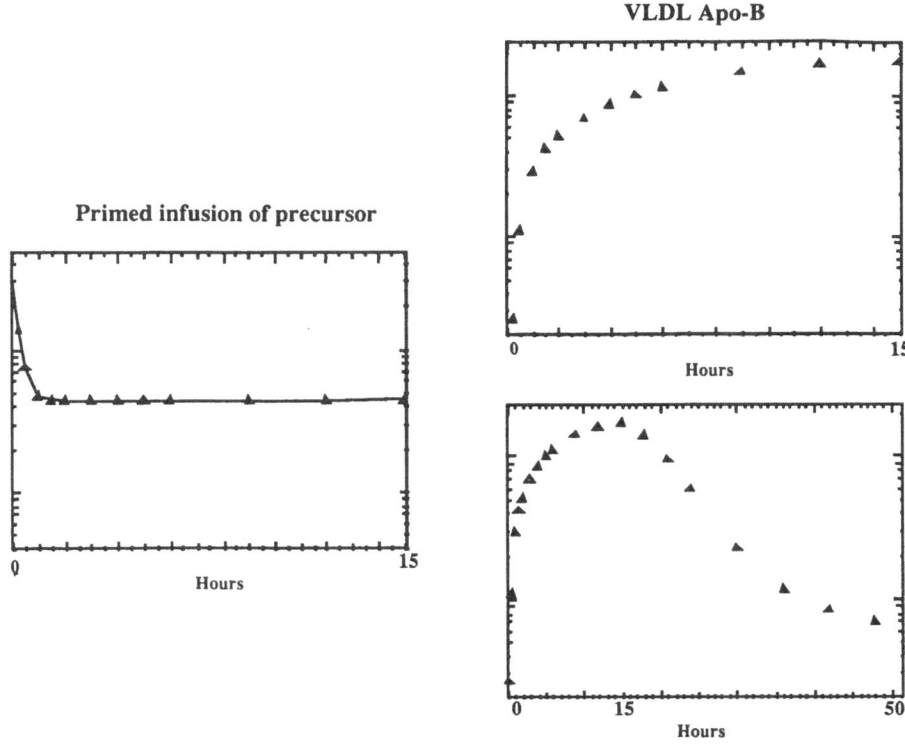

Figure 3. Endogenously labeled VLDL apo-B; see text for explanation.

It is of interest to note that all three models describe the data during the infusion period fairly well; it is only during the washout phase that models 1 and 2 are revealed to be inadequate.

How to the tracee FCR's estimated from the endogenous and exogenous labeling compare. The results are summarized below:

Table 3. Tracee FCR (hr^{-1})

	ENDOGENOUS	EXOGENOUS
Model 1	.056	.12
Model 2	.059	.14
Model 3	.085	.085

These FCR's will produce a different set of PR's, and will have the additional problem that, since V_I cannot be estimated from an endogenous labeling study, the researcher is forced to estimate a volume from which to calculate plasma VLDL apo-B mass.

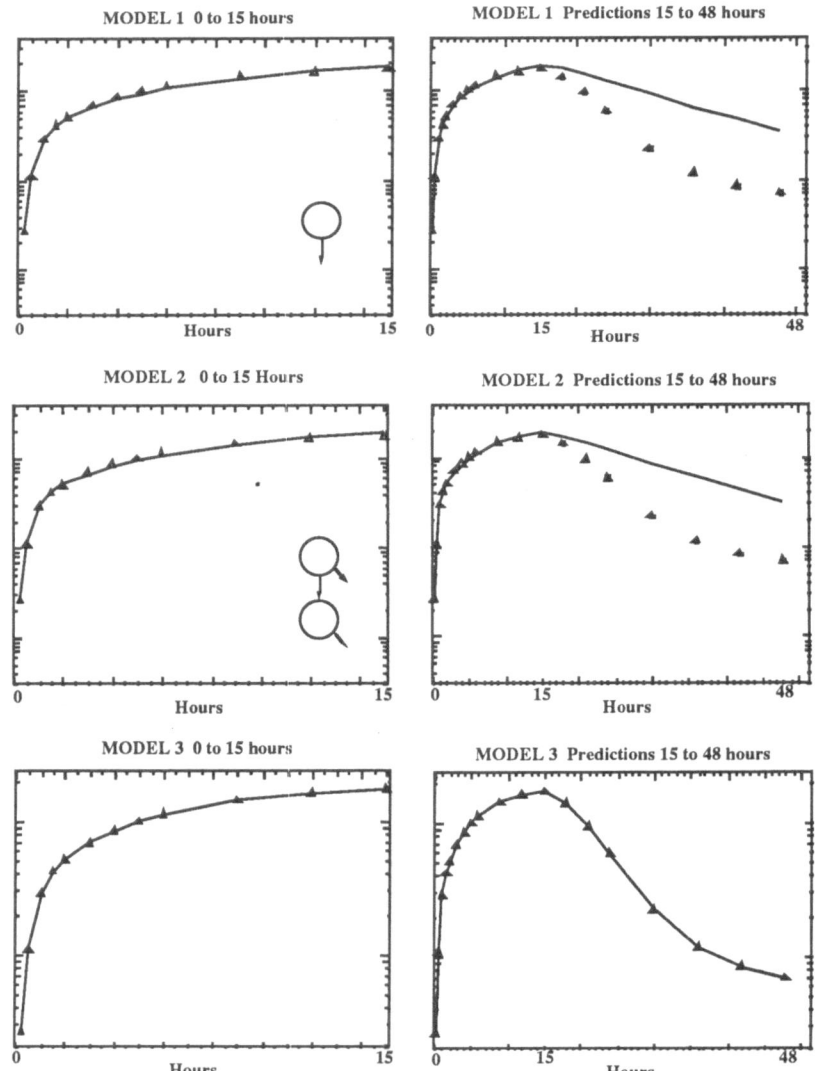

Figure 4. Prediction of models 1, 2 and 3 to the data shown in Figure 3, see text for explanation.

Summary from endogenously labeled VLDL apo-B turnover studies

In summary the following points can be made:

1. The FCR should not depend upon how the tracer is administered, and
2. It helps to follow the washout phase.

LDL TURNOVER STUDIES

We want to turn our attention for a few moments to LDL apo-B. Two questions of interest here are (i) what is LDL apo-B FCR and PR, and (ii) how much LDL apo-B is produced *de novo* versus arising from VLDL and/or IDL?

Figure 5 shows a typical LDL apo-B decay curve following a bolus injection of labeled LDL. Notice that the the duration of the study is 14 days as opposed to the 48 hours for the typical VLDL apo-B turnover study. Several models have been proposed to analyze LDL turnover data two of which are shown in Figure 6 (1,8).

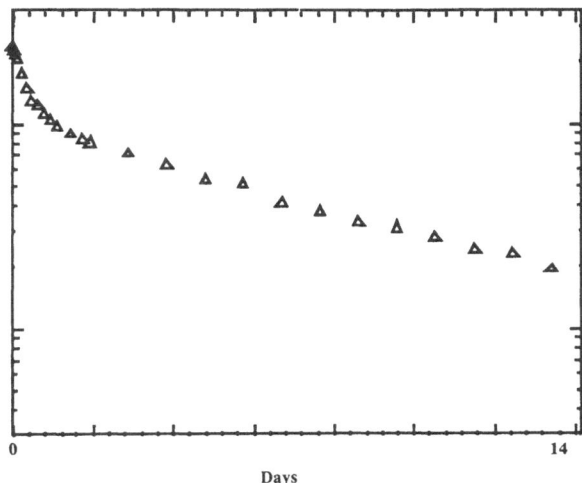

Figure 5. A typical LDL apo-B decay curve; see text for explanation.

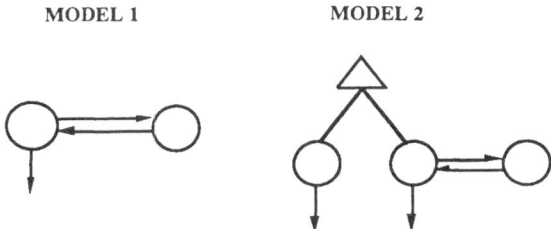

Figure 6. Two models used to analyze LDL apo-B decay; see text for explanation.

Model 1 is the so-called Matthews model (1); it is adequate for almost all plasma LDL apo-B decay curves. However, it assumes plasma LDL apo-B is kinetically homogeneous. This means the estimated FCR and PR will treat LDL apo-B as a single homogeneous collection of particles. Model 2, which was originally proposed by Goebel and Berman (1), provides for kinetic heterogeneity. However, this model or others involving heterogeneity cannot be identified from the plasma decay curve alone; it also requires urinary tracer data (8).

We do not wish to focus here on the heterogeneity issue beyond how it affects the interpretation of the FCR in perturbation studies. Figure 7 shows two simulated plasma LDL apo-B decay curves. Clearly the lower curve decays faster than the upper which means the lower one has a greater FCR. If this were, for example, a paired study to investigate the affect of a drug on the metabolism of LDL apo-B, an investigator would conclude the mechanism of action was due to an increased FCR. However, as shown in the model in the inset, these curves were generated by a model which accounts for plasma LDL heterogeneity. The only difference between the curves is the amount of label initially in compartments 1 and 2 which presumably reflects a change in the distribution of mass between the two subpopulations; none of the rate constants change. In particular, the FCR for both the tracer and tracee in pools 1 and 2 remain unchanged at 0.08/hr and 0.025/hr. For the LDL apo-B fraction as a whole however, the FCR of the tracer and tracee is increased. Although the affect of the drug is indeed increased clearance, is it through an anabolic rather than catabolic change.

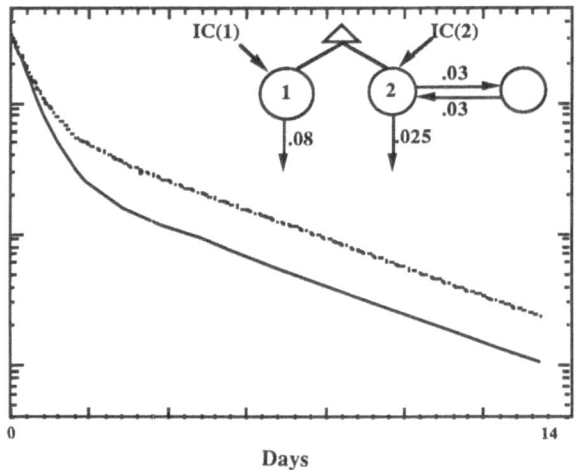

Figure 7. Simulated plasma LDL apo-B decay; see text for explanation.

Summary from LDL apo-B turnover studies

In summary the following points can be made:

1. Heterogeneity affects the interpretation of the FCR, and
2. It is highly recommended that urine samples be collected.

INTEGRATED APO-B MODELS

Finally, we want to turn out attention to apo-B metabolism through VLDL, IDL and LDL. Our focus will be on studies in which labeled VLDL or labeled VLDL and LDL are injected as a bolus; similar observations can be made when the label is introduced endogenously on precursors.

A simulated curve arising when exogenously labeled VLDL is injected and apo-B levels followed in VLDL, IDL and LDL is shown in Figure 8. Unlike the data shown previously, the units are shown on the ordinate. This is to underline the fact that it is the tracer movement through VLDL, IDL and LDL that is the basis for the kinetics. That is, one can plot these data in terms of specific activity or isotopic enrichment, for example, but when interpreting the data, one must be able to describe tracer movement since obviously it is not the specific activity or isotopic enrichment that is moving from VLDL to IDL and LDL.

208

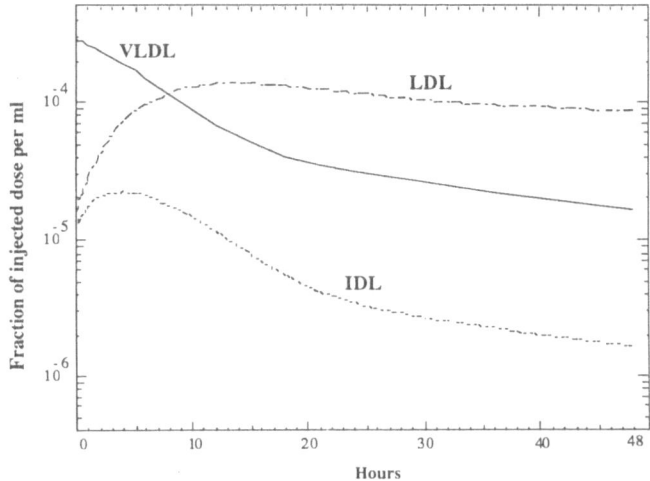

Figure 8. Simulated VLDL, IDL and LDL apo-B data following an "injection" of exogenously labeled VLDL apo-B; see text for additional explanation.

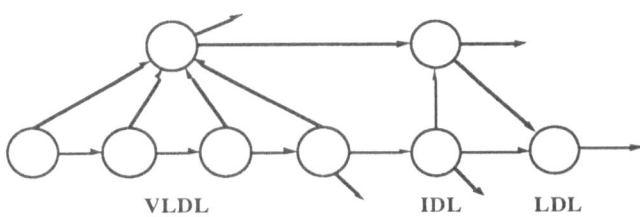

Figure 9. An integrated apo-B model; see text for explanation.

A model commonly used to interpret these data is shown in Figure 9 (2). It consists of the VLDL model 3 shown in Figure 2 together with a two compartment model for IDL and a one compartment model for LDL. This says that plasma LDL apo-B can be described by a single compartment which is in direct contrast with the models shown in Figure 6. The rate constant describing the loss from this compartment, the FCR, will be a combination of the true FCR and the exchange rate constant shown in model 2 of Figure 6; it will overestimate the FCR which will affect apo-B transfer estimates from VLDL. Why is there only a one compartment model describing LDL? One part of the answer is that the duration of the VLDL apo-B study is not long enough to accurately determine the kinetics of LDL apo-B. This can clearly be seen in Figure 10 where the VLDL data are plotted on a 14 day scale instead of the 50 hour scale; it shows that the characteristics of plasma LDL apo-B are simply not available.

The other part of the answer comes from the fact that when only labeled VLDL or precursor is injected, one does not know the structure of the LDL portion of the model; in short, it is a deconvolution problem (1,2). If one really wants to estimate apo-B transport through VLDL and LDL, they must coinject labeled VLDL and LDL. Such a study was recently conducted in baboons (9); the model used to analyze the data is shown below.

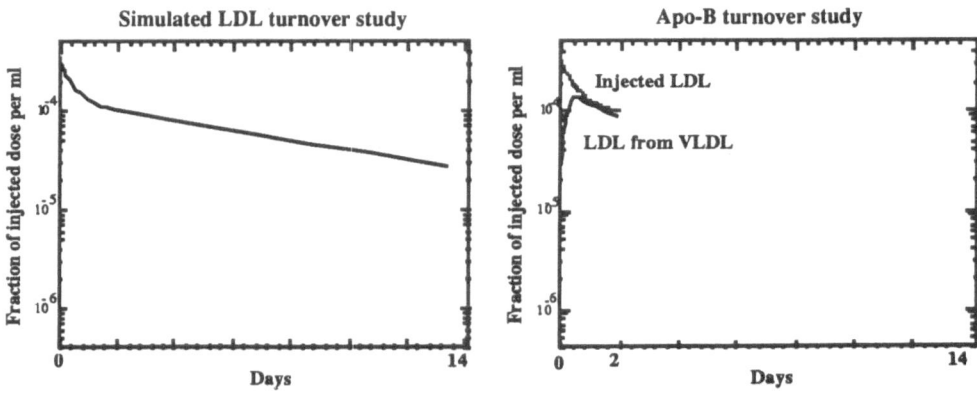

Figure 10. Simulated plasma LDL apo-B turnover, and simulated plasma VLDL and LDL apo-B turnover study; see text for explanation.

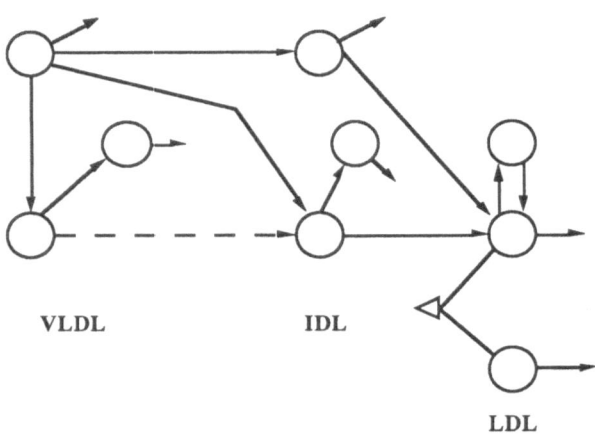

VLDL   IDL

LDL

Figure 11. Integrated apo-B model describing coinjected VLDL and LDL in baboons; see text for explanation.

Here, using plasma and urinary radioactivity following the LDL injection, an LDL model identical with model 2 in Figure 6 was derived to explain the data. To describe the VLDL, IDL and LDL data following the injection of VLDL, therefore, the LDL portion of the model was known. To describe these data, it was necessary only to convert radioactivity in to one plasma LDL compartment as shown in the Figure. The conclusion from this, which of course will require further testing, is that some plasma LDL particles arise from VLDL while others arise *de novo*. It is just this kind of information that is arising from more complex turnover protocols.

Summary from integrated apo-B turnover studies

In summary the following points can be made:

1. Apo-B transfer from VLDL to LDL cannot be estimated accurately using only labeled VLDL, and
2. It is highly recommended to coinject VLDL and LDL to obtain such estimates.

210

CONCLUSION

The purpose of the above discussion is to help the investigator considering using turnover studies to design the study to take maximal advantage of both experimental and modeling techniques available. Hopefully it will help in preventing mistakes and misinterpretations of tracer data.

The real point, however, is that the power of modeling comes more from the insights obtained during model development and testing than from using a model to estimate certain kinetic parameters.

BIBLIOGRAPHY

1.   M. Berman. Kinetic analysis of turnover data. Prog. Biochem. Pharmacol. 15:67 (1979).

2.   M. Berman, S. M. Grundy and B. V. Howard, eds. "Lipoprotein Kinetics and Modeling", Academic Press, New York (1982).

3.   C. A. Marzetta, D. M. Foster, and J. D. Brunzell. Relationships between LDL density and kinetic heterogeneity in subjects with normolipidemia and familial combined hyperlipidemia using density gradient ultracentrifugation. J. Lipid Res. 30:1307 (1989).

4.   E. Trezzi, C. Calvi, P. Roma, and A. L. Catapano. Subfractionation of human very low density lipoproteins by heparin-Sepharose affinity chromatography. J. Lipid Res. 24:790 (1983).

5.   D. R. Cryer, T. Matsushima, J. B. Marsh, M. Yudkoff, P. M. Coates, and J.A. Cortner. Direct measurement of apolipoprotein B synthesis in human very low density lipoprotein using stable isotopes and mass spectrometry. J. Lipid Res. 27:508 (1986).

6.   J. S. Cohn, D. A. Wagner, S. D. Cohn, J. S. Millar, and E. J. Schaefer. The measurement of very low density and low density lipoprotein apoB-100 and high density lipoprotein apoA-1 synthesis in human subjects using deuterated leucine: effect of fasting and feeding. J. Clin. Invest. 85:804 (1990).

7.   R. A. Shipley and C. E. Clark. "Tracer Methods for In Vivo Kinetics: Theory and Application", Academic Press, New York (1972).

8.   D. M. Foster, A. Chait, J.J .Albers, C. Harris and J.D. Brunzell. Evidence for kinetic heterogeneity among human low density lipoproteins. Metabolism. 35:685 (1986).

9.   R. S. Kushwaha, D. M. Foster, P. H. R. Barrett, and K. D. Carey. Effect of estrogen and progesterone on the metabolism of apoprotein B in baboons. Am. J. Physiol. 258: E172 (1990)

WHAT IS MEANT BY OVERPRODUCTION OF APO B-CONTAINING LIPOPROTEINS?

Scott M. Grundy and Gloria Lena Vega

University of Texas Southwestern Medical
Center at Dallas, Center for Human Nutrition
Departments of Clinical Nutrition
Biochemistry and Internal Medicine
5323 Harry Hines Boulevard
Dallas, Texas 75235-9052

Several mechanisms have been implicated in the development of hyperlipidemia. These include reduced activity of receptors for low density lipoproteins (LDL), deficiencies in lipoprotein lipase (LPL), abnormalities in the primary structures of apolipoproteins C-II, E, and B-100, and overproduction of lipoproteins containing apolipoprotein B-100 (apo B). All of these except the last are well documented. On the other hand, many researchers believe that overproduction of apo B-containing lipoproteins is one of the more common causes of hyperlipidemia. Therefore, it is worthwhile to consider the evidence for the existence of lipoprotein overproduction as a cause for hyperlipidemia. The present paper will examine the available data related to this question.

The term "overproduction" of apo B-containing lipoproteins implies to many investigators an excessive hepatic secretion of lipoproteins containing apo B. This further suggests to many an oversynthesis of apo B by the liver. Thus, if the liver were to oversynthesize apo B, this could lead to increased formation of lipoprotein particles containing apo B, which could be followed by an increased secretion of such particles into the circulation. Since it is widely accepted that some forms of hyperlipidemia are the result of dysregulation of hepatic apo B synthesis, we might review the data that is responsible for this concept.

The idea of hepatic oversynthesis of apo B was derived primarily from isotope kinetic studies in humans, i.e., studies indicating that in certain patients high flux rates of apo B-containing lipoproteins are associated with dyslipidemia. Indeed, high "production" rates for apo B have been reported in very low density lipoproteins (VLDL), intermediate density lipoproteins (IDL), and LDL. These tracer kinetics have been analyzed by complex multicompartmental models to obtain fractional catabolic rates (FCRs) for various lipoprotein fractions. The "production" rate then is calculated by multiplying the FCR by the plasma pool size of apo B in a particular lipoprotein fraction. In strict kinetic

Hypercholesterolemia, Hypocholesterolemia, Hypertriglyceridemia
Edited by C.L. Malmendier *et al.*, Plenum Press, New York, 1990

terminology, the preferred term for "production" rate is "transport" rate, which is a measure of the flux of apo B-containing lipoproteins through a given compartment.

Whether the absolute value for transport rate obtained by tracer kinetics is accurate have been a subject of considerable uncertainty. This question will be examined subsequently. A common practice is to equate estimated transport rates with rates of input of lipoproteins into a given compartment. Another common assumption is that the liver secretes VLDL particles directly into the circulation, and hence the finding of a high transport rate for VLDL apo B should reflect oversecretion of VLDL apo B by the liver. Although this is one possible explanation for a high transport rate for VLDL apo B, we must acknowledge that an elevated transport rate is not necessarily synonymous with increased hepatic secretion of VLDL apo B. For purposes of present discussion, three terms must be distinguished: (a) hepatic synthetic rate for apo B, (b) hepatic secretion rate for apo B-containing lipoproteins, and (c) plasma transport rate for a lipoprotein fractions (e.g. VLDL apo B or LDL apo B). An alternate term for "transport" rate is "flux" rate, whereas "input" rate or "production" rate should be used only if their meaning is clearly defined. Several questions related to these processes can be considered.

Which lipoproteins are secreted directly into plasma?

As indicated before, several studies indicate that some forms of hyperlipidemia are associated with increased transport rates for VLDL apo B, and this finding has been interpreted by some workers to mean that the liver can oversecrete VLDL apo B. This interpretation is based on the assumption that there are no metabolically active "compartments" between the liver and the circulating VLDL compartment. If not, that which appears in VLDL should be identical to that which is secreted by the liver. By carrying this reasoning one step further, it might be expected that the rate of flux of VLDL apo B approximates hepatic synthesis of apo B. If these relationships hold, VLDL particles isolated from plasma should correspond closely to "nascent" VLDL. In other words, the liver would secrete lipoprotein particles having the metabolic behavior of circulating VLDL. This idea will be challenged subsequently in this review.

When isotope kinetic studies examine the metabolism of VLDL apo B and LDL apo B simultaneously, the levels of LDL apo B are not explained by the input from VLDL apo B. This finding fostered the concept that the liver can secrete LDL particles directly into the circulation. In other words the notion has been put forth that the liver is able to synthesize LDL particles as well as VLDL particles. But does the liver really synthesize LDL particles and secrete them directly into the circulation, as suggested by isotope kinetic studies? Since at present no way exists to answer this question in humans we must turn to research in laboratory animals. Most studies in animal models, however, do not support the idea that the liver can directly secrete LDL-like particles. On the other hand, in cholesterol-fed animals, cholesterol-rich lipoproteins of LDL size have been reported to be secreted by the liver. Thus, the possibility cannot be completely excluded that direct input of LDL occurs in humans; but before this conclusion can be drawn, several difficult questions remain to be answered. For example, the cholesterol esters of LDL appear to be derived mainly from the activity of lecithin-cholesterol acyl transferase (LCAT), and it is doubtful that appreciable amounts of cholesterol esters are secreted with lipoproteins by the liver. In patients with LCAT deficiency, for example, LDL particles are not enriched in cholesterol esters as are normal LDL; if a major fraction of LDL are produced directly by the liver, one would expect to find many normal LDL particles in patients with LCAT deficiency, which is not the case. Thus, in our view, the possibility of direct secretion of "mature" LDL in humans should not be accepted on the basis of isotope kinetic studies. Corroborative evidence of other types is required before such a mechanism can be accepted. Although it is conceivable that the liver secretes LDL-sized nascent lipoproteins, they probably do not represent "mature" LDL. Consequently, circulating VLDL and LDL almost certainly represent lipoproteins that have been modified from nascent lipoproteins of a different composition.

What are states of high apo B flux?

Several conditions have been identified in which the flux of apo B-containing lipoproteins through one lipoprotein compartment or another is high. For instance, high flux rates for lipoprotein apo B can occur in either VLDL or LDL, or both. The causes of abnormal lipoprotein metabolism in these disorders can be either primary (familial) or secondary to other metabolic disorders. The lipoprotein profiles may be similar in these various conditions, but they are not necessarily identical; the various lipoproteins may differ depending on underlying metabolic defect(s). These conditions can be described briefly.

Familial combined hyperlipidemia (FCHL). This disorder is characterized by multiple lipoprotein phenotypes in a single family (1,2). Some family members have high triglycerides, others have high cholesterol levels, and still others have mixed hyperlipidemia. The disorder was originally proposed to be a monogenic disorder having variable phenotypic expression. However, the molecular defect responsible for FCHL has never been identified, and the possibility must be considered that FCHL often is a polygenic disorder. Still, several investigators have proposed that the underlying defect of FCHL is oversynthesis (or oversecretion) of apo B-containing lipoproteins by the liver (3-7). This hypothesis is based on the finding of enhanced flux of VLDL apo B and LDL apo B in patients classified as having FCHL, a result obtained from isotope kinetic studies.

Familial hypertriglyceridemia (FHTG). This condition represents the inheritance of elevated VLDL triglycerides as an isolated defect. Chait et al (3) reported that FHTG is due to overproduction of VLDL triglycerides without an increased production of VLDL apo B. Others (8) have reported that some forms of primary hypertriglyceridemia are due to defective lipolysis of triglyceride-rich lipoproteins. With neither of these abnormalities (3,8) is the flux of VLDL apo B necessarily increased. On the other hand, Kissebah et al (9) reported an increased flux of both VLDL triglycerides and VLDL apo B in patients with FHTG. The patients thus resembled hypertriglyceridemic individuals with FCHL, except that multiple lipoprotein phenotypes were not present in the families as a whole. Sniderman et al (10) have indicated that some patients with pure hypertriglyceridemia have elevated total concentrations of apo B, and these patients probably are among those having increased flux rates for VLDL apo B and LDL apo B (11).

Normolipidemic states. There is a subgroup of patients having high apo B-flux rates who have normal levels of total lipids. For example, Kesaniemi, Beltz and Grundy (12,13) reported that many patients with premature CHD have this pattern of lipoprotein kinetics. Increased rates of apo B flux were found in either VLDL or LDL, or both. Some of these individuals had increased plasma levels of total apo B, but this was not invariably the case. These particular patients cannot be said to have FCHL because they do not have hyperlipidemia, although other family members were found to have hyperlipidemia.

Intermittent hypertriglyceridemia. Other studies in our laboratory (14) have shown that normolipidemic patients with high flux rates for LDL-apo B often have intermittent hypertriglyceridemia. This observation suggests that a high rate of flux for LDL apo B may be secondary to abnormalities in the metabolism of plasma triglycerides. It is thus possible that even patients with normolipidemia (12,13) who have a high flux of apo B-containing lipoproteins might show intermittent hypertriglyceridemia if they were tested frequently.

Obesity. Several isotope kinetic studies (13,15,16) have shown that obese subjects often manifest high flux rates for VLDL apo B and LDL apo B. Elevated transport rates for VLDL apo B usually are accompanied by increased transport rates for VLDL triglycerides. The size and composition of VLDL particles is not abnormal in many obese subjects (13,15); this suggests that they secrete an increased number of normal VLDL particles. Although increased flux of LDL apo B occurs frequently in obese individuals, an elevation of LDL cholesterol is not always present. There are two reasons for this: (a) LDL particles may be somewhat depleted of cholesterol, and thus LDL-cholesterol levels are not necessarily raised when LDL-apo B levels are; and (b) removal rates of LDL may be normal or even increased.

Noninsulin dependent diabetes mellitus (NIDDM). High rates of flux of VLDL apo B and LDL apo B have been reported for patients with NIDDM (17,18). This response may be related in part to concomitant obesity since the majority of NIDDM patients are overweight. Howard et al (19) however reported that NIDDM does not increase production rates for of VLDL apo B, but it raises only the triglyceride content of newly secreted VLDL particles. On the other hand, Kissebah et al (18) found that NIDDM patients often have very high flux rates for LDL apo B. This latter finding at least raises the possibility that NIDDM induces oversynthesis of apo B-containing lipoproteins independently of obesity.

Nephrotic syndrome. Hyperlipidemia is a common feature of the nephrotic syndrome. The most common pattern of nephrotic hyperlipidemia is an elevated LDL cholesterol, although some patients show increases of VLDL triglycerides as well. Mechanisms for nephrotic hyperlipidemia have not been studied extensively in humans, but they have been examined in detail in laboratory animals. The available evidence indicates that loss of protein in the urine stimulates the synthesis of proteins in the liver, one group of which include apo B-containing lipoproteins. Liver perfusion studies in animals (20,21) reveal that livers from nephrotic animals have enhanced secretion of apo B-containing lipoproteins. Thus, nephrotic syndrome apparently represents one high-flux state in which there is a true oversecretion of lipoproteins by the liver.

What can be learned about hepatic apo B synthesis from VLDL-apo B turnover studies?

If the major apo B-containing lipoprotein secreted into plasma is VLDL, it might be expected that kinetic studies of VLDL apo B would reflect hepatic secretion rates of VLDL apo B. However, there are significant limitations to kinetic analysis of isotopic data for VLDL apo B, and hence to estimation of secretion rate. For example, multicompartmental models for apo B obviously represent an oversimplification of the true physiological state. There must be multiple sites of input and output along the delipidation chain for VLDL, and this adds a complexity to the VLDL - apo B model that is difficult to manage experimentally. Several subspecies of VLDL apo B having different metabolic fates undoubtedly exist, and it is impossible to trace all of these species. A recent study by Beltz et al (22) indicated that labeling VLDL proteins with an endogenous tracer (^{3}H-leucine) gives higher transport rates for VLDL apo B than does exogenous labeling of VLDL with radioiodine; still reasons presented in their paper (22) explain why even endogenous labeling may underestimate true hepatic secretion rates for VLDL apo B. Finally, a recent report by Ginsberg et al (23) revealed that radioiodine labeling of VLDL apo B does not label all subspecies of apo B identically, and this discrepancy further complicates kinetic analysis. Thus, there are serious problems with estimating transport rates for VLDL apo B based on kinetic analysis.

Perhaps an even greater problem for equating VLDL apo B flux rates with hepatic secretion rates for VLDL apo B is that newly secreted VLDL may be removed rapidly from the VLDL compartment, too rapidly to be traced accurately. There is growing evidence that not all newly secreted VLDL particles enter the relatively slow delipidation chain for VLDL, but instead some particles may be from the circulation within minutes (24-26). Particles of the latter type thus do not linger in the circulation in appreciable amounts for isolation and labeling, and because of their rapid decay, it is difficult to differentiate between physiological decay and artifact; this is to say, an appreciable fraction of newly secreted VLDL may have very-rapid turnover rates, and conventional isotope kinetics will be "blind" to their presence (27). The use of conventional kinetics therefore probably underestimates true hepatic secretion rates for VLDL apo B. The precise mechanisms for very rapid reuptake of newly secreted lipoproteins is unknown, but these lipoproteins may be removed by the putative "chylomicron remnant" receptor. For example, chylomicron remnants are cleared by the liver in a matter of minutes, not in hours typical of circulating VLDL.

What can be learned about apo B synthesis from LDL-apo B turnover studies?

Two patterns of high flux of LDL apo B have been reported. In one, a high input of LDL apo B is accompanied by a reduced fractional catabolic rate (FCR) for LDL (28). In the other, the high flux rate for LDL is associated with increased FCRs for LDL (29,30). Which of these patterns might be expected from hepatic oversynthesis of apo B-containing lipoproteins? In our view, the former is more likely for the following reason. If the total number of lipoprotein particles entering plasma is abnormally high, this should cause "overloading" of the LDL receptor pathway with newly secreted lipoproteins and their products, namely, VLDL, VLDL remnants, and LDL. Larger, triglyceride-rich lipoproteins have a greater affinity for LDL receptors than LDL itself because of the presence of apo E on the former. Hence they should have preference for clearance via LDL receptors. This is because particles having both apo E and apo B have greater affinity for LDL receptors than those with only apo B. In addition, if the input of apo B-containing particles is increased, more particles should be converted to LDL. Consequently, FCRs for LDL should be relatively low, and certainly not increased. This phenomenon of "saturation" of LDL receptors secondary to overproduction of LDL has been discussed in detail by Spady et al (31). On the other hand, the pattern of increased influx of LDL apo B and high FCRs for LDL seemingly is not compatible with true oversecretion of apo B-containing lipoproteins. Oversecretion of lipoproteins should produce low FCRs for LDL for the reasons described above. Instead, a different mechanism must account for <u>both</u> high input rates and high FCRs for LDL apo B. In our view, the most likely explanation for this high-flux pattern is a defect in direct removal of triglyceride-rich lipoproteins so that conversion of VLDL to LDL is enhanced. This mechanism would account for the high influx of LDL, and since fewer VLDL particles are removed by LDL receptors, more receptors should be available for uptake of LDL; this in turn should raise FCRs for LDL. Thus, the high-input, high-FCR pattern of LDL turnover apparently can be explained by a decrease in direct removal of VLDL particles, and not by hepatic oversecretion of apo B-containing lipoproteins (or oversynthesis of apo B).

<u>What is the mechanism for elevated plasma apo B concentrations in familial combined hyperlipidemia (FCHL)?</u>

Some investigators contend that patients with FCHL have oversynthesis of apo B causing increased hepatic secretion of apo B-containing lipoproteins (7). This mechanism could account for elevated levels of VLDL and LDL, and hence total apo B, in FCHL patients. Before this hypothesis can be accepted, however, other mechanisms must be explored. In fact, it may be helpful to differentiate between elevated concentrations of VLDL apo B and LDL apo B, both of which can occur in FCHL patients. It may be a mistake to assume that a common mechanism underlies high levels of both VLDL and LDL.

For instance, a defect in catabolism of VLDL apo B, such as a lipolytic defect, could produce the kinetic picture of "overproduction" of VLDL apo B. If the fraction of newly secreted lipoproteins that normally is removed rapidly from the circulation is decreased because of a lipolytic defect for VLDL triglycerides, more of these particles should enter the dilipidation cascade for VLDL, the compartment that is traced in isotope kinetic studies. Consequently, a high conversion of newly secreted lipoproteins to VLDL of the measured cascade would give the appearance of oversecretion of VLDL apo B, i.e., a lipolytic defect for VLDL triglycerides could masquerade as an oversecretion defect.

218

Likewise, a decrease in rapid removal of recently secreted lipoproteins should make more VLDL available for conversion to LDL. For reasons discussed before, however, an increased formation of LDL by this mechanism should be accompanied by increased availability of LDL receptors, and hence, high FCRs for LDL. But how can this mechanism account for elevated LDL-apo B concentrations? If there are more receptors available for LDL removal, should they not normalize LDL apo B concentrations? The answer probably is "yes" and may explain why some patients with FCHL do not have elevated LDL levels. If these patients should have a concomitant reduction in LDL receptors, however, they might develop hypercholesterolemia; a reduction in LDL receptors theoretically might have several causes, i.e., decline in LDL receptor activity with aging, excessive intakes of saturated fatty acids and cholesterol, inherited metabolic suppression of LDL receptor activity, or inherited defects in the gene encoding for LDL receptors. But in addition, since LDL particles have a lower affinity for LDL receptors than VLDL remnants, an increased conversion of VLDL to LDL itself should produce a rise in concentrations of apo B-containing lipoproteins even without a reduction in activity of LDL receptors. Thus, the increase in LDL apo B (and LDL cholesterol) concentrations of patients with FCHL may be multifactorial, and it need not be explained exclusively by enhanced secretion of lipoproteins.

The above considerations evoke new questions about the etiology of FCHL. First, they raise the possibility that FCHL may not be the result from hepatic oversynthesis of apo B or oversecretion of apo B-containing lipoproteins. And second, they suggest that this disorder may be multifactorial rather than monogenic in origin. Although we cannot exclude the possibility that a monogenic defect in lipoprotein metabolism could produce multiple-phenotype hyperlipoproteinemia, other explanations are possible. For example, if at least 10% of the general population has elevated VLDL triglyceride levels and another 10% has hypercholesterolemia, it might be expected that these two abnormalities would coexist and produce a mixed pattern of hyperlipidemia in about 1% of the population, which is the reported frequency of FCHL. Coinheritance of two forms of hyperlipidemia would suport a polygenic origin to FCHL.

What are mechanisms for other "high apo B-flux" states?

Several of the other dyslipidemic conditions listed before that have a high flux of apo B-containing lipoproteins by kinetic analysis likewise have increased FCRs for LDL apo B. These include persistent hypertriglyceridemia, transient hypertriglyceridemia, and normolipidemia. If our prior reasoning holds, these disorders should manifest (a) defective uptake of recently secreted lipoproteins, (b) increased conversion to VLDL and LDL, and (c) enhanced availability of LDL receptors, the latter being responsible for high FCRs for LDL. In none of these conditions is it necessary to postulate that hepatic synthesis and secretion of apo B-containing lipoproteins is increased.

On the other hand, what about the reported overproduction of lipoproteins in the obese state? Is the number of lipoprotein particles secreted by the liver increased in obese individuals? Three possibilities must be considered. First, obesity is known to enhance the synthesis of hepatic lipids (cholesterol and triglycerides), and synthesis of apo B could be increased in parallel. If so, the total number of apo B-containing lipoproteins entering the plasma could be increased. Second, the synthesis of apo B might not be increased, and rather, the triglyceride content of each lipoprotein particle is enhanced. This latter mechanism could explain an increased secretion of VLDL triglycerides without requiring an increased output of VLDL apo B. To date, neither of these mechanisms has been proven with certainty. Finally, a third possibility is worthy of consideration; the synthesis of apo B might not be increased in obese subjects, but an increased number of apo B molecules could be recruited to form lipoproteins. Limited evidence suggests that not all of newly synthesized apo B is utilized for formation of lipoprotein particles, and if not, the fraction utilized could be increased in obese individuals. These various mechanisms might be differentiated in animal models of obesity, but it is doubtful that they can be distinguished in humans with available methods. The limited number of isotope kinetic studies carried out in obese subjects do not reveal whether LDL kinetics are of the high-flux, high-FCR variety, a pattern that would speak against true oversecretion of lipoprotein particles.

Finally, is there a true oversynthesis of apo B in the nephrotic syndrome? Several studies in laboratory animals suggest that hepatic secretion rates of VLDL apo B are enhanced in nephrotic state, but it still remains to be proven that hepatic synthesis of apo B is increased. This question might be examined in animal models of the nephrotic syndrome; for example, in these animals, messenger RNA for apo B might be increased in the liver, which would strongly suggest an oversynthesis of apo B.

Conclusions

The considerations in this review raise the question of what is meant by the term "overproduction" of apo B-containing lipoproteins. There is little question that high flux rates for VLDL apo B and LDL apo B are present in several related forms of hyperlipidemia. It is now evident however that the term "production" as it is used to describe lipoprotein kinetic studies has been used to imply three different processes: (a) hepatic synthesis of apo B, (b) hepatic secretin of apo B-containing lipoproteins, and (c) flux of lipoproteins in different plasma compartments (i.e., VLDL, IDL, and LDL). Rates of plasma lipoprotein transport determined by isotope kinetic techniques are not necessarily identical to hepatic synthesis of apo B or hepatic secretion of apo B-containing lipoproteins. It is doubtful that human studies can be designed in the near future that will provide accurate measurements of these three parameters simultaneously. However, it may be possible through use of animal models of the "overproduction" state to determine the various mechanisms responsible for high rates of plasma lipoprotein flux.

REFERENCES

1. J. L. Goldstein, W. R. Hazzard, H. G. Schratt, E. L. Bierman, and A. G. Motulsky, Hyperlipidemia in coronary heart disease: I. Lipid levels in 500 survivors of myocardial infarction, J Clin Invest, 52:1533 (1973).

2. J. L. Goldstein, H. G. Schrott, W. R. Hazzard, E. L. Bierman, and A. G. Motulsky, Hyperlipidemia in coronary heart disease: II. Genetic analysis of lipid levels in 176 families and delineation of a new inherited disorder, combined hyperlipidemia, J Clin Invest, 52:1544 (1973).
3. A. Chait, J. J. Albers, and J. D. Brunzell, Very low density lipoprotein overproduction in genetic forms of hypertriglyceridaemia, Eur J Clin Invest, 10:17 (1980).
4. E. D. Janus, A. Nicoll, R. Wooton, P. R. Turner, P. J. Magill, and B. Lewis, Quantitative sutdies of very low density lipoprotein in normal controls and primary hyperlipidaemic states, Eur J Clin Invest, 10:149 (1980).
5. E. D. Janus, A. M. Nicoll, P. R. Turner, P. Magill, and B. Lewis, Kinetic bases of the primary hyperlipidaemias: Studies of apolipoprotein B turnover in genetically-defined subjects, Eur J Clin Invest, 10:161 (1980).
6. A. H. Kissebah, S. Alfarsi, and D. J. Evans, Low density lipoprotein metabolism in familial combined hyperlipidemia: mechanism of the multiple lipoprotein phenotype expression, Arteriosclerosis 24:199 (1976).
7. S. M. Grundy, A. Chait and J. D. Brunzell, Familial combined hyperlipidemia workshop, Arteriosclerosis 7:203 (1987).
8. F. L. Dunn, S. M. Grundy, D. W. Bilheimer, R. J. Havel and P. Raskin, Impaired catabolism of very low-density lipoprotein-triglyceride in a family with primary hypertriglyceridemia. Metabolism 34:316 (1985).
9. A. H. Kissebah, S. Alfarsi, and P. W. Adams, Integrated regulation of very low density lipoprotein triglyceride and apolipoprotein-B kinetics in man: normolipemic subjects, familial hypertriglyceridemia, and familial combined hyperlipidemia, Metabolism 30:856 (1981).
10. A. C. Sniderman, C. Wolfson, B. Teng, F. A. Franklin, P. S. Bachorik, and P. O. Kwiterovitch Jr, Association of hyperapolipoproteinemia with endogenous hypertriglyceridemia and atherosclerosis, Ann Intern Med 97:833 (1982).
11. B. Teng, A. D. Sniderman, A. K. Soutar, and G. R. Thompson, Metabolic basis of hyperapobeta lipoproteinemia: Turnover of apolipoprotein B in low density lipoprotein and its precursors and subfractions compared with normal and familial hypercholesterolemia, J Clin Invest, 77:663 (1986).
12. Y. A. Kesaniemi, and S. M. Grundy, Overproduction of low density lipoproteins associated with coronary heart disease, Arteriosclerosis, 3:40 (1983).
13. Y. A. Kesaniemi, W. F. Beltz, and S. M. Grundy, Comparisons of metabolism of apolipoprotein B in normal subjects, obese patients, and patients with coronary heart disease, J Clin Invest, 76:586 (1985).
14. G. L. Vega, W. Beltz, and S. M. Grundy, Low density lipoprotein metabolism in hypertriglyceridemic and normolipidemic patients with coronary heart disease, J Lipid Res, 26:115 (1985)
15. G. Egusa, W. F. Beltz, S. M. Grundy, and B. V. Howard, The influence of obesity on the metabolism of apolipoprotein B in man, J Clin Invest, 76:596 (1985).
16. Y. A. Kesaniemi, and S. M. Grundy, Increased low density lipoprotein production associated with obesity, Arteriosclerosis, 3:170 (1983).
17. A. H. Kissebah, S. Alfarsi, D. J. Evans, and P. W. Adams, Intergrated regulation of very low density lipoprotein triglyceride and apolipoprotein-B kinetics in non-insulin-dependent diabetes mellitus, Diabetes 31:217 (1982).

18. A. H. Kissebah, S. Alfarsi, D. J. Evans, and P. W. Adams, Plasma low density lipoprotein kinetics in non-insulin-dependent diabetes mellitus, _J Clin Invest_ 71:655 (1983).
19. B. V. Howard, W. G. H. Abbott, W. F. Beltz et al, Integrated study of low density lipoprotein metabolism and very low density lipoprotein metabolism in non-insulin-dependent diabetes, _Metabolism_ 36:877 (1987).
20. J. B. Marsh, and D. L. Drabkin, Experimental reconstruction of metabolic pattern of lipid nephrosis: Key role of hepatic protein synthesis in hyperlipemia, _Metabolism_, 9:946 (1960).
21. J. B. Marsh, Lipoprotein metabolism in experimental nephrosis, _J Lipid Res_, 25:1619 (1984).
22. W. F. Beltz, Y. A. Kesaniemi, N. H. Miller, W. R. Fisher, S. M. Grundy, and L. A. Zech, Studies on the metabolism of apolipoprotein B in hypertriglyceridemic subjects using simultaneous administration of tritiated leucine and radioiodinated very low density lipoprotein. _J Lipid Res,_ 31:361 (1990).
23. R. Ramakrishnan, Y. Arad, S. Wong, H. N. Ginsberg, Nonuniform radiolabeling of VLDL apolipoprotein B: implications for the analysis of studies of the kinetics of the metabolism of lipoproteins containing apolipoprotein B, _J Lipid Res_, 31:1031 (1990).
24. A. F. H. Stalenholf, M. J. Malloy, J. P. Kane, and R. J. Havel, Metabolism of apolipoprotein B-48 and B-100 of triglyceride-rich lipoproteins in normal and lipoprotein lipase-deficient humans, _Proc Natl Acad Sci USA_, 81:1839 (1984).
25. C. J. Packard, A. Munro, A. R. Lorimer, A. M. Gotto, and J. Shepherd, Metabolism of apolipoprotein B in large triglyceride-rich very low density lipoproteins of normal and hypertriglyceridemic subjects, _J Clin Invest_, 74:2178 (1984).
26. C. A. Marsetta, F. L. Johnson, L. A. Zeck, D. M. Foster, and L. L. Rudel, Metabolic behavior of hepatic VLDL and plasma LDL apo B100 in African green monkeys, _J Lipid Res_, 30:357 (1989).
27. W. F. Beltz, Y. A. Kesaniemi, B. V. Howard, and S. M. Grundy, Comparisons of metabolism of apolipoprotein B in normal subjects, obese patients, and patients with coronary heart disease, _J Clin Invest_, 76:586 (1985).
28. D. W. Bilheimer, N. J. Stone, and S. M. Grundy, Metabolic studies in familial hypercholesterolemia: Evidence for a gene-dosage effect in vivo, _J Clin Invest_, 64:524 (1979).
29. G. L. Vega and S. M. Grundy, Kinetic heterogeneity of low density lipoproteins in primary hypertriglyceridemia, _Arteriosclerosis_ 6:395 (1986).
30. G. L. Vega and S. M. Grundy, Studies on mechanisms for enhanced clearance of low density lipoproteins in patients with primary hypertriglyceridemia, _J Intern Med_, 226:5 (1989).
31. D. K. Spady, D. W. Bilheimer, and J. M. Dietschy, Rates of receptor-dependent and independent low density lipoprotein uptake in the hamster, _Proc Natl Acad Sci USA_, 80:3499 (1983).

KINETICS OF HETEROGENEOUS POPULATIONS OF PARTICLES

C. L. Malmendier, J-F. Lontie, C. Delcroix, and D. Y. Dubois

Fondation de Recherche sur l'Athérosclérose, Brussels
Belgium

Kinetics is a dynamic approach of metabolism.

After intravenous injection of labeled substances, it is possible by mathematical analysis of the plasma radioactivity decay curves and of the urinary excretion data, to build models, from the simplest to the most complex one, that allow to quantify metabolic parameters.

Most often kinetic studies have been undertaken with so-called "simple" lipoproteins because they were supposed at this time to consist of only one apoprotein (apo B for LDL) or mostly one apoprotein (apo A-I for HDL). The model most commonly used consists of a plasma compartment exchanging with one or two extraplasmatic compartments and only one entry and one output. This simplest model generally referred to as Matthews model (1) assuming an homogeneous pool can be used when only limited information is desired. The theory of compartimental modeling has been considerably developed by Berman and the availability of the most powerful computer programs SAAM and CONSAM allows the full description of a metabolic system (2-4). The development of a model permits to calculate kinetic parameters such as fractional catabolic rate, residence time, production rate.

The introduction of a new concept by P. Alaupovic (5) invited us to reconsider the problem of lipoprotein kinetics. The existence of discrete lipoprotein particles with definite apoprotein composition in each class of lipoproteins separated by ultracentrifugation according to their density, was based on the separation by immunological methods: affinity chromatography or immunoprecipitation. Does this heterogeneity correspond to a metabolic heterogeneity?

As we know that apoproteins are often associated in the lipoproteins or lipoprotein particles, it seemed interesting to compare the metabolism of 2 or more apoproteins simultaneously **in the same conditions and in the same subject**. Until now the use of two isotopes of iodine (131 and 125) permitted to follow simultaneously 2 apoproteins.

We will describe the different steps which were followed towards our goal: the understanding of in vivo metabolism of lipoprotein particles and its consequences in pathology.

Hypercholesterolemia, Hypocholesterolemia, Hypertriglyceridemia
Edited by C.L. Malmendier *et al.*, Plenum Press, New York, 1990

SUCCESSIVE STEPS TOWARDS LIPOPROTEIN PARTICLE'S METABOLISM

1. Differences in plasma curves of two apoproteins

When we compared 2 apoproteins such as apo A-I and C-I, A-I and SAA, or apo C-II and C-III, we observed different plasma radioactivity decay curves (Fig. 1).

Using the Matthews model independently for apo A-I and C-I (6) we just compared "globally" the classical parameters (FCR, residence time, and production rate) of one apoprotein to the other. The very different turnover rates reflect the differences in curves (much faster decay for apo C-I).

2. Parallelism between the slowest components of the curves of pairs of apoproteins

When the plasma decay curves of apo C-II and C-III for example, were compared - what seems logical as these apoproteins have been earlier said to have exactly the same metabolism (7) - it appears obvious that at least the last parts of the curves (from about day 7 to day 15 after injection) were parallel. When the use of this pair of apoproteins shows that the slowest exponentials were parallel, we consider reasonable to constraint the slopes of this component to be identical and thus to impose these two apoproteins to have at least part of their metabolism in common.

3. Evolution of U/P ratio with time

The urinary/plasma (U/P) ratio was always used as an indirect measurement of the FCR (8). Most authors assumed FCR calculated from plasma curve and U/P ratio to be the same or very close for at least the last 10 days of a 15 day kinetic experiment.

When it appears that the curve at least during the first 4 days was not constant with time, this observation was generally interpreted as the presence of a contaminant or denatured protein. To overcome this criticism,

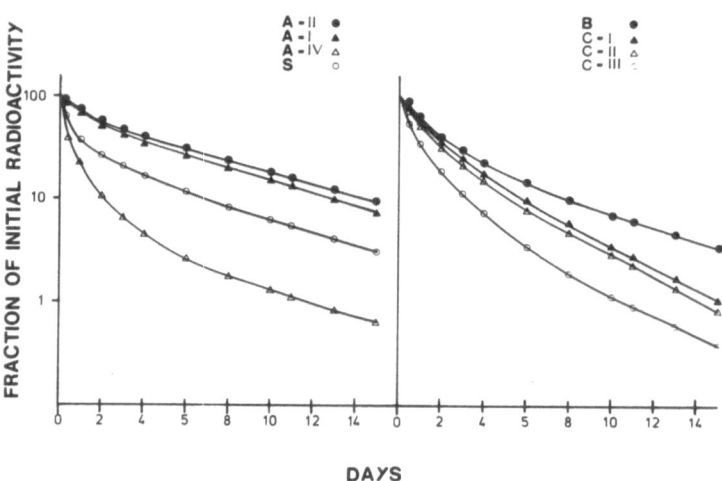

DAYS

Fig. 1. Comparison between plasma radioactivity decay curves of different apoproteins as a function of time. The terminal slopes of apo A-II, A-IV and SAA are parallel to that of apo A-I. The same is true for apo C-I and C-II compared to apo C-III.

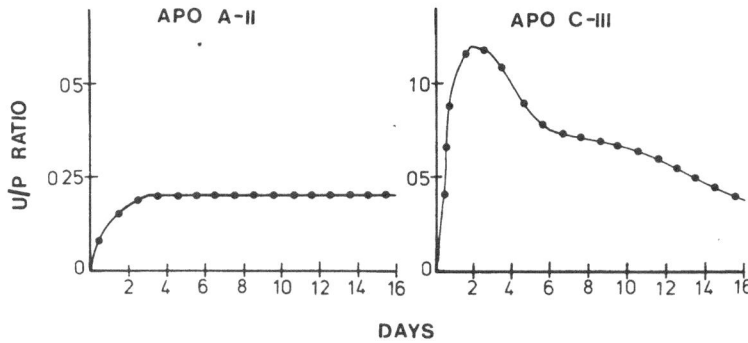

Fig. 2. A great variability exists in the profiles of urinary/plasma
ratio curves for the different apoproteins. Whereas apo A-II
U/P is rather constant with time, apo C-III curve shows a high
early peak followed by a decreasing slope.

we used as injected tracer the same labeled apoprotein in 3 different
forms: free apoprotein, lipoprotein in toto (VLDL or HDL) or apoprotein
associated in vitro to a lipoprotein (HDL or VLDL) and reisolated before
injection.

The results show that the "gross" parameter values were not much
affected by the form of material injected except for the apoprotein in the
free form, in what case a small portion (less than 10%) of the injected
material was eliminated very quickly (9). In fact the non-constancy of U/P
ratio with time was not artifactual but was a true indicator of a more
complex metabolism.

4. Concept of metabolic heterogeneity

If we accept that the material injected was in a "physiological" form,
the biphasic or not constant U/P ratio with time imposed the idea of
heterogeneity. An example of this heterogeneity is the varying extent of
change observed before and after treatment in hypertriglyceridemic patients
(10). The large differences in the U/P ratio curves from one patient to the
other and between pre- and post-treatment is a supplementary argument to
validate the variation with time of the U/P ratio and thus exclude an
artifactual phenomenon.

To overcome any artifact the kinetic studies have been performed since
then with a modified procedure: the pure labeled apoproteins were incubated
with autologous plasma and all labeled lipoproteins were separated by
ultracentrifugation at d 1.25 g/ml before reinjection. In this procedure,
labeled apoprotein not associated with lipoproteins was excluded.

5. Simultaneous fitting of plasma and urine data

Considering the variation with time of U/P ratio for apoprotein C-II
and C-III metabolism, we tried to fit urine excretion data. In this case,
we were unable to fit these data when plasma radioactivity was regarded as
a single homogeneous pool even when there is more than one output.
Thus, from this moment, in all kinetic experiments simultaneous fitting of
plasma and urine data became imperative, especially when the U/P ratio is
found not constant with time.

6. Re-estimation of LDL receptor pathway

We generally observed an inconstant U/P ratio for LDL kinetics. The question of kinetic heterogeneity of LDL has been already addressed by Berman *et al.* (11,12) and further developed by Foster *et al.* (13) who proposed several integrated models of increased complexity. Having in mind the lessons of apo C studies, we decided to re-estimate LDL metabolism via receptor and receptor-independent pathways. As an application we studied the effect of simvastatin on the metabolism of LDL in normolipidemic volunteers. With this aim in view native LDL and cyclohexanedione-modified LDL (LDL-CHD) not taken up by the receptors (14) were injected in absence and after one month of drug administration (15,16). A simple two-compartmental model was able to fit the plasma data of LDL-CHD. Separately an analogous model was used to fit the data of native LDL. As we know that LDL are catabolized by at least 2 pathways: one receptor-dependent and one receptor-independent (17) we tried, after having fixed the parameters of the LDL-CHD subsystem, to use for the receptor-dependent (R) subsystem a two-pool model similar to the one used for LDL-CHD data imposing as FCR for this subsystem the difference between total and LDL-CHD-FCR as used previously (18). Using this value we were unable to fit the data (Fig. 3 - upper left). To obtain a good fit the FCR of the proposed R subsystem had to be profoundly increased. Thus the summer (native or total LDL) does not represent the

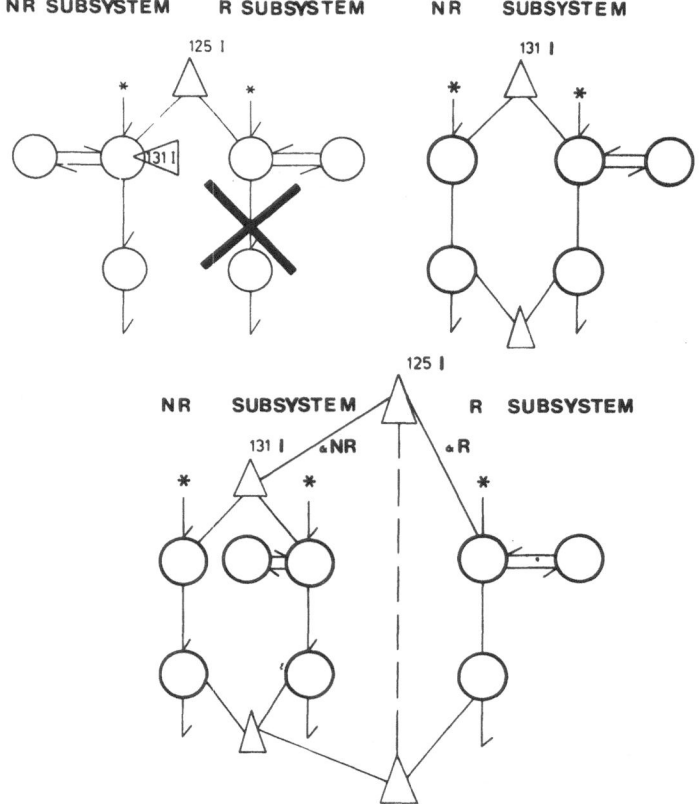

Fig. 3. Progressive stages in the building of the heterogeneous LDL model

Table 1. Kinetic Parameters of LDL and LDL-CHD

	FCR pools/day			percentage		PR mg/kg.day
	TOT	NR	R	NR	R	
MATTHEWS MODEL						
0	.315	.177	.138	56.2	43.8	9.25
+	.416	.194	.219	47.4	52.6	9.50
NEW MODEL						
0	.315	.177	.459	27.9	72.1	11.11
+	.416	.194	.582	19.4	80.6	11.56

arithmetical summation of 2 pathways but a combination in different proportions of 2 pathways with very different FCR.
This conclusion obliged us to build a new heterogeneous model taking into
account both plasma and urine data of both native and LDL-CHD for a simultaneous fit (Fig. 3 - lower part). The NR subsystem (^{131}I-CHD-LDL) must be
fitted first. This subsystem is already heterogeneous (Fig. 3 - upper
right). Then fixing all parameters of this NR subsystem a new subsystem was
built for the R pathway.

Using this constructed R subsystem the percentage of LDL following the
R pathway was much higher than the one erroneously calculated by difference
(18) but comparable to the one predicted by Brown and Goldstein (17).
The major modification produced by the drug was the increase in FCR of the
R pathway and the increase in the percentage of LDL following this pathway
(16).

7. Kinetics of Apoproteins C-II and C-III

An heterogeneous model was used also for describing the metabolism of
two apo C (C-II and C-III) present in the different lipoprotein classes. In
a similar manner as for LDL kinetics, the ^{131}I apo C-II data were first
fitted by an heterogeneous model (pathways 1 --> 7 and 2 --> 8), then
fixing the parameters of this apo C-II subsystem and assuming that a part
of apo C-III shares common metabolic pathways with apo C-II, the fitting of
^{125}I-apo C-III data necessitates additional pathways (3 --> 10 and 13 -->
11) specific to apo C-III (19). Simultaneous fitting of plasma and urine
data leads to the model illustrated in Fig. 4.

The metabolic implications of this model became manifest when normal
subjects'data were compared to date obtained in various pathologies (Tables
2 and 3). Clearly the metabolism of apo C-II and C-III is highly variable.
This may also result from the quick changes in the percentages of these
apo's in the different lipoproteins observable after modifications induced
by various stimuli such as diet, meals, exercise,...

8. Metabolic Heterogeneity in HDL Density Range

Experiments were performed with pairs of labeled apoproteins associated in different ratios into HDL:A-I and A-IV or A-I and A-II. In this
instance, two additional set of data were collected: 1. the comparison of
the metabolism of apoproteins distributed within different particles in one

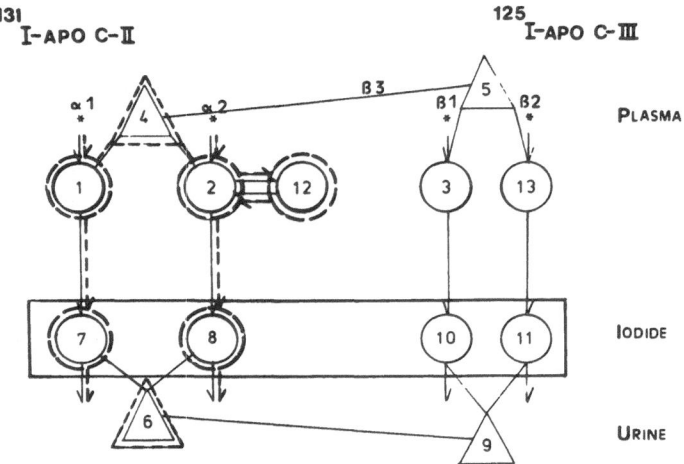

Fig. 4. Model used for apo C-II and C-III data fitting. α1 and α2 represent
fractions of apo C-II following each pathway; ß1 and ß2 percentages
of apo C-III following specific pathways, and ß3 part of the pool
sharing a common route with apo C-II.

Table 2. Estimated Rate Constants of Apolipoproteins C-II and C-III

	$L_{7,1}$	$L_{8,2}$	α_1	α_2	$L_{10,3}$	$L_{11,13}$	β_1	β_2	β_3
Normal	0.89	0.30	85.4	14.6	16.2	0.66	21.5	15.3	63.2
Hyper TG	1.53	0.42	85.6	14.4	----	1.03	----	39.4	60.6
Hypo ßLP	2.02	0.30	79.0	21.0	9.8	1.87	25.0	24.0	51.0
Renal Insuff.	0.25	0.36	35.0	65.0	----	0.41	----	28.3	71.7

L values are in pools/day and α and ß are percentages.

Table 3. Kinetic Parameters of Apolipoproteins C-II and C-III

	FCR		T		PR	
	C-II	C-III	C-II	C-III	C-II	C-III
	pools/day		day^{-1}		mg/kg x day	
Normal	0.70	0.87	2.24	1.43	0.73	11.10
Hypertriglyceridemia	1.11	0.95	1.42	0.82	2.04	5.83
HypoßLipoproteinemia	0.92	1.41	2.88	0.95	0.42	4.04
Renal insufficiency	0.45	0.43	8.82	3.61	0.43	2.47

class of lipoproteins (HDL) and 2. the quantitative comparison between radioactivity distribution and apoprotein distribution in the fractions eluted from affinity chromatography columns allowing separation of lipoprotein particles. Similar data have been already used in a simplified form for comparison of apoprotein S (or SAA) with apo A-I metabolism (20).

The building of the heterogeneous model shown in Fig. 5 follows a procedure identical to that described above for apo C-II and C-III.

There is an heterogeneity for apo A-I subsystem. Apoprotein A-IV is composed of one fraction having the same behavior as apo A-I and one fraction specific for apo A-IV (21).

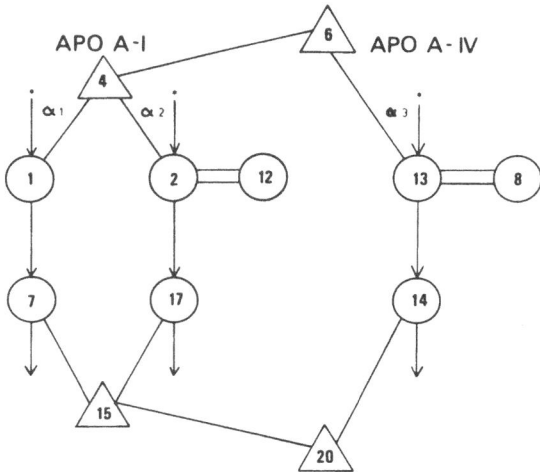

Fig. 5. Multicompartmental model describing apolipoproteins A-I and A-IV metabolism.

9. Metabolism of Lipoprotein Particles (LP-A-I and LP-A-:A-II)

Many authors have studied the metabolism of apo A-I and A-II by different approaches: labeling of total HDL and subsequent separation of A-I and A-II radioactivity by gel slicing (22), in vitro association of A-I and/or A-II with HDL (23) or free apoproteins A-I and A-II (24).

The aim of these studies was to point out a possible difference between apo A-I and A-II metabolism. Only slight differences were observed (22) and in some cases attributed to methodological problems (23). The first studies showing that apo A-I metabolism followed 2 pathways and thus was heterogeneous whereas apo A-II appeared homogeneous were published by Zech et al. (24) in complete independence with Alaupovic's concept of lipoprotein particles.

As it appeared that apo A-I may be present in particles with apo A-II (LP-A-I:A-II) or without apo A-II (LP-A-I) it looked essential to study the possible metabolic heterogeneity of these two particles. Already it was proposed that one particle may originate from intestine (LP-A-I) and the other (LP-A-I:A-II) from the liver. It is why we used for injection labeled LP-A-I prepared by incubation of ^{125}I-apo A-I with plasma devoid of apo A-II (after passage through an anti-A-II affinity column) and simul-

taneously with plasma lipoproteins labeled by incubation with ^{131}I-apo
A-II. It appears that 20 min after intravenous injection the distribution
between radioactivity in the two particles was comparable with distribution
of apo A-I and A-II in the plasma, i.e. about 2/3 of radioactivity of apo
A-I being linked in particles with apo A-II in normal subjects. All apo
A-II was linked to apo A-I. It was possible to build a model taking into
account two independent pathways for LPA-I and LPA-I:A-II (Fig. 6). The
parameter values of the two pathways were comparable to those of Zech *et
al.* FCR of P_1 (.163) (24) is similar to FCR of LP-A-I:A-II (.166) and FCR
of P_2 (.261) is in the same order of magnitude as FCR of LP-A-I (.32). The
same is true for the percentages following the two pathways (63% versus 69%
for the slow pathway). Thus FCR of LP-A-I:A-II is about the half of LP-A-I
(without A-II) particles and this particle represents 2/3 of the apo A-I
pool.

In conclusion, it seems that the two distinguishable routes of metabolism reported by Zech *et al.* (24) correspond to the metabolic pathways of
two clearly defined lipoportein particles.

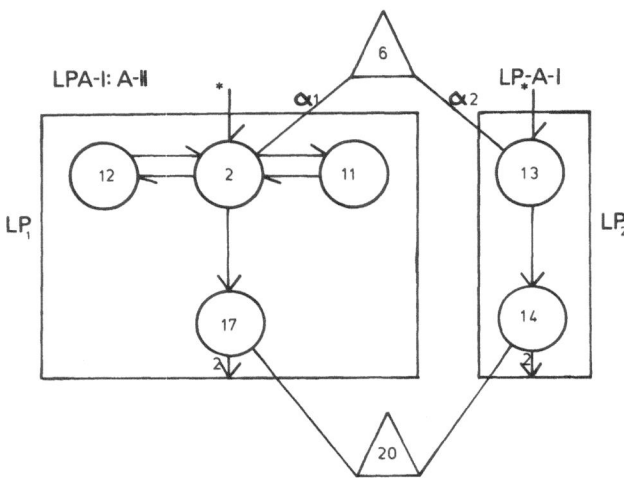

Fig. 6. Multicompartmental model for apolipoprotein A-I containing
lipoprotein particles (LP-A-I and LP-A-I:A-II)

CONCLUSIONS

Heterogeneity exists for the different density classes of lipoproteins. Since we know that many lipoprotein particles with different apoprotein composition are present in all lipoproteins, it is not surprising to
obtain results that necessitate the building of more complex models.

The non-constancy of U/P ratio with time implies heterogeneity. Thus
the collection of precise urine data becomes a necessity. The introduction
of an additional constraint: a simultaneous fitting of both plasma and
urinary curves to the model results in the considerable restriction of the
number of models giving an excellent fit of the data. As all kinetic
studies have been realized with pairs of apoproteins, it was possible not
only to build heterogeneous models but also to demonstate that these apoproteins had part of their metabolism in common.

The ultimate goal is to try to build a general model integrating all apoprotein kinetic data using pairs two by two. This methodology would allow quantifying the metabolic routes of discrete LP particles of definite apoprotein composition.

REFERENCES

1. C. M. E. Matthews, The theory of tracer experiments with iodine 131-labeled plasma proteins, <u>Phys. Med. Biol.</u>,2:36 (1957).
2. M. Berman, Kinetic analysis of turnover data, <u>Prog. Biochem. Pharmacol.</u>, 15:67 (1979).
3. M. Berman and M. Weiss, SAAM manual, US DHEW publ. N° (NIH) 78-180 (1978).
4. R. C. Boston, P. C. Greif, and M. Berman, CONSAM (Conversational Version of the SAAM Modeling Program, <u>in</u>: Lipoprotein kinetics and modeling, M. Berman, S. M. Grundy, and B. V. Howard, eds, Academic Press, New York (1982).
5. P. Alaupovic, The concepts, classification systems, and nomenclatures of human plasma lipoproteins, <u>in</u>: CRC handbook of electrophoresis. Lipoproteins: basic principles and concepts, L. A. Lewis and J. J. Opplt, eds, CRC Press, Boca Raton, FL, (1980).
6. C. L. Malmendier, J -F. Lontie, G. A. Grutman, and C. Delcroix, Metabolism of apolipoprotein C-I in normolipoproteinemic human subjects, <u>Atherosclerosis</u>, 62:167 (1986).
7. M. W. Huff, N. H. Fidge, P. J. Nestel, T. Billington, and B. Watson, Metabolism of C-apolipoproteins: kinetics of C-II, C-III1 and C-III2, and VLDL-apolipoprotein B in normal and hyperlipoproteinemic subjects, <u>J. Lipid Res.</u>, 22:1235 (1981).
8. S. A. Berson and R. S. Yalow, Quantitative aspects of iodide metabolism. The exchangeable organic pool, the rates of thyroidal secretion, peripheral degradation and fecal excretion of endogenously synthesized organically bound iodide, <u>J. Clin. Invest.</u>, 33:1533 (1954).
9. C. L. Malmendier, C. Delcroix, and J. P. Ameryckx, In vivo metabolism of human apoprotein A-I-phospholipid complexes. Comparison with human high density lipoprotein-apoprotein A-I metabolism, <u>Clin. Chim. Acta</u>,131:201 (1983).
10. C. L. Malmendier, J -F. Lontie, C. Delcroix, D. Y. Dubois, T. Magot, and L. De Roy, Apolipoproteins C-II and C-III metabolism in hypertriglyceridemic patients. Effect of a drastic triglyceride reduction by combined diet restriction and fenofibrate administration, <u>Atherosclerosis</u>, 77:139 (1989).
11. R. Goebel, M. Garnick, and M. Berman, A new model for low density lipoprotein kinetics: evidence for two labeled moieties, <u>Circulation</u>, 54 (Suppl. II):4 (1976).
12. M. E. Wastney, R. Riemke, C. L. Malmendier, and M. Berman, Heterogeneity of low-density lipoproteins: kinetics and modeling, M. Berman, S. M. Grundy, and B. V. Howard, eds, Academic Press, New York (1982).
13. D. M. Foster, A. Chait, J. J. Albers, R. A. Failor, C. Harris, and J. D. Brunzell, Evidence for kinetic heterogeneity among human low density lipoproteins, <u>Metabolism</u>, 35:685 (1986).
14. R. W. Mahley, T. L. Innerarity, R. E. Pitas, K. H. Weisgraber, J. H. Brown, and E. Gross, Inhibition of lipoprotein binding to cell surface receptors of fibroblasts following selective modification of arginyl residues in arginine-rich and B apoproteins, <u>J. Biol. Chem.</u>, 252:7279 (1977).
15. C. L. Malmendier, C. Delcroix, and J.-F. Lontie, Kinetics of a heterogeneous population of particles in low density lipoprotein apolipoprotein B, <u>Atherosclerosis</u>, 80:91 (1989).

16. C. L. Malmendier, J -F. Lontie, C. Delcroix, and T. Magot, Effect of simvastatin on receptor-dependent low density lipoprotein catabolism in normocholesterolemic human volunteers, Atherosclerosis, 80:101 (1989).

17. J. L. Goldstein and M. S. Brown, Atherosclerosis: the low-density lipoprotein receptor hypothesis, Metabolism, 26:1257 (1977). Alaupovic

18. J. Shepherd, H. R. Slater, and C. J. Packard, Low density lipoprotein receptor activity in man, in: Lipoprotein kinetics and modeling, M. Berman, S. M. Grundy, and B. V. Howard, eds., Academic Press, New York (1982).

19. C. L. Malmendier, C. Delcroix, J -F. Lontie, and D. Y. Dubois, Apolipoprotein C-II and C-III metabolism in a kindred of familial hypobeta-lipoproteinemia, Submitted to Metabolism, 1990.

20. C. L. Malmendier, J -F. Lontie, and C. Delcroix, In vivo metabolism of apolipoprotein S in humans. Comparison with apolipoprotein A-I metabolism, Clin. Chim. Acta, 170:169 (1987).

21. C. L. Malmendier, J -F. Lontie, L. Lagrost, D. Y. Dubois, and P. Gambert, In vivo metabolism of apolipoproteins A-IV and A-I in normo-lipidemic subjects, Submitted to J. Lipid Res., 1990.

22. C. B. Blum, R. I. Levy, S. Eisenberg, M. Hall, III, R. H. Goebel, and M. Berman, High density lipoprotein metabolism in man, J. Clin. Invest.,60:795 (1977).

23. J. Shepherd, C. J. Packard, A. M. Gotto, Jr., and O. D. Taunton, A comparison of two methods to investigate the metabolism of human apolipoproteins A-I and A-II, J. Lipid Res., 19:656 (1978).

24. L. A. Zech, E. J. Schaefer, T. J. Bronzert, R. L. Aamodt, and H. B. Brewer, Jr., Metabolism of human apolipoproteins A-I and A-II: compartmental models, J. Lipid Res., 24:60 (1983).

HDL METABOLISM IN HDL DEFICIENCY ASSOCIATED WITH FAMILIAL HYPER-

TRIGLYERIDEMIA: EFFECT OF TREATMENT WITH GEMFIBROZIL

Moti L. Kashyap and Keijiro Saku

Cholesterol Center, University of California at
Irvine, V. A. Medical Center
Long Beach, California, 90822, U.S.A.

ABSTRACT

Plasma high density lipoproteins (HDL) and their major proteins —
apolipoprotein (apo) AI and apo AII — are subnormal in most patients
with familial hypertriglyceridemia. However, the pathophysiology of
low plasma apo AI and apo AII is unclear. The kinetic parameters
(turnover) of HDL apo AI and apo AII were studied in six lean patients
with primary HDL deficiency associated with familial hypertriglyceridemia and normolipidemic healthy controls. The radioactivity decay
curve of ^{125}I labelled HDL was used for assessment of kinetics. Mean
plasma apo AI and apo AII were significantly lower (p<0.001) in
patients than normals (70.4± 2.7 v 106.9± 7.0; 24.2± 1.6 v 39.2±
0.9 mg/dl, respectively). The mean fractional catabolic rates (FCR)
obtained from plasma $^{125}I-HDL$, apo AI, apo AII radioactivity decay
curves and by Berson and Yalow's method (urine/plasma radioactivity
ratios) were significantly greater (p< 0.05) in patients than in
controls (0.387 v 0.299; 0.391 v 0.309; 0.361 v 0.275; 0.272 v 0.207/
d; respectively). The synthetic rates (SR of apo AI and apo AII were
significantly lower in patients than in controls (11.12 v 14.17 mg/
kg body weight/d. p < 0.05; 3.53 v 4.68 mg/kg body weight/d, p < 0.05,
respectively). Each patient was also investigated for HDL and
triglyceride metabolism immediately before and after 8 wk of
gemfibrozil (1200 mg/d) treatment. Gemfibrozil significantly increased plasma HDL cholesterol, apolipoprotein (apo) AI, and apo AII
by 36%, 29%, and 38% from baseline, respectively. Plasma TG decreased
by 54%. Gemfibrozil increased synthetic rates of apo AI and apo AII
by 27% and 34%, respectively, without changing the FCR. Stimulation
of apo AI and apo AII synthesis by gemfibrozil was associated with the
appearance in plasma of smaller (and heavier) HDL particles as assessed
by gradient gel electrophoresis and HDL composition. Postheparin
extra-hepatic lipoprotein lipase activity increased significantly by
25% after gemfibrozil, and was associated with the appearance in plasma
of smaller very low density lipoprotein particles whose apo CIII:CII
ratio was decreased. Hepatic triglyceride lipase was unchanged by
treatment.

These data indicate that the low HDL (and apo AI and AII) in these
patients is the result of increased catabolism and decreased
synthesis. Gemfibrozil increases the subnormal synthetic rate of apo
AI and AII without influencing the FCR. We suggest that increased

Hypercholesterolemia, Hypocholesterolemia, Hypertriglyceridemia
Edited by C.L. Malmendier *et al.*, Plenum Press, New York, 1990

233

turnover (flux, transport) of apo AI (whose major function is to efflux tissue cholesterol) by gemfibrozil may be significant in increasing reverse cholesterol transport in such patients with HDL deficiency.

Introduction and brief background

Apolipoprotein (apo) AI, major protein of high density lipoproteins (HDL) is a fascinating protein which has several properties. These include uptake of cholesterol from tissues (thus initiating the initial step in reverse cholesterol transport), (1) activation of the enzyme lecithin cholesterol acyl transferase (LCAT), (2) and other less well characterized properties of increasing cholesterol solubility in bile, (3) and in promoting fibrinolysis(4). Thus, it is not surprising that apo AI is more important as a discriminator of atherosclerotic cardiovascular disease than HDL cholesterol (5). Apo AII, the other major protein of HDL is important structurally in lipid binding. Apo AI is involved importantly in reverse cholesterol transport, a process in which cholesterol is taken up from tissues, esterified and transferred ultimately to the liver for disposal (6). In theory, this process can be stimulated at various steps, the most important of which is elevation of HDL level. The mechanism by which HDL and apo AI levels are increased in plasma is important because, theoretically, this can impact on reverse cholesterol transport. For example, if apo AI levels are increased by blocking catabolism, plasma levels will be elevated but tissue cholesterol removal rate *in vivo* may not be increased to the same extent as a situation in which plasma apo AI levels are increased because of increased synthetic (i.e. transport) rate of apo AI (6). Thus, defining the mechanism by which apo AI levels are increased or decreased in plasma may be important in our understanding of the relationship between low HDL levels and atherosclerosis.

Study aims

The aim of this study was, (a) to assess the *in vivo* kinetic parameters of apo AI and apo AII in a group of patients with primary HDL deficiency associated with low plasma apo A levels and familial hypertriglyceridemia, (b) to study the relationship between apo A kinetics and post heparin lipoprotein lipase and hepatic triglyceride lipase activities, and (c) to assess the mechanism by which gemfibrozil, a drug that raises HDL levels, influences the *in vivo* kinetics of apo AI and apo AII, and (d) to determine the significance of the data in terms of pathophysiology of HDL deficiency and its theoretical relationship to reverse cholesterol transport. These studies have been reported in part previously (7,8).

Methods and results

The patients selected were well defined patients with very low levels of HDL cholesterol (mean 22mg percent) and apo AI and AII levels below the tenth percentile of reference values. All patients had hypertriglyceridemia (type V phenotype) which was familial in nature. All patients were lean with a mean desirable body weight = 106% and none of the patients were > 15% of desirable body weight. After recruitment into the study, each patient was completely evaluated, and once the diagnosis was established they were treated with a diet. After stabilization, they were switched to an isocaloric diet for at least six weeks before the first admission to the Clinical Research Center. During this period, the six patients and six normolipidemic healthy controls were investigated for turnover of apo AI and AII, post-heparin lipolytic activities, and isolation of VLDL and HDL for chemical characterization. After this baseline study was over, each patient was treated with gemfibrozil (600 mg bid) for 8 weeks. Each patient was then readmitted and the detailed research studies were repeated. For

the turnover studies autologous HDL was labelled with ^{125}I, injected
into each patient, and blood samples obtained periodically for twelve
days. Kinetic parameters were based upon radioactivity decay curves
of apo AI and AII isolated from HDL of each sample subjected to SDS
polyacrylamide gel electrophoresis.

Mean apo AI and AII levels in patients (70, 24 mg/dl) were significantly
lower than controls (107,39 mg/dl) respectively. Analysis of the decay
curves of apo AI and AII showed that the patients had significantly
steeper decay curves indicating an increase in the FCR. Compared to
controls, patients had significantly higher FCR (0.39 and 0.36 for apo
AI and AII in patients versus .31 and .27 in control respectively).
Also, calculations of the synthetic rate indicated that patients had
significant reductions in synthetic rate compared to controls (11.1
and 3.5 mg/kg body weight/per day for apo AI and AII respectively in
patients vs 14.2 and 4.7 mg/per kg body weight/per day in controls.

Thus, the reduction in the apo AI and AII levels in these patients was
the result of decreased synthesis and increased fractional catabolic
rates. Assessment of the post-heparin total, extra-hepatic lipoprotein lipase and hepatic triglyceride lipase indicated no significant
differences between patients and normal subjects. However, evaluation
of the relationship between the post heparin lipases and the kinetic
parameters revealed a significant positive correlation between
hepatic triglyceride lipase and the fractional catabolic rate.

Effect of gemfibrozil therapy on apo AI and AII kinetics and post heparin lipolytic activities

After 8 weeks of gemfibrozil therapy, the most dramatic changes were
reductions in total plasma triglycerides from a mean of 449 mg/dl to
189 mg/dl. LDL-Cholesterol rose from 87 to 115 mg/per dl. Apo AI levels
increased significantly from 70 to 91 mg/per dl and apo AII levels
likewise increased from 24 to 33 mg/per dl.

Radioactivity decay curves of both apo AI and AII were almost
superimposable in most patients indicating no change in the fractional
catabolic rates. Since plasma levels of apo AI and AII were increased,
the synthetic rate of these apoproteins were significantly elevated.
The absolute apo AI synthetic rate (i.e. transport rate) increased from
12.1 to 15.2 mg/per kg body weight/day and apo AII absolute synthetic
rate increased from 3.9 to 5.1 mg/per kg body weight/day. Thus,
gemfibrozil increased the transport rate of apo AI and AII without
affecting the FCR.

Evaluation of HDL composition revealed that the major difference
before and after gemfibrozil therapy in isolated HDL-2 and HDL-3 was
a reduction in the core triglyceride content of HDL particles. Also,
the ratio of the core components (TG plus CE) to surface components
(protein + cholesterol + phospholipid), an index of particle size,
revealed that the mean size of both isolated HDL-2 and HDL-3 were
reduced, the difference being statistically significant in HDL-2
particles. Isolated HDL-2 and HDL-3 were subjected to gradient gel
electrophoresis. Scans of these electrophoretograms revealed a shift
in the peak towards smaller molecular weight size. Assessment of the
molecular diameter and molecular weight of the two major peaks of HDL
in these fractions revealed significant reduction in both these
parameters after gemfibrozil therapy.

Measurement of post heparin total, extra hepatic lipoprotein lipase
and hepatic triglyceride lipase indicated significant increase in the
lipoprotein lipase activity from 19.2 µM free fatty acid/hour/ml, to

23.2 µM free fatty acid/hour/ml. Hepatic triglyceride lipase activity remained unaltered.

Conclusions

A. Low apo AI and apo AII levels of HDL deficiency associated with familial hypertriglyceridemia resulted from
      (a) decreased synthetic (transport) rate and
      (b) increased fractional catabolic rate

B. In these patients, gemfibrozil therapy significantly increased the subnormal transport rates of apos AI and AII (to the normal range) but did not affect their fractional catabolic rates.

C. Hepatic triglyceride lipase activity correlated positively with fractional catabolic rate. Both hepatic triglyceride lipase activity and FCR were unchanged after gemfibrozil therapy indicating that the drug did not increase apo AI and apo AII levels via changes in the hepatic lipase

D. We hypothesize that, because apo AI is major mediator of tissue cholesterol efflux, increasing its transport rate (i.e. synthetic rate or turnover) will be expected to enhance reverse cholesterol transport. This maybe the common antiatherogenic mechanism of the few other agents (e.g. estrogens, ethanol, fenofibrate) that have been found to increase the synthetic rate of apo AI (9 - 11).

References

1. Fielding, C.J., and P.E. Fielding. 1982. Cholesterol transport between cells and body fluids. *Med. Clin. N. Am.* 66:363-373.

2. Golmset, J.A. 1968. The plasma lecithin:cholesterol acyltransferase reaction. *J. Lipid Res.* 9:155-167.

3. Kibe, A., R.T. Holzbach, N.F. LaRusso, and S.J.T. Mao. 1984. Inhibition of cholesterol crystal formation by apolipoproteins in supersaturated model bile. *Science (Wash. DC)*. 255:514-516.

4. Saku, K., M. Ahmad, P. Glas-Greenwalt, M.L. Kashyap. 1985. Activation of fibrinolysis by apolipoproteins of high density lipoproteins in man. *Thromb. Res.* 39:1-8.

5. Mendoza, S.G., A. Zerpa, H. Carrasco, O. Colmenares, A. Rangel, P.S. Gartside, and M.L. Kashyap. 1983. Estradiol, testosterone, apolipoproteins, lipoprotein cholesterol, and lipolytic enzymes in men with premature myocardial infarction and angiographically assessed coronary occlusion. *Artery*. 12:1-23.

6. Kashyap, M.L. 1989. Basic considerations in the reversal of atherosclerosis: significance of high-density lipoprotein in stimulating reverse cholesterol transport. *Am. J. Card.* 63:56H-59H.

7. Saku, K., P.S. Gartside, B.A. Hynd, S.G. Mendoza, M.L. Kashyap. 1985. Apolipoprotein AI and AII metabolism in patients with primary high-density lipoprotein deficiency associated with familial hypertriglyceridemia. *Metabolism*. 34:754-764.

8. Saku, K., P.S. Gartside, B.A. Hynd, M.L. Kashyap. 1985. Mechanism of action of gemfibrozil on lipoprotein metabolism. *J. Clin. Invest.* 75:1702-1712.

9. Schaefer, E.J., Foster, D.M., Zech L.A., et al. 1983. The effects of estrogen administration on plasma lipoprotein metabolism in premenopausal females. *J Clin Encocrinol Metab*. 57:262-267.

10. Malmendier, C.L., Delcroix, C. 1985. Effect of alcohol intake on high and low density lipoprotein metabolism in healthy volunteers. *Clin Chim Acta*. 152:281-288.

11. Malmendier, C.L., Delcroix, C. 1985. Effect of fenofibrate on high and low density lipoprotein metabolism in heterozygous familial hypercholesterolemia. *Atherosclerosis*. 55:161-169.

RECENT ADVANCES IN LIPOPROTEIN METABOLISM AND THE GENETIC DYSLIPOPROTEINEMIAS

H.B. Brewer, Jr., D.J. Rader, J.M. Hoeg, A. Mann, and G.Tennyson

Molecular Disease Branch, National Heart, Lung, and Blood Institute, National Institutes of Health Bethesda, MD USA

During the last two decades major advances have occurred in our understanding of lipoprotein metabolism in normal subjects, and patients with dyslipoproteinemias. Of particular interest has been the elucidation of the role of plasma apolipoproteins, lipoproteins receptors, transfer factors, and lipolytic enzymes in the metabolism of plasma lipoproteins. An improved understanding of the regulation of lipoprotein metabolism, and the intracellular as well as extracellular transport of cholesterol has permitted the determination of molecular defects in patients with dyslipoproteinemias. This new information has been used to development improved methods for diagnosis and treatment of patients at risk for the development of hyperlipidemia and premature cardiovascular disease. This report will summarize recent studies in our laboratory on three apolipoproteins, apoA-I, apoB, and Lp(a).

HDL and ApoA-I

HDL have been of particular interest over the last decade due to the relationship of HDL and premature cardiovascular disease. High levels of HDL are associated with a decreased risk, whereas reduced levels increase the risk of early heart disease (1-5). HDL has been proposed to modulate the development of atherosclerosis by the process termed reverse cholesterol transport (6). In this still hypothetical process HDL is proposed to transport excess cellular cholesterol in peripheral tissues back to the liver for removal from the body.

One of the major questions which remains in our understanding of the role of HDL in the development of atherosclerosis is the importance of heterogeneity of lipoprotein particles within HDL. The physiological role of the different lipoproteins within HDL has not been definitively established. To effectively analyze the physiological functions of the different lipoproteins within HDL requires the development of effective methods to separate the different lipoproteins within HDL. Some of the major techniques employed to characterize and isolate HDL lipoprotein particles have include hydrated density (7),

Hypercholesterolemia, Hypocholesterolemia, Hypertriglyceridemia
Edited by C.L. Malmendier *et al.*, Plenum Press, New York, 1990

237

electrophoresis (8-11), gradient gel electrophoresis (11), and apolipoprotein composition (12,13). The characterization of lipoproteins into lipoprotein particles based on apolipoprotein composition has become particularly useful. The major lipoprotein particles within HDL based on apolipoprotein composition are LpA-I, LpA-I,A-II, LpA-I,E, and LpA-IV. In this report we will focus on lipoprotein particles containing both apoA-I and apoA-II (LpA-I,A-II), or apoA-I only particles (LpA-I).

Recent studies have suggested it is the reduction in LpA-I, not LpA-I,A-II which is correlated with the risk of development of premature atherosclerosis (14). *In vitro* studies utilizing the mouse adipocyte cell line, OB 1771 cells, have shown that LpA-I, but not LpA-I,A-II particles are able to facilitate cholesterol eggress from the adipocyte (15). In addition, liposomes containing increasing amounts of apoA-II relative to apoA-I showed a dose response decrease in cholesterol efflux. Thus the presence of apoA-II or lipoprotein particles containing apoA-II inhibited cholesterol efflux from the adipocyte. These studies have been interpreted as indicating that there is an important physiological difference between LpA-I and LpA-I,A-II particles within HDL. The combined results from these studies suggest that the plasma level of the subfraction within HDL that may be related to the development of atherosclerosis is LpA-I, and not LpA-I,A-II particles.

In order to provide additional information on the potential difference in physiological function of LpA-I and LpA-I,A-II we have performed kinetic studies on the individual HDL lipoprotein particles (16). ^{131}I-LpA-I and ^{125}I-LpA-I,A-II were simultaneously injected into normal volunteers, and the LpA-I and LpA-I,A-II particles separated by affinity chromatography. LpA-I particles were catabolized at a faster rate than LpA-I,A-II (16, Fig. 1). In addition, analysis of the radiolabelled apoA-I in the LpA-I and LpA-I,A-II particles separated by affinity chromatography indicated that LpA-I particles are slowly converted to LpA-I,A-II particles during the 14 day study. These metabolic studies support the physiological difference between LpA-I and LpA-I,A-II particles and lend support to the concept that a full understanding of the role of HDL in cholesterol homoeostasis will require a detailed analysis of the functional role of individual lipoprotein particles within HDL.

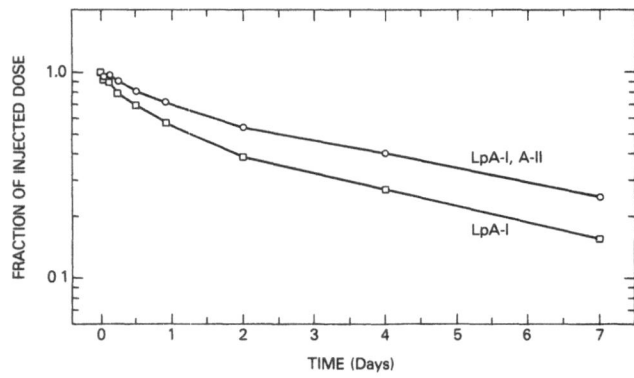

Figure 1

ApoB

Two B apolipoproteins, apoB-100 (Mr 512 kDa) and apoB-48 (Mr 250 kDa) are major apolipoprotein constituents of VLDL/LDL and chylomicrons, respectively. In man, the liver synthesizes primarily apoB-100, and both apoB-48 and apoB-100 are synthesized by the intestine (17,18). ApoB-100 serves as the classic ligand for the LDL receptor, however, apoB-48 has not been shown to interact with a specific cellular receptor. ApoB-100 and apoB-48 are the product of a single apoB gene on chromosome 2. ApoB-100 and apoB-48 are translated from an apoB mRNA by a remarkable RNA editing mechanism in which the glutamine codon (CAA) at amino acid 2143 of apoB-100 undergoes editing with a conversion of a C to U converting the CAA codon to a UAA stop codon (17, 19-21). Translation of the native and edited apoB mRNA results in the synthesis of the 4536 amino acid apoB-100 and the 2152 amino acid B-48 apolipo- proteins. Thus apoB mRNA editing provides a unique mechanism for the synthesis of two large proteins from a single gene.

The degree of apoB mRNA editing in the intestine and liver differs in different species. The percentage of apoB mRNA edited in the rat, rabbit, chicken, and man is shown in Table I.

TABLE I

Analysis of ApoB mRNA Editing in Different Species

	ApoB-48 (%)	ApoB-100 (%)
Liver		
Human	1	99
Rat	60	40
Rabbit	1	99
Intestine		15
Human	85	15
Rat	85	15
Rabbit	87	13
Chicken	--	100

The large percentage (up to 80%) of apoB mRNA edited in the rat liver is in marked contrast to the virtual absence of editing in the rabbit and human liver tissues. The teleological reason for the increase in editing in the rat liver is unknown, however, the relative increase in the percentage of VLDL containing apoB-48 results in an increased rate of removal of VLDL-LDL particles leading to a decreased plasma level of LDL. The increased rate of removal of apoB-48 equivalent lipoproteins as compared to apoB-100 lipo- proteins results in more effective removal of hepatic secreted VLDL in the rat species. The advantages of increased clearance of apoB containing lipoproteins and the lower plasma level of LDL in the rat is not as yet known, however, it would be anticipated to reduce the risk of atherosclerosis.

The degree of unedited apoB mRNA in the human intestine is approximately 15 percent. Therefore approximately 15 percent of human chylomicrons may contain apoB-100 instead of apoB-48. The metabolism of the apoB-100 containing chylomicrons may be significantly different than the apoB-48 chylomicrons, and the apoB-100 containing lipoproteins may be potentially more atherogenic. There exists the interesting possibility that selected patients with disorders of chylomicron metabolism and premature cardiovascular disease may have a defect in the editing process resulting in the biosynthesis of predominantly apoB-100 containing chylomicrons. The identification of patients with defects in apoB editing will be of great interest, and provide new additional insights into potential mechanisms responsible for premature cardiovascular disease.

The identification of this novel apoB mRNA editing process has focused research on mechanisms that may potentially modulate the degree of editing in a particular tissue. Our laboratory (22) and others (23) have recently reported that the intact rat liver provides a unique hepatic model for investigating mRNA editing. To date no effective _in vitro_ system which efficiently edits a significant fraction of apoB mRNA has been reported. We have therefore developed an _in vitro_ model system using several rat hepatocyte cell lines which could be used to ultimately define the minimal cognate apoB mRNA template require for editing and provide a model system for analysis of ligands which modulate the extent of apoB mRNA editing. A novel primer extension analysis procedure has also been developed based on the inclusion of a single terminating dideoxynucleotide (AdG) and three extending deoxynucleotides (dA, dC, dT) to rapidly survey for apoB transcript editing.

Several different rat hepatocyte cell lines were evaluated for apoB editing. The percentage of edited and unedited mRNA which code for apoB-48 and apoB-100 equivalent apolipoproteins is shown in Table II.

TABLE II

Cell Line	Percentage mRNA ApoB-48	ApoB-100
Fao	12.3	88.7
Clone 9	7.1	92.9
BRL	8.8	91.2
H411E	7.0	93.0
H411EC3	11.5	88.5
Rat Liver	60.0	40.0
Rat Intestine	81.2	18.8

The Fao and H411EC3 cell lines effectively synthesized and secreted apoB isoproteins into the cell culture media. These two cell lines will now be utilized to analyze the physiological as well as regulatory ligand systems which may modulate the apoB editing process.

Lp(a)

Lp(a) is now widely recognized as an important independent risk factor for the development of premature cardiovascular disease (24-27). Plasma Lp(a) levels in the population range from undectable to greater than 120 mg/dl with levels higher than 30 mg/dl resulting in at least a two fold increase in the relative risk of premature coronary artery disease (26).

Lp(a) is a LDL-like particle with a unique apolipoprotein, apo(a), covalently linked by a disulfide bridge to apoB-100 (28-31). The structure of apo(a) bears a striking homology to plasminogen (32). Several different apo(a) alleles have been recognized in the population, and the apo(a) protein coded for by these alleles differs in size due to variable repeats of the kingle 4 domain of plasminogen (33-34). Lp(a) concentrations in individuals are inherited in an autosomal codominant manner at the apo(a) structural gene locus and the size of the polymorphic apo(a) isoprotein is inversely correlated with the plasma level (35).

Plasma Lp(a) levels are also moderated by a second gene locus, the LDL receptor gene (36). The plasma level of Lp(a) is elevated approximately three fold in heterozygous individuals with familial hypercholesterolemia (36). This observation suggests that the LDL receptor may play a role in the catabolism of Lp(a), however, this has been controversial (37).

In order to gain additional insights into role of the LDL receptor in the metabolism of Lp(a) we have investigated the in vivo catabolism of Lp(a) in man by simultaneously injecting radiolabelled Lp(a) and LDL particles in a homozygous familial hypercholesterolemic patient and two normal subjects (38).

The catabolism of ^{131}I-LDL and ^{125}I-Lp(a) in a normal subject are illustrated in Figures 2, 3. The catabolism of Lp(a) was slower (residence time 3.43) when compared to LDL (residence time 2.42 days). In the homozygous familial hypercholester-olemic patient with a total absence of the LDL receptor, Lp(a) and LDL are catabolized at very similar rates with residence times of 3.93 and 4.29 days, respectively. In addition, Lp(a) is catabolized slower in the familial hypercholesterolemic patient than in the control. These combined results establish a significant role of the LDL receptor in the in vivo catabolism of Lp(a).

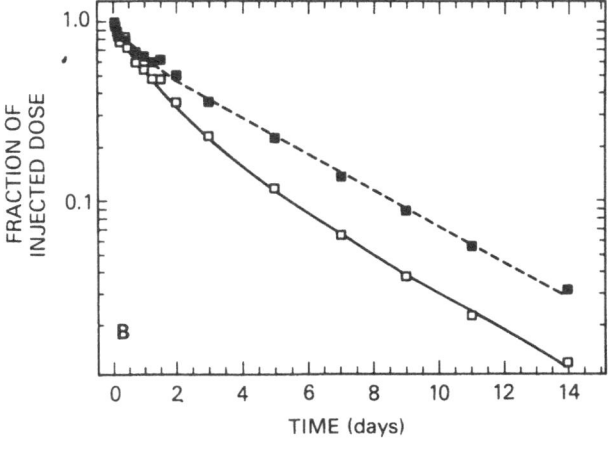

Figure 2

LDL (□)and Lp(a) (■) catabolism in a normal subject

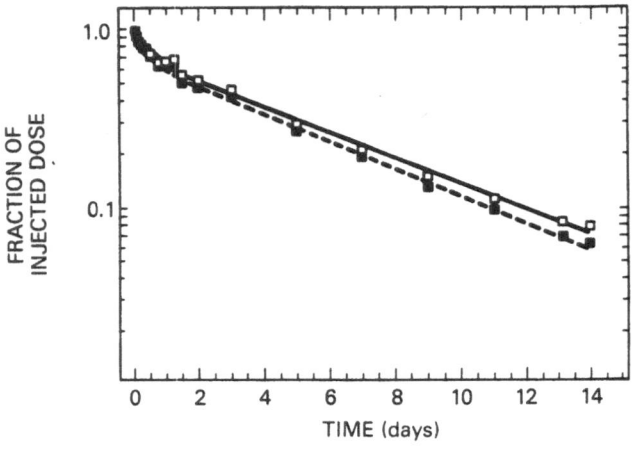

Figure 3

LDL (▢)and Lp(a) (■) catabolism in a homozygous familial hypercholesterolemic patient

Summary

The elucidation of the structure and function of the plasma apolipoproteins has provided the unique opportunity to understand the physiological pathways for the transport and cellular metabolism of the plasma lipoproteins. The complexity of the individual density classes of plasma lipoproteins has been revealed by a detailed analysis of the apolipoprotein composition of the individual lipoprotein particles. In addition, the elucidation of the molecular defects in patients with dyslipoproteinemias has now permitted the understanding of the defects at the level of the apolipoprotein gene. The ability to define the genetic defect in individuals at risk for the development of premature cardiovascular disease provides the unique opportunity to now identify these individuals at an earlier age, and to initiate therapy to prevent the development of early heart disease.

References

1. T. Gordon, W.P. Castelli, M.C. Hjortland, W.B. Kannel, and T.R. Dawber *Am. J. Med.* 62:707-714 (1977).
2. G. Heiss, N.J. Johnson, S. Reiland, L.E. Davis, and H.A. Tyroler *Circulation* 62:IV-116-IV-136 (1980).
3. W.P. Castelli, J.T. Dougle, T. Gordon, C.G. Hames, M.C. Hjortland, S.B. Hulley, A. Kagan, and W.J. Zukel *Circulation* 55:767-772 (1977).
4. P. Avogaro, B.G. Bittolo, G. Cazzaloto, and E. Rosa *Atherosclerosis* 37:69-76 (1980).
5. J.L.H.L. Third, J. Montag, M. Flynn, J. Freidel, P. Laskarzewski, and C.J. Glueck *Metabolism* 33:136-146 (1984).

6. J.A. Glomset _J. Lipid Res._ 9:155-167 (1968).
7. R.J. Havel, H.A. Eder, and J.H. Bragdon _J. Clin. Invest._ 34:1345-1353 (1955).
8. D.L. Sprecher, L. Taam, and H.B. Brewer, Jr. _Clin. Chem._ 30:2084-2092 (1984).
9. G. Utermann, G. Ferissner, G. Franceschini, J. Hass, A. Steinmetz _J. Biol. Chem._ 257:501-507 (1982).
10. H.J. Menzel, R.G. Kladetzky, and G. Assmann _J. Lipid Res._ 23:915-922 (1982).
11. J.J. Albers, G.R. Warnick, and A.V. Nichols. Clinical and Metabolic Aspects of High Density Lipoproteins. ed. N.E. Miller and G.J. Miller. Elsevier 1984 pp. 381-414.
12. P. Alaupovic. _Protides of Biol. Fluids Proc. Colloq._ 19:9-19 (1972).
13. J. Osborne and H.B. Brewer, Jr. _Adv. Prot. Chem._ 31:253-337 (1977).
14. P. Puchois, A. Kandoussi, P. Fievet, J.L. Fourrin, M. Bertrand, E. Koren, J.C. Fruchart. _Atherosclerosis_ 68:35-40 (1987).
15. R. Barbaras, P. Puchois, J.C. Fruchart, and G. Aihaud. _Biochem. Biophys. Res. Comm._ 142:63-69 (1987).
16. D.J. Rader _Circulation_
17. K. Higuchi, A.V. Hospattankar, S.W. Law, N. Meglin, J. Cortright, and H.B. Brewer, Jr. _Proc. Natl. Acad. Sci._ 85:1772-1776 (1988).
18. J.M. Hoeg, D.D. Sviridov, G.E. Tennyson, S.J. Demosky, Jr., M.D. Meng, D. Bojanovski, I.G. Safonova, V.S. Repin, M.B. Kuberger, N.V. Smirnov, K. Higuchi, R.E. Gregg, H.B. Brewer, Jr. _J. Lipid Res._ In press.
19. L.M. Powell, S.C. Wallis, R.J. Pease, Y.H. Edwards, T.S. Knott, and J. Scott. _Cell_ 50:831-840 (1987).
20. A.V. Hospattankar, K. Higuchi, S.W. Law, N.M. Meglin, and H.B. Brewer, Jr. _Biochem. Biophys. Res. Comm._ 148:279-285 (1987).
21. S.H. Chen, G. Habib, C.Y. Yang, Z.W. Gu, B.R. Lee, S.A. Weng, S.R. Silberman, S.J. Cai, J.P. Deslypere, M. Rosseneu, A.M. Gotto, Jr., W.H. Li, and L. Chan _Science_ 238:363-366 (1987).
22. G.E. Tennyson, C.A. Sabatos, K. Higuchi, N. Meglin, and H.B. Brewer, Jr. _Proc. Natl. Acad. Sci._ USA 86:500-504 (1989).
23. N.O. Davidson, L.M. Powell, S.C. Wallis, and J. Scott _J. Biol. Chem._ 263:13482-13485 (1988).
24. K. Berg _Acta Path_ 59:369-382 (1963).
25. G.M. Kostner, P. Avogaro, G. Cazzolato, E. Marth, G. Bittolo-Bon, and G.B. Qunici _Atherosclerosis_ 38:51-61 (1981).
26. V.W. Armstrong, P. Cremer, E. Eberle, A. Menke, F. Schulze, H. Wieland, H. Kreuzer, and D. Seidel _Atherosclerosis_ 62:249-257 (1986).
27. G.G. Rhoads, G. Dahlen, K. Berg, N.E. Morton, and A.L. Dannenberg. _J. Am. Med. Assoc._ 256:2540-2544 (1986).
28. G.M. Kostner. In Low Density Lipoproteins, ed. C.E. Day and R.S. Levy, Plenun Press, New York. (1976) pp. 229-269.
29. J.W. Ganbatz, C. Heideman, A.M. Gotto, J.D. Morrisett, and G.H. Dahlen _J. Biol. Chem._ 254:4582-4589 (1983).
30. G. Utermann and W. Weber _FEBS Lett._ 154:357-361 (1983).
31. G.M. Fless, C.A. Rolih, and A.M. Scanu _J. Biol. Chem._ 259:11470-11478 (1984).
32. J.W. McLean, J.E. Tomlenson, W.J. Keeang, D.L. Eaton, E.Y. Chen, G.M. Fless, A.M. Scanu, and R.M. Lawn _Nature_ 330:132-137 (1987).
33. M.L. Koschinsky, U. Beisiegel, D. Henne-Bruns, D.L. Eaton, and R.M. Lawn _Biochem._ 29:640-644 (1990).

34. G. Utermann <u>Science</u> 246:904-910 (1990).
35. G. Utermann, H.J. Mengel, H.G. Kraft, H.C. Duba, H.G. Kemmler, and C. Seitz <u>J. Clin. Invest.</u> 80:458-465 (1987).
36. G. Utermann, F. Hoppichles,H. Dieplinger, M. Seed, G. Thompson, and E. Boerwinkle <u>Proc. Natl. Acad. Sci. USA</u> 86:4171-4174 (1989).
37. K. Maatmann-Moe and K. Berg <u>Clin. Genet.</u> 20:352-362 (1981).
38. A.W. Mann, H.O. Kraft, R.E. Gregg, D.J. Rader, J.R. Schaefer, L.A. Zech, J.M. Hoeg, J. Davignon, and H.B. Brewer, Jr. <u>J. Clin. Invest.</u> In press.

KINETIC STUDIES OF THE ORIGIN OF APOLIPOPROTEIN (APO) B-100 IN LOW DENSITY LIPOPROTEINS OF NORMAL AND WATANABE HERITABLE HYPERLIPIDEMIC (WHHL) RABBITS

Richard J. Havel and David M. Shames
Cardiovascular Research Institute and Department
of Medicine, University of California, San
Francisco, California, U.S.A.

Numerous studies in humans, as well as in experimental animals, support the general conclusion that low density lipoproteins (1.02 <d<1.06 g/ml; LDL) are a product of the metabolism of hepatogenous triglyceride-rich precursor particles of lower density, mainly by the action of lipolytic enzymes (lipoprotein lipase and, perhaps, hepatic lipase) on component triglycerides and phospholipids (1). From kinetic studies in humans, the concept has developed that this process occurs progressively, by conversion of large very low density lipoproteins (VLDL) to lipoproteins of progressively higher density in a lipolytic cascade (2).

Liver perfusion experiments with normal animals have generally, but not always, shown that VLDL are the primary secretory products that contain apolipoprotein B, the protein common to VLDL and LDL (3). "Direct" production of LDL has been found under some circumstances, but these observations can be explained by "washout" of trapped lipoproteins from the space of Disse or by conversion from VLDL in the perfusate following secretion (3). Cholesteryl ester-enriched particles, secreted from cholesterol-loaded livers, may have an increased density approaching that of LDL (4), but these can hardly be considered as normal secretory products. Likewise, abnormal lipoproteins in the density range of LDL are produced durnig cholestasis (5). In the rat, virtually all lipoproteins isolated from Golgi fractions of the liver float at a density of 1.01 g/ml, provided that care is taken to exclude contamination with endosomes containing lipoproteins destined for liposomal degradation (Hamilton, R.L., Havel, R.J., unpublished data). The "nascent" VLDL thus isolated from Golgi-rich fractions of normal rat livers have a similar "core" composition to plasma VLDL (but their "surface" components are quite different. The nascent particles contain several fold less unesterified cholesterol, several fold more phosphatidyl enthanolamine, and considerably less sphingomyelin than plasma VLDL. Unlike plasma VLDL the nascent particles contain proapolipoprotein (apo) A-I and considerably less of the C apoproteins than plasma VLDL. These observations indicate that plasma VLDL represent particles that have become modified by post-secretory events.

Hypercholesterolemia, Hypocholesterolemia, Hypertriglyceridemia
Edited by C.L. Malmendier *et al.*, Plenum Press, New York, 1990

Although rat plasma VLDL are converted to LDL, this species is a difficult one in which to study this process because most VLDL are removed by the liver as partially degraded VLDL "remnants" (1). Furthermore, the rat liver, unlike that of other mammals, secretes VLDL-containing apo B-48, normally the primary form of apo B produced by the intestine, as well as apo B-100, which in most mammals is the sole form of apo B secreted by the liver.

We have undertaken studies of the metabolism of lipoproteins containing apo B in rabbits during the last eight years. In this species, in which hepatogenous VLDL contain apo B-100 and not apo B-48, a major perturbation of lipoprotein metabolism can be studied in animals with a mutant form of the LDL receptor, the Watanabe heritable hyperlipidemic (WHHL) rabbit. In our early experiments, we showed that in normal, New Zealand white (NZW) and WHHL rabbits alike, apo B-100 is secreted from perfused livers virtually exclusively as VLDL (d<1.006 g/ml) and none as LDL or even as particles with intermediate density (IDL, 1.006<d<1.019 g/ml) (6). The rabbit also differs from the rat by the presence of an active process, mediated by the "cholesteryl ester transfer protein" (CETP) which redistributes cholesteryl esters and triglycerides among plasma lipoprotein particles (7). Thus, although rabbits normally convert considerably less VLDL to LDL than do humans, this species does resemble the human in important aspects of the metabolism of lipoproteins containing apo B-100.

We have carried out detailed kinetic analysis of the metabolism of apo B-100 in blood plasma of NZW and WHHL rabbits. The results of these investigations have been communicated in four primary reports (8-11). In all cases, radioiodinated autologous or homologous VLDL, IDL and LDL have been injected into animals fed comparable diets and data obtained from groups of such animals have been averaged. In early studies of the effects of fasting upon the metabolism of apo B-100 in NZW rabbits, we showed that the nutritional state modifies not only the rates of removal of lipoproteins containing apo B-100, but also the composition of the particles (12). When IDL from fed rabbits, which contain few cholesteryl esters, were injected into fasted rabbits, whose IDL and other lipoproteins were enriched in this lipid, rapid and extensive conversion of IDL to particles with the density of LDL was shown (unpublished data). This conversion could, however, be demonstrated as well when tracer amounts of radioiodinated IDL from fed animals were mixed with plasma of fasted animals, and incubated at 37° C. The "conversion" in this case resulted from reciprocal net transfers of cholesteryl esters and triglycerides between the tracer IDL and the cholesterol-enriched lipoproteins from the plasma of the fasted animals, resulting in a substantial increase in the particle density. No such "conversion" was observed in the reciprocal case, because the tracer VLDL, enriched in cholesteryl esters, presumably became more triglyceride-enriched during incubation, and remained at a density below 1.019 g/ml. An opposite "conversion" of IDL to VLDL has been reported by Minnich et al. during incubation of IDL from normal rabbits with VLDL from hypertriglyceridemic rabbits (13).

Our kinetic experiments took a form different from that ordinarily used in studies of the metabolism of lipoproteins

246

containing apo B. Our injected lipoproteins (VLDL, IDL, and LDL) were obtained by repeated ultracentrifugations at the appropriate densities and then mixed with plasma of the recipients to promote reacquisition of any surface components that may have been lost during preparation. After the lipoproteins were injected, however, we routinely analyzed the concentration of radiolabelled apo B in total plasma rather than in the isolated lipoproteins (VLDL, IDL and LDL). For this purpose, a method to analyze radioiodine in apo B of unfractionated plasma was developed (14). In addition, we separated from plasma those particles containing apo B-100 that also contained apo E (B,E particles) from those that lacked apo E (B particles) and measured separately the radioactivity contained in each population. Separately, we measured the concentration of apo B-100 in B,E and B particles within the VLDL, IDL and LDL density ranges to provide the steady state data needed to estimate transport rates on the basis of multicompartmental modeling. Using a form of "brute force" deconvolution available to us in the SAAM program (15,16), the whole plasma data could be separated mathematically into kinetic components corresponding to VLDL, IDL, and LDL.

The results of these experiments revealed metabolic as well as structural heterogeneity related to the presence of apo E on VLDL particles in both NZW and WHHL rabbits. We and others have pointed out that kinetic heterogeneity can also be related to the differing size of VLDL particles secreted from the liver (17,18). Such heterogeneity is evident in VLDL isolated from liver perfusates from several species, including rabbits (6). We therefore also carried out experiments in which large VLDL (mainly greater than 450Å in diameter) rather than total VLDL were injected. As in humans, we found that these particles were rapidly removed from the blood of NZW rabbits and none was converted to LDL. In WHHL rabbits, many but not all of the large VLDL were metabolized at a comparable rate and likewise were not converted to LDL. From these combined experiments, we developed a model for the metabolism of apo B-100 in large VLDL (with rapid rates of metabolism, designated "fast"), smaller VLDL (with slower rates of metabolism, designated "slow"), IDL and LDL in NZW and WHHL rabbits. Our model allowed for no independent production of IDL or LDL because of our observation that virtually all apo B-100 is secreted from the liver in the form of VLDL particles (6).

The steady state solutions of the model suggested that the rates of secretion of VLDL apo B-100 into blood plasma were similar in NZW and WHHL rabbits. This result is also consistent with our perfused liver data. The metabolism of VLDL apo B-100 differed, however, in the two groups of animals. In NZW rabbits, most apo B-100 was removed from the blood as VLDL and only about 7% was eventually converted to LDL and removed as such. In WHHL rabbits, VLDL were removed more slowly as such and considerably more was converted to IDL and eventually to LDL (about 24%). The increased conversion of apo B-100 secreted as VLDL to LDL, coupled with a reduced fractional rate of removal of VLDL, IDL, and LDL, accounted for the greater than twenty-fold increase in the concentration of LDL apo B-100 in plasma of WHHL rabbits, as well as the lesser increases in the concentration of apo B-100 in VLDL and IDL.

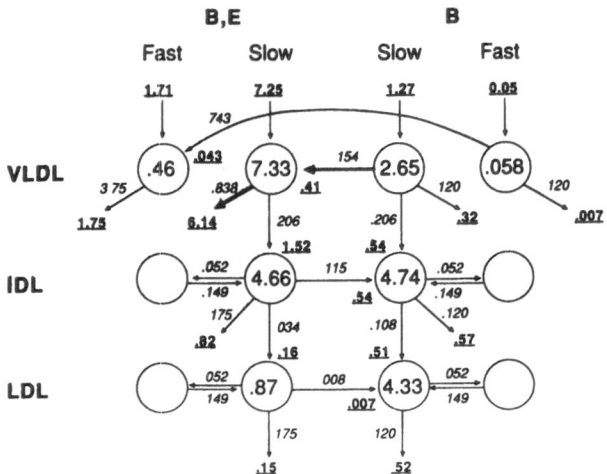

Figure 1. Multicompartmental model of the metabolism of apo B-100 in NZW rabbits developed from combined analysis of data on radioactivity in apo B-100 in whole plasma (8-10) and after density gradient ultracentrifugal separation of VLDL, IDL and LDL (11). The concentrations (mg/dl) of B,E and B particles of lipoprotein classes in the intravascular compartment are shown within the circles. Empty circles represent extravascular compartments for IDL and LDL. Transport rates (mg·dl⁻¹hr⁻¹) are underlined. Small, italicized numbers are rate constants (hr⁻¹). From D.M. Shames et al. in J. Lipid Res. 31:753 (1990), with permission of the publisher.

In those experiments in which total VLDL were injected into NZW and WHHL rabbits, we also measured in the distribution of radioiodine in apo B-100 in lipoprotein fractions separated by density gradient ultracentrifugation (11). The data thus obtained were combined with the data, summarized above, from analysis of apo B-100 in whole blood plasma. The results of the density gradient analysis were inconsistent with those obtained from analysis of the whole plasma data. To reconcile the two sets of data for NZW rabbits, it was necessary to add one new pathway in the VLDL portion of the model (whereby some B particles were converted to B,E particles) and to assume that there was appreciable cross-contamination between the fractions from the density gradients that were designated as IDL and LDL (Fig. 1). The latter was not surprising because it is well established that IDL and LDL do not exist as discrete ultracentrifugal populations in rabbits (1). For WHHL rabbits, it was necessary to add a new compartment for VLDL B,E particles with a fractional rate of irreversible removal intermediate between the "fast" and "slow" components defined from the earlier analyses, as well as to assume some cross-contamination between VLDL and IDL as well as between IDL and LDL (Fig. 2). These modifications of the models yielded only slight differences in the calculated rate of production of apo B-100 in VLDL and in the conversion of VLDL to IDL and IDL to

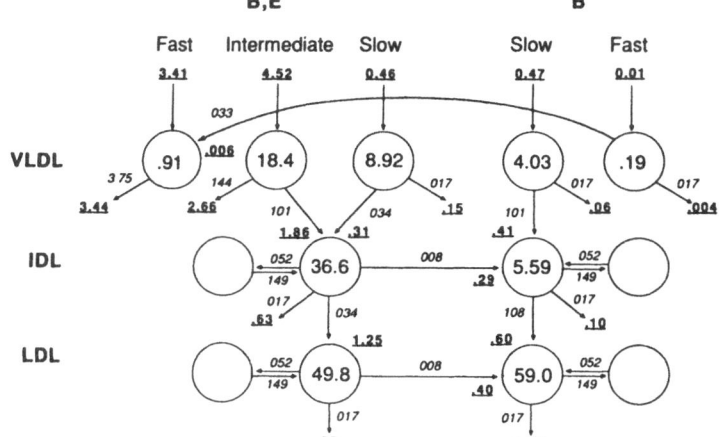

Figure 2. Multicompartmental model of the metabolism of apo B-100 in WHHL rabbits developed from combined analysis of data on radioactivity in apo B-100 in whole plasma (9-10) and after density gradient ultracentrifugal separation of VLDL, IDL and LDL (11). The concentrations (mg/dl) of B,E and B particles of lipoprotein classes in the intravascular compartment are shown within the circles. Empty circles represent extravascular compartments for IDL and LDL. Transport rates (mg·dl^{-1}hr^{-1}) are underlined. Small, italicized numbers are rate constants (hr^{-1}). From D.M. Shames et al. in J. Lipid Res. 31:753 (1990), with permission of the publisher.

LDL, as compared with the models based upon analysis of whole plasma data alone. The combined results therefore supported the usefulness of our approach based upon analysis of data obtained in whole plasma, which permitted the ready separation of B from B,E particles, and this made possible the appreciation and quantification of a novel form of structural and metabolic heterogeneity of these lipoproteins. Our results also indicated that ultracentrifugal separations, even when carried out in density gradients, can yield confusing results, owing to even small amounts of cross-contamination between kinetically distinct lipoprotein classes.

Although our current model of the metabolism of lipoproteins containing apo B-100, like all such constructs, provides an incomplete picture of the complexity of their metabolism, we believe that the calculated rates of production of apo B-100, and of its removal and conversion to lipoproteins of higher density, provide reasonable estimates of the overall processes that we have studied. We also believe that they provide a rational explanation for the effects of LDL receptor deficiency on these processes that have been poorly understood heretofore. In addition, our data have provided, for the first time, quantitative estimates of the effects of the presence of apo E upon the metabolic fate of lipoprotein particles that contain apo B-100.

It is, of course, of interest to speculate on the relevance of our observations for human lipoprotein metabolism. Are human VLDL (and, possibly IDL and LDL) metabolically heterogenous owing to their content of apo E? We have carried out analyses of the content of apo E in human plasma lipoproteins and have confirmed reports that VLDL particles that do not bind heparin contain little or no apo E (19,20). Such particles contain approximately 20% of the apo B-100 of normal human plasma VLDL. In order to minimize loss of apo E from VLDL during ultracentrifugation, we have separated these VLDL B particles on heparin-agarose columns directly from whole blood plasma and have subsequently isolated the VLDL by ultracentrifugation, as was done previously by Fielding and Fielding (20). We have found, however, that B particles that bind to heparin are also present in human VLDL (Campos, E., Havel, R.J., unpublished data). These particles, which are separated by sequential chromatography on anti apo E affinity columns and heparin-agarose columns, also contribute to about 20% of normal human VLDL. Thus, only about 60% of normal human VLDL particles contain apo E. Similar results have been obtained in humans with primary hypertriglyceridemia. Both forms of B particle bind very poorly to the LDL receptor and it is therefore reasonable to assume that they are removed slowly from the blood as such. It is likely, therefore, that the metabolic heterogeneity of VLDL that we have documented in rabbits also exists in humans.

A final issue of considerable interest and controversy is the extent to which apo B-100 enters human blood plasma at densities greater than that of VLDL. A number of investigators have obtained kinetic data that they have interpreted to show "direct" production of LDL. This has been a particularly common interpretation of kinetic data from studies of humans with homozygous familial hypercholesterolemia (1). By contrast, our hepatic perfusion experiments show that virtually all apo B is secreted from the liver of WHHL as well as NZW rabbits (6). This observation accords with the detailed in vivo kinetic data that we have obtained in the normal and mutant rabbits (8-11). This discrepancy between our findings in the rabbit and those obtained in other experimental animals and humans could be related to species variation in the metabolism of apo B. On the other hand, interpretations of in vivo kinetic data that lead to the hypothesis of significant de novo production of LDL could be in error. As an approach to this question, we have reanalyzed our in vivo kinetic data in NZW and WHHL rabbits subject to the hypothesis that significant de novo production of LDL does occur. We speculated that if our own in vivo kinetic data in the rabbit (where we have independent evidence that no de novo production of LDL exists), can be interpreted to "show" the presence of de novo production of LDL, then it might be possible that the kinetic data of others could be reinterpreted to "show" the absence of de novo production of LDL.

We have carried out such an analysis in a very simple way. Using our most recent models of apo B-100 metabolism in both B,E and B particles in NZW and WHHL rabbits (11), which are based on a large number of in vivo and hepatic perfusion experiments, we simulated a set of theoretical data corresponding to the concentration of radiolabelled apo B in plasma VLDL, IDL and LDL following injection of radiolabelled

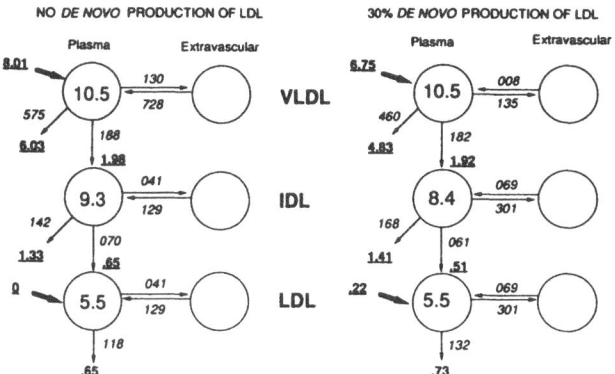

Figure 3. Simplified multicompartmental models proposed to fit the simulated data points generated from our most recent model of apo B-100 metabolism in NZW rabbits assuming zero (left) and 30% *de novo* production of LDL (right). The concentrations (mg/dl) of apo B-100 in VLDL, IDL and LDL in the intravascular compartment are shown within the circles. Empty circles represent extravascular compartments except for VLDL (see text). Transport rates (mg·dl^{-1}hr^{-1}) are underlined and fractional turnover rates (hr^{-1}) are italicized.

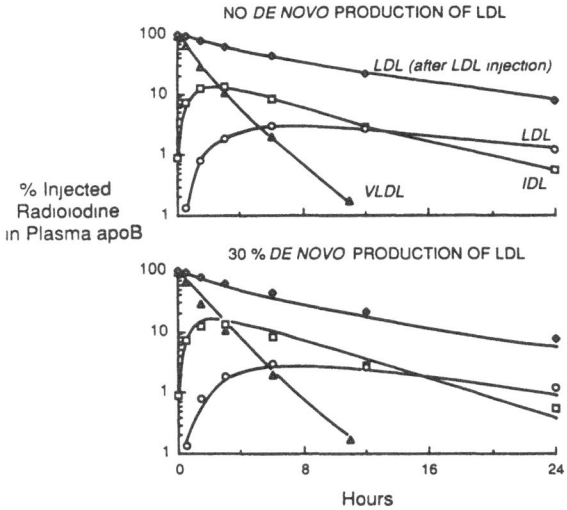

Figure 4. Simulated apo B-100 data points in VLDL (Δ), IDL ($\square$) and LDL (O) were generated from our most recent model of apo B-100 metabolism in NZW rabbits using tracer VLDL as the initial condition. Simulated apo B-100 data points in LDL ($\Diamond$) were also generated using tracer LDL as the initial condition. Solid lines through the data points represent the fits of the models of Fig. 3 to the simulated data points.

VLDL as the initial condition. We also simulated the apo B
response in plasma LDL following injection of radiolabelled LDL
as the initial condition. We then attempted to fit these
theoretical data, generated from a model without *de novo*
production of LDL, with a simple model where we assumed a range
of values for de novo production of LDL in NZW and WHHL
rabbits. A simple model was used rather than the complex
models of Fig. 1 and 2 in order to evaluate the hypothesis
being tested in the most straightforward manner.

The compartmental system used to fit the theoretical data
in NZW rabbits is shown in Fig. 3 and the results of our
analyses are also shown as the estimated values (rate
constants, transport rates and pool sizes) in Fig. 3 and as the
fits to the simulated data points in Fig. 4. The compartmental
structure exhibits a single plasma pool of apo B in LDL and IDL
with extravascular equilibration. Although no extravascular
penetration of the VLDL is known to exist, an extravascular
VLDL compartment was used in this model for simplicity to
explain the biexponential plasma disappearance of injected
VLDL. The differences in rate constants for this model shown
in the two halves of the figure reflect the different fits of
the model to the simulated data points subject to the
constraints of either zero or 30 percent de novo production of
LDL.

The fits to the simulated data at 30% direct production
of LDL do not differ appreciably from the fits with no direct
production of LDL. This result is perhaps not surprising given
the fact that according to this model only 8% of apo B-100 in
VLDL is converted to LDL, with direct production of LDL
accounting for only 3% of total apo B-100 production.

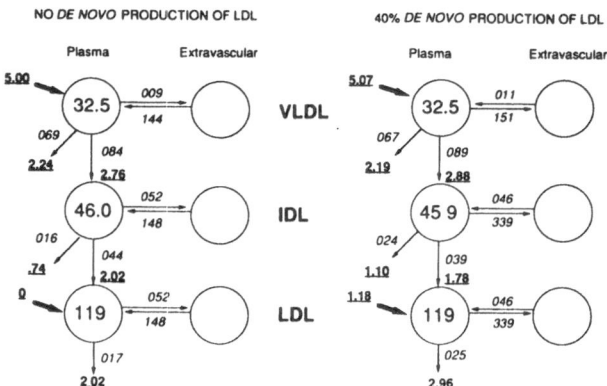

Figure. 5. Simplified multicompartmental models
proposed to fit the simulated data points generated from our
most recent model of apo B-100 metabolism in WHHL rabbits
assuming zero (left) and 40% *de novo* production of LDL (right).
The concentrations (mg/dl) of apo B-100 in VLDL, IDL and LDL in
the intravascular compartment are shown within the circles.
Empty circles represent extravascular compartments except for
VLDL (see text). Transport rates (mg·dl^{-1}hr^{-1}) are underlined
and fractional turnover rates (hr^{-1}) are italicized.

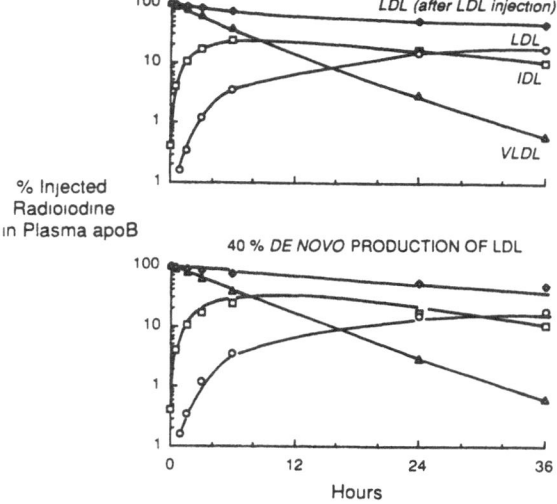

Figure 6. Simulated apo B-100 data points in VLDL (Δ), IDL (□) and LDL (O) were generated from our most recent model of apo B-100 metabolism in WHHL rabbits using tracer VLDL as the initial condition. Simulated apo B-100 data points in LDL (◇) were also generated using tracer LDL as the initial condition. Solid lines through the data points represent the fits of the simplified models of Fig. 5 to the simulated data points.

Figure 5 shows the compartmental system for WHHL rabbits and the estimated values with direct production of LDL set to zero or to 40%. As shown in Fig. 6, the fits to simulated data points exhibited by the model with direct production set to 40% differ little from those generated from the compartmental system with no direct LDL production. In the former case, 35% of apo B-100 in VLDL is converted to LDL, with direct production accounting for 19% of total apo B-100 production.

It seems incongruous, initially that this model is so insensitive to such a large variation in direct production of LDL. However, the reason becomes apparent when one considers that only a small fraction of the area that would be encompassed by the total specific activity-time curve for LDL (time zero to infinity) is represented by the data obtained over the 24 and 36 hour periods after injection in the NZW and WHHL rabbits respectively (Figs. 4 and 6). Thus significant discrepancies in this area that would become evident at longer times after injection may not be observed when only a small fraction of the total LDL area is represented by the data. We conclude that in such analyses, failure to take account of the extrapolative consequences of the data set beyond the last datum can lead to erroneous interpretations, in this case apparent direct production of LDL of up to 30% and 40% in NZW and WHHL rabbits, respectively. Clearly, the severity of this problem is inversely related to the final slope of the line derived from the product-activity data. In our studies, it has been possible to set this value of *de novo* production of LDL to zero because we had obtained independent evidence that direct production of LDL does not occur in NZW or WHHL rabbits.

CONCLUSIONS

1.   VLDL in rabbit blood plasma are structurally and metabolically heterogenous. This heterogeneity is related to differences in particle size and content of apo E. Data in humans confirm this structural heterogeneity and strongly suggest related metabolic heterogeneity.

2.   By utilizing independent information on the nature of nascent lipoproteins secreted from rabbit liver and by taking into account the structural and metabolic heterogeneity of VLDL, the hyperlipoproteinemia associated with LDL receptor deficiency in rabbits can be explained by reduced direct removal of VLDL, IDL and LDL from the blood, together with increased formation of IDL and LDL from their precursors. LDL receptor deficiency has no evident effect upon the secretion of apo B-100 into blood plasma.

3.   All apo B-100 enters blood plasma of normal and LDL receptor-deficient rabbits as VLDL. No particles resembling IDL or LDL are secreted. Analyses of kinetic data with simple models can, however, yield false estimates of "direct" LDL-production in rabbits. These can be shown to result, at least in part, from failure to obtain tracer data over a long enough period of time after injection to define adequately the shape of the curve that necessarily must be extrapolated beyond the last datum. Although we have employed a simple compartmental system to analyze our data, similar conclusions may be drawn for integral equation analysis. The current analysis raises the question whether estimates of direct production of LDL in humans may suffer from similar limitations.

This research was supported by a grant from the U.S. Public Health Service (HL-14237: Arteriosclerosis SCOR).

REFERENCES

1. R.J. Havel, The formation of LDL: mechanisms and regulation. J. Lipid. Res. 25:1570 (1984).
2. M. Berman, Kinetic analysis and modeling: theory and application to lipoproteins, in: "Lipoprotein Kinetics and Modeling," M. Berman, S.M. Grundy, B.V. Howard, eds., Academic Press, Inc., London (1982).
3. R.L. Hamilton, Hepatic secretion of nascent plasma lipoproteins, in: "Plasma Protein Secretion by the Liver," H. Glaumann, T. Peters, Jr., and C. Redman, eds., Academic Press Inc., London (1983).
4. L.S.S. Guo, R.L. Hamilton, R. Ostwald and R.J. Havel. Secretion of nascent lipoproteins and apolipoproteins by perfused livers of normal and cholesterol-fed guinea pigs, J. Lipid Res. 23:543 (1982).
5. T.E. Felker, R.L. Hamilton, J.-L. Vigne and R. J. Havel, Properties of lipoproteins in blood plasma and liver perfusates of rats with cholestasis, Gastroenterology 83:652 (1982).

6. C.A. Hornick, T. Kita, R.L. Hamilton, J.P. Kane and R.J. Havel, Secretion of lipoproteins from the liver of normal and Watanabe heritable hyperlipidemic rabbits, Proc. Natl. Acad. Sci. U.S.A. 80:6096 (1983).

7. Y.C. Ha, P.J. Barter. Differences in plasma cholesteryl ester transfer activity in sixteen vertebrate species, Comp. Biochem. Physiol. 71B:265 (1982).

8. N. Yamada, D.M. Shames, J B. Stoudemire and R. J. Havel, Metabolism of lipoproteins containing apolipoprotein B-100 in blood plasma of rabbits: heterogeneity related to the presence of apolipoprotein E, Proc. Natl. Acad. Sci. U.S.A. 83:3479 (1986).

9. N. Yamada, D.M. Shames and R.J. Havel. Effect of LDL receptor deficiency on the metabolism of apo B-100 in blood plasma: kinetic studies in normal and Watanabe heritable hyperlipidemic (WHHL) rabbits, J. Clin. Invest. 80:507 (1987).

10. N. Yamada, D.M. Shames, K. Takahashi, and R.J. Havel, Metabolism of apolipoprotein B-100 in large very low density lipoproteins of blood plasma: kinetic studies in normal and Watanabe heritable hyperlipidemic rabbits, J. Clin. Invest. 82:2106 (1988).

11. D.M. Shames, N. Yamada and R.J. Havel, Metabolism of apo B-100 in lipoproteins separated by density gradient ultracentrifugation in normal and Watanabe heritable hyperlipidemic rabbits. J. Lipid Res., 31:753 (1990).

12. J.B. Stoudemire, G. Renaud, D.M. Shames and R.J. Havel, Impaired receptor-mediated catabolism of low density lipoproteins in fasted rabbits, J. Lipid Res. 25:33 (1984).

13. A. Minnich, D.B. Nordestgaard, D.B. Zilversmit, A novel explanation for the reduced LDL cholesterol in severe hypertriglyceridemia. J. Lipid Res. 30:347 (1989).

14. N. Yamada and R.J. Havel, Measurements of apolipoprotein B radioactivity in whole blood plasma by precipitation with isopropanol, J. Lipid Res. 27:910 (1986).

15. M. Berman and M.F. Weiss, 1978, SAAM Manual, DHEW Publication No. (NIH) 78-180, National Institutes of Health, Bethesda, Maryland.

16. M. Berman, W.F. Beltz, P.C. Greif, R. Grabay, and R.C. Boston, 1983, "CONSAM User's Guide," U.S. Department of Health and Human Services, Public Health Service, National Institutes of Health, Bethesda, Maryland.

17. A.F.H. Stalenhoef, M.J. Malloy, J.P. Kane and R.J. Havel, Metabolism of apolipoproteins B-48 and B-100 of triglyceride-rich lipoproteins in normal and lipoprotein lipase-deficient humans, Proc. Natl. Acad. Sci. U.S.A. 81:1839 (1984).

18. C.J. Packard, A. Munzo, A.R. Lorimer, A.M. Gotto, and J. Shepherd, Metabolism of apolipoprotein B in large triglyceride-rich very low density lipoproteins of normal and hypertriglyceridemic subjects, J. Clin. Invest. 74:2178 (1984).

19. F.A. Shelburne and S.H. Quarfordt, The interaction of heparin with an apoprotein of human very low density lipoprotein, J. Clin. Invest. 60:944 (1977).

20. P.E. Fielding, C.J. Fielding, An apo-E free very low density lipoprotein enriched in phosphatidyl ethanolamine in human plasma, J. Biol. Chem. 261:5233 (1986).

ALTERATIONS IN CHOLESTEROL METABOLISM IN THE GENETICALLY HYPER-

CHOLESTEROLEMIC RICO RAT: AN OVERVIEW

K. Ouguerram, T. Magot, and C. Lutton

Laboratoire de Physiologie de la Nutrition
UPS, Bât. 447
URA 0646 CNRS, 91405 ORSAY Cedex, France

INTRODUCTION

The selection of spontaneously hypercholesterolemic rats in a large animal breeding unit and the maintenance of this characteristic over numerous successive generations were performed by Zucker[1], Kolestsky[2], Imai et al.[3] and Müller[4]. We were particularly interested by this last strain: by as early as 1979, RICO rats (rats with increased cholesterol) were 100% hypercholesterolemic after the fifteenth generation of consanguineous breeding. These animals are normotriglyceridemic and non-obese, in contrast with ''Zucker'' rats. They were isolated from heterozygous normocholesterolemic rats from this animal breeding unit (CIBA-GEIGY) which we shall refer to as SW. A single study, published in 1979, reported some of the characteristics of the cholesterol metabolism of these animals[4]. Since 1983, we have therefore conducted an exhaustive study of the characteristics of the cholesterol metabolism of this strain of rats.

PHYSIOLOGICAL CHARACTERISTICS

a. General physiological characteristics

The weight gain as well as the daily food ingestion of RICO rats were similar to those of normocholesterolemic heterozygotes killed at the age of 7 months (Table I). On the other hand, their liver weight was significantly higher than that of controls ($3.39 \pm 0.10\%$ of body weight versus $3.04 \pm 0.08\%$, n=6). Moreover, the intestine was longer in RICO rats (129 ± 3 cm) than in SW rats (113 ± 4 cm) and its weight was also greater (11.4 ± 0.8 g versus 7 ± 0.5 g), corresponding to $2.02 \pm 0.12\%$ and $1.45 \pm 0.09\%$ of the rat's body weight, respectively. It is noteworthy that the daily water intake of SW and RICO rats was much higher than that of Wistar rats from the Janvier breeding stock raised in our laboratory, and reached 40 to 50 ml/day versus 25-30 ml for the latter.

Several other physiological parameters such as plasma glucose, insulin and thyroid hormone levels and the glycogen concentration in the liver are presented in Table II. No significant differences were observed between the two groups of rats for any of these parameters.

Hypercholesterolemia, Hypocholesterolemia, Hypertriglyceridemia
Edited by C.L. Malmendier *et al.*, Plenum Press, New York, 1990

TABLE I
Body and organ weight and daily food ingestion
in SW and RICO rats

		SW	RICO
Body weight	(g)	559 + 18*	568 + 7
Liver weight	(g)	17.0 + 0.5	20.4 + 0.8[++b]
Intestine weight	(g)	7.7 + 0.5	11.4 + 0.8[++b]
Diet Ingestion	g/day	19.7 + 0.9	20.3 + 0.3

*Results are expressed as Mean + SEM (n=6)
  The rats were killed at 7 months)

TABLE II
Liver glycogen and plasma glucose, insulin
and thyroid hormones in SW and RICO rats

		SW	RICO
PLASMA			
Glucose	(µg/ml)	1000 + 38*	962 + 50
Insulin	(µU/ml)	112 + 17	111 + 16
Thyroxin T_4	(ng/ml)	43.7+ 2.8	44.1+ 2.7
Triodothyronine T_3	(ng/ml)	0.9+ 0.1	0.8+ 0.1
LIVER			
Glycogen concentration			
(mmole/g fresh weight)		228 + 21	188 + 30
(mmole/g dry weight)		715 + 32	598 + 42

*Results are expressed as mean + SEM (n=6) except for liver glycogen
(n=3).

b. Extrinsic characteristics of the cholesterol system

The size of the mobile cholesterol pool and the rates of the mobile
cholesterol turnover processes measured by the isotope equilibrium
method[5] are illustrated in Figure 1 for the two groups of rats.

The values for the mobile cholesterol turnover processes in SW rats
were identical to those obtained previously in Wistar rats from the
Janvier breeding unit[6] and in Wistar rats raised in our laboratory over
many generations. More specifically, the daily rate of cholesterol
degradation into bile acids (BA) and the internal cholesterol secretion
(IS) were equal to 15 and 16 mg/day respectively in adult rats fed a
semi-purified diet containing 53% sucrose and 10% lipids. In contrast,
the mobile cholesterol system of the RICO rat was characterized by a
substantial increase (about 50%) in the cholesterol present in the
mobile pool (433 mg versus 350 mg) and a simultaneous increase in one of
the forms of cholesterol influx (internal secretion) as well as its
efflux (cholesterol excretion and transformation into bile acids).

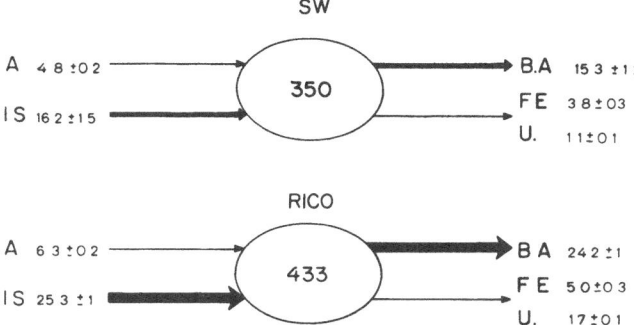

Fig. 1. Rates of mobile cholesterol turnover processes in SW and RICO rats

As the isotope equilibrium method cannot provide any information about the movements of cholesterol between the various parts of the system, and is also unable to express the efflux processes in terms of parameters (or catabolic rates), we followed the course of the specific activity (SA) of cholesterol in the plasma after intravenous injection of tritiated cholesterol for a period of 4 months. Under these conditions, the application of the occupancy principle provided the total influx rate of the system together with the elimination parameters of the system (fractional catabolic rate). Compartmental analysis (Figure 2) of the results provided parameters for the movements of cholesterol between the plasma and the tissue cholesterol pool according to the two-pool model classicaly used in the rat[7]: k21 represents the transfer of the rapidly exchangeable cholesterol pool towards the slowly exchangeable cholesterol pool and therefore corresponds to a movement of plasma cholesterol towards the tissue cholesterol pool; k12 describes the movement of cholesterol from these organs towards the plasma and k01 is the parameter of irreversible elimination from the first compartment, i.e. its fractional catabolic rate. R10 and R20 are the rates of influx of cholesterol into the 2 compartments of the system.

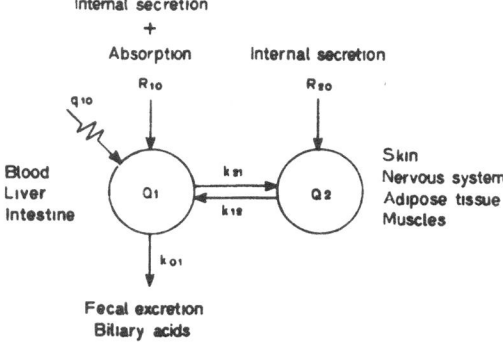

Fig. 2. 1. Bicompartmental model of body cholesterol metabolism.
1 and 2 rapidly and slowly exchangeable cholesterol pool.

The mass Q1 of the pool rapidly exchangeable with plasma was significantly higher in the RICO rat than in the SW rat (210 ± 5 mg versus 129 ± 10 mg), while the mass Q2, was not different between the two strains of rats. The values for k12 and k01 were also the same in the 2 groups. In contrast, the parameter k21, which describes the movement of plasma cholesterol towards the slowly exchangeable tissue cholesterol pool, was significantly lower in the RICO rat (0.18 ± 0.02 day^{-1}) than in the SW rat (0.26 ± 0.02 day^{-1}). The rate of influx of cholesterol into the system (R10+R20), was higher in the RICO rat (29 mg/day) than in the SW rat (18.5 mg/day). These values, obtained by compartmental analysis, were identical to those obtained by application of the occupancy principle.

It should be noted that the sum of Q1+Q2, obtained here by application of the occupancy principle, was higher than the value for the total body mass of exchangeable cholesterol determined by application of the isotope equilibrium method (Figure 1). The difference observed between the values obtained by the two methods was due to the fact that, with the isotope equilibrium method, performed over a period of 8 weeks, only a partial equilibrium was reached which underestimated (by about 20%) the mass of slowly exchangeable cholesterol.

c. Origin of synthesized cholesterol

In order to investigate whether the stimulation of internal cholesterol secretion observed in the RICO rat is of hepatic or non-hepatic origin, the radioactivity of the sterols in the liver, intestine and carcass (rest of the animal without the liver and intestine) was measured 70 minutes after subcutaneous injection of 100 µCi of 1-^{14}C acetate (Figure 3).

In the control SW rats, the radioactivity of the intestinal sterols was 4 times higher than that of the liver sterols, in agreement with previous data obtained in Wistar rats fed this semi-purified base diet[8]. In RICO rats, the radioactivity of the intestinal sterols was multiplied by a factor of 1.8 and that of the liver sterols was multiplied by a factor of 3, in comparison with the controls. Thus the ratio of the radioactivity of the intestine liver sterols, which reached a value of 4 in the control rats, was only 2.4 in RICO rats. In the latter group, the radioactivity of the liver sterols increased by a factor of 2.8 for free cholesterol and by a factor of 7.5 for cholesterol esters. A similar level of radioactivity was found in the sterols in the carcass in RICO and control rats.

We can conclude from these experiments that the increased internal secretion of cholesterol observed in RICO rats was due to stimulation of intestinal, but more particularly hepatic cholesterol synthesis.

d. Intestinal cholesterol synthesis

The mean cholesterol specific activity of intestinal mucosa, in RICO and control rats 70 minutes after a subcutaneous injection of 100 µCi of 1-^{14}C acetate, is shown in Figure 4. Independently of the segment of small intestine (duodenum or terminal ileum), an increasing gradient of cholesterol synthesis was observed from the cell located in the upper part of the villus towards those cells located more deeply near the crypts. Furthermore, a decrease in the cholesterol radioactivity of the enterocytes was observed from the 1st to the 3rd quarter, followed by an increase in the terminal quarter of the small intestine. Stimulation of intestinal cholesterol synthesis has been demonstrated in RICO rats. This stimulation of cholesterol synthesis by enterocytes does not appear to be specific to their location in the small intestine (duodenum, jejunum, ileum) or in the villus. The general profile of this synthesis was the same in the RICO and SW rat and in the Janvier Wistar rat with or without cholesterol in its diet[8-9].

An original observation concerns the decrease in the cholesterol/protein ratio in the RICO rat enterocyte due to an increase by more than 20% in its protein content (RICO: 24.6±1.2 mg/mg DNA, n=46; SW: 20.9 mg/mg DNA, n=47, p<0.001).

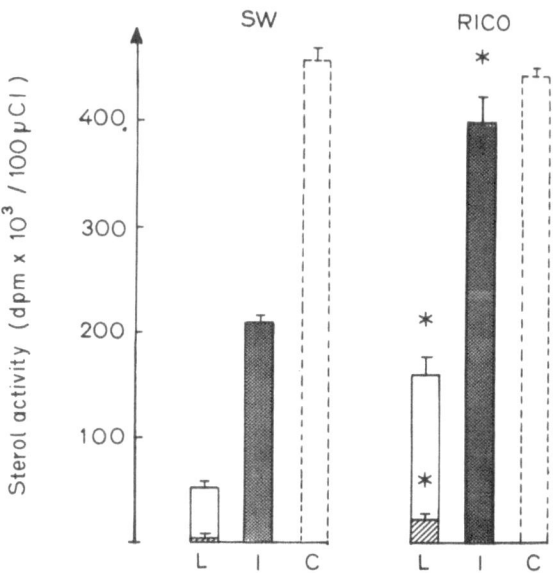

Fig. 3. Sterol radioactivity (dpm/100 µCi) in liver (L), small intestine (I) and carcass (C), 70 minutes after a subcutaneous administration of 100 µCi of 1-^{14}C acetate in the SW or RICO rat.

In the light of these findings, we wanted to determine whether there was any difference in the secretion of cholesterol into intestinal lymph between the two groups of rats. The rate of secretion of total cholesterol into intestinal lymph was measured after cannulation of the thoracic duct. For a lymph flow rate of the order of 1 to 1.4 ml/hour, the quantity of cholesterol secreted into the lymph was almost twice as high in the RICO rat (580 ± 20 µg.hr^{-1}) than in the SW rat (330 ± 10 µg.hr^{-1}). The increase in the cholesterol flow rate in the chyle was also detected in the chylomicrons (x 1.7), VLDL (x 1,5) as well as the lipoproteins with a density>1.006 (x 1.4).

These results confirm earlier observations demonstrating that the incorporation of ^{14}C-acetate into intestinal sterols and therefore intestinal cholesterol synthesis is increased in the RICO rat.

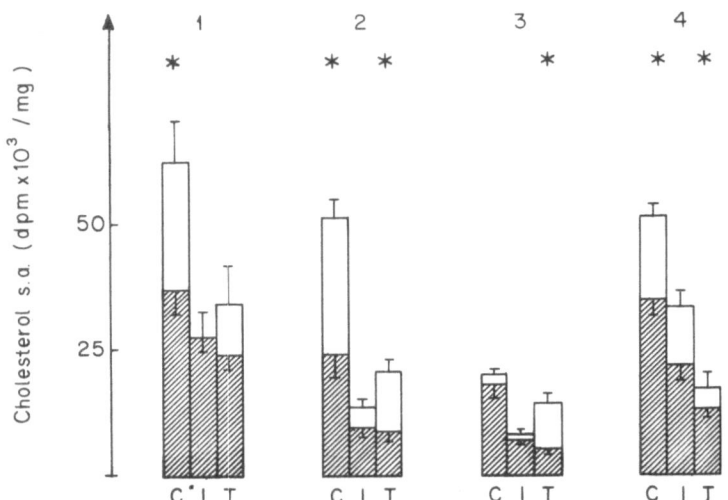

Fig. 4. Mean specific activity of intestinal mucosa sterols (dpm per mg of cholesterol) in SW or RICO rat, 70 minutes after a subcutanenous administration of 100 μCi of |1-^{14}C|acetate. The cells were collected after the small intestine was divided into four equal quarters (1st, 2nd, 3rd, 4th). T, top cells (fractions 1-3, Weiser's method): I: intermediate cells (fractions 4-6, Weiser's method); C, crypt cells (fractions 7-9, Weiser's method).

DISTRIBUTION OF PLASMA CHOLESTEROL AND APOLIPOPROTEINS

This distribution was studied in SW and RICO rats receiving a semi-purified base diet (containing 0.05% cholesterol, SW and RICO groups) or supplemented with 0.5% cholesterol for 3 weeks (groups SW$_{CH}$ and RICO$_{CH}$).

The cholesterol concentration (indicating the fraction of free cholesterol in brackets) in the plasma lipoprotein fractions and the sum of total cholesterol are shown for the four groups of rats in Table III. The total plasma cholesterol was twice as high in the RICO rat than in the SW rat. The raised plasma cholesterol was due to a rise in free cholesterol (x 1.55), but more importantly to a rise in cholesterol esters (x 2.25). After removal of the chylomicrons and the VLDL by sequential ultracentrifugation, the separation of 2 ml of plasma on a density gradient (Figure 5) demonstrated the distribution of cholesterol and total apolipoproteins on the first eighteen 0.5 ml fractions isolated successively. The rise in the cholesterol and protein concentration in the RICO rat essentially concerned LDL1 1.006<d<1.050, LDL2 (or HDL1) 1.050<d<1.063 and HDL2 1.063<d<1.095. The hypercholesterolemia of the RICO rat therefore corresponds to hyperlipoproteinaemia of the lipoproteins in the density zone: 1.020-1.095. However, the significant increase in the cholesterol/protein ratio of these lipoproteins in the RICO rat demonstrates that the hypercholesterolemia of this rat is also partly due to an overload of cholesterol in its lipoproteins. In contrast, the RICO rat does not present any hypertriglyceridemia[10].

TABLE III

Concentration (µg/ml plasma) of a) cholesterol, indicating the fraction of free cholesterol in brackets, in the plasma lipoprotein fractions in rats fed the base diet (SW and RICO groups) and in rats fed the base diet supplemented with 0.5% cholesterol (SW$_{CH}$ and RICO$_{CH}$ groups).

	SW	RICO	SC$_{CH}$	RICO$_{CH}$
Chylomicrons	12± 3 (54%)	15± 3 (47%)	82±17[a] (55%)	185±20[b] (42%)
VLDL	63± 7 (54%)	83± 8 (45%)	790±28[a] (30%)	760±35 (25%)
LDL1	110± 6 (30%)	267± 4[a] (30%)	92±11 (64%)	228±27[b] (76%)
LDL2	190±44 (28%)	352±22[a] (26%)	145±18 (70%)	507±37[b] (51%)
HDL	355±51 (50%)	559±29[a] (48%)	294± 7 (39%)	424±10[b] (40%)

Mean ± SEM (n=6)   a: p ≤ 0.05 RICO or SW$_{CH}$ versus SW
b: p ≤ 0.05 RICO$_{CH}$ versus SW$_{CH}$

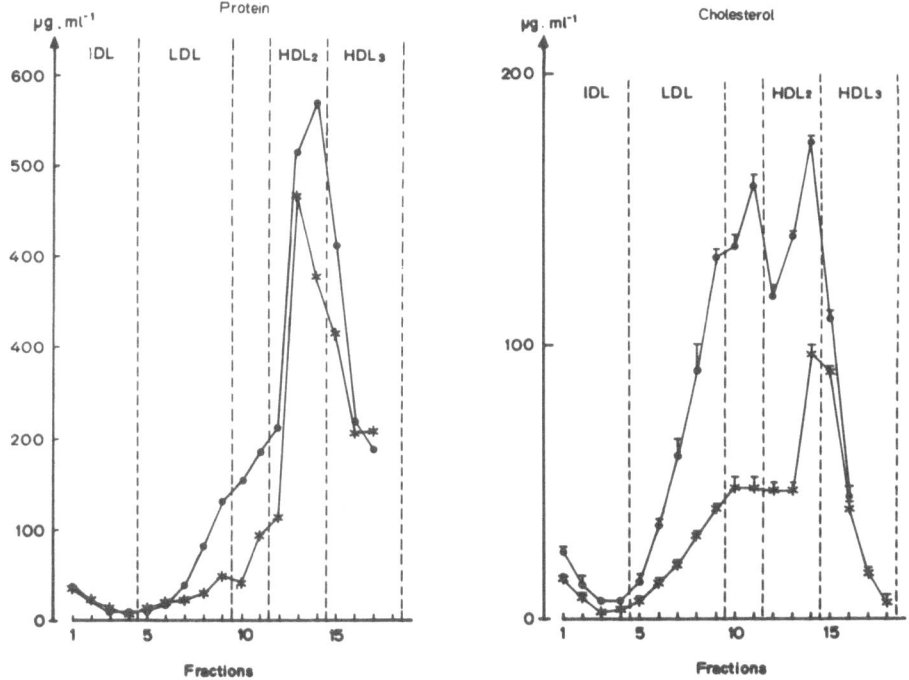

Fig. 5. Protein and cholesterol distribution in 18 lipoprotein fractions from 2 ml of plasma after separation of chylomicrons and VLDL in SW (*) or RICO (●) rat (n = 4)

IDL : 1.006 < d < 1.020
LDL$_1$: 1.020 < d < 1.050
LDL$_2$: 1.050 < d < 1.063
HDL$_2$: 1.063 < d < 1.125
HDL$_3$: 1.125 < d < 1.21

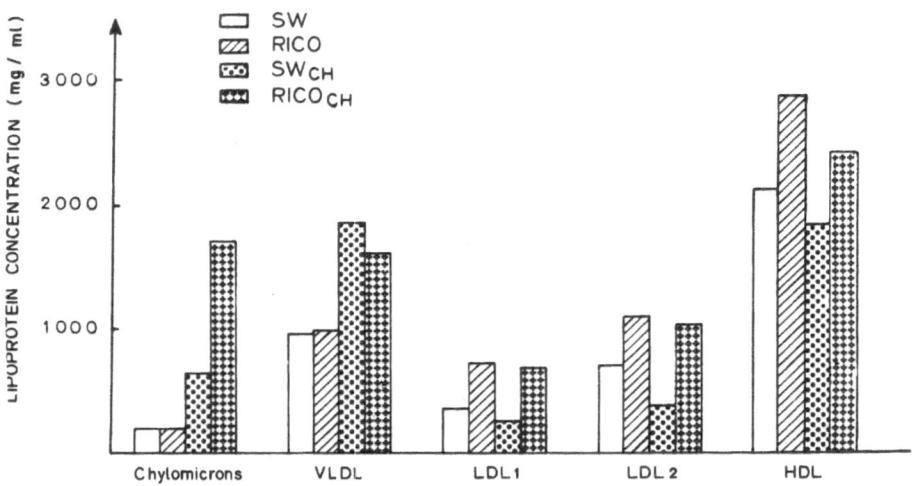

Fig. 6. Plasma lipoprotein profile from SW and RICO rats fed the base diet with or without 0.5% cholesterol.

The addition of 0.5% of cholesterol to the diet increased the plasma cholesterol concentration by 60% in the $RICO_{CH}$ rats and by 52% in the SW_{CH} rats (Table III). With the exception of the VLDL, the total plasma lipoprotein concentration was higher in the $RICO_{CH}$ rats than in the SW_{CH} rats (Figure 6). This was especially the case for the chylomicron, LDL1 and LDL2 fractions, which were 1.6 times higher in the RICO rats than in the SW rats.

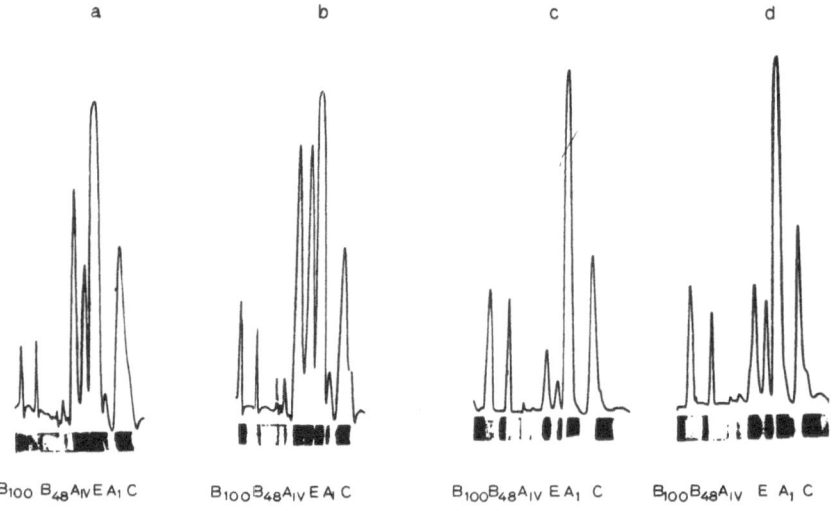

Fig. 7. SDS 5 and 10% polyacrylamide gels and densitometric scanning of total lipoproteins from SW and RICO rats fed the base diet with or without 0.5% cholesterol (a) SW (b) RICO (c) SW_{CH} (d) $RICO_{CH}$. The bands corresponding to $apoB_{100}$, B_{48}, A_{IV}, E, A_I and C (C_I + C_{II} + C_{III}) are indicated.

SDS-PAGE analysis of total plasma lipoproteins, as illustrated in Figure 7, clearly showed apo E enrichment in RICO versus normocholesterolemic SW rats. A higher quantity of apo E was also seen in various lipoproteins from the RICO rat compared to those from the SW rat (particularly the chylomicrons, VLDL, LDL1 and LDL2) (for detailed results, see Ref. 9). The higher content of apo E and the lower content of apo B48 has also been observed in the chylomicrons isolated from thoracic duct lymph.

LIPOPROTEIN LIPASE AND HEPATIC LIPASE ACTIVITIES

We have seen that the major characteristics of the RICO rat are an enhanced hepatic and intestinal cholesterol synthesis and an increased mobile cholesterol pool as compared to the normocholesterolemic control (SW). With a semi-purified diet, the hypercholesterolemia of the RICO rat is mainly due to an increased cholesterol content in the d>1.006 lipoprotein fractions. When 0.5% cholesterol was added to the diet, hypercholesterolemia was induced in both strains of rats, but cholesterol enrichment was located in the lighter lipoproteins (VLDL and chylomicrons) which contained 60% (SW) and 45% (RICO) of the plasma cholesterol, respectively. Under these conditions, cholesterol enrichment of VLDL (9 fold) was comparable for both strains of rats, but that of chylomicrons was greater in the RICO rats (12 fold) than in the SW rats (7 fold), suggesting a longer delay in chylomicron removal in the first strain. In the light of these results, we consequently studied hepatic lipase (HL) and lipoprotein lipase (LPL) activies in the two groups of animals[11].

The plasma post-heparin LPL and HL activities in SW and RICO rats fed the control or the 0.5% cholesterol diet are described in Table IV. Plasma post-heparin LPL was significantly lower (35% decrease) in RICO rats, while no inter-group difference in heparin-released HL could be detected.

To confirm the absence of any difference in HL activity, the secretion of the enzyme by isolated hepatocytes was measured in vitro. HL, stabilized by increasing doses of heparin (Figure 8a), was comparable for SW and RICO rats after 150 minutes of incubation. The time course of secreted HL activity (Figure 8b) was also similar, with a slight tendency towards a lower activity in RICO hepatocytes. HL activities measured in liver homogenates were identical for SW and RICO rats (1.56±0.13 and 1.65±0.12 units/g, respectively).

These results show that, with a standard diet, only a defect in post-heparin LPL (-35%) could be detected. HL activity was strictly comparable in RICO and SW rats whether the enzyme was measured in post-heparin plasma, in liver homogenates or in incubated hepatocytes. Uptake of chylomicron remnants by isolated hepatocytes was also identical, demonstrating the absence of any abnormality at the cellular level (apo E and/or apo B,E receptors or other processes). However, since heterologous remnants (from Wistar rats) were used in this experiment, the possibility of a different behaviour in response to remnants prepared from RICO rats cannot be excluded. In order to confirm this hypothesis, we studied the in vivo clearance with homologous chylomicrons.

TABLE IV

PLASMA POST—HEPARIN LPL and HL ACTIVITIES IN SW AND RICO RATS FED THE CONTROL AND THE CHOLESTEROL DIETS

After blood aliquots were taken for cholesterol determination, heparin (20 U/100 g) was injected IV and blood samples were collected 2 min later. LPL and HL activities were assayed as described in ref. 11. Results expressed as mU/ml are means+SEM (n=number of rats). Values of RICO versus SW rats were significantly different at a: $p<0.05$ and b: $p<0.001$ levels. Effects of cholesterol diets were significant at 1: $p<0.05$.

	SW	RICO
Post—heparin LPL		
Control diet	485 $\pm$ 71 (8)	316 $\pm$ 43 (10)[a]
+ 0.5% cholesterol	488 $\pm$ 30 (6)	287 $\pm$ 18 (6)[b]
Post—heparin HL		
Control diet	1306 $\pm$ 89 (5)	1369 $\pm$ 57 (6)
+ 0.5% cholesterol	1108 $\pm$ 32 (6)[1]	1161 $\pm$ 86 (5)

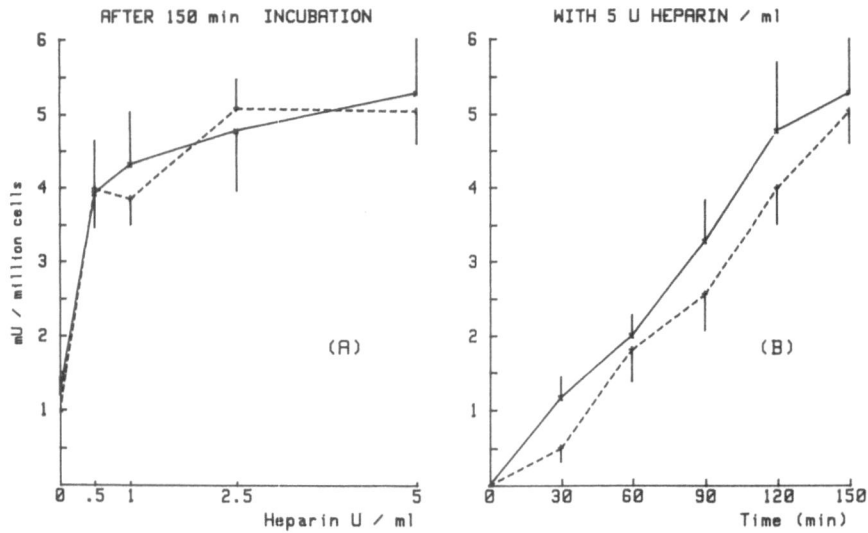

Fig. 8. HL activity secreted in the medium during incubation of isolated hepatocytes from SW and RICO rats fed the standard diet. Cells (10^7 per flask) were incubated at 37°C in Krebs medium for 150 min with increasing concentrations of heparin (Fig. 1a) or with 5 U heparin/ml as a function of time (Fig. 1b). At indicated times, the medium was separated from the cell pellet by brief centrifugation and stored at -20°C until HL activity was determined. Values expressed as mU/10^6 cells are means $\pm$ SEM for 5 independent experiments. No significant difference was found between SW x——x and RICO x- - --x hepatocytes.

CATABOLISM OF CHYLOMICRON CHOLESTEROL ESTERS

a. Turnover of cholesterol esters in autologous chylomicrons

Autologous chylomicrons, essentially labelled on the cholesterol esters, were injected into RICO and SW rats. The plasma cholesterol esters were isolated at various times (3, 5, 8 and 15 minutes) and the kinetics of clearance of their radioactivity was determined (Figure 9).

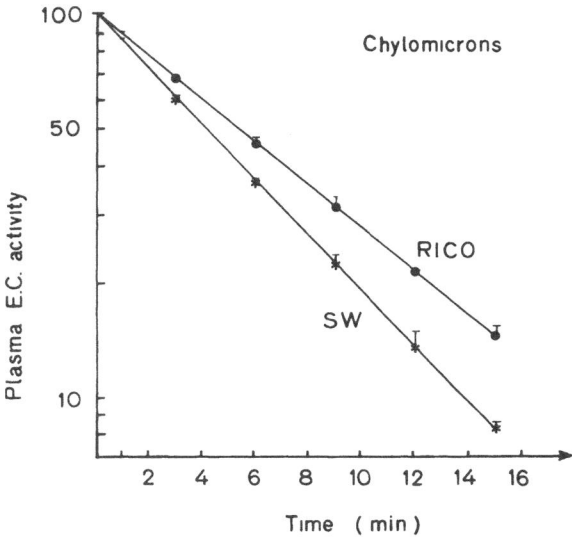

Fig. 9. Removal from blood plasma of autologous |3H C.E.| chylomicrons in SW (*) and RICO (●) rats.
Blood samples were obtained from a tail vein. Each point represents the mean value from five rats. Bars indicate ± SEM

The fractional catabolic rate was significantly lower in the RICO rats (7.7 ± 0.6 hr^{-1}) than in the SW rats (10.2 ± 0.7 hr^{-1}). Thus, 13% and 17% of the cholesterol esters of plasma chylomicrons disappeared each minute in the RICO rats and SW rats, respectively. The rate of production of cholesterol esters in the chylomicrons, calculated from the masses measured in experimental rats, was higher in the RICO rats than in the SW rats: 1.9 ± 0.15 mg.hr^{-1} versus 1.1 ± 0.07 mg.hr^{-1}, respectively. The decrease in the fractional catabolic rate in RICO rats may be related to a property (composition, diameter, ...) of the chylomicron itself and/or to a defect in its catabolic system (action of LPL, ...). In order to answer this question, we studied the catabolism of chylomicron cholesterol esters in RICO rats following the administration of chylomicrons from SW rats and in SW rats following the administration of chylomicrons from RICO rats (heterologous chylomicrons).

b. Cholesterol ester turnover in heterologous chylomicrons

When heterologous chylomicrons were used, the clearance of cholesterol esters was faster in the RICO rat (9.5 ± 0.7 hr^{-1}) than in the SW rat (7.8 ± 0.6 hr^{-1}). Regardless of the recipient, the fractional catabolic rate obtained when the donor was a RICO rat was therefore significantly lower than that obtained for any situation in which the donor was an SW rat.

As we have seen, the relative apolipoprotein composition of lymphatic chylomicrons essentially differs by an increase in apo E (6 fold) and a decrease in apo B48.

In the RICO rats, the fractional catabolic rate of chylomicron cholesterol esters was significantly lower than that found in the SW rat. We have shown above that the LPL activity was lower in the RICO rat. The activity of this enzyme is essential for the formation of remnants, subsequently bound by the liver by specific apo E receptor[12]. The fact that chylomicrons from SW rats injected into RICO rats demonstrated normal catabolism indicates that the RICO rat does not present any abnormality in the receptors for chylomicron remnants. Our findings also tend to suggest that the low fractional catabolic rate of chylomicrons in RICO rats is not due to saturation of this system, but is rather related to the nature of the particle itself. The apparent contradiction between the relative apolipoprotein E enrichment of the chylomicrons in RICO rats and their low fractional catabolic rate supposes the existence of an abnormal form of apolipoprotein E, as has already been observed in certain human diseases[13]. However, determination of the various isoforms of apo E, performed by two-dimensional electrophoresis (2DE) did not reveal any differences in the proportions of the 4 isoforms present in the chylomicrons or VLDL of SW and RICO rats. The fact that the apo CII/apo CIII ratio of these fractions was lower in the RICO rat than in the SW rat is coherent with the decreased lipoprotein lipase activity observed in the hypercholesterolemic rat.

CATABOLISM OF CHOLESTEROL ESTERS IN LYMPHATIC AND PLASMA VLDL

a. Lymphatic VLDL

After injection of autologous lymphatic VLDL labelled with tritiated cholesterol, the clearance of cholesterol esters obeyed mono-exponential kinetics (unpublished results). Although the mean fractional catabolic rate of VLDL was lower in the RICO rat than in the SW rat, the difference was not significant (4.3 ± 0.6 hr^{-1} versus 5.1 ± 0.4 hr^{-1}).

b. Plasma VLDL

When the experiment was repeated with VLDL isolated from plasma (plasma VLDL), the fractional catabolic rate (FCR) was considerably lower (3.4 to 3.6 hr^{-1}) but did not differ significantly between RICO and SW rats. In these experiments[14], we observed that the production of lymphatic (intestinal) VLDL was higher in the RICO rat than in the SW rat. The same applies for the plasma transformation of VLDL into LDL, as, in the RICO rat compared with the SW rat, two to three times the amount of radioactivity was detected in the LDL 30 minutes after the injection of VLDL labelled with ^{3}H-cholesterol esters.

In order to differentiate between the clearance of cholesterol esters and that of the particle itself, the catabolism of HDL was studied after labelling of the cholesterol esters and after labelling of the apolipoproteins.

a. Catabolism of HDL cholesterol esters

After labelling of autologous HDL with ^{14}C-cholesteryl linoleyl-ether[15], the kinetics of clearance of the radioactivity were followed for 24 hours. The FCR, determined by the occupancy principle, was identical in the 2 groups of animals ($8.9+0.3\%.hr^{-1}$ and $8.8+0.5\%.hr^{-1}$, for the RICO rats and SW rats, respectively).

24 hours after the injection, the uptake by the organs, expressed as a percentage of the dose injected, was compared for the 2 groups of animals (Table V). The total uptake of radioactivity by all of the organs examined represented 70% of the total radioactivity injected for the 2 groups of animals. The liver was by far the major site of this uptake, as it was 20 times more active than the adrenal glands or small intestine.

b. Turnover and tissue uptake of ^{14}C-sucrose labelled HDL

The turnover and tissue uptake of HDL (d: 1.095-1.21) were compared in SW and RICO rats by using a constant infusion method of ^{14}C-sucrose labelled HDL[16]. Again, the HDL clearance rate was not significantly lower in the RICO rat ($320+22$ $\mu l.hr^{-1}$ per 100 g of rat body weight) than in the SW rat ($366+24$ $\mu l.hr^{-1}$ per 100 g of rat body weight). The same applied to the fractional catabolic rate which was equal to $6.6+0.3\%.hr^{-1}$ when determined for a 24 hour experiment[14].

Table V

Uptake ^{14}C cholesteryl linoleyl ether of HDL by whole organs expressed as a percentage of the quantity injected, measured 24 hours after the injection (n = 5)

	SW	RICO
Adrenals	3.3 + 0.2	0.1 + 0.0[a]
Liver	68 + 3.1	68 + 2.8
Kidneys	0.3 + 0.0	0.4 + 0.1
Jejunum	2.4 + 0.3	1.5 + 0.1[b]
Colon	0.2 + 0.0	0.2 + 0.00
Caecum	0.2 + 0.0	0.2 + 0.0
Testes	0.4 + 0.0	0.2 + 0.0[b]

Difference in relation to SW rats:

a : P<0.001 ; b : P<0.01 ; c : P<0.02

CATABOLISM OF ^{14}C-SUCROSE LABELLED LDL

As the catabolism of LDL cholesterol esters in the rat is totally dependent on that of the particle, as demonstrated by the fact that the cholesterol esters of these lipoproteins disappear at an identical rate to that of the proteins[17,18], we compared the catabolism of ^{14}C-sucrose labelled LDL (1.020<d<1.050). This density range was selected in order to eliminate IDL (1.066<d<1.019)[19] and to minimise possible contamination by apo E-rich HDL with d<1.050[20,21].

As the RICO rat has a very marked increase in LDL_1 cholesterol and in the number of these particles in the plasma, we studied the kinetics of autologous and heterologous ^{14}C-sucrose labelled LDL (1.020<d<1.050) in SW and RICO rats over a period of 24 hours.

Experiments concerning LDL catabolism in SW and RICO rats were described in another part of this volume. The LDL FCR was significantly lower in the RICO rat (-30%; $5.5+0.6\%.hr^{-1}$) than in the SW rat ($7.9+0.8\%.hr^{-1}$). The rate of production of LDL (equal to the rate of degradation) was $0.4+0.05$ $mg.hr^{-1}$ in the RICO rat versus $0.19+0.03$ $mg.hr^{-1}$ in the SW rat, which, in terms of cholesterol esters, corresponds to fluxes of $0.24+0.04$ $mg.hr^{-1}$ and $0.12+0.01$ $mg.hr^{-1}$, respectively. The fractional catabolic rate of heterologous LDL administered to SW rats ($5.4+0.5\%.hr^{-1}$) or RICO rats ($5.5+0.6\%.hr^{-1}$) was identical to that of the homologous LDL administered to the RICO rat ($5.5+0.6\%.hr^{-1}$), but significantly lower than that of the SW rats which received their own LDL ($7.9+0.8\%.hr^{-1}$).

The total plasma clearance of LDL (FCR x plasma volume) was $1762+148$ $\mu l.hr^{-1}$ in the SW rats versus only $1.189+131$ $\mu l.hr^{-1}$ in the RICO rats. Amongst the various organs studied, the liver, small intestine and kidney constituted, in order of importance, the major organs of LDL binding (expressed per whole organ) in the SW and RICO rat. When the organ binding of LDL was compared in the 2 groups of SW and RICO rats, we found that it was reduced by 40% for the liver and by 30% for the adrenal glands in the RICO rat. These differences were further accentuated for the liver when the binding was expressed per gram of organ.

CONCLUSIONS

Figure 10 illustrates all of the principal data obtained concerning the movements of cholesterol and lipoproteins in the RICO rat in comparison with the normocholesterolemic SW rats.

The hypercholesterolemia of the RICO rat is related to a decreased rate of catabolism of chylomicrons and LDL, but more especially to an excessive production of these 2 types of lipoproteins.

We have observed that the RICO rat has a 25% reduction in the FCR of chylomicron cholesterol esters and a 30% reduction in the catabolism of LDL. We know that chylomicrons, transformed into ''remnants'' by the action of LPL (whose activity is reduced by 25% in the RICO rats), are actively bound by high affinity specific apo E receptors in the liver[22,23]. The contradictory increase in the apo E concentration in the RICO rat, already observed in the lymphatic chylomicrons, may suggest the presence of an isoform which is poorly recognised by the hepatic receptors. However, our results failed to demonstrate such a defect, which could be situated at the level of the molecular structure of the apo E, but which does not appear to be related to a modification in the hepatic receptor itself. As in the case of the chylomicrons, the plasma clearance of LDL is essentially hepatic: the efficiency of this process in the liver of the RICO rat is decreased by 40% in comparison with that of the normocholesterolemic rat. This reduction in the FCR could be due to a modification related to the nature of the lipoprotein (for example, the cholesterol/protein ratio, the apo E content ... are higher) but also related to the receptors themselves (decreased number or saturation of the receptors). Apo B/E receptor quantification experiments need to be conducted in order to determine which type of abnormality is responsible for hyper-LDL lipoproteinemaemia in the RICO rat.

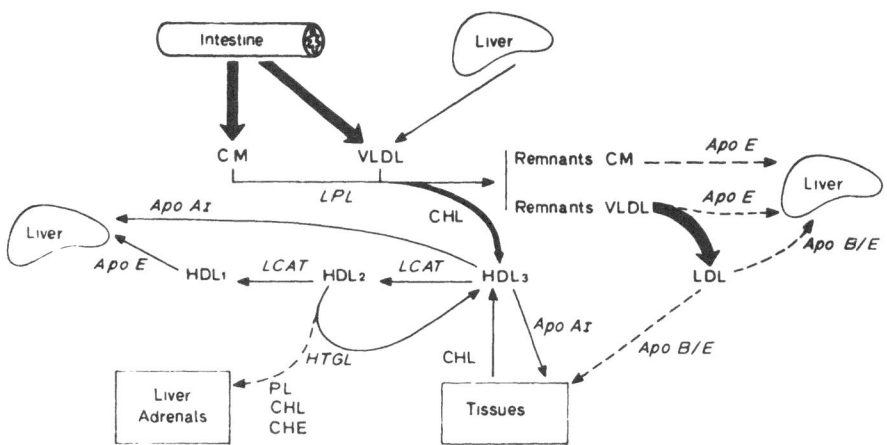

Fig. 10. Metabolism of lipoprotein cholesterol in the RICO rat.

The increased production of chylomicrons and LDL in the RICO rat is an essential factor responsible for hypercholesterolemia in this strain. The excessive secretion of these lipoproteins is apparently not due to an alteration in triglyceride or carbohydrate metabolism (diabetes ...) or secondary to hypothyroidism. The synthesis of cholesterol by the intestine (x 2), liver (x 4) and the intestinal secretion of chylomicrons and VLDL (x 1.5) are higher in the RICO rat than in the normocholesterolemic rat. In the rat, a very small proportion ($\leq 2\%$) of the VLDL is normally transformed into LDL in the plasma, as the great majority is bound by the liver. After labelling the cholesterol esters of the VLDL, 2 to 3 times more VLDL were slowly transformed into LDL in the RICO rat than in the SW rat, contributing to an increased production of LDL. It remains to be seen whether the liver of the RICO rat is also able to secrete larger quantities of LDL than the liver of the SW rat[24,25]. All of these data demonstrate the value of this genetically hypercholesterolemic animal for the study of hypolipidemic drugs, particularly those designed to decrease the plasma concentrations of chylomicrons and LDL.

REFERENCES

1. **ZUCKER, T.F., and ZUCKER, L.M.**
   Hereditary obesity in the rat associated with high serum fat and cholesterol.
   Proc. Soc. exp. Biol. Med. **110** (1962) 165-171.

2. **KOLETSKY, S.**
   Obese spontaneously hypertensive rats - A model for study of atherosclerosis.
   Exp. Molec. Pathol. **19** (1973) 53-60.

3. **IMAI, Y., MATSUMURA, H., MIYAJIMA, H., and OKA, K.**
   Serum and tissue lipids and glomerulonephritis in the spontaneously hypercholesterolemic (SHC) rat, with a note on thye effects of gonadectomy.
   Atherosclerosis **27**(1977) 165-178.

4. **MULLER, K.R., LI, J.R., DINH, D.M., SUBBIAH, M.T.R.**
   The characteristics and metabolism of a genetically hypercholesterolemic strain of Rats (RICO).
   Biochim. Biophys. Acta, **574** (1979)334-343.

5. **LUTTON, C., CHEVALLIER, F.**
   Vitesses des processus de renouvellement du cholestérol contenu dans son espace de transfert, chez le rat.
   III- Modifications et étude critique de la méthode d'équilibre isotopique.
   Biochim. Biophys. Acta, **255** (1972) 762-779.

6. **MATHE, D., LUTTON, C.**
   Le Cholestérol. Aspects dynamiques et métaboliques.
   J. Physiol., **79** (1984) 41-97.

7. **MAGOT, T., FREIN, Y., CHAMPARNAUD, G., CHERUY, A., LUTTON, C.**
   Origin and fate of rat plasma cholesterol in vivo. Modelling of cholesterol movements between plasma and organs.
   Biochim. Biophys. Acta, **921** (1987) 587-594.

8. **LUTTON, C., PERRODIN, M., CARDONA-SACLEMENTE, L.E., SEROUGNE, C.**
   In vivo cholesterol synthesis by the rat digestive tract. III. Evaluation of modulating factors.
   Reprod. Nutr. Dev., **26** (1986) 1241-1253.

9. **CARDONA-SANCLEMENTE, L.E., VERNEAU, C., MATHE, D., LUTTON, C.**
   Cholesterol metabolism in the genetically hypercholesterolemic rat (RICO).
   I. Measurement of turnover processes.
   Biochim. Biophys. Acta **919** (1987) 205-212.

10. **CARDONA-SANCLEMENTE, L.E., FEREZOU, J., and LUTTON, C.**
    Cholesterol metabolim in the genetically hypercholesterolemic (RICO) rat. II. A study of plasma lipoproteins and effect of dietary cholesterol.
    Biochim. Biophys. Acta **960** (1988) 382-389.

11. **SULTAN, F., CARDONA-SANCLEMENTE, L.E., LAGRANGE, D., LUTTON, C., GRIGLIO, S.**
    Lipoprotein lipase and hepatic lipase activities in a hypercholesterolemic RICO strain of Rat.
    Biochem. J., **226** (1990) 349-353.

12. COOPER, A.D., SHREWSBURY, M.A., ERICKSON, S.K.
Comparison of binding and removal of remnant of triglyceride-rich
lipoproteins of intestinal and hepatic origin by rat liver in vitro.
Am. J. Physiol. **243** (1982) 389-395.

13- UTERMANN, G., KINDERMANN, I., KAFFARNIK, H. STEINMETZ, A.
Apolipoprotein E phenotypes and hyperlipidemia.
Hum. Genet., **65** (1984) 232-236.

14- OUGUERRAM, K.
Etude du métabolisme du cholestérol des lipoprotéines plasmatiques
chez le rat normocholestérolémique et génétiquement hypercholestéro-
lémique.
Thèse de Doctorat, Orsay, 1989.

15- ROBERTS, D.C.K., MILLER, N.E., GROOK, D., CORTESE, C., LA VILLE, A.,
MASANA, L., LEWIS, B.
An alternative procedure for incorporating radiolabelled cholesteryl
ester into human plasma lipoproteins in vitro.
Biochem. J. **226** (1985) 319-322.

16- LUTTON, C., OUGUERRAM, K., SAUVAGE, M., MAGOT, T.
Turnover of ^{14}C sucrose HDL and uptake by organs in the normal or
genetically hypercholesterolemic (RICO) rats using a constant
infusion method.
Reprod. Nutr. Dév., **30** (1990) 97-101.

17- STEIN, Y., HALPERIN, G., STEIN, O.
The fate of cholesteryl linoleyl ether and cholesteryl linoleate in
the intact rat after injection of biologically labeled human low
density lipoprotein.
Biochim. Biophys. Acta **663** (1981) 569-574

18- STEIN, O., STEIN, Y., COETZEE, G.A., VAN DER WESTHUYZEN, D.R.
Metabolic fate of low density lipoprotein and high density
lipoprotein labeled with an ether analog of cholesteryl ester.
Klin. Wochenschenschr. **62** (1984) 1151-1156.

19- EISENBERG, S., OLIVECRONA, T.
Very low density lipoprotein. Fate of phospholipids, cholesterol and
apolipoprotein C during lipolysis in vitro.
J. Lipid Res., **20** (1979) 614-623.

20- MAHLEY, R.W., HOLCOMBE, K.S.
Alteration of the plasma lipoproteins and apoproteins following
cholesterol feeding in the rat.
J. Lipid Res. **18** (1977) 314-324.

21- WEISGRABER, K.H., MAHLEY, R.W., ASSMAN, G.
The rat arginine rich lipoprotein and its redistribution following
injection of iodinated lipoproteins into normal and
hypercholesterolemic rats.
Atherosclerosis **29** (1977) 121-140.

22- HUI D.Y., INNERARITY T.L., MAHLEY R.W.
Lipoprotein binding to canine hepatic membranes. Metabolically
distinct apo E and apoB, E receptors.
J. Biol. Chem. **256** (1981) 5646-5655.

23- **VAN BERKEL T.J.C., KRUIJT J.K., VAN GENT T., VAN TOL A.**
Saturable high affinity binding uptake and degradation of rat plasma
lipoproteins by isolated parenchymal and non-parenchymal cells from
rat liver.
Biochim. Biophys. Acta **665** (1981) 22-23.

24- **FAINARU M., FELKER T.E., HAMILTON R.L., HAVEL R.J.**
Evidence that a separate particle containing β-apoprotein is present
in high-density lipoprotein from perfused rat liver.
Metabolism **26** (1977) 999-1004.

25- **FIDGE N.H., POULIS P.**
Metabolic heterogeneity in the formation of low density lipoprotein
in the rat. Evidence for independent production of a low density
lipoprotein subfraction.
J. Lipid Res. **19** (1978) 342-349.

HYPERTRIGLYCERIDEMIA

GENETIC VARIATION AT THE LIPOPROTEIN LIPASE GENE

ASSOCIATES WITH CORONARY ARTERIOSCLEROSIS

J.C. Chamberlain[1], J.A. Thorn[1], R. Morgan[2], A. Bishop[2]
J. Stocks[1], A. Rees[2], K. Oka[3], and D.J. Galton[1]

1. Department of Human Genetics and Metabolism, Medical
   Professorial Unit, St Bartholomew's Hospital
   London, EC1A 7BE, UK

2. University of Wales College of Medicine, Heath Park
   Cardiff, Wales

3. Medlantic Research Foundation, Washington DC, USA

INTRODUCTION

The aggregation of coronary artery disease (CAD) within families is well established 14,13,10 and concordance rates for the disease are known to be higher in mono - than dizygotic twins, 2,3.

The use of restriction fragment length polymorphisms (RFLP's) to detect DNA sequence variation around genes considered to be implicated in the aetiology of any inherited disease has been extensively applied to CAD. Amongst the many candidates for the development of CAD and hyper-triglyceridaemia, our previous work has emphasised a role for those genes encoding the enzymes responsible for lipolysis, particularly lipoprotein-lipase (LPL) 5, 15.

LPL is a glycoprotein synthesised primarily by a parenchymal tissue and secreted and bound to glycosaminoglycans found on the luminal surfaces of capillary endothelia. It is known to catalyse the removal of triglyceride-rich lipoproteins such as VLDL and chylomicrons, from the plasma. During this clearance surface components are transferred from the triglyceride-rich particles to high density lipoproteins 11.

Using a cDNA LPL clone as a hybridisation probe, a polymorphic restriction site has been described at the LPL gene 7 and one of the two alleles identified by this RFLP has been found to associate with both primary hypertriglyceridaemia and premature CAD 5 and 15. This present study was designed to test this association by comparison of allelic frequencies between a group of patients with angiographically defined CAD and a control group with an angiographically defined absence of CAD.

Hypercholesterolemia, Hypocholesterolemia, Hypertriglyceridemia
Edited by C.L. Malmendier *et al.*, Plenum Press, New York, 1990

MATERIALS AND METHODS

Subjects

One hundred and twenty Welsh Caucasian subjects were selected from amongst 1,371 individuals undergoing coronary angiography. All had fasting plasma glucose measurements 6.5 mmol/l. Sixty were older than forty years and had minimal CAD, the remaining sixty were aged less than sixty-five years and had severe 2 or 3 vessel disease. No subject had received any hypolipidaemic therapy or dietary advice. Fasting plasma total cholesterol, triglyceride and HDL-cholesterol were measured and details are summarised in table 1. Venous blood was taken from all subjects for genotyping .

DNA Analysis

DNA was isolated as previously described 12 and 8 microgrammes digested with 8 units of the restriction endonuclease Hind III according to manufacturers instructions (Gibco U.K). The digests were electro-phoresed and blotted onto "Hybond N" filters for subsequent hybridisation with 5 x 10 c.p.m./ml labelled probe. Bands were visualised by autoradiography with "Hyperfilm-MP."

Lipid Analysis

Fasting plasma triglycerides and total cholesterol were measured by fully enzymatic methods (Boehringer Mannheim FRG). HDL -cholesterol was measured by the heparin - Mn Cl_2 precipitation method.

Statistical Analysis

Genotype frequencies in the various groups as shown in Table 2 were analysed by 2 x 3 contingency tables using chi-squared analysis. Allelic frequencies were compared by a Z-test.

RESULTS

Lipoprotein Lipase Gene Polymorphism

The Hind III polymorphic site has been reported as lying between exons 8 & 9. 8 It produces a two allele polymorphism with an invariant band at 4.5Kb and a band at either 17.5 Kb in the absence of the restriction site (H1 allele) or 8.7 Kb in its presence (H2 allele).

Genotypes in Subjects with Negligible and Severe CAD

The genotype distribution and allelic frequencies for the Hind-III RFLP in severe and minimal CAD subjects are shown in Table 2. There is a trend to favour genotypes containing the H2 allele in the severe CAD group as compared to the control group, although this does not quite reach significance (p < 0.10 chi-squared). The frequency of the H2 allele is higher in this Welsh control group than there has been previously described for either control or random populations 5, 15, 7, 6. The proportion of subjects with at least one H2 allele is significantly raised in the severe CAD group as compared with the control group (p < 0.05).

Fasting plasma levels for both triglyceride and total cholesterol are seen to be higher in the severe CAD than in the control group, whilst the HDL-cholesterol levels are lower.

276

DISCUSSION

Our data demonstrate that an allele of an RFLP found at the LPL gene locus is seen more frequently in a population with severe angiographically defined CAD than in a similar group with angiographically minimal CAD.

Epidemiological studies have previously suggested that hypertriglyceridaemia may not be an independent risk factor in the development of CAD once allowance has been made for HDL levels [9, 4] but the use of multivariate analysis for such inter-related variables may be questionable [1]. Our data would suggest that the catabolism of triglyceride-rich lipoproteins, rate-limited by LPL, may be closely associated with the development of coronary atherosclerosis in genetically predisposed individuals.

We would suggest that the H2 allele of the Hind III RFLP at the LPL gene is acting as a linkage marker for an aetiological mutation, close to or at this locus, which influences the clearance of triglyceride-rich particles from the plasma and the metabolism of HDL. These effects may then predispose the individual to the development of coronary atherosclerosis.

ACKNOWLEDGEMENTS

We gratefully acknowledge the financial assistance provided by the British Heart Foundation, the Medical Research Council and the Wellcome Trust.

TABLE 1

CLINICAL DETAILS OF STUDIED GROUPS

Group	n	Age Years	BMI	Triglyceride mmol/l	Plasma:- Cholesterol mmol/l	HDL Cholesterol mmol/l
Controls	60	54.4	26.0	1.84	5.76	0.88
		± 8.7	± 3.1	± 0.97	± 1.00	± 0.21
Severe	60	53.9	26.9	2.45	6.45	0.81
CAD		± 6.7	± 2.9	± 1.38	± 1.16	± 0.19

Values are
± Standard Deviation

TABLE 2

ALLELES OF LPL PRODUCED BY HIND-III DIGESTION OF GENOMIC DNA FROM SUBJECTS WITH SEVERE OR MINIMAL CAD:

			Number of Subjects Represented by Genotype (%)	
Group	n	H1H1	H1H2	H2H2
Controls	60	7(11.7)	21(35.0)	32(53.3)
Severe CAD	60	1 (1.7)	22(36.7)	37(61.6)*

* Controls as Severe CAD $p < 0.10$ (chi-squared 2 x 3)

		Allelic Frequencies		Proportion of Subjects with at least one allele	
Group	No. of chromo- somes (n)	H1	H2	No. of Subjects (n)	H1
Control	120	0.29	0.71	60	0.467	0.883
Severe CAD	120	0.19	0.81	60	0.383	0.983*

* Control as Severe CAD H1 ns; H2 $p < 0.05$ (Z-test)

REFERENCES

1. Abbott RD, Caroll RJ. Interpreting multiple logistic regression coefficients in prospective observational studies. Am J. Epidemiol 1984; 119 : 830 - 836.
2. Berg K. Genetics of coronary heart disease. In: Steinberg AE, Brown AG, Motulsky AR, Childs B Eds. Prog Med Genet 1983; 35 - 90.
3. Berg K. Twin studies of coronary heart disease and its risk factors. Actor Med Gemell 1984; 33: 349 - 361.
4. Castelli COP. The triglyceride issue: a view from Framingham. Am Heart J 1986; 112: 432 - 437.
5. Chamberlain JC, Thorn JA, Oka K et al. DNA polymorphisms at the lipoprotein lipase gene. Associations in normal and hyper-triglyceridaemic subjects. Atherosclerosis 1989; 75: 85 - 91.
6. Fischer L, Fitzgerald GA, Lawn RM. Two polymorphisms in the human lipoprotein lipase gene. Nucleic Acids Res 1987; 15: 7657.
7. Heinzmann G, Ladias J, Antonarakis S et al. RFLP for the human lipoprotein lipase gene: Hind-III. Nucleic Acids Res 1987: 15; 6763.
8. Kirchgessner TG, Chuat JG, Heinzmann G et al. Organisation of the human lipoprotein lipase gene and evolution of the lipase gene family. Proc Natl Acad Sci USA 1989; 86: 9647 - 9651.

9. Hulley SB, Rosenmann AM, Bawol RD, Brand RJ. Epidemiology as a guide to clinical decisions. The associations between triglyceride and coronary heart disease. NEJ Med 1980; 302: 1383 - 1389.
10. Nora JJ, Lortscher RM, Spangler RD, Nora AH, Kimberling WJ. Genetic-epidaemiologic study of early onset ischaemic heart disease. Circulation 1980; 61: 503 - 508.
11. Patsch JR, Prasad S, Gotto AM Jr, Patsch W. HDL2 relationships of the plasma levels of this lipoprotein species to its composition, to the magnitude of postprandial lipaemia and to the activities of lipoprotein lipase and hepatic lipase. J Clin Invest 1987: 80: 341 - 347.
12. Rees A, Shoulders CC, Stocks J, Galton DJ, Baralle FE. DNA polymorphisms adjacent to human apolipoprotein A-I gene: relation to hypertriglyceridaemia. Lancet 1983; i: 444 - 446.
13. Rissanen AM. Familial aggregation of coronary heart disease in a high incidence area (North Karelia, Finland). Br Heart J 1979; 42: 294 - 303.
14. Slack J, Evans KA. The increased risk of death from ischaemic heart disease in first-degree relatives of 121 men and 96 women with ischaemic heart disease. J Med Genet 1966; 3: 239 - 259.
15. Thorn JA, Chamberlain JC, Stocks J, Galton DJ. RFLP's at the lipoprotein and hepatic lipase gene loci in coronary athero-sclerosis. Proceedings of the Fifth European Symposium on Metabolism. Atherosclerosis 1989; in press.

THE LIPID SURFACE OF TRIGLYCERIDE-RICH PARTICLES CAN MODULATE

(APO)PROTEIN BINDING AND TISSUE UPTAKE

Donald M. Small, Susanne Bennett Clark, Anna Tercyak
John Steiner, Donald Gantz, and Arie Derksen

Department of Biophysics, Boston University School of
Medicine, 80 East Concord Street, Boston, MA 02118

Plasma lipoproteins are aggregates varying in size from large chylomicrons to small HDL_3. They are composed of complex combinations of apoproteins and lipids. The triglyceride-rich lipoproteins, secreted by the intestine as chylomicrons or by the liver as VLDL, contain a core which is rich in triacylglycerols. Small but varying amounts of cholesterol esters also are contained in the core. The surface lipids are extremely complex and include a variety of phospholipids, including phosphatidylcholines, phosphatidylethanolamines and sphingomyelins, as well as glycosphingolipids such as cerebrosides and gangliosides. Also present in the surface is cholesterol and small but significant amounts of triacylglycerols and cholesterol esters. In addition the surface contains insoluble and non-exchangeable apolipoproteins, specifically B_{100} or B_{48}, and exchangeable soluble apolipoproteins such as A-I, A-II, A-IV, C-I, C-II, C-III, and E. The determinants of triglyceride-rich lipoprotein composition are: 1) the metabolism of the cell that secretes the nascent chylomicron or VLDL, 2) the physical exchanges of lipids and apoproteins that occur in plasma, 3) the transfer proteins mediated exchange between core molecules such as triacylglycerols and cholesterol ester and surface molecules such as phospholipids, and 4) the action of lipolytic enzymes such as lipoprotein lipase, hepatic lipase and lecithin-cholesterol acyltransferase, to produce metabolically important lipid products including fatty acids, monoacylglycerols, diacylglycerols, and lysophosphatides. Thus, the lipid composition of lipoproteins is extremely complex and changing and potentially this complexity can affect the adsorption and desorption of different apolipoproteins which in turn can affect the metabolism.

Over the past several years we have been examining the partitioning of various lipid molecules between the core and surface of emulsion particles created to model triglyceride-rich lipoproteins. The summary of our findings concerning the partitioning of cholesterol, cholesterol esters, triacylglycerols, fatty acids, diacylglycerols and monoglycerols is shown in Figure 1. These conclusions have been made from experiments utilizing physical isolation of core and surface,

Hypercholesterolemia, Hypocholesterolemia, Hypertriglyceridemia
Edited by C.L. Malmendier *et al.*, Plenum Press, New York, 1990

chemical determination of core and surface composition and use of carbon-13 labeled lipids in high-field nuclear magnetic resonance (NMR) studies. The chemical shift of ^{13}C carboxyl and carbonyl groups identify the location of the molecule in core or surface and help establish its conformation (1).

Minor Molecules in Lipoprotein-like Emulsion Surface

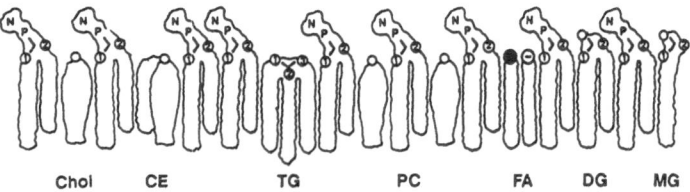

Chol   CE   TG   PC   FA   DG   MG

FIGURE 1. Conformation and Properties of Minor Lipids in Phosphatidylcholine Monolayer of a Lipoprotein-like Emulsion Particle. Chol=cholesterol, CE=cholesterol ester, TG=triacylglycerols, PC=phosphatidylcholine, FA=fatty acids (partly ionized), DG=diacylglycerols, MG=monoacylglycerols. Free cholesterol increases the viscosity and decreases the fluidity in an expanded monolayer. Cholesterol ester can be incorporated to about 2.8 mol% relative to PC. It has a hairpin-like configuration as shown. Increasing the free cholesterol partly pushes cholesterol ester out of the surface and into the core. Triacylglycerols (triolein, tripalmitin) have a solubility of about 3 mol% relative to PC. The conformation shows that all three chains lay side-by-side and parallel to the chains of the PC. The -1 and -3 carbonyl sn-glycerol groups are more hydrated and closer to the water interface than the sn-2 carbonyl. Increasing free cholesterol progressively pushes triolein out of the monolayer so that at 1C:1PC virtually no triolein remains. Fatty acid is about half ionized at pH 7.4 and exists as an "acid-soap". Its solubility is virtually unlimited in this state and it is not altered by increasing amounts of free cholesterol. Diacylglycerol has limited but relatively substantial solubility and monoacylglycerols have unlimited solubility.

Briefly, free cholesterol can partition into the surface up to a molar ratio of 1C:1PC and even higher if supersaturated. The distribution coefficient of cholesterol between core and surface on an equivalent mass basis is about 11:1 surface to core (1). In the surface it increases the surface pressure and viscosity, decreases the surface fluidity and free volume, and forces some other molecules out of the interface. For instance, cholesterol ester (2,3) and triacylglycerols (3-7) are soluble to about 2.8 to 3 mol% in a phosphatidylcholine interface. However, when free cholesterol is introduced into the surface these molecules are progressively pushed out (8,9) so that at equimolar cholesterol to phosphatidylcholine, triacylglycerols (9) are entirely displaced from the monolayer covering the model lipoprotein. Fatty acids partition readily into the interface (10-12). The partition coefficient of protonated fatty acid between surface and core is 7:1, whereas all of the ionized fatty acid partitions into the interface. The apparent pKa of fatty acids in the surface of emulsion particles and triglyceride-rich lipoproteins is about 7.4, indicating that at physiologic pH half of the fatty acid is ionized and half is protonated (10). The half-ionized fatty acids (acid soap complex) behave as a swelling amphiphilic molecule (13) and partitions to a nearly unlimited extent into the surface of the lipoprotein and in doing so imparts a negative charge. The incorporation of increasing amounts of cholesterol does not displace fatty acid from the interface (11). Diacylglycerols also partition mainly into the interface and have rather high solubilities, up to about 15 to 20 mol% depending on the temperature and the nature of the phospholipids in the surface (14). Their conformation in terms of the glycerol backbone is like that of phosphatidylcholine. The glycerol backbone is perpendicular to the plane of the bilayer with the-OH group hydrogen bonded to the aqueous interface (14). Monoacylglycerols behave like swelling amphiphiles and partition into the interface with unlimited solubility.

It is well known that fatty acid composition of lipoproteins can be modulated by the fatty acids in the diet (15). In fact, dietary fats high in saturated fatty acids produce chylomicrons with triacylglycerols high in saturated fatty acids (16,17). These triacylglycerols may be packaged as metastable undercooled liquids in the cores of triacylglycerol-rich lipoproteins (16,18, 19). When cooled below their crystallization temperature these triacylglycerols may crystallize distorting the shape and the metabolic behavior of the triacylglycerol particle if reinjected into an animal (16,17,20). The lipid composition of the diet may also influence the phospholipid chain composition and thus render the surface more or less saturated (21).

We reasoned that such changes in surface and core composition might lead to changes in the physical properties of the respective parts and thus to changes in the adsorption of apolipoproteins and ultimately to changes in metabolism of the particles. As an attempt to modulate surface characteristics we have generated a series of triglyceride-rich emulsion particles with different physical properties (Figure 2). Utilizing cholesterol to modulate the fluidity we have made particles using different phosphatidylcholines containing a sufficient amount of cholesterol to produce surfaces which are fluid, viscous, very viscous, mainly solid, or surfaces at the transition between viscous and solid states (22). Based on NMR studies utilizing ^{13}C enriched triolein in phosphatidylcholine vesicles we believe that triolein partitions into the low cholesterol interface at about 2-2.5 mole percent with respect to DPPC (23) and about 2 mole percent with respect to DSPC even when the DSPC is in the solid state (J.A. Hamilton and D.M. Small, unpublished).

We have begun to study the binding of apoproteins and other proteins from the plasma to these surfaces. The metabolic fate of these particles has been followed in intact rats and compared to native chylomicrons.

In particles made with egg yolk phosphatidylcholine (EYPC), cholesterol and triolein, increasing the cholesterol limits the amount of apoprotein, such as apoA-I and apoE, that can bind to the surface (24). In particles with low cholesterol (cholesterol/phospholipid mole ratio in surface = 1/7) resembling the cholesterol content in nascent chylomicrons (1, 7) the surface is fluid. Particles resembling chylomicron remnants have higher cholesterol, i.e. approaching 1 cholesterol per 1 phospholipid, and the surfaces are much more viscous. The binding of apoA-I or apoE-3 shows a similar binding constant (Kd ~ 9 x 10^{-7} M) but the maximum amount of protein which can bind to the surface (N) is about 5 times greater with particles with low cholesterol (24). The addition of small amounts of fatty acid partly restore binding of apoA-1 (25). Preliminary studies using human apoC-II indicate that apoC-II binds well to the low cholesterol particles but appears to be virtually excluded from the high cholesterol particles.

When the EYPC-triolein particles are incubated briefly with rat plasma, which of course contains not only apolipoproteins but many other proteins, the results are more complex. After incubation the particles are washed by centrifuging through a water layer and analyzed. The apoC's are the major proteins bound to the low cholesterol EYPC surface but they are excluded from the high cholesterol

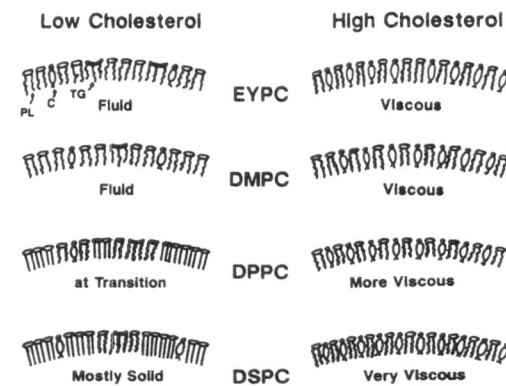

FIGURE 2. The Physical State at 38°C of the Surfaces of Emulsions Made with Different Phosphatidylcholines and Low or High Cholesterol. Low cholesterol refers to about 1 cholesterol per 7 phospholipids, and high cholesterol refers to about 1 cholesterol to 1-2 phospholipids. EYPC=egg yolk phosphatidylcholine, DMPC=dimyristoyl phosphatidylcholine, DPPC=dipalmitoyl phosphatidylcholine, DSPC=distearoyl phosphatidylcholine. Other abbreviations as in Figure 1. In the low cholesterol series the chains are either fluid, at the transition between viscous and solid (DPPC) or mostly solid (DSPC). High cholesterol increases the viscosity of the surface and abolishes the transition to the solid state. However, with the longer chained phosphatidylcholines, especially DSPC the surfaces become very viscous and tightly packed.

EYPC surface. These high cholesterol particles, modeling chylomicron remnants, bind a 66 Kd protein, perhaps albumin, as the major protein and apoE as a major apoprotein by immunoblots. DSPC high or low cholesterol particles bind apoA-I and apoA-IV as major apoproteins and a 53 Kd plasma protein as the major protein. Particles at the gel-liquid crystal transition (DPPC-low cholesterol) bound A-IV as their major protein (see table).

In an attempt to understand the metabolism and tissue uptake of the particles with the varying surfaces, particles were doubly labeled with a trace of cholesterol ester and also in the triolein and injected as a bolus into intact unanesthetized rats. The disappearance of radioactivity from triacylglycerol reflects both the lipolysis of triacylglycerol and particle uptake, whereas the disappearance of the cholesterol ester (which is not hyrolyzed, nor transferred) represents only the tissue uptake (26-28). Thus, when the disappearance rate of the two markers are the same, there is no lipolysis. After 10 minutes

TABLE 1

Physical State of Surface	LIPOLYSIS	rate pools/min	to liver 10 min	liver/ spleen	Pr/PL[b]	Major[c]	Major[d] apoproteins
EMULSION		**REMNANT (CE) REMOVAL**			**PROTEINS BOUND**		
FLUID							
EYPC-loC	rapid	fast 0.12	43%[a]	43	0.10	apo C's	apo C's
DMPC-loC	fairly rapid	fast 0.13	39%	8.7			
VISCOUS							
EYPC-hiC	v.slow	v.fast 0.18	56%	6.6	0.01	66kd	apo E
DMPC-hiC	0	slow 0.04	20%	5.0			
DPPC-hiC	0	v.slow 0.005	10%	2.9			
DSPC-hiC	0	v.slow 0.003	6%	2	0.03	53kd	apo A-I[e] apo A-IV
AT GEL/FLUID TRANSITION							
DPPC-loC	0	slow 0.02	24%	5.3	0.11	apo A-IV	apo A-IV
MAINLY SOLID							
DSPC-loC	0	v.fast 0.22	56%	8.6	0.02	53kd	apo A-I[e] apo A-IV
CHYLOMICRONS	rapid	fast 0.14	40-50%	25-50			

[a] Mean values as % of injected dose. Each emulsion was infused into 4-10 rats.
[b] Protein/Phospholipid ratio on particles.
[c] Estimated by SDS PAGE.
[d] Estimated by SDS PAGE and immunoblots.
[e] Estimated by SDS PAGE stained with colloidal Coomassie Blue.
   (S.B. Clark, A. Derksen and D.M. Small, 1990, unpublished results)

the animals were sacrificed and the counts in liver and spleen and that remaining in plasma were analyzed. A summary of these experiments, carried out by Drs. Bennett Clark and Derksen is given in the table along with the in vitro binding experiments from rat plasma, shown on the right of the table. For comparison the disappearance of chylomicrons is given at the bottom.

Chylomicrons were rapidly hydrolyzed and the remnant uptake was rapid. Most went to the liver, the liver to spleen ratio being approximately 25-50. The liver to spleen ratio indicates in a general way how much of the uptake is through the reticulo-endothelial system. The lower the ratio the more important the reticulo-endothelial component of uptake. The emulsions with low cholesterol having a fluid surface mimic nascent chylomicron removal quite faithfully. Initially they bind the apoC's, they are rapidly hydrolyzed and rapidly taken up by the liver. The unsaturated PC chain (EYPC) particles appear to distribute better to the liver than the saturated DMPC. The high cholesterol EYPC emulsion, as shown in the past (26-28), mimics the uptake of chylomicron remnants. Hydrolysis is very slow, if present at all, whereas the uptake by the liver is rapid, and there is very little spleen uptake. These emulsions bind a 66 Kd protein as their main protein, but apoE is the main apoprotein. In the saturated chain PC series with high cholesterol the viscosity increases as the chain lengthens from DMPC to DSPC. These particles are not hydrolysed. As the chain length increases the uptake becomes progressively slower until with DSPC it is extremely slow and probably mainly taken up by the reticulo- endothelium system as noted from the low liver to spleen ratio. These emulsions bind a major protein of 53 Kd from plasma as well as A-IV and A-I as the major apoproteins identified by immunoblots and Coomassie Blue stains respectively. When the surface is at the gel to liquid crystal transition (DPPC-low cholesterol) hydrolysis is blocked and the uptake is rather slow. When the surface is solid (DSPC-low cholesterol) the uptake is very fast, most of it going to the liver, even though the major protein and apoproteins bound appear quite similar to that of the highly viscous surface (DSPC-high cholesterol) which has the slowest removal. The reason for the striking difference in uptake between these two particles is not clear from the apoproteins bound, but it could be that specific cells (e.g. Kupffer cells) within the liver actually recognize a solid surface as compared to a highly viscous one and rapidly remove these particles.

In summary these studies suggest that the lipids which are present at the interface determine the physical properties of the surface of these model lipoproteins, which in turn determine the proteins which can bind. Presumably the surfaces then determine the extent to which lipolysis can occur and which receptors remove the particles from the plasma compartment. Although bound apoC's are important for lipolysis and bound apoE is important for hepatic uptake, other apoproteins such as apoA-IV may play important roles in modulating the metabolism and uptake of certain triglyceride-rich particles.

ACKNOWLEDGEMENTS

This work was supported by National Institutes of Health research grant HL-26335 and training grant HL-07291. We also thank Irene L. Miller for typing the manuscript.

BIBLIOGRAPHY

1. Miller, K.W. and D.M. Small. 1987. Structure of triglyceride-rich lipoproteins: An analysis of core and surface phases. in New Comprehensive Biochemistry, Vol. 14, Plasma Lipoproteins (ed. A.M.Gotto,Jr.), Elsevier Science Publ. B.V. (Biomed. Div.) pp. 1-75.

2. Hamilton, J.A. and D.M. Small. 1982. Solubilization and localization of cholesteryl oleate in egg phosphatidylcholine vesicles:A carbon-13 NMR study. J. Biol. Chem. (Communication) - 257:7318-7321.

3. Hamilton, J.A., K.W. Miller and D.M. Small. 1983. Solubilization of triolein and cholesteryl oleate in egg phosphatidylcholine vesicles. J. Biol. Chem. - 258:12821-12826.

4. Hamilton, J.A. and D.M. Small. 1981. Solubilization and localization of triolein in phosphatidylcholine bilayers:A ^{13}C NMR study. Proc. Natl. Acad. Sci. USA - 78(no.11): 6878-6882.

5. Miller, K. W. and D.M. Small. 1982. The phase behavior of triolein, cholesterol, and lecithin emulsions. J. Colloid & Interface Sci. 89(2):466-478.

6. Miller, K.W. and D.M. Small. 1983. Triolein-cholesteryl oleate-cholesterol-lecithin emulsions: Structural models of triglyceride-rich lipoproteins. Biochemistry - 22:443-451.

7. Miller, K.W. and D.M. Small. 1983. Surface-to-core and interparticle equilibrium distribution sof triglyceride-rich lipoprotein lipids. J. Biol. Chem. 258:13772-13784.

8. Spooner, P.J.R., J.A. Hamilton, D. Gantz, and D.M. Small. 1986. The effect of free cholesterol on the solubilization of cholesteryl oleate in phosphatidylcholine bilayers: A ^{13}C-NMR study. Biochim. et Biophys. Acta - 860:345-353.

9. Spooner, P.J.R. and D.M. Small. 1987. Effect of free cholesterol on incorporation of triolein in phospholipid bilayers. Biochemistry 26:5820-5825.

10. Spooner, P.J.R., S. Bennett Clark, D.L. Gantz, J.A. Hamilton, and D.M. Small. 1988. The ionization and distribution behavior of oleic acid in chylomicrons and chylomicron-like emulsion particles and the influence of serum albumin. J. Biol. Chem. - 263(3):1444-1453.

11. Ekman, S., A. Derksen and D.M. Small. 1988 The partitioning of fatty acid and cholesterol between core and surfaces of phosphatidylcholine-triolein emulsions at pH 7.4. Biochim. Biophys. Acta - 959(3):343-348.

12. Spooner, P.J.R., D.L. Gantz, J. A. Hamilton, and D.M. Small. 1990. The distribution of oleic acid between chylomicron-like emulsions, phospholipid bilayers and serum albumin. A model for fatty acid distribution between lipoproteins, membranes, and albumin. J. Biol. Chem. - in press.

13. Small, D.M. 1986. The Physical Chemistry of Lipids from Alkanes to Phospholipids. Handbook of Lipid Research Series, Vol. 4, , ed. D. Hanahan, Plenum Press. pp. 1-672.

14. Hamilton J.A., S. Bhamidipati, D.R. Kodali and D.M. Small. 1989. The conformation and molecular mobility of 1,2-diacylglycerols in phospholipid bilayers. J. Biol. Chem. - in press.

15. Breckenridge, W.C. 1978. Stereospecific Analysis of Triacylglycerols. in Handbook of Lipid Research Series, Vol. 1, ed: A. Kuksis, Plenum Press, Chapt. 4, pp. 197-232.

16. Bennett Clark, S., D. Atkinson, J.A.Hamilton, T. Forte, B. Russell, E.B. Feldman, and D.M. Small. 1982. Physical studies of d § 1.006 g/ml lymph lipoproteins from rats fed palmitate-rich diets. J. Lipid Res. - 23:28-41.

17. Feldman, E.B., B.S. Russell, R. Chen, J. Johnson, T. Forte, and S. Bennett Clark. 1983. Diet saturated fatty acid content affects lymph lipoprotein composition and configuration in the rat. J. Lipid Res. 24:967-976.

18. Parks, J.S., D. Atkinson, D.M. Small and L.R. Rudel. 1981. Physical characterization of lymph chylomicra and very low density lipoproteins from nonhuman primates fed saturated dietary fat. J. Biol. Chem. - 256:12992-12999.

19. Hamilton, J.A., D.M. Small and J. Parks. 1983. ^{1}H NMR studies of lymph chylomicra and very low density lipoproteins from nonhuman primates. J. Biol. Chem. - 258:1172-1179.

20. Puppione, D.L., S.T. Kunitake, R.L. Hamilton, M.L. Phillips, V.N. Schumaker and L.D. Davis. 1982. Characterization of unusual intermediate density lipoproteins. J. Lipid Res. 23:283-290.

21. Patton, G.M., S. Bennett Clark, J.M. Fasulo and S.J. Robins. 1984. Utilization of individual lecithins in intestinal lipoprotein formation. J. Clin. Invest. - 73:231-240.

22. Small, D.M., J.W. Steiner, A. Derksen, S. Bennett Clark. 1988. Thermal transitions of phosphatidylcholines on the surface of lipoprotein-like emulsion particles. Biophys. J. 53:211a.

23. Hamilton, J.A. 1989. Interactions of triglycerides with phospholipids: Incorporation into the bilayer structure and formation of emulsions. Biochemistry 28:2514-2520.

24. Derksen, A. and D.M. Small. 1989. Interaction of apoA-1 and apoE-3 with triglyceride-phospholipid emulsion containing increasing cholesterol concentrations. Model of triglyceride-rich nascent and remnant lipoproteins. Biochemistry 28:900-905.

25. Derksen, A., S. Ekman and D.M. Small. 1989. Oleic acid allows more apoprotein A-1 to bind with higher affinity to large emulsion particles saturated with cholesterol. J. Biol. Chem. - 264(12):6935-6940.

26. Redgrave, T.G. and R.C. Maranhao. 1985. Metabolism of protein-free lipid emulsion models of chylomicrons in rats. Biochimica et Biophysica Acta - 835:104-112.

27. Maranhao, R.C., A.M. Tercyak and T.G. Redgrave. 1986. Effects of cholesterol content on the metabolism of protein-free emulsion models of lipoproteins. Biochim. et Biophys. Acta. - 875:247-255.

28. Bennett Clark, S. and A. Derksen. 1987. Phosphatidylcholine composition of emulsions influences triacylglycerol lipolysis and clearance from plasma. Biochim. Biophys. Acta - 920:37-46.

STRUCTURE AND METABOLISM OF LOW DENSITY LIPOPROTEINS FROM

NORMAL AND HYPERTRIGLYCERIDEMIC SUBJECTS

Barry J. McKeone, Josef R. Patsch, and Henry J. Pownall

Department of Medicine, Baylor College of Medicine and
The Methodist Hospital, Houston, TX 77030

INTRODUCTION

Human plasma lipoproteins are complex assemblies of lipids and proteins whose identities are operationally defined by the densities at which they are isolated as the high, low, and very low density lipoproteins, HDL, LDL and VLDL, respectively. Lipoproteins transport glycerol and sphingolipids as well as free and esterified cholesterol along with apoproteins which may, depending on their identity, activate lipolytic enzymes or serve as ligands for receptor-mediated endocytosis. In addition to these important proteins, human plasma also contains a number of proteins that transport monomeric lipids between lipoprotein classes. Most notable among these are cholesterol ester transfer protein and phosphatidylcholine transfer protein. Lipolytic enzymes, lipoprotein receptors,
and lipid transfer protein are the major endogenous factors that modify the concentration, composition, and structure of plasma lipoproteins. The major exogenous factors are found in the diet and include the amount and kinds of fatty acids that are consumed.

Over the past two decades, the LDL-receptor, which binds to a ligand on LDL, has emerged as an important regulator of cholesterol synthesis and of plasma LDL concentrations (Goldstein and Brown, 1989). Most plasma LDL is removed via hepatic receptors that recognize a ligand within either the B-protein of LDL or the apoE of some triglyceride-rich lipoproteins. Many of the LDL-receptor defects summarized by Brown and Goldstein (1989) are associated with elevated plasma LDL-cholesterol levels and premature coronary artery disease. Recently, it has been found that defective association of LDL with its receptor can be due to mutations in apoB-100, the sole LDL protein (Soria et al., 1989; Talmud et al., 1989; Young et al., 1988). In addition, there is some evidence that changes in the neutral lipid composition of LDL are associated with modified receptor binding (Kleinman et al., 1985). The purpose of this investigation was to quantify the effects of plasma triglyceride levels on the composition, structure and function of LDL.

EXPERIMENTAL

Normal and hypertriglyceridemic subjects were recruited from the laboratory staff in Houston and the Lipid Clinic in Innsbruck. Informed consent was obtained from all participants. Following a 12 h fast, each subject gave 100 ml blood, from which approximately 50 ml of plasma was isolated. A portion of the sample was used to measure plasma levels of total cholesterol, HDL_2- and HDL_3-cholesterol, and triglycerides. The LDL from each subject

Hypercholesterolemia, Hypocholesterolemia, Hypertriglyceridemia
Edited by C.L. Malmendier *et al.,* Plenum Press, New York, 1990

was isolated by flotation between the densities of 1.063 and 1.019 g/ml; a second flotation into the same density range was used to wash the sample. Protein (Lowry et al. 1951), phospholipid (Bartlett, 1959), triglyceride (Boehringer Kit), and free and total cholesterol (Boehringer Kit) were measured on each LDL sample. Methyl ester fatty acid composition was measured by gas chromatography using a capillary column. Individual fatty acids were quantified according to their respective integrated areas and the peaks were identified by comparison with methyl ester standards.

LDL transition temperatures were measured in Tris buffered saline, pH = 7.4, in a MicroCal MC-2 differential scanning calorimeter at a scan rate of 20 deg/min. Peak melting temperatures were determined using an algorithm supplied by the manufacturer.

Uptake of LDL by cells in culture was conducted with the human hepatoma line, HepG2 and with human skin fibroblasts using [^{125}I]LDL. The binding of the monoclonal antibodies to LDL was measured by an enzyme-linked immunoassay. MB47, which binds close to the receptor ligand of LDL, was kindly provided by Dr. Linda Curtiss (Young et al., 1986).

In vitro hypertriglyceridemic LDL was prepared by incubation of normal LDL with a high concentration of VLDL for various times. The rate of triglyceride transfer from VLDL to LDL and HDL was quantified by measuring the change in the triglyceride content of the latter as a function of time at 37 deg C.

RESULTS

Plasma Triglyceride Levels and LDL Composition

The free cholesterol, protein, and phospholipid composition of the LDL from all subjects fell within the normal range. However, the relative amount of triglyceride and cholesteryl ester varied according to the severity of the hypertriglyceridemia. When expressed in terms of the ratio of LDL-triglyceride to LDL-(triglyceride + cholesteryl ester) there is a clear correlation between the ratio and the plasma triglyceride levels. Figure 1 contains a plot in which these data have been fitted to an adsorption isotherm in which the asymptote represents the maximum capacity of LDL for triglyceride and the binding constant represents the relative affinity of triglycerides for the LDL particle. According to this analysis, LDL is saturated when it contains approximately 70% of its neutral lipid as triglyceride and the plasma concentration at which half the saturation level is reached is 500 mg/dl. The fit of these data to this isotherm shows that LDL-triglyceride composition is a predictable function of total plasma triglycerides, and that subjects with elevated plasma triglycerides carry a large amount of triglyceride in the LDL fraction, which can be removed from the plasma compartment by receptor-mediated endocytosis. In this case, in addition to a substantial amount of cholesterol, the cell takes in fatty acid as a part of the triglyceride molecule. This process could contribute to the pool of fatty acid available for membrane biogenesis and beta-oxidation. When supplied in excess, these fatty acids are likely to form fat droplets within the cell.

LDL Melting Temperatures

The neutral lipid core of LDL undergoes a transition from an ordered liquid crystalline phase to an isotropic liquid over a range of temperatures that extends from 15 to 35 deg. Although the fatty acid composition of the cholesteryl esters plays a small role in determining the transition temperature, the major determinant is the triglyceride content (Deckelbaum et al., 1977). Figure 2 (lower panel) contains some representative calorimetric curves obtained from LDL having various triglyceride contents; these curves show that the melting point of the LDL decreases as the triglyceride content increases. Figure 2 (upper panel) summarizes these data for all of the subjects for whom transitions were measurable. In some of the very hypertriglyceridemic subjects, the transitions were broadened to the extent that a peak melting temperature was not readily discernible. Nevertheless, it is clear that the transition temperature of LDL decreases as the LDL-triglyceride increases. Fatty acid analysis of the cholesteryl esters of the LDL showed no significant difference between those that with high melting points and those with low melting points, so we conclude that the major determinant of the core melting of the LDL is the triglyceride content. Since the triglyceride content of LDL is a function of the total plasma triglyceride concentration, we infer that subjects with elevated plasma triglycerides will have low LDL melting points.

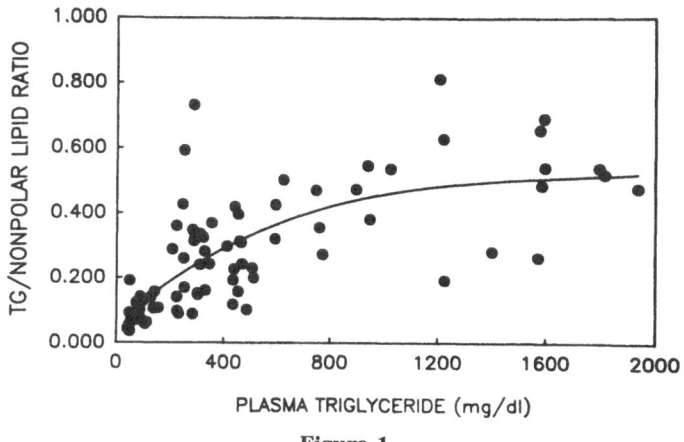

Figure 1

LDL Binding to Cultured Cells

We have measured the uptake of LDL by HepG2 cells and fibroblasts as a function of the triglyceride content of the LDL. As previously reported (Kleinman et al., 1985, 1987), increasing the triglyceride content of LDL lowers its uptake by fibroblasts (Figure 3). By contrast, in HepG2 cells, more of the hypertriglyceridemic than normal LDL was taken up. These data suggest that the determinants for LDL uptake by fibroblasts and HepG2 cells differ and that normal and hypertriglyceridemic LDL are taken up at independent tissue sites. To identify other LDL structural differences induced by increased triglyceride content, we measured the circular dichroism (CD) and tested each LDL with a series of monoclonal antibodies. Over the range where CD data could be collected we found no difference in the structure of normal and hypertriglyceridemic LDL. Moreover, none of 6 different monoclonal antibodies, including MB-47, (a gift from Dr. Linda Curtiss) distinguished between normal and hypertriglyceridemic LDL. This is unexpected because MB-47 has been shown to bind at or close to the LDL-receptor ligand (Young et al., 1986).

DISCUSSION

The LDL-receptor plays an important role in cholesterol homeostasis and individuals who lack this receptor commonly have premature atherosclerosis. The effects of hypertriglyceridemia are not as clear. Although there are higher circulating levels of lipoproteins, the correlation between early CAD and the severity of the hyperlipidemia is not as clear. There may be several reasons for this. First, as our data suggest, the increased triglyceride content decreases the melting point of the core lipids. As has been shown in vitro, the rate at which the lipid is catabolized by cells increases as the triglyceride content of LDL is increased (Adelman et al., 1984; Glick et al., 1983). Second, based on the preferential uptake of triglyceride enriched LDL that we observed with HepG2 cells, we speculate that these modified LDL may be switched away from peripheral to hepatic sites of catabolism. The strong correlation that we observed between the severity of hyperlipidemia and the triglyceride content of LDL suggests this mechanism most likely to operates in the most lipemic subjects. Further tests to identify the source of the receptor ligand modulation are continuing in this laboratory. Finally, the correlation observed between the plasma triglyceride levels and LDL-triglyceride content suggests that the triglyceride of VLDL, which contains most of the total plasma triglyceride, is exchangeable with that of LDL. These data can readily be fitted to an adsorption isotherm which demonstrates that, at a steady state, of LDL-triglyceride is a predictable function of plasma triglycerides.

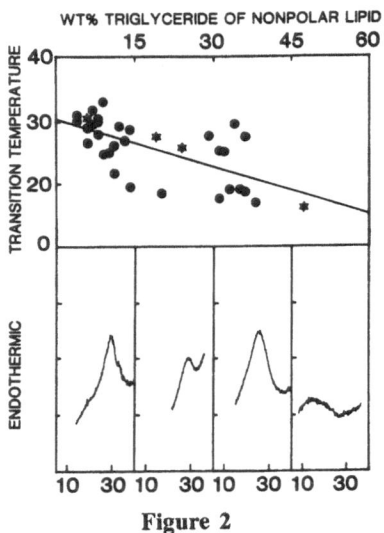

Figure 2

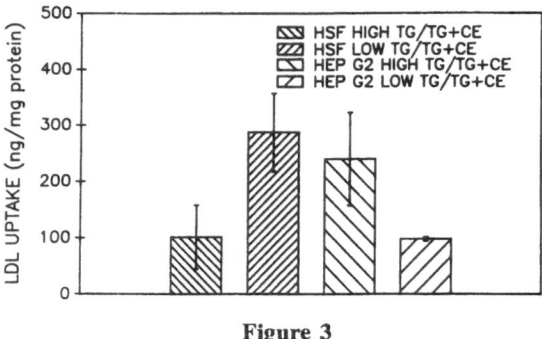

Figure 3

REFERENCES

Adelman, S. J., Glick, J. M., Phillips, M. C., and Rothblat, G. H., 1984, Lipid composition and physical state effects on cellular cholesteryl ester clearance, J. Biol. Chem., 259:13844-13850.

Bartlett, G. R. 1959, Phosphorus assay in column chromatography, J. Biol. Chem., 234:466-468.

Deckelbaum, R. J., Shipley, G. G., and Small, D. M., 1977, Structure and interactions of lipids in human low density lipoproteins, J. Biol. Chem., 252:744-754.

Glick, J. M., Adelman, S. J., Phillips, M. C., and Rothblat, G. H., 1983, Cellular cholesteryl ester clearance, J. Biol. Chem., 258:13425-13430.

Goldstein, J. L., and Brown, M. S., 1989, Familial hypercholesterolemia in: "The Metabolic Basis of Inherited Disease," C. R. Scriver, A. L. Beaudet, W. S. Sly and D. Valle eds., McGraw Hill, NY, Vol. I Chapter 48.

Kleinman, Y., Eisenberg, S., Oschry, Y., Gavish, D;., Stein, O., and Stein, Y., 1985, Defective metabolism of hypertriglyceridemic low density lipoprotein in cultured human skin fibroblasts, J. Clin. Invest. 75:1796-1803.

Kleinman, Y., Schonfeld, G., Gavish, D., Oschry, Y., and Eisenbeg, S., 1987, Hypolipidemic therapy modulates expression of apolipoprotein B epitopes on low density lipoproteins. Studies in mild to moderate hypertriglyceridemic patients, J. Lipid Res., 28:540-548.

Lowry, O. H., Rosebrough, N. J., Farr, A. L., Randall, R.J., 1951, Protein measurement with the Folin phenol reagent J. Biol. Chem., 193:265-275.

Soria, L. F., Ludwig, E. H., Clarke, H. R. G., Vega, G. L., Grundy, S. M., and McCarthy, B. J., 1989, Association between a specific apolipoprotein B mutation and familial defective apolipoprotein B-100, Proc. Nat. Acad. Sci. (USA), 86:587-591.

Talmud, P., King-Underwood, L., Krul, E., Schonfeld, G., and Humphries, S. 1989, The molecular basis of truncated forms of apolipoprotein B in a kindred with compound heterozygous hypobetalipoprotenemia, J. Lipid Res., 30:1773-1779.

Young, S. G., Northey, S. T., and McCarthy, B. J., 1988, Low plasma cholesterol levels caused by a short deletion in the apolipoprotein B gene, Science, 241:591-593.

Young, S. G., Witztum, J. L., Casal, D. C., Curtiss, L. K., and Bernstein, S. 1986, Conservation of the low density lipoprotein receptor-binding domain of apoprotein B, Arteriosclerosis, 6:178-188.

THE APOB 100-APO(a) COMPLEX:

RELATION TO TRIGLYCERIDE-RICH PARTICLES

Angelo M. Scanu and Gunther Fless

The University of Chicago
Departments of Medicine, Biochemistry and Molecular Biology
Chicago, Illinois USA

INTRODUCTION

Lipoprotein(a) is one of the plasma lipoproteins with general characteristics of LDL but having as a protein moiety apoB100 linked by disulfide bridge(s) to apolipoprotein(a) or apo(a), the specific glycoprotein marker of Lp(a) (1-3). Thus, the protein moiety of Lp(a) is the apoB100-apo(a) complex. We will here examine the physico-chemical properties of this complex and try to relate them to its function.

A STRUCTURE

ApoB100 is an apolipoprotein of about 550 kDa, made mainly in the liver and representing the major component of the circulating apoB pool contrary to apoB48 which is made mainly in the intestine. Both apoB100 and apoB48 are coded by the same gene localized in chromosome 2 (4). However, because of an intestine-specific RNA editing protein apoB48 represents only the first NH_2 terminal half of apoB100 with a molecular weight of 206,000. Recent studies by Coleman et al (5) suggest that the two cysteines with a potential attachment to apo(a) are the terminal portion of apoB100 (positions 3734 and 4190), a notion supported by recent unpublished studies by this laboratory on human truncated apoB mutants carried out in collaboration with Dr. Steve Young of the Gladstone Foundation in San Francisco. ApoB$_{100}$ as obtained by the delipidation of LDL is not water-soluble and thus not amenable to physico-chemical characterization probably due to its highly aggregated state (6).

Apo(a) is a glycoprotein that contains about 30% carbohydrate by weight. The polypeptide chain is polymorphic in size, from 300 to 800 kDa; the size isoforms are coded by alleles of the apo(a) gene located in the long arm of chromosome 6 (7, 8). Apo(a) has a strong structural similarity to plasminogen (9, 10). One of its striking features is to contain several kringle 4 repeats (15 to 37) followed by one kringle V and a protease region both highly homologous to the corresponding domains in plasminogen. The number of kringle 4 domains determines the size of the apo(a) isoform as indicated by the recent studies in man (11) and in the baboon (12). Contrary to apo B100, apo(a) is readily water-soluble, a property which is likely related, at least in part, to its high degree of glycosylation.

In terms of the ApoB100-apo(a) complex its properties have been determined only in preparations obtained by removing the lipid moiety from Lp(a) by organic solvents in vitro. Although a lipid-free or an essentially lipid-free apoB100-apo(a) complex may be present in the circulating plasma, its very low abundance and its possible artifactual derivation have limited the interest in its analysis. Contrary to

Hypercholesterolemia, Hypocholesterolemia, Hypertriglyceridemia
Edited by C.L. Malmendier *et al.*, Plenum Press, New York, 1990

295

apoB100, the delipidated apoB100-apo(a) complex is water-soluble, a property which is likely due to apo(a) (6). How does apo(a) render apoB100 water-soluble is unclear but different factors may be operating either alone or in combination: 1) high glycosylation of apo(a); 2) capacity of apo(a) to prevent apoB100 from aggregating; 3) effect of apo(a) on the conformation of apoB100. Regardless of the mechanism, the fact that the apoB100-apo(a) complex is water-soluble permits this complex to potentially occur in solution both in a lipid-free or in a lipidated form.

B FUNCTION

One of the main functions of apoB100 is to act as the ligand for the LDL receptor and thus in key regulation of intracellular cholesterol metabolism (13). Mutations in the ligand binding region of apoB100 are known to cause clinical situations that phenotypically resemble familial hypercholesterolemia associated with an LDL-R deficiency (13).

In terms of apo(a) at this time we know little about its function since in its native state it has not been isolated from the plasma or other body fluids. The available information relates to the reduced and carboxylated product obtained from the treatment of Lp(a) with reducing agents like β-mercaptoethanol or dithiothryotol, both of which may affect the stability of the three disulfides of the kringle domains. Haijar et al (14) have reported that reduced "free" apo(a) binds to the high affinity plasminogen sites located at the cell surface as effectively as whole Lp(a). However, since the properties of the preparations used were not defined, the conclusions must be considered tentative. Only very recently a recombinant apo(a) product was used in binding studies in a mouse macrophage cell line (15). In such a system the recombinant product behaved in the same way as native Lp(a). However, even in this case the physico-chemical characteristics of the product used were not reported.

The apo B100-apo(a) complex once incubated in vitro with artificial lipid emulsions, binds avidly to them; as a consequence of this complexation, the recombinant particles assume a heavier density. An interaction with the apoB100-apo(a) complex has also been demonstrated with either artificial or natural chylomicrons as well as with very low density lipoproteins (6). From these observations, it is apparent that the apoB100-apo(a) complex can affiliate with either triglyceride-rich or cholesteryl ester-rich particles thus potentially setting a basis for a complex pattern of Lp(a) heterogeneity both in normo- and hypertriglyceridemias (see below) (6-16). An important consideration to keep in mind is that any particle that acquires the apoB100-apo(a) complex changes density and hydrodynamic properties due to its large mass and high viscosity of this complex.

Hypertriglyceridemic subjects, regardless of the levels of plasma Lp(a), have only 2-3% of the total mass of apoB100-apo(a) associated with tryglyceride-rich particles, the largest portion being bound to (CE)Lp(a). There are at least three explanations for it: 1) the liver produces (TG)Lp(a) in lesser amounts than (CE)Lp(a); 2) the permanence time of (TG)Lp(a) is much shorter than that of (CE)Lp(a) possibly due to the fact that the former particles are taken up and degraded by the apoE-mediated scavenger receptor pathway a fate not exhibited by the apoE-free (CE)Lp(a) particles; 3) (TG)Lp(a) upon hydrolysis by lipoprotein lipase, releases the apoB100-apo(a) complex that becomes lipid-poor and potentially more readily degradable by plasma proteases. Regardless of the mechanisms involved, there appear to be two pools of apoB100-apo(a) particles in the plasma, probably attended by distinct functional properties and metabolic fate. This area deserves more attention in future studies.

CONCLUDING REMARKS

As this time we know little about the physiological role of Lp(a). Epidemiological studies have shown that Lp(a) particles when present in the circulation at levels above 30 mg/dl are associated with an increased prevalence of atherosclerotic cardiovascular disease, by mechanisms yet to be established. In all of the published studies the attention has been directed at (CE)Lp(a) particles and possible mechanism(s) for their atherogeneity. Recently a pro-thrombotic role for Lp(a) has been postulated on the basis of in <u>vitro</u> studies showing that apo(a) can compete with key physiological funtions of plasminogen with which apo(a) has a strong structural similarity (16, 17-20). In this context, consideration should be given to (TG)Lp(a) particles whether present in the post-prandial state or in endogenous hypertriglyceridemias since triglycerides have been associated with an increased prevalence of thrombotic events (21, 22). The potential connection between Lp(a) pathogenicity and hypertriglyceridemias deserves exploration.

ACKNOWLEDGEMENTS

The original work cited in this review was supported by Program Project NIH-NHLBI grant 18577. The Author is indebted to Ms. Sue Hutchison for valuable help in preparing the manuscript and Dr. Luciani Scandiani for editorial assistance.

REFERENCES

1. A.M. Scanu, L. Scandiani. Lipoprotein(a) structure, biology and clinical relevance. Adv. in Intern. Med., In press (1990).
2. A.M. Scanu, G.M. Fless. Lipoprotein(a) heterogeneity and biological relevance. J.Clin. Invest. In press (1990).
3. G. Utermann. The mysteries of Lipoprotein(a). Science 246:904-910 (1989).
4. C.Y. Yang, L. Chan, A.M. Gotto Jr. The complete structures of human apolipoprotein B-100 and its messenger RNA. In Plasma lipoproteins. A.M. Gotto, Jr. editor. Elsevier, Amsterdam, New York, Oxford. 77-93 (1987).
5. R.D. Coleman, T.W. Kim, A.M. Gotto, Jr., C.Y. Yang. Determination of systeine on low-density lipoproteins using the fluorescent probe, 5 -iodoacetamidofluoresceine. Biochim. Biophys. Acta. 1037:129-132 (1990).
6. G.M. Fless, D.J. Pfaffinger, J.D. Eisenbart, A.M. Scanu. Solubilty, immunochemical, and lipoprotein binding properties of apoB-100-apo(a), the Protein moiety of Lipoprotein(a). J. Lipid. Res. In press (1990).
7. S.L. Frank, I. Klisak, R.S. Sparkes, T. Mohandas, J.E. Tomlinson, J.W. McLean, R.M. Lawn, A.J. Lusis. The apolipoprotein(a) gene resides on human chromosome 6q26-27, in close proximity to the homologous gene for plasminogen. Hum. Genet. 79:352-356 (1988).
8. G. Lindhal, E. Gersdorf, H.J. Menael, C. Duba, H. Cleve, S. Humphries, G. Utermann. The gene for the Lp(a)-specific glycoprotein is closely linked to the gene for plasminogen on chromosome 6. Hum. Genet. 81:149-152 (1989).
9. D.L. Eaton, G.M. Fless, W.J. Kohr, J.W. McLean, Q.Xu, C.G. Miller, R.M. Lawn, A.M. Scanu. Partial amino acid sequence of apolipoprotein(a) shows that it is homologous to plasminogen. 84:3224-3228 (1987).
10. J.W. McLean, J.E. Tomlinson, W. Kuang, D.L. Eaton, E.Y. Chen, C.M. Fless, A.M. Scanu, R.M. Lawn. cDNA sequence of human apolipoprotein(a) is homologous to plasminogen. Nature 330:132-137 (1987).
11. R.M. Lawn, J.E. Tomlinson, J.W. McLean, D.L. Eaton. Molecular biology of apolipoprotein(a). In Lipoprotein(a): 25 years of progress. A.M. Scanu, editor. Academis Press Inc., New York, In press (1990).
12. J.E. Hixson, M.L. Britten, G. S. Manis, D.L. Rainwater. Apolipoprotein(a)[Apo(a)] glycoprotein isoforms result from size differences in apo(a) mRNA in baboons. J. Biol. Chem. 264:6013-6016 (1989).
13. J.E. Goldstein, M.S. Brown. In The Metabolic Basis of Inherited Disease, eds. C.R. Scriver, A.L. Beaudet, W.S. Sly, D. Valle. MCGraw-Hill Co, NY, 6:1215 -1250 (1989).
14. K.A. Hajjar, D. Gavish, J.L. Breslow, R.L. Nachman. Lipoprotein(a) modulation endohelial cell surface fibrinolysis and its potential role in atherosclerosis. Nature. 339:303-305 (1989).
15. L.M. Powell, G. Rice, R. Lawn, E. Eaton. Binding of Lp(a) and recombinant apo(a) to the mouse macrophage cell line P388D.1. Atherosclerosis, 701a (1989).

16. G.M. Fless. Heterogeneity of particles containing the apoB-apo(a) complex. In Lipoprotein(a): 25 years of progress. A.M. Scanu, editor. Academic Press Inc. NY, In press (1990).

17. J. Loscalzo, G.M. Fless, A.M. Scanu. Lp(a) and the fibrinolytic system. In Lipoprotein(a): 25 years of progress. A.M. Scanu, editor. Academic Press Inc. NY, In press (1990).

18. E.F. Plow, L.A. Miles. Relationship between plasminogen receptors and Lp(a). In Lipoprotein(a): 25 years of progress. A.M. Scanu, editor. Academic Press Inc. NY, In press (1990).

19. M. Gonzalez-Gronow, J.M. Edelberg, S.V. Pizzo. Further characterization of the cellular plasminogen binding site; evidence that plasminogen 2 and lipoprotein *a* compete for the same site. Biochemistry, 28:2374-2377 (1989).

20. J.M. Edelberg, M. Gonzalez-Gronow, S.V. Pizzo. Lipoprotein(a) inhibition of plasminogen activation by tissue-type plasminogen activator. Throm. Res. 57:155-162 (1990).

21. R.S. Elkeles, R. Chakrabarti, M. Vickers, Y. Stirling, T.W. Meade. Effect of treatment of hyperlipidaemia on haemostatic variables. Br. Med. J. 281:973-974 (1980).

22. H.C.R. Simpson, J.I. Mann, T.W. Maede, R. Chakrabarti, Y. Stirling, L. Woolf. Hypertriglyceridaemia and hypercoagulability. Lancet, i:786-790 (1983).

APOB-CONTAINING LIPOPROTEIN PARTICLES AS RISK FACTORS

FOR CORONARY ARTERY DISEASE

P. Alaupovic, D.H. Blackenhorn, C. Knight-Gibson, M. Tavella,
J.-M. Bard, D. Shafer, E.T. Lee and J. Brasuell

Lipoprotein and Atherosclerosis Research Program, Oklahoma
Medical Research Foundation, Oklahoma City, Oklahoma,
Department of Medicine, University of Southern California
School of Medicine, Los Angeles, California, and USPHS
Indian Hospital, Claremore, Oklahoma, USA

INTRODUCTION

The function of lipid transport is to facilitate the transfer of exogenous and endogenous triglycerides and cholesterol from their sites of
absorption or formation to their sites of storage and utilization. This
function is carried out by triglyceride-rich lipoproteins of intestinal and
hepatic origin through a series of enzymic conversion reactions resulting
in release of triglyceride fatty acids and generation of cholesterol-rich
remnant lipoproteins. Under normal, steady-state conditions, the input and
output of triglyceride-rich lipoproteins are balanced with little or no
change in their plasma levels. However, an increased influx and/or a decreased efflux of triglyceride-rich lipoproteins from the plasma compartment have been identified as main pathophysiologic mechanisms leading to
hypertriglyceridemia[1]. The increased formation and/or decreased removal of
cholesterol-rich lipoproteins results in hypercholesterolemia which may or
may not accompany typically hypertriglyceridemic states. Deranged lipid
transport processes are of great clinical significance, because they are
identified as one of the main factors responsible for the genesis and development of atherosclerosis[2-4].

Hypertriglyceridemia is a common disorder of lipid transport occurring
either as a primary or secondary manifestation of the underlying disease.
However, in contrast to plasma cholesterol levels, the independent contribution of plasma triglyceride concentrations to atherogenesis remains a
controversial issue. Most case-control studies have shown a strong univariate association between plasma triglyceride levels and coronary heart
disease[4]. However, with some exceptions[5], prospective studies have provided no evidence that plasma triglyceride levels are an independent predictor of coronary heart disease after adjustments for the other risk
factors including plasma cholesterol, HDL-cholesterol and hypertension[4,6].
On the other hand, a recent Swedish prospective study has shown that in
women, aged 38-60, plasma levels of triglyceride are a more accurate predictor of ischemic heart disease and mortality than plasma cholesterol[7].
These results are in agreement with previous studies suggesting that plasma
triglyceride levels are better predictors of ischemic heart disease than
cholesterol levels in women aged 50 and older[8,9]. Results of the Paris
prospective study have indicated that in subjects with serum cholesterol

Hypercholesterolemia, Hypocholesterolemia, Hypertriglyceridemia
Edited by C.L. Malmendier *et al.*, Plenum Press, New York, 1990

299

levels lower than 220 mg/dl, serum triglyceride levels are an independent predictor for coronary heart disease death even after adjustments for all other risk factors[10]. Several studies of patients with angiographically-documented coronary artery disease have demonstrated the significance of VLDL- and IDL-cholesterol and LDL- and IDL-triglyceride levels as indicators of severity of coronary atherosclerosis[11,12]. It has been shown in a recent comparative study that hypertriglyceridemic men with angiographically documented coronary artery disease had a higher coronary score than hypercholesterolemic men[13]. In general, hypertriglyceridemia with or without associated hypercholesterolemia occurs more frequently in patients with coronary artery disease than hypercholesterolemia[14,15]. It appears, thus, that certain subsets of the population with hypertriglyceridemia have a predisposition to coronary artery disease, but this association is not evident in population studies.

In hypertriglyceridemia, impaired catabolism of triglyceride-rich lipoproteins may lead to the formation of remnant particles only partially depleted of triglycerides but enriched in cholesterol[16]. It has been suggested on the basis of clinical[11,12,16,17] and metabolic[18-21] studies that these compositionally modified triglyceride-rich particles may have atherogenic potential similar, if not equal, to that of cholesterol-rich and triglyceride-poor lipoproteins. The apparent dichotomy between the epidemiologic approach and clinical and metabolic studies in the assessment of lipid or lipoprotein involvement in atherogenesis may stem in part from the customary use of cholesterol and triglyceride as the sole markers of a complex macromolecular system of lipoprotein particles. These two lipids, albeit occurring in varying proportions, are integral constituents of all lipoproteins; however, they are not specific markers of lipoprotein particles[22] or determinants of their various biologic properties[23]. This may also apply to the atherogenicity of lipoprotein particles in that all potentially atherogenic lipoproteins contain ApoB and all potentially nonatherogenic lipoproteins ApoA-I as their characteristic protein constituents[24]. Furthermore, it is generally considered that ApoB-containing lipoproteins differ in their atherogenic potential which increases with decreasing size and increasing density of lipoprotein particles[2,8,25]; however, as mentioned before, recent evidence suggests that cholesterol-enriched IDL particles may have atherogenic potential similar, if not equal, to that of cholesterol-rich LDL particles[11,12,16,17,26]. If supported by further evidence, the concept of relative atherogenicity of ApoB-containing lipoprotein particles may be of considerable theoretical and clinical significance.

The purpose of this paper is to provide some additional information regarding the possible differences among ApoB-containing lipoproteins as risk factors for coronary artery disease. It is based conceptually on the theory of lipoprotein families and experimentally on the recent results of the Cholesterol Lowering Atherosclerosis Study (CLAS)[27,28] and a study on the ApoB-containing lipoprotein particles in patients with non-insulin dependent (type II) diabetes mellitus.

ApoB-Containing Lipoprotein Particles

Results from several laboratories have shown that the compositional and metabolic heterogeneity of operationally defined lipoprotein classes is due to the presence of discrete lipoprotein particles of similar sizes and densities but distinct apolipoprotein composition. We have proposed on the basis of these findings that apolipoproteins be used as the new criterion for identifying and classifying plasma lipoproteins[22,29]. As major apolipoproteins, ApoA (ApoA-I + ApoA-II) and ApoB form two groups of lipoproteins separable either by concanavalin A affinity chromatography or immunoaffinity chromatography of whole plasma, with either polyclonal antibodies to ApoB or monoclonal antibodies that bind equally to all ApoB-containing lipoproteins[27]. Fractionation of ApoB-containing lipoproteins in VLDL, LDL-1 and LDL-2 or in whole plasma by sequential immunoprecipitation or immunoaffinity chromatography on immunosorbers with antibodies to ApoA-II, ApoE and ApoC-III, respectively, results in the separation of four major lipoprotein families identified as lipoprotein A-II:B:C-I:C-II:C-III:D:E (LP-A-II:B complex), lipoprotein B:C-I:C-II:C-III:E (LP-B:C:E), lipoprotein B:C-I:C-II:C-III (LP-B:C) and lipoprotein B (LP-B)[29-32]. The LP-B:C:E particles may also contain varying amounts of lipoprotein B:E (LP-B:E) separable, if necessary, by immunoaffinity chromatography on an anti-ApoC-III immunosorber. All ApoB-containing lipoprotein families are polydisperse systems of particles heterogeneous with respect to size, density and lipid/apolipoprotein ratios, but homogeneous with respect to their qualitative apolipoprotein composition. The lipid composition of ApoB-containing lipoproteins is characterized by a considerable degree of specificity. The LP-B particles, regardless of their density properties, contain cholesterol esters as their main neutral lipid. On the other hand, LP-A-II:B complex, LP-B:C:E and LP-B:C particles contain triglycerides as their main neutral lipid, irrespective of their density properties. The LP-B:E particles contain cholesterol esters and triglycerides in similar proportions. The apolipoprotein composition of these latter lipoprotein particles changes with increasing densities: the relative content of ApoB increases and the relative contents of ApoC-peptides and ApoE decrease. The polydisperse character of ApoB-containing lipoproteins is further demonstrated by measurement of their concentration profile in major lipoprotein density classes. Although in normolipidemic and dyslipoproteinemic subjects the LP-B particles are present mainly in LDL subfractions, they are also detectable in VLDL, especially in patients with phenotypes III and V and in patients with chronic renal failure[29,31,32]. The LP-B:C:E, LP-B:C and LP-A-II:B complex are present mainly in VLDL and LDL-1, but, in many dyslipoproteinemic states, relatively high concentrations of these particles are also found in LDL-2. Determination of lipoprotein particle profiles in patients with primary and secondary hyperlipoproteinemias[29,32] has demonstrated that lipid transport disorders are characterized by quantitative rather than qualitative differences in five major ApoB-containing lipoprotein particles. Hypercholesterolemic states contain cholesterol ester-rich LP-B particles as the main ApoB-containing lipoprotein, whereas hypertriglyceridemic states contain triglyceride-rich LP-B:C:E, LP-B:C and LP-A-II:B complex as the predominant lipoprotein forms of ApoB. More recent findings have indicated that discrete ApoB-containing lipoprotein particles may differ in their metabolic properties. For example, the HepG2 cells seem to secrete mainly, if not exclusively, triglyceride-rich LP-B and LP-B:E particles[33] suggesting that these lipoprotein families also serve, at least in part, as precursors for the extracellular formation of LP-B:C:E, LP-B:C and LP-A-II:B complex. The LP-B:C:E particles were found to be a more efficient substrate for lipoprotein lipase than LP-A-II:B complex[30,32]. The LP-B:E particles bind to LDL-receptors with greater affinity than LP-B particles[33], which, in turn, bind with greater affinity to LDL-receptors than LP-B:C particles (unpublished results). There is, however, some uncertainty as to whether these chemically and, possibly, metabolically distinct ApoB-containing lipoproteins also differ with respect to their atherogenic properties. Is ApoB the only determinant of this pathophysiologic capacity or are there other modulators and factors enhancing or diminishing this capacity? To examine this question we shall first analyze data provided by the CLAS study[27,28].

The Cholesterol Lowering Atherosclerosis Study

The purpose of the CLAS study, conducted by Blankenhorn and his associates was to determine whether a prolonged, aggressive lowering of LDL-cholesterol with concomitant increase in HDL-cholesterol will retard or reverse the growth of atherosclerotic lesions in native coronary arteries and bypass grafts[27]. It was a randomized, placebo-controlled, angiographic trial involving 162 non-smoking men with progressive atherosclerosis and previous coronary bypass surgery who were treated with combined colestipol hydrochloride and niacin therapy for 2 years. Global angiographic change including both native arteries and bypass grafts after 2 treatment years was the end point. A detailed description of patients, dietary and drug treatment, coronary evaluation and statistical methodology has been reported previously[27]. The drug treatment reduced significantly the percentage of subjects who developed new lesions in both native coronary arteries ($p < 0.03$) and grafts ($p < 0.04$). There were 61% nonprogressors in the drug group and 40% in the placebo group; the corresponding percentages for progressors were 38% in the drug group and 59% in the placebo group. To compare the predictive power of apolipoproteins and lipids[28], the measurements included cholesterol, triglyceride, HDL-cholesterol, non-HDL-cholesterol (the sum of VLDL and LDL cholesterol), and apolipoproteins A-I, B and C-III. In addition, ApoC-III was determined in heparin-Mn^{++} supernates ("HDL") and heparin-Mn^{++} precipitates ("VLDL + LDL") as previously described[28,34]. The ratio of ApoC-III in heparin-Mn^{++} supernate/ApoC-III in heparin-Mn^{++} precipitate (ApoC-III-ratio) has been introduced as a biochemical link or marker for the reciprocal relationship between the HDL and VLDL particles. The conceptual basis and validation of ApoC-III-ratio has been previously described[34]. It suffices to state that ApoC-III-ratio is highly significantly ($p < 0.001$) negatively correlated with plasma and VLDL triglyceride levels and equally significantly positively correlated with levels of HDL-cholesterol. The higher the ApoC-III-ratio the lower the concentration of triglyceride-rich lipoproteins and vice versa. Thus, the ApoC-III ratio is an excellent marker for the distribution of ApoC-III between triglyceride-rich lipoproteins and triglyceride-poor HDL particles, and a useful means for assessing the efficiency of catabolic processes responsible for the degradation and uptake of triglyceride-rich lipoproteins.

A comparison on-trial values between the placebo and drug groups showed significant differences ($p < 0.001$) for all lipids, ApoA-I, ApoB, ApoC-III-supernate and ApoC-III-precipitate but not for plasma ApoC-III[28]. In the drug group there were statistically significant decreases in total cholesterol, non-HDL-cholesterol, triglyceride, LDL-cholesterol, ApoB and ApoC-III-precipitate, and statistically significant increases in HDL-cholesterol, ApoA-I, ApoC-III-supernate and ApoC-III-ratio. The univariate analysis of risk factors comparing coronary progressors and nonprogressors showed that in the placebo group progressors had higher levels of total cholesterol, non-HDL-cholesterol, triglyceride, LDL-cholesterol, ApoB and ApoC-III than nonprogressors. In the drug group progressors had significantly lower levels of ApoC-III-supernate than nonprogressors as the only on-trial predictor of the probability of coronary progression. The multivariate logistic regression analysis showed that in the placebo group the significant independent predictor of coronary progression was non-HDL-cholesterol and in the drug group ApoC-III-supernate: each 34 mg/dl increase in non-HDL-cholesterol was associated with a twofold increase in risk of coronary progression, whereas each 1.9 mg/dl increase in ApoC-III-supernate was associated with a twofold decrease in the risk of progression. These results have indicated the importance of triglyceride-rich lipoproteins in the progression of atherosclerosis in both placebo and drug groups and the significance of ApoC-III ratio as a marker for the potential contribution of triglyceride-rich lipoproteins as a risk factor. It should be pointed out, however, that drug treatment also had a significant effect on the levels of LDL-cholesterol and ApoB, and that higher levels of ApoC-III-supernate in the absence of drug therapy were not protective.

To determine the effect of drug treatment on major ApoB-containing lipoproteins, we have measured LP-B, LP-B:C:E and LP-B:C particles in a limited number of patients on and off drug treatment[35]. The results have shown a highly uniform and significant decrease (40-45%) in the levels of LP-B particles, but no effect on the concentrations of LP-B:C:E (including LP-A-II:B complex and LP-B:E) and LP-B:C particles. These results have suggested that cholesterol-rich LP-B particles have a significant atherogenic potential as demonstrated by increased percentages of nonprogressors in the drug group and progressors in the placebo group. However, patients who also had increased concentrations of triglyceride-rich LP-B:C:E and LP-B:C particles, as demonstrated by decreased levels of ApoC-III in HDL and increased levels of ApoC-III in VLDL and LDL (decreased ApoC-III-ratio), showed progression of atherosclerotic lesions in their native arteries and grafts despite decreased levels of LP-B particles. Thus, LP-B:C:E, LP-A-II:B complex, LP-B:E and LP-B:C particles also carry a certain degree of risk for coronary artery disease that may or may not be equal to that of LP-B particles.

The significance of triglyceride-rich lipoproteins in the progression of atherosclerotic lesions has also been supported recently by results of the Helsinki Heart Study[36] which showed that a 5 year treatment of dyslipidemic men with gemfibrozil resulted in modest decreases in the levels of total cholesterol (10%) and LDL-cholesterol (11%), a marked decrease in triglycerides (35%) and a moderate increase in HDL-cholesterol (11%). This effect on plasma lipids was accompanied by a 34% reduction in the incidence of definite coronary heart disease events. The results were interpreted as a further confirmation of the importance of LDL as a risk factor, but, more importantly, as evidence for the protective effect of HDL-cholesterol against coronary heart disease in men. However, it should be recalled that patients with phenotypes IIB and IV who had a minimal decrease in the levels of LDL-cholesterol, but a significant decrease in the levels of triglyceride, also had the greatest reduction in incidence of end points. Our study on the effect of gemfibrozil on ApoB-containing lipoproteins showed that the administration of this drug to patients with phenotypes V (37) and IV (unpublished results) caused significant reductions of triglyceride-rich LP-B:C:E and LP-B:C particles accompanied by no change or a slight increase in the levels of LP-B particles. This specific effect of gemfibrozil has provided at least a partial explanation for the relatively modest effect of drug treatment observed in patients with phenotype IIA whose high levels of LP-B particles had not been reduced sufficiently to produce a favorable effect comparable to those seen in patients with phenotypes IIB and IV.

When interpreted in light of the lipoprotein family concept, findings of the CLAS and Helsinki Heart Study suggest that both the cholesterol-rich LP-B and triglyceride-rich LP-B:C:E, LP-A-II:B complex and LP-B:C particles contribute to the development and progression of atherosclerotic lesions. It appears on the basis of circumstantial evidence that the relative atherogenicity of LP-B particles may be greater than those of triglyceride-rich ApoB-containing lipoproteins. One of the most compelling arguments supporting this conclusion is the fact that very high concentrations of LP-B particles in patients with familial hypercholesterolemia[29,31] are associated, especially in homozygotes, with a fulminating coronary heart disease in the early life.

If one subscribes to the generally accepted view that LP-B particles are more atherogenic than other ApoB-containing lipoproteins, the next question to be raised is whether these latter lipoprotein families differ from one another in that capacity or are equal as risk factors for coronary artery disease. In the absence of a direct method for assessing the relative atherogenicity of lipoprotein families, we shall address this question by presenting indirect evidence provided by a study of patients with non-insulin dependent (type II) diabetic mellitus.

Triglyceride-rich Lipoproteins and Macrovascular Disease in Patients with Non-Insulin Dependent Diabetes Mellitus

The selection of type II diabetics for a study of triglyceride-rich lipoprotein particles was prompted by a number of reports demonstrating that the high incidence of atherosclerotic vascular disease in diabetic patients is associated with an increased concentration of triglyceride-rich lipoproteins as the most characteristic risk factor[38,39]. The determination of triglyceride-rich lipoprotein particle profile was performed as a part of an extensive study examining the possible relationship between lipoprotein abnormalities and macrovascular disease in 428 Northeastern Oklahoma Indians with non-insulin dependent (type II) diabetes mellitus selected randomly from among diabetic patients regularly attending the USPHS Indian Hospital in Claremore, Oklahoma. Two hundred sixteen asymptomatic Northeastern Oklahoma Indians attending the Indian Hospital for routine medical examination served as controls. The study only included patients with established diabetes of at least six month duration and fasting hyperglycemia (plasma glucose levels of over 140 mg/dl on more than one occasion). Records of all patients and controls included complete medical history, standard physical and laboratory data and, in case of patients, the type of treatment. All patients and control subjects signed written informed consents. The protocols were approved by the institutional Human Experimentation Committee at the Oklahoma Medical Research Foundation.

Whereas the measurements of plasma cholesterol, triglyceride, VLDL-, LDL- and HDL-cholesterol and apolipoproteins A-I, A-II, B, C-III, C-III-supernate, C-III-precipitate and E were performed in all patients and controls, the isolation of major lipoprotein density classes and determination of LP-B, LP-B:C:E and LP-B:C particles were performed in 8 randomly selected normolipidemic and 8 hyperlipidemic diabetics and in the same number of nondiabetic controls by previously described procedures[29-34].

Briefly, the results showed that the lipid profile of diabetic patients was characterized by significantly higher levels of plasma triglyceride and VLDL-cholesterol and lower levels of HDL-cholesterol than nondiabetic controls; however, there was no difference in the levels of LDL-cholesterol. Diabetics also had significantly higher levels of plasma ApoB, ApoC-III and ApoE and lower levels of ApoA-I and ApoA-II than control subjects. The determination of ApoC-III in heparin-Mn^{++} supernates and precipitates showed that diabetic patients had significantly lower ApoC-III-ratio than nondiabetics (0.94 vs 1.74, $p < 0.0001$) suggesting an impaired degradation and/or overproduction of triglyceride-rich lipoproteins. The incidence of hypertriglyceridemia in this diabetic population was 37% (TG $\geq$ 200 mg/dl), but that of hypercholesterolemia was only 11% (TC $\geq$ 250 mg/dl). The estimation of the mass of major lipoprotein density classes showed that both hypertriglyceridemic and normotriglyceridemic diabetics had a several-fold (3-10) increase in the mass of VLDL and a 50% reduction in the mass of HDL in comparison with nondiabetics, but no difference in the mass of LDL.

Angina represented the major manifestation of macrovascular disease affecting 24% of this diabetic population. The most significant variables differentiating affected and non-affected diabetics were the ApoC-III-ratio (0.7 ± 0.6 vs 1.0 ± 1.0 mean ± SD, $p < 0.02$), VLDL-cholesterol (39 ± 42 vs 28 ± 25 mg/dl, $p < 0.02$) and HDL-cholesterol (39 ± 14 vs 42 ± 13 mg/dl, $p < 0.03$). The non-fatal myocardial infarction occurred in 10.2% of diabetics; the most significant variables in this subpopulation of patients included HDL-cholesterol (36 ± 11 vs 42 ± 13 mg/dl, $p < 0.001$), VLDL-cholesterol (45 ± 44 vs 28 ± 27 mg/dl, $p < 0.01$) and plasma triglyceride (249 ± 199 vs 204 ± 202, $p < 0.05$). These results showed that atherosclerotic complica-

Table 1. VLDL Constituents of Normolipidemic and Hyperlipidemic
Type II Diabetics With and Without Vascular Disease[a]

Lipoprotein Constituents	Normal Triglycerides		Increased Triglycerides	
	No Vascular Disease (n=2)	Vascular Disease (n=4)	No Vascular Disease (n=2)	Vascular Disease (n=4)
ApoC-III (%)	28.8	<u>41.1</u>	37.2	<u>53.8</u>
ApoE (%)	29.2	<u>47.3</u>	39.2	<u>57.3</u>
LP-B:C (mg/dl)	1.1	5.0	5.3	25.6
LP-B:C:E (mg/dl)	4.3	6.0	10.6	13.6
$\frac{LP-B:C}{LP-B:C:E}$ (ratio)	0.25	<u>0.83</u>	0.41	<u>1.8</u>

[a] The cutoff point for triglycerides was 150 mg/dl.

tion of type II diabetes are associated with some of the typical constituents of triglyceride-rich rather than cholesterol-rich lipoproteins. It was, therefore, of considerable interest to determine the concentrations of major ApoB-containing lipoprotein particles and to compare them with those isolated from nondiabetic controls. Qualitatively, ApoB-containing lipoprotein particles identified in major lipoprotein density classes of diabetic patients were the same as those of nondiabetic controls. However, there were some significant differences in the concentrations of lipoprotein particles between patients and controls. The levels of LP-B particles were slightly higher in VLDL, LDL-1 and LDL-2 of diabetic than nondiabetic subjects, but the statistical significance was only reached in LDL-1. The most characteristic feature of the lipoprotein particle profile of diabetic patients was the significantly increased concentrations of LP-B:C (11.3 vs 1.2 mg/dl, $p < 0.05$) and LP-B:C:E + LP-A-II:B complex (9.1 vs 3.2 mg/dl, $p < 0.01$) in VLDL. Moreover, the increase in the concentration of LP-B:C particles was proportionally greater than that of LP-B:C:E particles resulting in higher LP-B:C/LP-B:C:E ratios in VLDL of diabetic than nondiabetic subjects. Diabetics also had higher ratios of these two lipoprotein families in LDL-1 and LDL-2. Furthermore, both hypertriglyceridemic and normotriglyceridemic diabetics had higher levels and ratios of LP-B:C and LP-B:C:E particles than normal controls.

A comparison of plasma lipid and apolipoprotein patterns between diabetic patients with and without clinically diagnosed vascular disease has shown that the typical abnormalities in the composition and concentration of triglyceride-rich lipoproteins are more pronounced in affected than non-affected patients. It was, therefore, of considerable interest to explore the possible relationship between lipoprotein abnormalities already detected in diabetic patients and the presence of vascular disease. Results of this exploratory study are presented in Table 1. In the normotriglyceridemic group, patients with vascular disease had higher proportions of total plasma ApoC-III and ApoE in VLDL, and higher levels and ratios of LP-B:C and LP-B:C:E particles than patients without the disease. These same differences were also observed between hypertriglyceridemic diabetics with and without the vascular disease. However, the most interesting aspect of this preliminary study was the observation that higher proportions of total plasma ApoC-III and ApoE and higher LP-B:C/LP-B:C:E ratios (due to a significant increase in LP-B:C but not LP-B:C:E particles) in VLDL were characteristic of all diabetic patients with vascular disease regardless of their plasma triglyceride levels (Table 1). Patients with vascular disease also had higher LP-B:C/LP-B:C:E ratios in LDL-2 irrespective of their plasma or LDL-2 triglyceride levels. These results based on a small number of selected diabetic patients ought to be confirmed and expanded before any definite conclusions can be reached regarding the possible association between the increased concentration of LP-B:C particles, or a certain ratio of LP-B:C/ LP-B:C:E particles, and the presence of coronary and/or peripheral artery disease in type II diabetes. Nevertheless, these findings suggest that LP-B:C particles may represent a greater risk for coronary atherosclerosis than LP-B:C:E particles, but that this potential may only be realized at or above a certain threshold value for the LP-B:C/LP-B:C:E ratio. Since LP-B:C:E particles measured in this study also contained the LP-A-II:B complex, it is not yet known what effect this latter lipoprotein family may have on the atherogenic potential of LP-B:C particles, especially in view of the fact that LP-A-II:B complex is a less efficient substrate for lipoprotein lipase than LP-B:C:E particles[32]. Finally, the determination of LP-B:C and LP-B:C:E levels in several primary and secondary hyperlipoproteinemias has shown that LP-B:C/LP-B:C:E ratios are higher in dyslipoproteinemic states known to have higher (chronic renal failure, phenotype IV, diabetes mellitus, type II) rather than lower (phenotype V, glycogen storage disease, type I) incidence of coronary artery disease (unpublished results).

As pointed out in recent commentaries[25,40], there is still a considerable uncertainty regarding the atherogenic potential of various lipoproteins, the mode of their interaction with arterial wall cells, and the methodology for their identification and quantification. One of the reasons for this uncertainty may be due to the customary use of total cholesterol and, more recently, LDL-cholesterol as the markers for the atherogenicity of lipoprotein particles. The calculated values for LDL-cholesterol include a substantial part of partially delipidized triglyceride-rich lipoproteins and, thus, provide an erroneous impression that only the cholesterol-rich, triglyceride-depleted lipoproteins are included in this measurement. The use of lipoprotein density classes may also be misleading because, as shown in this and other studies, these lipoprotein preparations consist of several polydisperse ApoB-containing lipoproteins of distinct chemical and metabolic properties and, possibly, different atherogenicity. Consequently, the composition of lipoprotein density classes such as VLDL or LDL may differ considerably among individuals depending on the percentage content of discrete ApoB-containing lipoprotein particles. For this reason, the measurement or mapping of discrete lipoprotein particles may reflect more accurately the metabolic status of lipid transport in individual subjects. Moreover, if the concept of relative atherogenicity of ApoB-containing particles survives the necessary testing and confirmation, mapping of these particles may also provide a more rational estimation of their potential as risk factors for coronary artery disease. Whereas increased levels of LP-B particles may represent the major risk factor in some cases, elevated concentrations of LP-B:C or LP-A-II:B complex particles may carry the major risk in others. In the majority of cases, however, a combination of these discrete lipoproteins may have the greatest atherogenic potential. In one of the first studies to correlate ApoB-containing lipoprotein particles with the severity of coronary artery disease in 67 normotensive, nondiabetic patients with angiographically documented coronary artery disease and normal or moderately increased plasma cholesterol (215 ± 34 mg/dl) and triglyceride (140 ± 86 mg/dl),

ApoC-III-ratio and LP-B:C + LP-B:C:E particles showed the most significant correlation with the coronary score[41]. When considered in conjunction with recent findings indicating a rather specific effect of hypolipidemic drugs on ApoB-containing lipoproteins,[35,37] the identification of atherogenic potential of lipoprotein particles may become a very useful means for diagnosing and treating dyslipoproteinemic states.

CONCLUSIONS

1. Both the cholesterol ester-rich and triglyceride-rich ApoB-containing lipoprotein particles are potential risk factors for coronary artery disease.

2. It appears on the basis of circumstantial evidence that cholesterol ester-rich LP-B particles may have a more pronounced atherogenic potential than the triglyceride-rich LP-B:C, LP-A-II:B:C:D:E (LP-A-II:B complex) or LP-B:C:E particles.

3. Among the triglyceride-rich lipoprotein families, LP-B:C particles might be more atherogenic than LP-A-II:B complex and LP-B:C:E particles.

4. The measurement of ApoC-III in heparin-Mn^{++} supernate and heparin-Mn^{++} precipitate (ApoC-III-ratio) is a useful marker for the presence of potentially atherogenic triglyceride-rich ApoB-containing lipoprotein particles.

5. The concept of relative atherogenicity of ApoB-containing lipoproteins needs to be confirmed and expanded by additional studies on the association between these lipoproteins and the presence and severity of coronary artery disease.

ACKNOWLEDGEMENTS

We wish to thank Ms. M. French for editing and typing the manuscript.

This study was supported in part by Grants HL-23181 and HL-23619 from the U.S. Public Health Service and by the Oklahoma Medical Research Foundation and the Upjohn Company.

REFERENCES

1. S.M. Grundy, Pathogenesis of hyperlipoproteinemia, J. Lipid Res. 25:1611 (1984).
2. R.B. Wallace, and R.A. Anderson, Blood lipids, lipid-related measures, and the risk of atherosclerotic cardiovascular disease, Epidemiol. Rev. 9:95 (1987).
3. K.M. Anderson, W.P. Castelli, and D. Levy, Cholesterol and mortality: 30 years of follow-up from the Framingham study, JAMA 257:2176 (1987).
4. M.A. Austin, Plasma triglyceride as a risk factor for coronary heart disease: the epidemiologic evidence and beyond, Am. J. Epidemiol. 129:249 (1989).
5. L.A. Carlson, and L.E. Böttiger, Serum triglycerides, to be or not to be a risk factor for ischaemic heart disease? Atherosclerosis 39:287 (1981).
6. S.B. Hulley, R.H. Rosenman, R.D. Bawol, and R.J. Brand, Epidemiology as a guide to clinical disorders. The association between triglyceride and coronary heart disease, N. Engl. J. Med. 302:1383 (1980).

7. L. Lapidus, C. Bengtsson, O. Lindquist, J.A. Sigurdsson, and E. Ryleo, Triglycerides - main lipid risk factor for cardiovascular disease in women? *Acta* *Med*. *Scand*. 217:481 (1985).

8. W.B. Kannel, W.P. Castelli, T. Gordon, and P.M. McNamara, Serum cholesterol, lipoproteins and the risk of coronary heart disease. The Framingham study, *Ann*. *Intern*. *Med*. 74:1 (1971).

9. S. Heyden, G. Heiss, C.G. Hames, and A.G. Bartel, Fasting triglycerides as predictors of total and CHD mortality in Evans County, Georgia, *J*. *Chron*. *Dis*. 33:275 (1980).

10. F. Cambien, A. Jacqueson, J.L. Richard, J.M. Warnet, P. Ducimetière, and J.R. Claude, Is the level of serum triglyceride a significant predictor of coronary death in "normocholesterolemic" subjects? *Am*. *J*. *Epidemiol*. 124:624 (1986).

11. R. Tatami, H. Mabuchi, K. Ueda, R. Ueda, T. Haba, T. Kametani, S. Ito, J. Koizumi, M. Ohta, S. Miyamoto, A. Nakayama, H. Kanoya, H. Oiwake, A. Genda, and R. Takeda, Intermediate-density lipoprotein and cholesterol-rich very low density lipoprotein in angiographically determined coronary artery disease, *Circulation* 64:1174 (1981).

12. M.F. Reardon, P.J. Nestel, I.H. Craig, and R.W. Harper, Lipoprotein predictors of the severity of coronary artery disease in men and women, *Circulation* 71:881 (1985).

13. C.H. Breier, J.R. Patsch, V. Mühlberger, H. Drexel, E. Knapp, and H. Braunsteiner, Risk factors for coronary artery disease: a study comparing hypercholesterolaemia and hypertriglyceridaemia in angiographically characterized patients, *Eur*. *J*. *Clin*. *Invest*. 19:419 (1989).

14. J.L. Goldstein, W.R. Hazzard, H.G. Schrott, E.L. Bierman, and A.G. Motulsky, I. Lipid levels in 500 survivors of myocardial infarction, *J*. *Clin*. *Invest*. 52:1533 (1973).

15. E.J. Schaefer, J.R. McNamara, J. Genest, and J.M. Ordovas, Clinical significance of hypertriglyceridemia, *Semin*. *Thromb*. *Hemost*. 14:143 (1988).

16. J.P. Kane, G.C. Chen, R.L. Hamilton, D.A. Hardman, M.J. Malloy, and R.J. Havel, Remnants of lipoproteins of intestinal and hepatic origin in familial dysbetalipoproteinemia, *Arteriosclerosis* 3:47 (1983).

17. L.A. Simons, T. Dwyer, J. Simons, L. Bernstein, P. Mack, N.S. Poon, S. Balasubramaniam, D. Baron, J. Branson, J. Morgan, and P. Roy, Chylomicrons and chylomicron remnants in coronary artery disease: a case-control study, *Atherosclerosis* 65:181 (1987).

18. D.B. Zilversmit, Atherogenesis: a postprandial phenomenon, *Circulation* 60:473 (1979).

19. C.H. Florén, J.J. Albers, and E.L. Bierman, Uptake of chylomicron remnants causes cholesterol accumulation in cultured human arterial smooth muscle cells, *Biochim*. *Biophys*. *Acta* 663:336 (1981).

20. S. Eisenberg, Lipoprotein abnormalities in hypertriglyceridemia: significance in atherosclerosis, *Am*. *Heart*. *J*. 113:555 (1987).

21. S.H. Gianturco, and W.A. Bradley, Lipoprotein-mediated cellular mechanisms for atherogenesis in hypertriglyceridemia, *Semin*. *Thromb*. *Hemost*. 14:165 (1988).

22. P. Alaupovic, Conceptual development of the classification systems of plasma lipoproteins, *Protides* *Biol*. *Fluids* *Proc*. *Colloq*. 19:9(1972).

23. B.J. Dolphin, Lipoprotein metabolism and the role of apolipoproteins as metabolic programmers, *Can*. *J*. *Biochem*. *Cell* *Biol*. 63:850 (1985).

24. J.D. Brunzell, A.D. Sniderman, J.J. Albers, and P. Kwiterovich, Jr., Apolipoproteins B and A-I and coronary artery disease in humans, *Arteriosclerosis* 4:79 (1984).

25. D. Steinberg, Lipoproteins and atherosclerosis: some unanswered questions, *Am*. *Heart* *J*. 113:626 (1987).

26. R. Krauss, Relationship of intermediate and low-density lipoproteins subspecies to risk of coronary artery disease, *Am*. *Heart* *J*. 113:578 (1987).

27. D.H. Blankenhorn, S.A. Nessim, R.L. Johnson, M.E. Sanmarco, S.P. Azen, and L. Cashin-Hemphill, Beneficial effect of combined colestipol-niacin therapy on coronary atherosclerosis and coronary venous bypass grafts, JAMA 257:3233 (1987).

28. D.H. Blankenhorn, P. Alaupovic, E. Wickham, H.P. Chin, and S.P. Azen, Prediction of angiographic change in native human coronary arteries and aortocoronary bypass grafts. Lipid and nonlipid factors, Circulation 81:470 (1990).

29. P. Alaupovic, W.J. McConathy, J. Fesmire, M. Tavella, and J.M. Bard, Profiles of apolipoproteins and apolipoprotein B-containing lipoprotein particles in dyslipoproteinemias, Clin. Chem. 34:B13 (1988).

30. P. Alaupovic, C.S. Wang, W.J. McConathy, D. Weiser, and D. Downs, Lipolytic degradation of human very low density lipoproteins by human milk lipoprotein lipase: the identification of lipoprotein B as the main lipoprotein degradation product, Arch. Biochem. Biophys. 224:226 (1986).

31. P. Alaupovic, M. Tavella, and J. Fesmire, Separation and identification of ApoB-containing lipoprotein particles in normolipidemic subjects and patients with hyperlipoproteinemias, Adv. Exp. Med. Biol. 210:7 (1987).

32. P. Alaupovic, M. Tavella, J.M. Bard, C.S. Wang, P.O. Attman, E. Koren, C. Corder, C. Knight-Gibson, and D. Downs, Lipoprotein particles in hypertriglyceridemic states, Adv. Exp. Med. Biol. 243:289 (1988).

33. N. Dashti, P. Alaupovic, C. Knight-Gibson, and E. Koren, Identification and partial characterization of discrete apolipoprotein B-containing lipoprotein particles produced by human hepatoma cell line HepG2, Biochemistry 26:4837 (1987)

34. P. Alaupovic, David Rubinstein Memorial Lecture: the biochemical and clinical significance of the interrelationship between very low density and high density lipoproteins, Can. J. Biochem. 59:565 (1981).

35. M. Tavella, P. Alaupovic, D. Blankenhorn, and H.P. Chin, Specific effect of combined colestipol and niacin therapy on apolipoprotein B-containing particles, Arteriosclerosis 7:515a (1987).

36. V. Manninen, O. Elo, H. Frick, K. Haapa, O.P. Heinonen, P. Heinsalmi, P. Helo, J.K. Huttunen, P. Kaitaniemi, P. Koskinen, H. Mäenpää, M. Mälkönen, M. Mänttäri, S. Norola, A. Pasternack, J. Pikkarainen, M. Romo, T. Sjöblom, and E. A. Nikkilä, Lipid alterations and decline in the incidence of coronary heart disease in the Helsinki Heart Study, JAMA 260:641 (1988).

37. C. Corder, M. Tavella, and P. Alaupovic, Effect of gemfibrozil on discrete ApoB-containing lipoproteins in patients with type V hyperlipoproteinemia, Arteriosclerosis 7:515a (1987).

38. P. Alaupovic, and P.R. Blackett, The dyslipoproteinemias of diabetes mellitus, in Human Plasma Lipoproteins, J.C. Fruchart and J. Shepherd, eds., Walter de Gruyter Press, Berlin•New York, pp. 173-206 (1989).

39. A. Fontbonne, E. Eschwège, F. Cambien, J.-L. Richard, P. Ducimetière, N. Thibult, J.-M. Warnet, J.-R. Claude, and G.-E. Rosselin, Hypertriglyceridiamia as a risk factor of coronary heart disease mortality in subjects with impaired glucose tolerance or diabetes. Results from the 11-year follow-up of the Paris prospective study, Diabetologia 32:300 (1989).

40. R.J. Havel, Role of triglyceride-rich lipoproteins in progression of atherosclerosis, Circulation 81:694 (1990).

41. E. Koren, C. Corder, G. Mueller, J. Centurion, J. Fesmire, L. Hunt, and P. Alaupovic, Coronary artery disease and triglyceride enriched lipoproteins, FASEB J. 2:A1596 (1988).

IN VITRO BINDING AND IN VIVO UPTAKE OF CHYLOMICRON REMNANTS AFTER THEIR HYDROLYSIS BY HEPATIC LIPASE

Fabrice SULTAN, Dominique LAGRANGE and Sabine GRIGLIO

INSERM U.177. Physiopathologie de la Nutrition

15 rue de l'Ecole de Médecine, 75270 Paris Cedex 06

INTRODUCTION

Chylomicron remnants (CM-RM), resulting from intravascular hydrolysis of chylomicrons by the endothelium bound lipoprotein lipases (LPL), are rapidly taken up by the liver. It appears clearly that this uptake is mostly receptor mediated[1-3] and that CM-RM are taken up as whole particles. Hepatic lipase (HL), which exerts both a triacylglycerol lipase and phospholipase Al activity, may be involved in with CM-RM uptake. Indeed, its inhibition by a specific anti-HL antibody leads to an accumulation of apo B-48 rich-particles in the plasma of rats[4,5]. We have previously shown that CM-RM prealably hydrolyzed by HL and then added to isolated hepatocytes are taken up at a higher rate than non-treated remnants (6). Moreover, we showed that following hydrolysis by HL, there was no modification in remnant apo E levels (unpublished results). Since apo E is the ligand for remnant binding to their hepatic receptors[1-3], we examined herein, whether HL-treatment of CM-RM could despite unchanged apo E levels, increase 1) the binding of such particles to liver membranes and 2) their plasma clearance and hepatic uptake in vivo.

MATERIALS and METHODS

CM-RM preparation and hydrolysis by HL

The procedure of CM-RM preparation was essentially as described[6]. CM-RM were incubated with HL for 2 h at 37°C as described[7]. HL-treated and control remnants (0.4 mg protein) were incubated with 0.5 unit HL and 0.15M NaCl respectively and ultracentrifuged[6]. HL was purified from post-heparin liver perfusates of six to twelve rats as detailed in [8].

Chylomicron remnant labeling

Radioiodination of remnants was performed using the modified iodine monochloride method[9] and resulted in specific activities of about 100 cpm/ng protein. Following filtration on a PD-10 column (Pharmacia) and extensive dialysis to remove free iodine, 61% of the label was bound to protein after TCA precipitation. Calculations were done on radioactivity of the protein moiety.

Hypercholesterolemia, Hypocholesterolemia, Hypertriglyceridemia
Edited by C.L. Malmendier *et al.*, Plenum Press, New York, 1990

311

Binding of labeled CM-RM to liver plasma membranes

Liver plasma membranes were prepared from 6 male Wistar rats (150-200g) as described in [10]. Binding assays were carried out at 37°C in plastic Eppendorf centrifugation tubes in a final volume of 200 µl. Incubation medium contained 50 µg membrane protein and buffer (50mM Tris/HCl pH 7.5, 25mM NaCl and 1mM $CaCl_2$). Total binding of ^{125}I labeled CM-RM and non-specific binding (assessed with a 20-fold excess of unlabeled CM-RM), were measured after a 60 min incubation. The suspension was then transferred into microfuge tubes and the membranes separated from the medium by centrifugation through a mineral oil layer (dibutylphthalate/dinonylphthalate 20:11, v/v, final density d = 1.025 g/ml). Pellets were counted in a Beckman gamma counter.

In vivo plasma clearance and hepatic uptake of CM-RM

Male Wistar rats of 200-250 g were injected with 60 µg protein of ^{125}I labeled HL-treated or control remnants and blood aliquots (200 µl) were drawn from the jugular vein into tubes containing disodium EDTA (0.1% wt/v). Animals were killed 6 to 7 min later and the livers were then immediately washed throught a perfusion with 0.15 M NaCl at 10°C. Plasma and liver aliquots were counted.

RESULTS

Characterization of HL-treated and control CM-RM

Composition of control remnants was 91.4 % triacylglycerol, 1 % esterified cholesterol, 0.7 % free cholesterol, 4.4 % phospholipids, 2.5 % protein. In this work, CM-RM were particularly rich in triacylglycerol as compared to other preparations[6] possibly because parent chylomicrons also were so (95 % TG). The degree of hydrolysis was estimated by the loss of phospholipids and triacylglycerols from HL-treated as opposed to control remnants and 35 % hydrolysis of phospholipids, but not hydrolysis of triacylglycerols was obtained under the present conditions.

Binding of CM-RM to rat liver membranes

[125]I - labeled remnants bound to liver membranes in a saturable manner (Fig. 1). Saturation for specific binding occurred at a concentration superior to 40 µg protein/ml for HL-treated remnants and of 20 to 25 µg protein/ml for control remnants. Scatchard plots of data (Fig. 2) was consistent with CM-RM binding to a single class of high affinity sites. Nevertheless, a Scatchard linearization may mask a multiplicity of binding sites. Capacity of the binding sites (Bmax) was more than 2-fold increased for HL-treated remnants (1.7 µg/mg membrane protein) as compared to untreated remnants (0.74 µg/mg). Dissociation constant (Kd) was 4.6 µg/ml for HL-treated and 1.9 µg/ml for control remnants. These values are well in the range of those obtained by Nagata et al.[12] and suggest and increase of the number of CM-RM receptors rather than an increase in the affinity of CM-RM for the receptors.

In vivo experiments

When injected into rats, [125]I -labeled CM-RM disappeared very rapidly from the plasma, showing a clear-cut biphasic decay (Fig.3). The initial disappearence curves show clearly a faster clearance of HL-treated remnants. The radioactivity recovered in the liver was significantly higher for HL-treated CM-RM (64 ± 3.1 % vs 52 ± 0.7 %, p < 0.01). These results show a a stimulatory effect of HL treatment on CM-RM uptake by the liver.

312

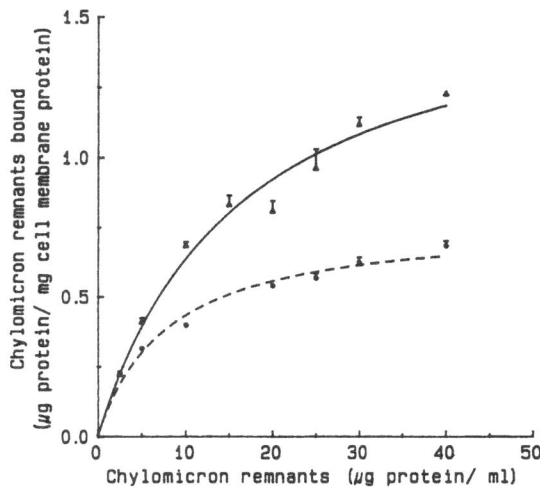

Figure 1. Binding of ^{125}I chylomicron remnants to liver plasma
membranes. Incubations were carried out at 37°C for 1h,
with 50 μg membrane protein and the indicated amounts of
remnant proteins. Control incubations contained a 20-
fold protein excess of non-labeled remnants in order to
determine non-specific binding. Untreated chylomicron
remnants (●- - -●) and HL-treated remnants (▲——▲) of
the same batch of native chylomicron remnants were used
and determinations made in triplicates. Mean ± SEM is
shown.

DISCUSSION

It is confirmed here that HL, when acting on CM-RM, modifies these
particles in such a way that they are taken up at faster rates by the liver.
The HL-treated remnants displayed indeed a higher binding to liver membranes
and also an enhanced in vivo uptake by the liver although their apoprotein
composition was identical to that of control remnants (unpublished results).
While a increased binding of HL-treated remnants to liver membranes was
measured, a decreased binding affinity of these particles was simultaneously
observed as shown by the increment in Kd. It has been shown that CM-RM bind
to rat liver membranes to a site different from the LDL receptor, as demonstrated by the uneffectiveness of an anti-LDL receptor antibody to modify
CM-RM binding[12]. By contrast, intact hepatocytes seem to bind and internalize these remnants by an LDL receptor[12]. Thus the question as to whether
mainly LDL receptor[12-14] or a separate remnant receptor[3,15,16] or both, are
involved in CM-RM uptake remains to be further investigated. This point is
important for interpretation of our present results.
In vivo, CM-RM treated with HL were more efficiently cleared from the plasma
than control ones, whether they were labeled with ^{14}C cholesteryl ether
(unpublished results) or with ^{125}I -iodine as shown in the present work.
It is now well established that CM-RM are taken up as whole particles[17-18].

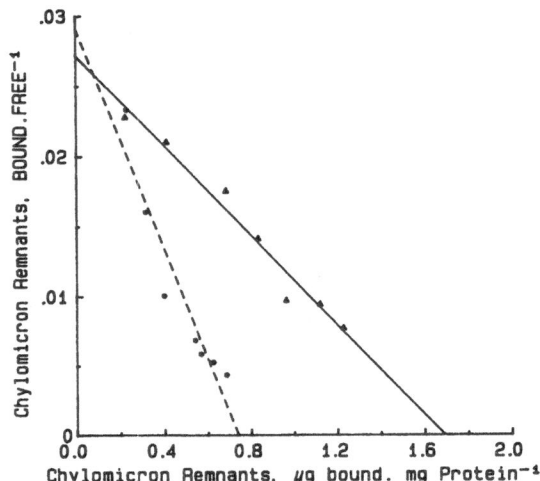

Figure 2. Data of the figure 2 wee plotted as bound/free vs bound
chylomicron remnant concentration. Untreated remnants
(•- - - -•) (Kd = 1.9 µg chylomicron remnants/ml, maximum
bound equals 0.74 µg chylomicron remnants/mg protein) ;
HL-treated remnants (▲——▲) (Kd = 4.6 µg chylomicron
remnants/ml, maximum bound equals 1.7 µg chylomicron
remnants/mg protein).

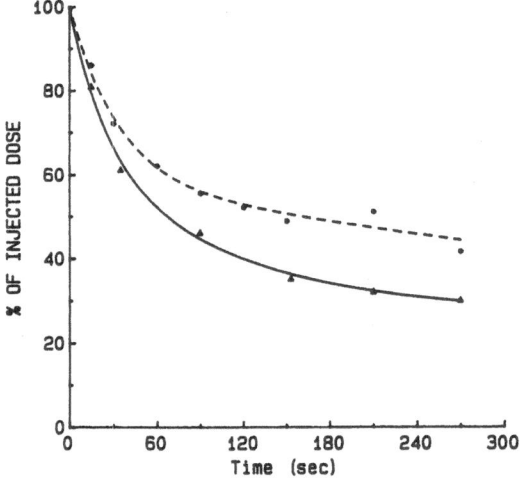

Figure 3. Plasma disappearance of chylomicron remnants either untreated
(•- - - -•) or HL-treated (▲——▲). Radioiodinated remnants were
injected at t = 0 and blood aliquots drawn every 30 sec as
described in Material and Methods. The disappearance kinetics
were obtained from the mean of 4 determinations.

However, the kinetics of plasma disappearance were different when obtained with ^{14}C -cholesteryl ester labeled remnants or with ^{125}I -labeled ones. Indeed, the second phase of radioactivity plateaued after 2 min with ^{125}I -labeled remnants whereas the decay of ^{14}C -cholesteryl ether radioactivity continued after that period. Due to the absence of cholesteryl ester exchange between plasma lipoproteins in the rat[19], ^{14}C -cholesteryl ether remains in the remnants and reflects their catabolism. In contrast, apolipoproteins undergo exchanges between lipoproteins in the rat plasma. Such a transfer has been shown in vitro for apo E, from lipoproteins of the size of IDL towards HDL when the former were hydrolyzed by human post-heparin HL[20]. Thus, ^{125}I -labeled apoproteins, with the exception of apo B-48, may be transferred from CM-RM which have short half-lives to HDL which have much longer half-lives. Total plasma ^{125}I -iodine radioactivity is then no more the reflect of remnant catabolism only. Nevertheless, the initial increased disappearance observed in the plasma for HL-treated remnants is in accordance with the increased uptake of ^{125}I measured in the liver. Concerning the mechanism whereby HL favors CM-RM uptake, it has been recently shown that CM treated with this enzyme were taken at faster rates by the perfused rat liver, provided that more than 25 to 30% of their phospholipids were hydrolyzed[21]. We reach the same conclusion with CM-RM in which 35% phospholipids were hydrolyzed by HL. Interestingly, no hydrolysis of their triacylglycerol content could be detected under these conditions. It appears then, that phospholipid hydrolysis alone is sufficient for HL-treated remnants to be better recognized by their hepatic receptors. Apo E, which is the specific ligand for remnant binding to these receptors[1-3], may exhibit new epitopes involved in the binding process after modification of apo E phospholipid environment.

In conclusion, we showed in the present study that HL increased CM-RM binding to liver membranes. While binding affinity seems to be decreased, a higher number of binding sites appeared to be available. This resulted in an enhanced uptake of CM-RM by the whole liver in vivo. The accurate mechanism of HL effect, remains to be further investigated.

ACKNOWLEDGMENTS

We are grateful to Dr. M. Caron (INSERM U.181) for liver membrane preparation and we thank Dr. M.I. Malewiak for helpful discussion of the manuscript.

REFERENCES

1. E.E.T. Windler, Y.S. Chao and R.J. Havel. Regulation of the hepatic uptake of triglyceride -rich lipoproteins in the rat. Opposing effects of homologous apolipoprotein E and individual C apolipoproteins, J. Biol. Chem. 255 : 8303 (1980).

2. A.D. Cooper, S.K. Erickson, S.K., R. Nutik and M.A. Shrewsbury. Characterization of chylomicron remnants binding to rat liver membranes, J. Lipid Res. 23 : 42 (1982).

3. R.W. Mahley, D.Y. Hui, T.L. Innerarity and U. Beisigel. Chylomicron remnant metabolism. Role of hepatic lipoprotein receptors in mediaring uptake, Arteriosclerosis 9 (1) : 14 (1989).

4. P.B. Daggy and A. Bensadoun. Enrichment of apolipoprotein B-48 in the LDL density class following in vivo inhibition of hepatic lipase, Biochim. Biophys. Acta 877 : 252 (1986).

5. F. Sultan, D. Lagrange, H. Jansen and S. Griglio. Inhibition of hepatic lipase activity impairs chylomicron remnant removal in rats, Biochim. Biophys. Acta, 1042 : 150 (1990).

6. F. Sultan, D. Lagrange, X. Le Liepvre and S. Griglio. Chylomicron remnant uptake by freshly isolated hepatocytes. Effect of heparin and of triacylglycerol lipase, Biochem. J., 258 : 587 (1989).

7. A. Van Tol, T. Van Gent and H. Jansen. Degradation of HDL by heparin releasable liver lipase, Biochem. Biophys. Res. Commun. 94 : 101 (1980).

8. K. Schoonderwoerd, W.C. Hülsmann and H. Jansen. Regulation of liver lipase. Evidence for several regulatory sites studied in corticotrophin-treated rats, Biochim. Biophys. Acta 754 : 279 (1983).

9. D.W. Bilheimer, S. Eisenberg and R.I. Levy. The metabolism of VLDL proteins. Preliminary in vitro and in vivo observations, Biochim. Biophys. Acta 260 : 212 (1972).

10. T.K. Ray. A modified method for the isolation of the plasma membrane from rat liver. Biochim. Biophys. Acta 196 : 1 (1970).

11. D.H. Lowry, N.J. Rosebrough, A.L. Farr and R.J. Randal. Protein measurement with the folin phenol reagent. J. Biol. Chem. 193 : 265 (1951).

12. Y. Nagata, J. Chen. and A.D. Cooper. Role of LDL-receptor dependent and independent sites in binding and uptake of chylomicron-remnants in rat liver. J. Biol. Chem. 263 : 15151 (1988).

13. E. Jensen, C.H. Floren and A. Nilsson. Cell density dependent uptake of chylomicron-remnants in rat hepatocyte monolayers. Effects of compactin and mevalonic acid. Biochim. Biophys. Acta 917 : 74 (1987)

14. E.E.T. Windler, J. Greeve, W. Daerr and H. Greten. Binding of rat chylomicrons and their remnants to the LDL receptor and its role in remnant removal. Biochem. J. 252 : 553 (1988).

15. J. Hertz, V. Hamann, S. Rogne, O. Myklebost, H. Gausepohl and K.K. Stanley. Surface location and high affinity for calcium of a 500-kd liver membrane protein closely related to the LDL-receptor suggest a physiological role as lipoprotein receptor, EMBO J., 7 : 4119 (1988).

16. S. Jeackle, S.E. Brady and R.J. Havel. Membrane binding sites for plasma lipoproteins on endosomes from rat liver. Proc. Natl. Acad. Sci. 86 : 1880 (1989).

17. A.D. Cooper. The metabolism of chylomicron-remnants by isolated perfused rat liver. Biochim. Biophys. Acta 488 : 464 (1977).

18. P.M. Lippiello, P.J. Sisson and M. Waite. The uptake and metabolism of chylomicron-remnants by rat liver parenchymal and non-parenchymal cells in vitro. Biochem. J. 232 : 395 (1985).

19. Y. Oschry and S. Eisenberg. Rat plasma lipoprotein : reevaluation of a lipoprotein system in an animal devoid of cholesteryl ester transfer activity. J. Lipid. Res. 23 : 1099 (1982).

20.   J.C. Gibson and W.V. Brown. Effect of lipoprotein lipase and hepatic triglyceride lipase activity on the distribution of apolipoprotein E among the plasma lipoproteins. Atherosclerosis 73 : 45 (1988).

21.   J. Borensztajn, G.S. Getz and T.J. Kotlar. Uptake of chylomicron remnants by the liver : further evidence for the modulating role of phospholipids. J. Lipid Res. 29 : 1987 (1988).

THE ENDOCYTOSIS OF LIPOPROTEINS BY THE LIVER AND THEIR INTRACELLULAR PATHWAY IN COMPARISON TO OTHER LIGANDS

Stefan Jäckle, Bodo Levkau, Thomas Lorenzen, Franz Rinninger, Wolfgang Daerr, Heiner Greten, and Eberhard Windler

Medizinische Kernklinik und Poliklinik
Universitäts-Krankenhaus Eppendorf
Martinistraße 52
D-2000 Hamburg 20, FRG

Receptor-mediated endocytosis is well-recognized as a general mechanism used by many cells to take up biologically important molecules. The variety of ligands taken up by receptor-mediated endocytosis includes hormones (e.g. epidermal growth factor (EGF) and insulin), transport proteins carrying nutritional substances (e.g., lipoproteins and transferrin), asialoglycoproteins (ASGPs) and immunoglobulins (Goldstein, Anderson et al. 1979). The terminal catabolism of most plasma lipoproteins and asialoglycoproteins occurs chiefly in the liver by means of receptor-mediated endocytosis (Havel and Hamilton 1988).

The liver has little capacity to take up nascent very low density lipoproteins (VLDL) and nascent chylomicrons, as they are initially found in the blood (Windler, Chao et al. 1980). Chylomicrons and VLDL are metabolized by lipoprotein lipase on the surface of capillary endothelial cells in several extrahepatic tissues. This enzyme catalyzes the hydrolysis of most triglycerides, leading to the formation of remnants. The remnants lose their affinity for C apoproteins, which are returned to high density lipoproteins (HDL), but retain their affinity for apoprotein E, which is then competent to bind to receptors (Windler, Chao et al. 1980; Windler and Havel 1985; Windler, Preyer et al. 1986). Chylomicron remnants and a major fraction of VLDL remnants are rapidly taken up into the liver (Windler, Chao et al. 1980) (Fig.1).

LDL and VLDL remnants, with the possible exception of very large VLDL remnants (Havel, Yamada et al. 1987), can interact with the LDL receptor (Windler, Kovanen et al. 1980). Evidence that chylomicron remnants are removed by a process distinct from that mediated by the LDL receptor comes from several types of observations (Havel and Hamilton 1988). First, the rate of hepatic uptake of chylomicrons in humans and rabbits that are grossly deficient in the LDL receptor seems to be normal (Kita, Goldstein et al. 1982). Second, interventions that affect the number of functional hepatic LDL receptors have little effect on chylomicron removal (Angelin, Raviola et al. 1983; Arbeeny and Rifici 1984). However, chylomicron remnants do bind to the LDL receptor on liver membranes and LDL can inhibit the uptake of chylomicron remnants into the perfused rat liver (Kita, Goldstein et al. 1982, Windler, Greeve et al. 1988).

Hypercholesterolemia, Hypocholesterolemia, Hypertriglyceridemia
Edited by C.L. Malmendier *et al.*, Plenum Press, New York, 1990

The pathway of endocytosis of lipoproteins has been studied by autoradiography in livers of rats in which radioiodinated plasma lipoproteins have been injected intravenously. Between 3 and 15 min after injection, lipoproteins are concentrated in endosomal vesicles near the cell surface (Chao, Jones et al. 1981). This hepatocytic endosomal compartment has been described in studies of receptor mediated endocytosis of EGF and asialoglycoproteins (Wall and Hubbard 1980; Dunn and Hubbard 1984); it represents the compartment of uncoupling of receptor and ligand (CURL) (Geuze, Slot et al. 1983; Geuze, Slot et al. 1987). At later time points after injection, LDL and chylomicron remnants accumulate in endosomes near the Golgi/nuclear pole of the cell (Chao, Jones et al. 1981; Jones, Hradek et al. 1984). They were shown to contain internal bilayer vesicles, which identified them as multivesicular bodies (MVBs).

We isolated three endosomal fractions from livers of estradiol-treated and untreated rats. These endosomal fractions had characteristic physical and ultrastructural properties, but the lipid composition and major proteins of their membranes were similar and differed from those derived from the Golgi apparatus (Belcher, Hamilton et al. 1987; Jaeckle and Havel 1990). Injected radioiodinated ligands (lipoproteins, EGF, transferrin) accumulate first in the fraction of intermediate density (CURL) and later in the low density fraction (MVBs).

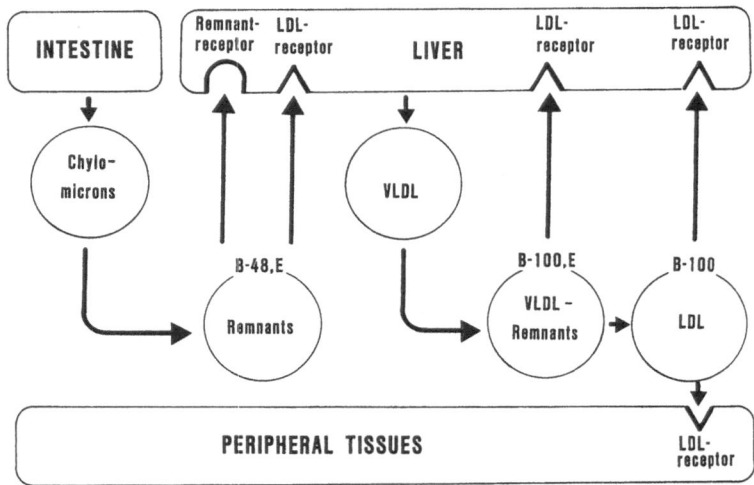

Fig 1   Model of the lipoprotein metabolism in plasma.

Chylomicrons are secreted from the small intestine. Most trriglycerides are hydrolyzed in extrahepatic tissues by lipoprotein lipase. The resulting remnant particles are taken up into rat hepatocytes after binding to the chylomicron or LDL receptor. VLDL are secreted from hepatocytes; after hydrolysis of most of their triglycerides VLDL remnants are taken up by the hepatic LDL receptor or are converted to LDL.

The endosomes of high density were composed of membranes resembling the appendages of the two vesicular fractions, which are highly enriched in LDL and asialoglycoprotein receptors (Belcher, Hamilton et al. 1987). The concentration of EGF receptors, which are delivered to lysosomes and degraded together with EGF (Dunn, Connoly et al. 1986; Lai, Cameron et al. 1989), remains low in this fraction (Jaeckle, Miranda-Brady et al. 1990). These observations support the role of endosomes of high density as a receptor recycling compartment (RRC). The current concept of receptor mediated endocytosis of lipoproteins and asialoglycoproteins in hepatocytes is shown in Fig 2.

To compare the intracellular pathway of chylomicron remnants with those of LDL and asialoglycoproteins, we injected these ligands and isolated three endosomal fractions at specified times thereafter. Like the tested lipoproteins asialofetuin was also rapidly taken up by the liver. LDL, ß-VLDL, and chylomicron remnants and asialofetuin were concentrated in CURL and MVBs (Jaeckle, Brady et al. 1989).

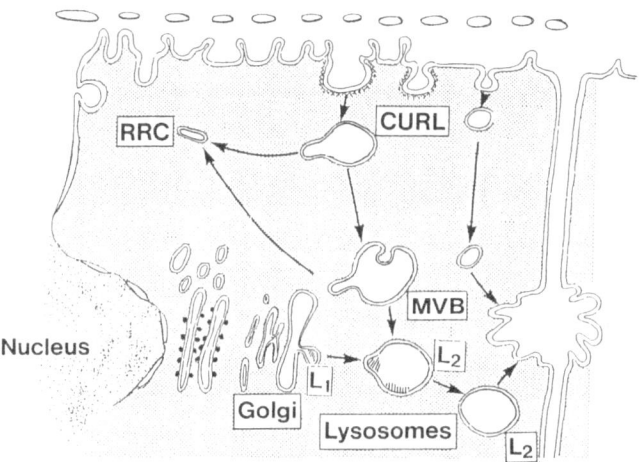

Fig 2   Current concept of the receptor mediated endocytosis in hepatocytes.
(modified from Havel and Hamilton, 1988)

Ligands are bound by receptors that cluster in clathrin coated pits; the pits pinch off and become vesicles, which fuse together. A proton pump lowers the pH within the endosomes, which in many cases causes dissociation of ligands from receptors in the compartment of uncoupling of receptor and ligands (CURL). The receptor-rich extensions pinch off and recycle receptors back to the plasma membrane (RRC: receptor recycling compartment). Some kinds of receptors as the EGF receptor are sequestered into the vesicle lumen by invagination generating multivesicular bodies (MVBs) in the Golgi/lysosomal region of the cell. The MVBs fuse with primary lysosomes (Ly 1) resulting in the formation of secondary lysosomes (Ly 2), where degradation occurs.

At 2.5 min after injection the concentration of ^{125}I was higher in CURL than in MVBs. However, between 2.5 and 15 min the concentration of ^{125}I increased much more in MVBs, supporting our previous findings that the CURL fraction is a precursor of MVBs (Belcher, Hamilton et al. 1987; Jaeckle and Havel 1990; Jaeckle, Miranda-Brady et al. 1990). The concentration of ^{125}I in the RRC fraction remained at considerably lower levels as is the case for all ligands which are determined for lysosomal degradation.

Endosomal membranes contain many endocytotic receptors at high concentrations including the receptors for EGF, transferrin, and mannose-6-phosphate, as well as the LDL receptor, asialoglycoprotein receptor, and the LDL-receptor related protein (Belcher, Hamilton et al. 1987; Lund, Takahashi et al. 1989; Jaeckle and Havel 1990; Jaeckle, Miranda-Brady et al. 1990;). We have therefore sought to determine the contribution of the LDL receptor and other receptors, which may interact with lipoproteins and are present in these endosomal membranes, to the hepatic uptake of remnant lipoproteins.

Specific binding of chylomicron remnants and ß-VLDL to endosomal and plasma membranes could be almost completely inhibited by unlabeled LDL and by a polyclonal antibody against the bovine LDL receptor (Jaeckle and Havel 1990). In ligand blots chylomicron remnants and ß-VLDL can bind to the LDL receptor (Soutar, Harders-Spengel et al. 1986; Windler, Greeve et al. 1988; Jaeckle, Brady et al. 1989). Treatment of rats with estradiol is shown to increase the number of hepatic LDL receptors about tenfold. The number of binding sites for LDL, ß-VLDL, and chylomicron remnants on endosomal membranes were increased to almost the same extent.

The effect of estradiol on the removal of chylomicron remnants was also investigated in vivo. In preliminary experiments small dosages of ^{3}H-chylomicron remnants, ^{125}I-ß-VLDL, or ^{125}I-LDL were injected into estradiol-treated and untreated rats. The decay of ^{125}I-LDL could be substantially increased by treatment with estradiol. However, ß-VLDL were almost completely taken up by the liver within 2 min indistinguishable in estradiol-treated and in untreated rats. Evidently the decay is maximal for very small dosages of ß-VLDL that do not saturate their binding sites; thus the induction of the LDL receptor does not lead to a further increase in the uptake velocity. Yet, after injection of increasing amounts of ß-VLDL, their plasma half life increased significantly, demonstrating the saturability of the hepatic uptake mechanism for remnants. The decay of chylomicron remnants could also be dramatically decreased by simultaneous injection of large amounts of ß-VLDL. Under these conditions, we could demonstrate a significant stimulation of the hepatic uptake of chylomicron remnants by the treatment with estradiol. After saturation of the binding sites, the number of receptors seems to become rate limiting.

In the human hepatic tumor cell line Hep-G2 lipoproteins seem to be taken up to some extent by the asialoglycoprotein receptor (Windler, Greeve et al. 1990). Yet, using rat endosomal membranes, there was no interaction of lipoproteins with the asialoglycoprotein-receptor detectable. Neither was the binding of ^{3}H-chylomicron remnants or ^{125}I-ß-VLDL inhibited by excess asialofetuin, nor the binding of ^{125}I-asialofetuin by excess ß-VLDL. In vivo the removal of ^{3}H-chylomicron remnants or ^{125}I-ß-VLDL from the circulation of rats were greatly delayed by simultaneous injection of excess ß-VLDL but not of excess asialofetuin. Vice versa the decay of radiolabeled asialofetuin was greatly inhibited by excess asialofetuin but not by excess ß-VLDL.

Thus, in the rat the asialoglycoprotein receptor does not appear to play a major role for the removal of chylomicron remnants, which contrasts with previous results obtained with Hep-G2 cells and human liver membranes. However, the results support our previous hypothesis that the LDL-receptor is the single most important receptor for the catabolism of chylomicron remnants and ß-VLDL in the rat.

References

Angelin, B., C. A. Raviola, T. L. Innerarity and R. W. Mahley. (1983). "Regulation of hepatic lipoprotein receptors in the dog; rapid regulation of apolipoprotein B,E receptors, but not of apolipoprotein E receptors, by intestinal lipoproteins and bile acids." J.Clin.Invest. 71: 816-831.

Arbeeny, C. M. and V. A. Rifici. (1984). "The uptake of chylomicron remnants and very low density lipoprotein remnants by the perfused rat liver." J.Biol.Chem. 259: 9662-9666.

Belcher, J. D., R. L. Hamilton, S. E. Brady, C. A. Hornick, S. Jaeckle, W. J. Schneider and R. J. Havel. (1987). "Isolation and characterization of three endosomal fractions from the liver of estradiol-treated rats." Proc.Natl.Acad.Sci.USA. 84: 6785-6789.

Chao, Y. -s., A. L. Jones, G. T. Hradek, E. E. T. Windler and R. J. Havel. (1981). "Autoradiographic localization of the sites of uptake, cellular transport, and catabolism of low density lipoproteins in the liver of normal and estrogen-treated rats." Proc.Natl.Acad.Sci.USA. 78: 597-601.

Dunn, W. A. and A. L. Hubbard. (1984). "Receptor-mediated endocytosis of epidermal growth factor by hepatocytes in the perfused rat liver:ligand and receptor dynamics." J.Cell Biol. 98: 2148-2159.

Dunn, W. A., T. P. Connoly and A. L. Hubbard. (1986). "Receptor-mediated endocytosis of epidermal growth factor by rat hepatocytes:receptor pathway." J.Cell Biol. 102: 24-36.

Geuze, H. J., J. W. Slot, G. J. A. M. Strous, H. F. Lodish and A. L. Schwartz. (1983). "Intracellular site of asialoglycoprotein receptor-ligand uncoupling:doublelable immunoelectron microscopy during receptor-mediated endocytosis." Cell. 32: 277-287.

Geuze, H. J., J. W. Slot and A. L. Schwartz. (1987). "Membranes of sorting organelles display lateral heterogeneity in receptor distribution." J.Cell Biol. 104: 1715-1723.

Goldstein, J. L., R. G. W. Anderson and M. S. Brown. (1979). "Coated pits, coated vesicles, and receptor-mediated endocytosis." Nature. 279: 679-685.

Havel, R. J., N. Yamada and D. M. Shames. (1987). "Role of apolipoprotein E in lipoprotein metabolism." Am.Heart J. 113: 470-474.

Havel, R. J. and R. L. Hamilton. (1988). "Hepatocytic lipoprotein receptors and intracellular lipoprotein catabolism." Hepatology. 8: 1689-1704.

Jaeckle, S., S. E. Brady and R. J. Havel. (1989). "Membrane binding sites for plasma lipoproteins on endosomes from rat liver." Proc.Natl.Acad.Sci.USA. 86: 1880-1884.

Jaeckle, S., S. Miranda-Brady and R. J. Havel. (1990). "Dynamics of epidermal growth factor and its receptor in three hepatic endosomal fractions." submitted for publication.

Jaeckle, S. and R. J. Havel. (1990). Unpublished data

Jones, A. L., G. T. Hradek, C. Hornick, G. Renaud, E. E. T. Windler and R. J. Havel. (1984). "Uptake and processing of remnants of chylomicrons and very low density lipoproteins by rat liver." J.Lipid Res. 25: 1151-1158.

Kita, T., J. L. Goldstein, M. S. Brown, Y. Watanabe, C. A. Hornick and R. J. Havel. (1982). "Hepatic uptake of chylomicron remnants in WHHL rabbits: a mechanism genetically distinct from the low density lipoprotein receptor." Proc.Natl.Acad.Sci.USA. 79: 3623-3627.

Lai, W. H., P. H. Cameron, I. Wada, J. -. Doherty, D. G. Kay, B. I. Posner and J. J. M. Bergeron. (1989). "Ligand-mediated internalization, recycling, and downregulation of the epidermal growth factor receptor in vivo." J.Cell Biol. 109: 2741-2749.

Lund, H., K. Takahashi, R. L. Hamilton and R. J. Havel. (1989). "Lipoprotein binding and endosomal itinerary of the low density lipoprotein receptor-related protein in rat liver." Proc.Natl.Acad.Sci.USA. 86: 9318-9322.

Soutar, A. K., K. Harders-Spengel, D. P. Wade and B. L. Knight. (1986). "Detection and quantitation of low density lipoprotein (LDL) receptors in human liver ligand blotting, immunoblotting, and radioimmunoassay; LDL receptor protein content is correlated with plasma LDL cholesterol concentration." J.Biol.Chem. 261: 17127-17133.

Wall, D. A. and A. L. Hubbard. (1980). "The galactose-specific recognition system of mammalian liver: the route of ligand internalization in rat hepatocytes." Cell. 21: 79-93.

Windler, E., Y. -. Chao and R. J. Havel. (1980a). "Determinants of hepatic uptake of triglyceride-rich lipoproteins and their remnants in the rat." J.Biol.Chem. 255: 5475-5480.

Windler, E., Y. -. Chao and R. J. Havel. (1980b). "Regulation of the hepatic uptake of triglyceride-rich lipoproteins in the rat; opposing effects of homologous apoprotein E and individual C apoproteins." J.Biol.Chem. 255: 8303-8307.

Windler, E. E. T., P. T. Kovanen, Y. -. Chao, M. S. Brown, R. J. Havel and J. L. Goldstein. (1980). "The estradiol-stimulated lipoprotein receptor of rat liver; a binding site that mediates the uptake of rat lipoproteins containing apoprotein B and E." J.Biol.Chem. 255(21): 10464-10471.

Windler, E. and R. J. Havel. (1985). "Inhibitory effects of C apolipoproteins from rats and humans on the uptake of triglyceride-rich lipoproteins and their remnants by the perfused rat liver." J.Lipid Res. 26: 556-565.

Windler, E. E. T., S. Preyer and H. Greten. (1986). "Influence of lysophosphatidylcholine on the C-apoprotein content of rat and human triglyceride-rich lipoproteins during triglyceride hydrolysis." J.Clin.Invest. 78: 658-665.

Windler, E. E. T., J. Greeve, W. H. Daerr and H. Greten. (1988). "Binding of rat chylomicrons and their remnants to the hepatic low-density-lipoprotein receptor and its role in remnant removal." J.Biol.Chem. 252: 553-561.

Windler, E., J. Greeve, B. Levkau, V. Kolb-Bachofen, W. Daerr and H. Greten. (1990). "The human asialoglycoprotein-receptor: a possible binding site for LDL and chylomicron remnants." submitted for publication.

CONCOMITANT INHIBITION OF VLDL TRIGLYCERIDE AND APOPROTEIN SECRETION BY

HEPATOCYTES OF RATS ADAPTED TO A HIGH-FAT DIET

O.L. Francone, A.D. Kalopissis and G. Griffaton
INSERM U.177
15 rue de l'Ecole de Médecine, 75270 Paris 6°. France

INTRODUCTION

Hepatic VLDL synthesis and secretion is under nutritional control, being increased 2-fold by a sucrose-rich diet. Conversely, it is decreased 2-fold by a high-fat diet (1,2), despite the fact that fat-fed rats have elevated plasma non esterified fatty acids (3), which are generally considered a stimulatory factor of hepatic VLDL production. The aim of this study was to investigate whether hepatic VLDL apoprotein (apo) secretion was also decreased by fat-feeding in the rat.

METHODS

Wistar rats were adapted for 3 weeks to the control or the high-fat diet (30% lard, by wt). 150×10^6 freshly isolated hepatocytes (4) were incubated in Krebs saline in the presence of 1.5% albumin (fatty acid-free), glucose (20 mM), amino acids, oleate (0.3mM), aprotinin, penicillin G (100 U/ml) and streptomycin (100 µg/ml). L-[3,4,5-^{3}H] leucine (0.75 mCi) was added simultaneously with cold leucine (5 mM) to measure protein synthetic rates (5). After 1,2 or 3 h incubations, the cells were separated from the media. Total cellular protein synthesis was determined after precipitation with TCA 5%. The incubation media were extensively dialyzed to eliminate free amino acids and ultracentrifuged at 100 000 g at 15° C for 24 h to obtain VLDL. After dialysis, VLDL triglyceride content was determined and total VLDL apoproteins were counted. Then, VLDL were lyophilized, delipidated with diethyl ether and subjected to electrophoresis on discontinuous polyacrylamide-SDS slab gels (composed of 5% acrylamide in the upper part and 8.25% acrylamide in the lower part of the gel), so as to obtain the apo-B's and the other smaller apoproteins on the same gels. Gels were sliced into 5mm-sections, digested with 0.25 ml NCS at 50° C for 2 h and their radioactivity was determined. Blank gels counted 80 dpm or less.

RESULTS AND DISCUSSION

Cellular protein content was 1.1-1.2 mg/10^6 cells, irrespective of the diet. Protein synthesis (assessed by incorporation of ^{3}H-leucine into total hepatocyte proteins) increased linearly and was comparable with both diets throughout the 3 h incubations (Fig.1). After fat-feeding, secretion of total VLDL apoproteins (Fig. 2) was slightly lower after 2h and was decreased by 31% after 3h (p<0.05).

Hypercholesterolemia, Hypocholesterolemia, Hypertriglyceridemia
Edited by C.L. Malmendier *et al.*, Plenum Press, New York, 1990

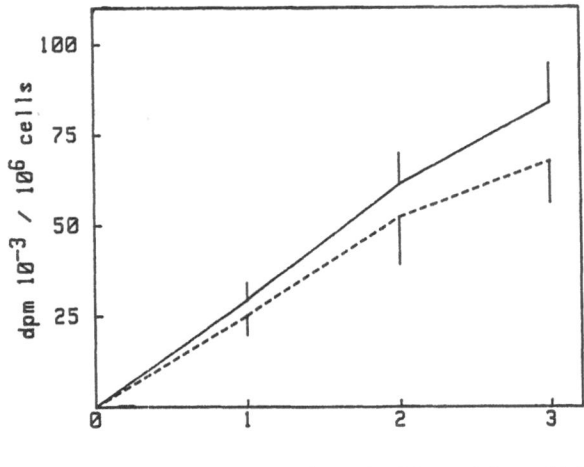

Fig. 1. Incorporation of [3]-H leucine into total
cellular proteins.
C diet (solid line); HF diet (dashed line)

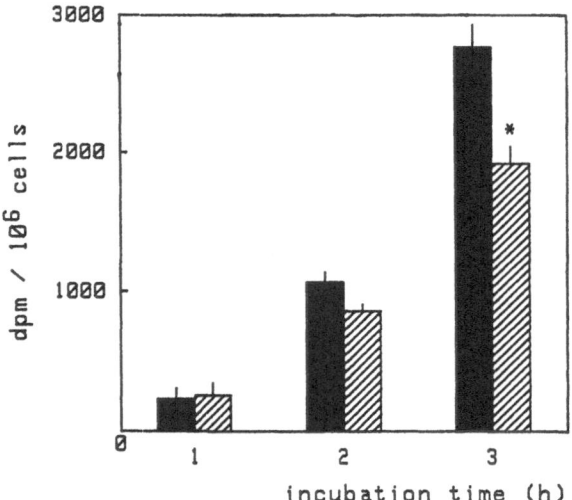

Fig. 2. [3]-H leucine incorporation into total VLDL
apoproteins.
C diet (full bars); HF diet (hatched bars).

Table 1.   Secretion of VLDL-triglyceride (TG) and
            total apoprotein (apo) after 3h-incubations.

Diet	VLDL-TG nmol/10^6 cell	VLDL-apo dpm/10^6 cell	VLDL-apo/ cell prot.
C	24.1 ± 2.0 (3)	2765.2 ± 157.9 (4)	3.3 %
HF	14.9 ± 1.5 (3)	1919.3 ± 120.4 (4)	2.8 %
	$p < 0.05$	$p < 0.05$	

C, control diet ; HF, high-fat diet.

Table 1 summarizes the results after 3 h incubations. Adaptation to the
high-fat diet induced a 40% decreased secretion of VLDL triglyceride and a
31% decreased output of radioactive total VLDL apoproteins into the medium.
Interestingly, the ratio of total VLDL apoproteins to total cellular proteins
was also lower with the high-fat diet.

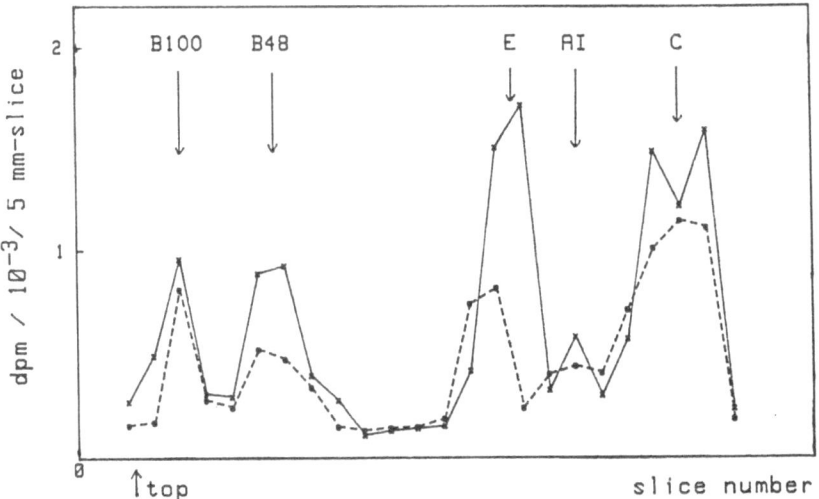

Fig. 3. ^{3}-H leucine incorporation into individual VLDL apoproteins.
        Legends are as in Fig. 1.

    Fig. 3. shows individual VLDL apoproteins separated by SDS-PAGE electro-
phoresis. Apo-AI was the least labelled apoprotein, with both diets. Apo-C's
were the most highly labelled apoproteins, accounting for 32% of total VLDL
apoproteins with the control diet and for 42% with the high-fat diet. Fat-
feeding decreased the secretion of practically all VLDL apoproteins, but in
varying degree. The greatest decreases (approximately 50%) were measured for
apo-B48 and apo-E.
    Apo E is synthetized mainly in the liver and secreted in three forms :
on VLDL, on lipoproteins smaller than VLDL and also as a peptide associated
with phospholipid. In our experiments we have measured only apo-E attached
to VLDL particles which was decreased by 60% after adaptation of the rats to
the high-fat diet. However, total apo-E secretion rates by hepatocytes of
fat-fed rats may not be decreased.

On the other hand, apo-B100 and apo-B48 are constitutive VLDL apoproteins and are both synthetized in the rat liver - in contrast to the human liver which synthetizes only apo-B100. Several papers have reported that apo-B48 production is closely related to hepatic VLDL-triglyceride production in the rat. Using rat hepatocytes in primary culture, Davis et al. (6) have shown that fasting greatly decreased secretion of VLDL-triglyceride and apo-B48, but not secretion of apo-B100. On the contrary, obese Zucker rats (which have a 3-fold higher hepatic VLDL secretion than their lean littermates) also display a 3-fold higher secretion of apo-B48 as compared to apo-B100(7).

Thus, the decreased secretion of apo-B48 induced by the high-fat diet seems to correlate well with the decreased VLDL triglyceride secretion. It raises the interesting question of a specific regulatory role of hepatic apo-B48 on VLDL synthesis and secretion in the rat.

In conclusion, this study showed that adaptation of the rats to the high-fat diet resulted in decreased secretion of all VLDL apoproteins, in varying degrees. VLDL apo-B48 and apo-E were the most decreased apoproteins. Since the high-fat diet affected neither hepatocyte protein content nor total protein synthesis, our results are consistent with a specific inhibition of VLDL apoprotein synthesis and secretion by this diet. Finally, the concomitant inhibition of VLDL triglyceride secretion (by 40%) and of VLDL apoprotein secretion (by 31%) suggests that fat-feeding may decrease the number of VLDL particles secreted by the rat liver.

REFERENCES

1. A.D. Kalopissis, S. Griglio, M.I. Malewiak, R. Rozen and X. Le Liepvre. Very low density lipoprotein secretion by isolated hepatocytes of fat-fed rats. Biochem. J. 198 : 373 (1981).
2. G.F. Gibbons and C.R. Pullinger. Regulation of hepatic very low density lipoprotein secretion in rats fed on a diet high in unsaturated fat. Biochem. J. 243 : 487 (1987).
3. A.D. Kalopissis, A. Girard, and S. Griglio. Diurnal variations of plasma lipoproteins and liver lipids in rats fed starch sucrose or fat. Horm. Metab. Res. 11 : 118 (1979).
4. P.O. Seglen. Preparation of rat liver cells. Exptl. Cell. Res. 74 : 450 (1972).
5. R.C. Feldhoff, J.M. Taylor, and L.S. Jefferson. Synthesis and secretion of rat albumin in vivo, in perfused liver, and in isolated hepatocytes. J. Biol. Chem. 252 : 3611 (1977).
6. R.A. Davis, J.R. Boogaerts, R.A. Borchardt, M. Malone-Mc Neal and J. Archambault-Schexnayder. Intrahepatic assembly of VLDL. Varied synthetic response of individual VLDL apolipoproteins to fasting. J. Biol. Chem. 260 : 14137 (1985).
7. M.J. Azain, N. Fukuda, F.-F. Chao, M. Yamamoto and J.A. Ontko. Contributions of fatty acid and sterol synthesis to triglyceride and cholesterol secretion by the perfused rat liver in genetic hyperlipemia and obesity. J. Biol. Chem. 260 : 174 (1985).

MOLECULAR GENETICS OF APOC-II AND LIPOPROTEIN LIPASE DEFICIENCY

S.S. Fojo, J.L. de Gennes[1], U. Beisiegel[2], G. Baggio[3]
A.F.H. Stalenhoef[4], J.D. Brunzell[5], and H.B. Brewer, Jr

NIH, Bethesda, MD, USA
[1]Groupe d'Endocrinologie-Metabolism, Paris, France
[2]Univ. Klinik Eppendorf, Hamburg,FRG
[3]Univ. of Padova, Padova, Italy
[4]St. Radboud Hospital, Nijmegen, The Netherlands
[5]Univ. of Washington, Seattle, Washington

INTRODUCTION

Apolipoprotein (apo) C-II plays a central role in normal triglyceride metabolism as cofactor for the enzyme lipoprotein lipase (LPL). Patients with a deficiency of either apoC-II or LPL have marked derangements in triglyceride metabolism which include an elevation of plasma triglycerides, fasting chylomicrons and VLDL (1). Clinical features of this syndrome, which is inherited as an autosomal recessive trait, include hepatosplenomegaly, eruptive xanthomas and an increased risk of pancreatitis. The diagnosis of apoC-II or LPL deficiency is established by finding a reduced post-heparin lipoprotein lipase activity which in apoC-II deficiency is corrected by the addition of normal apoC-II containing plasma.

We have studied the underlying molecular defects in the apoC-II and LPL genes of patients with the familial hyperchylomicronemia syndrome. Several point mutations have been identified which lead to the synthesis of dysfunctional proteins and ultimately, to a deficiency of apoC-II or LPL in these patients.

Materials and Methods

Clinical Data: The clinical features of the study patients with a deficiency of apoC-II have been previously described (2-5). ApoC-II levels in these individuals ranged from undetectable to 0.2 ng/dl (nl-5 mg/dl) while post-heparin LPL activity in the presence of an exogenous source of apoC-II was normal. The LPL deficient patient from the Bethesda kindred (6) had a pre- and post-heparin LPL mass of 256 and 393 ng/ml, respectively, but no detectable LPL activity in pre- or post-heparin plasma.

Genomic Libraries. WBC genomic DNA of the proband from the Hamburg and Nijmegen kindreds with apoC-II deficiency was partially digested with Mbo I and used to prepare a genomic library by infecting E.coli P2492 cells with EMBL-3 recombinant phage as described (3). The library was screened for apoC-II cDNA clones by using an apoC-II cDNA probe.

Hypercholesterolemia, Hypocholesterolemia, Hypertriglyceridemia
Edited by C.L. Malmendier *et al.*, Plenum Press, New York, 1990

329

<u>Sequence Amplification of the ApoC-II gene by Taq DNA</u>
<u>Polymerase</u>.   One g of genomic DNA from apoC-II deficient subjects was
amplified by the automated polymerase chain reaction (PCR) technique (2) as
described previously.   Amplified DNA was digested with SstI and XbaI and
subcloned into M13 vector DNA for sequencing.

<u>Reverse Transcription and Polymerase Chain Reaction (PCR) Amplification</u>
<u>of LPL cDNAs</u>.   LPL cDNA from the LPL-deficient subject was synthesized by
incubation of 1   g of monocyte-derived macrophage total RNA with 15 units of
Molony murine leukemia virus reverse transcriptase (Pharmacia LKB) and two
32 bp LPL primers (each at 0.2 mM) containing the restriction enzyme sites
for HindIII and BamHI, which spanned bases 129-1665 of the LPL cDNA (7).
After incubation at 37°C for 2 hr the newly generated cDNA was amplified by
the automated PCR technique.   Amplified DNA was digested with the appropriate
restriction enzymes and subcloned into pGEM-3Z vector DNA for sequencing.

<u>DNA Sequencing</u>.   Double-stranded DNA sequencing of LPL cDNA cloned into
pGEM-3Z vector DNA was performed by the dideoxynucleotide chain-termination
method of Sanger et al. (8).

<u>Plasmid Construction and Transfection of Mammalian Cells</u>.   The parent
plasmid (pCMV) utilized in the transfection studied contains the
cytomegalovirus (CMV) promoter and transcriptional start site as well as the
polyadelylation signal from the Simian virus 40 (SVC40) cloned into pU-18
(6).   The constructs, pCMV-NL and pCMV-PT, containing the normal and
patient's cDNA have been previously described (6).   Standard techniques were
used for all cloning procedures and plasmid isolations were performed by the
CsCl double-banding method (9).   DNA transfections were performed by the
calcium phosphate coprecipitation method (10) with the exception that the
medium was supplemented with 2 units of heparin sulfate (Elkins-Sinn, Cherry
Hill, NJ) per ml.

<u>Results</u>

Table I summarizes the molecular defects which have been identified in
the apoC-II gene of 4 patients with a deficiency of apoC-II from different
kindreds by sequence analysis.   Single base substitutions result in the
introduction of premature stop codons and the synthesis of truncated non-
functional proteins in apoC-II$_{Padova}$, apoC-II$_{nijmegen}$ and apoC-II$_{Paris}$.   A donor
splice site mutation in the second intron of the apoC-II gene leads to
abnormal splicing of the apoC-II$_{Hamburg}$ mRNA, while a mutation of the
initiation methionine codon results in a met to a val substitution that
prevents normal initiation of translation of apoC-II$_{Paris}$.   Analysis of PCR

TABLE I

MUTATIONS IN ApoC-II DEFICIENCY

MUTANT CLASS	ApoC-II level	Kindred
Nonfunctional mRNA		
Nonsense mutants (codon 18)	Absence	Nijmegen
Frameshift (codon 37)	Deficiency	Padova
RNA Processing Mutants		
IVS-II donor splice site	Deficiency	Hamburg
position 1 (G -> C)		
Translational Start Site (A -> G)	Absence	Paris

amplified DNA from family members demonstrated that despite the absence of a family history of consanguinity (except in the Hamburg kindred) all probands were true homozygotes for their respective mutations (data not shown).

Figure 1 is a schematic representation of apolipoprotein C-II and illustrates the location of the mutations within the protein. All 4 defects an are clustered near the amino terminal end of apoC-II, although 2 previously described variants, apoC-II$_{Toronto}$ (11) and ApoC-II$_{StMichael}$ (12), have mutations near the carboxyl-terminal end of apoC-II.

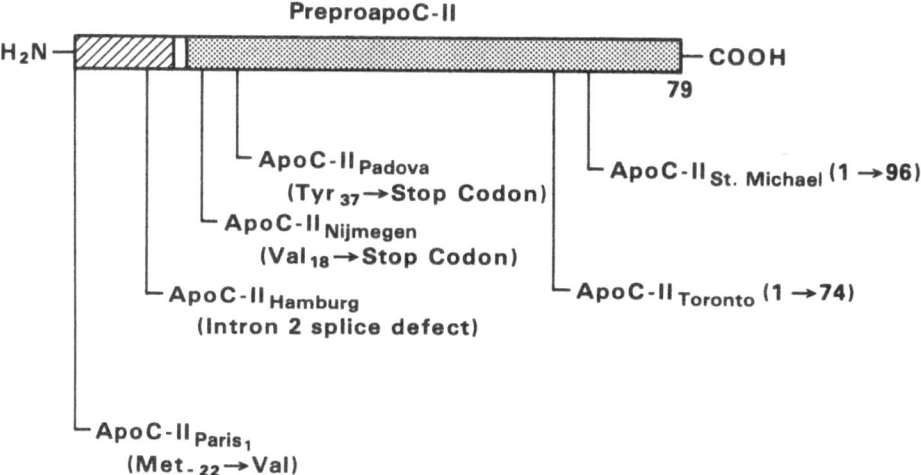

Figure 1. Schematic representation of apoC-II. Arrows illustrate the location of the defects in the apoC-II mutants.

Figure 2A illustrates the results of sequence analysis of the LPL cDNA of the proband from the Bethesda kindred. A G to A mutation at position 781 within the fifth exon of LPL that results in the substitution of ala-176 to thr was identified. Autoradiographs of sequence gels from normal and the LPL-deficient proband illustrating the G to A substitution are shown in Figure 2B.

In order to establish that the ala 176 to thr substitution results in the expression of an enzymatically inactive LPL$_{Bethesda}$ enzyme, expression plasmids encoding for both the normal and mutant LPL were transfected into COS-7 cells. The medium was then assayed for LPL mass and activity. The results of the expression of normal and patient LPL cDNA in COS-7 cells are summarized in Table II. Medium harvested from COS-7 cells transfected with the pCMV-PT construct that contained an LPL concentration (by ELISA) similar to that of the pCMV-NL (normal) construct had no detectable LPL activity whereas successful expression of an enzymatically active LPL was achieved with the pCMV-NL (normal) plasmid. These studies establish that the ala $_{176}$ -> thr substitution is the mutation that results in the loss of enzymic activity in LPL$_{Bethesda}$.

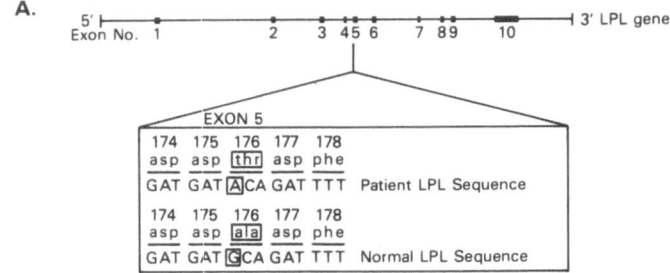

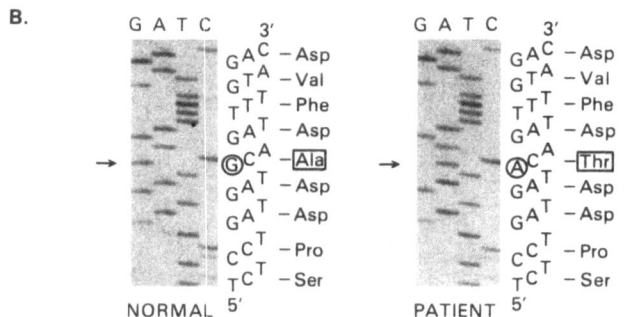

Figure 2. (A) Schematic representation of the LPL gene.
Exons are illustrated by solid bars interrupted by lines
that represent introns. The G -> A substitutions in the
patient's LPL sequence is highlighted by a box. (B) Auto-
radiograms of sequencing gels of DNA from a normal subject
and the proband. The G -> A substitution is indicated by
the arrow.

Table II. LPL activity and mass in the medium of transfected COS-7 cells

Plasmid	LPL mass, (ng/ml)	LPL activity, (neq of FFa/ml/min)
pCMV	1.56 ± 0.55	0.18 ± 0.07
pCMV-NL	26.80 ± 1.22	4.46 ± 0.15
pCMV-PT	24.37 ± 1.91	0.12 ± 0.04

Values are expressed as mean ± SD. The difference between the mean
post-heparin lipolytic activity and HL activities was used to determine
the LPL activity (mean ± SD) in each fraction.

Discussion

We have studied the underlying molecular defects that result in a deficiency of apoC-II in 4 different patients from independent kindreds. Sequence analysis of the patient's apoC-II gene identified 4 unrelated mutations suggesting that there is no evidence for a founder gene effect as the cause of apoC-II deficiency. Despite the absence of a family history of consanguinity in most kindreds, all patients were identified as true homozygotes for their respective mutations. These substitutions, were clustered near the amino and carboxyl terminal end of apoC-II.

In the $LPL_{Bethesda}$ gene a G to A mutation at position 781 leads to an ala 176 to thr substitution in a highly conserved region near the proposed interfacial recognition site of LPL. Transient expression analysis of the mutant LPL cDNA in COS-7 cells demonstrated the functional significance of the $LPL_{Bethesda}$ mutation as the expressed mutant protein had no detectable enzymatic activity.

The elucidation of the genetic defects in patients with a deficiency of apoC-II and LPL enhances our understanding of the molecular derangements that can result in the familial hyperchylmicromenia syndrome and provides new insights into the structure and function of apoC-II and LPL.

References

1. J. D. Brunzell. in The Metabolic Basis of Inherited Disease, eds. C. R. Scriver, A. L. Sly, W. S. & D. Valle (McGraw-Hill, New York), 6th Ed., 1165-1180 (1989).
2. S. S. Fojo, P. Lohse, C. Parrott, G. Baggio, C. Gabelli, F. Thomas, J. Hoffman, and H. B. Brewer Jr. J. Clin. Invest. 84:1215-1219 (1989).
3. S. S. Fojo, A. F. H. Stalenhoef, K. Marr, R. E. Gregg, S. Ross, and H. B. Brewer Jr. J. Biol. Chem. 263:17913-17916 (1988).
4. S. S. Fojo, U. Beisiegel, U. Beil, K. Higuchi, M. Bojanovski, R. E. Gregg, H. Greten, and H. B. Brewer Jr. J. Clin. Invest. 82:1489-1494 (1988).
5. S. S. Fojo, J. L. deGennes, J. Chapman, C. Parrott, P. Lohse, S. S. Kwan, J. Truffert, and H. B. Brewer Jr. J. Biol. Chem. 264:20839-20842 (1989).
6. O. U. Beg, M. S. Meng, S. I. Skarlatos, L. Previato, J. D. Brunzell, H. B. Brewer Jr., and S. S. Fojo, Proc. Natl. Acad. Sci. USA. In press.
7. K. L. Wion, T. G. Kirchgessner, A. J. Lusis, and M. C. Schotz, Science. 235:1638-1641 (1987).
8. F. Sanger, A. R. Coulson, B. G. Barrell, A. J. H. Smith, and B. A. Roe, J. Mol. Biol. 143:161-178 (1980).
9. T. Maniatis, E. F. Fritsch, and J. Sambrook, in A Laboratory Manual (Cold Spring Harbor Lab., Cold Spring Harbor, NY), Mol. Clon. 229-246.
10. N. Rosenthal, Methods Enzymol. 152:704-720 (1987).
11. P. W. Connelly, G. F. Maguire, T. Hofmann, and J. A. Little, Proc Natl Acad. Sci. USA. 84:270-273 (1987).
12. P. W. Connelly, G. F. Maguire, T. Hofmann, and A. J. Little, J. Clin. Invest. 80: 1597-1606 (1987).

WHAT FACTORS REGULATE THE ACTION OF LIPOPROTEIN LIPASE?

Thomas Olivecrona (1), Gunilla Bengtsson-Olivecrona (1)
Magnus Hultin (1), Jonas Peterson (1), Senén Vilaró (2)
Richard J. Deckelbaum (3), Yvon A. Carpentier (4) and
Josef Patsch (5)

(1) Medical Biochemistry and Biophysics, University of Umeå
Sweden (2) Cell Biology, University of Barcelona, Spain
(3) Pediatric Gastroenterology, Columbia University, New
York, USA (4) Clinical Nutrition, Free University of Brussels
Belgium and (5) Medicine, University of Innsbruck, Austria

Triglyceride transport is an efficient process. Meals containing as much as 100 g triglycerides are usually absorbed and disposed within a few hours with only a moderate rise in the plasma triglyceride concentration.[1] Most of the lipoprotein triglycerides are unloaded through hydrolysis by lipoprotein lipase at endothelial sites in extra-hepatic tissues.[2] It is often stated that this step is rate-limiting for triglyceride transport, and that it directs the tissue distribution of lipid uptake.[3] Inherent in this concept are two assumptions that we will discuss in this paper.

The first assumption is that endothelial LPL is dependent on local production in nearby parenchymal cells. This is certainly true in part. For instance, there are several situations where a decrease in adipose tissue LPL activity can be ascribed in part to a decrease of LPL mRNA within the tissue. Examples are fasting,[4] TNF[4] and dioxins (Bengtsson-Olivecrona, unpublished observation).

The binding of LPL to endothelial cells is moderately strong. The reported dissociation constants are 0.14 µM or more (reviewed in 5). Hence, LPL probably moves rapidly along the endothelial wall from one binding site to the next, carried by blood. Because of avid uptake in the liver the concentration of LPL in the general circulation is kept low (Fig. 1, reviewed in 5). The system is not in equilibrium, but there must be a

Hypercholesterolemia, Hypocholesterolemia, Hypertriglyceridemia
Edited by C.L. Malmendier *et al.*, Plenum Press, New York, 1990

335

concentration gradient from areas with high LPL synthesis to other parts of the vascular bed, and there is net flow of LPL to the liver.

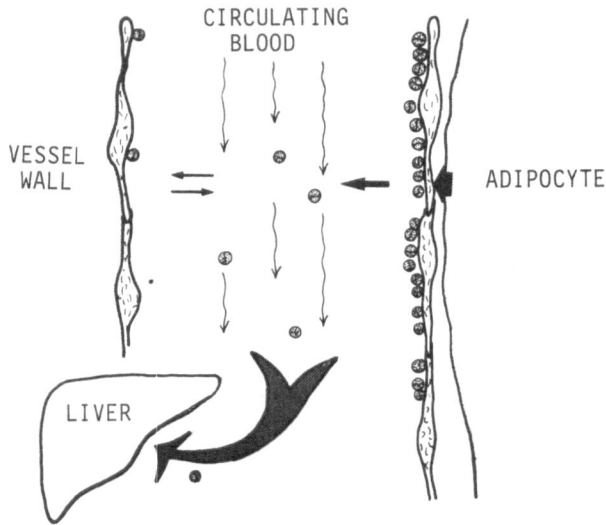

Fig. 1. Transport of LPL. The enzyme is produced by parenchymal cells, illustrated here by an adipocyte to the right. LPL is released from these cells and moves to binding sites at the luminal side of endothelial cells in adjacent capillaries. The lipase then slowly moves along the endothelial surface from one binding site to the next, carried by blood. When it enters larger vessels and the general circulation it can spread to binding sites in other tissues, including those which do not produce LPL. This spreading is counteracted by avid uptake of LPL in the liver. Hence, the system is not equilibrium but there is continuous supply of new LPL molecules from sites of synthesis, resulting in high concentration at these capillary segments, and there is continuous net uptake and degradation of LPL by the liver. (From Olivecrona and Bengtsson-Olivecrona in Borensztajn ed. Lipoprotein lipase, Evener, Chicago, 1987, pp. 15-58).

To put these ideas to an experimental test, a study was carried out in which _in situ_ hybridization was used to determine where LPL is synthesized, and immunohistochemistry was used to localize the enzyme protein.[6] In agreement with previous studies there was immunofluorescence over the dominant parenchymal cells in adipose tissue and muscles, i.e. adipocytes/

preadipocytes and myocytes, and also at the endothelial surface of all vessels. In situ hybridization showed LPL mRNA in the same parenchymal cells, but not in the endothelial cells. LPL at the capillary endothelium could come from nearby parenchymal cells, but the enzyme found in larger vessels must have come via the blood stream. In the aorta, there was intense immunoreaction for LPL, but only a few cells in the deeper layers of the vessel wall reacted for LPL protein, or LPL mRNA, indicating that local synthesis of the enzyme was sparse.

In some tissues, e.g. the lungs, only scattered cells gave positive reactions for LPL protein and LPL mRNA.[7] Nonetheless, there was rather homogeneous labeling of the endothelial surfaces, indicating that the lipase must have spread rapidly over the entire vascular bed and did not remain confined to areas of LPL-producing cells.

There were also sites with no evidence for local LPL synthesis, but strong reaction for LPL protein at the vascular surfaces. A particularly striking example of this were the glomeruli in the kidney.[6] Here, all the LPL must have come from blood.

The second assumption is that LPL-catalyzed hydrolysis is the rate-limiting step for triglyceride transport, and that the underlying tissue can assimilate the fatty acids as rapidly as the lipase can provide them. One early piece of evidence for this came from kinetic modeling of chylomicron metabolism which one of us (T.O.) was privileged to perform with the late Dr. Mones Berman at the National Institutes of Health in Bethesda, MD. These studies led to the conclusion that only a minor fraction of the fatty acids from labeled chylomicrons recirculated through the plasma FFA pool.[8] We are happy to join in the tribute to Mones at the present symposium.

The question remaining to be answered was whether the ability of the underlying tissue to assimilate fatty acids is indeed unlimited. To answer this question, Peterson et al. created an experimental overloading of the triglyceride clearing system by infusion of Intralipid at a high rate to fasted volunteers.[9] The dose given, 0.3 g/kg body weight per hour, approximated the potential rate of endothelial LPL hydrolysis, but was far above immediate energy needs. The infusion caused a progressive rise of plasma triglycerides in all study subjects, and increases of plasma free fatty acids (FFA) which varied among individuals. Plasma LPL activity also rose. This rise was correlated to FFA, but not to triglycerides, both with respect to the levels reached and to the time course. In ongoing studies similar results have been obtained in rats infused with lipid emulsions, and in humans after oral fat loads.

Earlier in vitro studies (discussed in 5) have shown that LPL binds fatty acids and that the resulting complexes have much reduced affinities for lipid/water interfaces, for apolipoprotein CII and for heparin-like polysaccharides. Moreover, Saxena et al. have recently shown that physiologic concentrations of fatty acids dissociate LPL from cultured endothelial cells.[10] This suggests the following sequence of events: Fatty acids can be released by LPL at endothelial sites more rapidly than the underlying tissue is capable of transporting and/or metabolizing them. The result is spillage of fatty acids into blood, and formation of LPL-fatty acid complexes which leads to dissociation of LPL from endothelial heparan sulfate which in turn leads to cessation of ongoing hydrolysis. The implication is that there exist situations in which LPL is present in excess and the real limitation is transport and metabolism of fatty acids by the underlying tissue.

The starting point for our discussion was the traditional view according to which endothelial LPL activity is goverened by local production of the enzyme, and the resulting activity governs the uptake of fatty acids from hydrolyzed lipoprotein triglycerides. This is certainly true in part but the data discussed in this article indicate substantial transport in blood of both LPL and the fatty acids that it generates. The magnitude of these transport processes can evidently be quite different in various metabolic situations. A challenging task for future research will be to explore the regulation of these transport processes.

ACKNOWLEDGEMENTS

This study was supported by grants from the Swedish Medical Research Council (B31-727) and from the Swedish Margarine Industry Fund for research in nutrition to T.O. and G.B.-O., by grant from the Fonds Belgé de la Recherche Scientifique Medicale (9002-86) to Y.A.C., by a grant from the National Institutes of Health (P50 21006) to R.J.D., by grants from Comisión Asesora de Investigatión y Ciencia, grant AR87 (CIRIT) from the Generalitat de Catalunya to S.V., and by grants HL-27341 from the National Institutes of Health and 5-46106 from the Austrian Funds on Förderung der Wissenschaftlichen Forschung to J.P.

REFERENCES

1. Patsch JR, Karlin JB, Scott LW, Smith LC, Gotto AM (1983) Inverse relationship between blood levels of high density lipoprotein sub-fraction 2 - and magnitude of postprandial lipemia. Proc Natl Acad Sci USA 80:1449-1453

2. Robinson DS (1987) Lipoprotein lipase - past, present and future. In Lipoprotein Lipase (Bòrensztajn J, ed), pp 1-13, Evener Publishers, Chicago

3. Nestel PH (1987) The regulation of lipoprotein metabolism. In Plasma Lipoproteins (Gotto AM Jr, ed), New Comprehensive Biochemistry, vol 14, pp 153-182, Elsevier, Amsterdam

4. Enerbäck S, Semb H, Tavernier J, Bjursell G, Olivecrona T (1988) Tissue-specific regulation of guinea pig lipoprotein lipase; effects of nutritional state and of tumor necrosis factor on mRNA levels in adipose tissue, heart and liver. Gene 64:97-106

5. Olivecrona T, Bengtsson-Olivecrona G (1989) Heparin and lipases. In Heparin (Lane D and Lindahl U, eds) pp 335-361, Edward Arnold, London

6. Camps L, Reina M, Llobera M, Vilaró S, Olivecrona T (1990) Lipoprotein lipase: cellular origin and functional distribution. Am J Physiol. In press

7. Camps L, Reina M, Llobera M, Olivecrona T, Vilaró S (1990) Lipoprotein lipase in lungs, spleen and liver: synthesis and distribution. In preparation

8. Olivecrona T (1962) Kinetics of fatty acid transport. An experimental study in the rat. Thesis, University of Lund, Sweden

9. Peterson J, Bihain BE, Bengtsson-Olivecrona G, Deckelbaum R, Carpentier YA, Olivecrona T (1990) Fatty acid control of lipoprotein lipase: A link between energy metabolism and lipid transport. Proc Natl Acad Sci 87:909-913

10. Saxena U, Witte LD, Goldberg IJ (1989) Release of endothelial cell lipoprotein lipase by plasma lipoproteins and free fatty acids. J Biol Chem 264:4349-4355

PATHOGENESIS OF ATHEROSCLEROSIS

CYTOTOXICITY OF REMNANTS OF TRIGLYCERIDE-RICH LIPOPROTEINS:

AN ATHEROGENIC INSULT?

B. Hong Chung, and Jere P. Segrest

Department of Medicine, University of Alabama Medical Center
University of Alabama at Birmingham, Birmingham, Alabama

INTRODUCTION

One of the more widely accepted hypotheses (1) to explain the development of atherosclerosis is that some form of functional and/or structural injury occurs to endothelial cells that line the lumen of the artery. This injury disrupts the permeability barrier and may lead to an infiltration of plasma components (lipoproteins) into the artery and an adherence of circulating blood cells (monocytes and platelets) to the injured endothelium. The conversion of monocyte-derived macrophages to lipid laden foam cells has been thought to play a central role in the development of atherosclerotic lesions. Although the importance of endothelial cell injury in the initiation of the atherogenic process has been demonstrated by the development of atherosclerotic lesions in the artery following mechanical (balloon catheter) injury of the endothelium (2), the factor(s) responsible for the injury of arterial endothelium *in vivo* is not currently clear. Ross and Harker (3) suggested that hypercholesterolemia might cause endothelial cell damage, and so set off a sequence of reactions, which results in an atherosclerotic lesion (4). Although there is currently no strong evidence that hypercholesterolemia indeed leads to endothelial cell damage *in vivo*, several investigators have reported a cytotoxic effect of LDL and VLDL on cultured endothelial cells (5-9). The cytotoxic effect of LDL has been indicated to be a result of lipid peroxidation (6-7). Lipid peroxides have been identified in the sera of healthy men, hypercholesterolemic and hypertriglyceridemic subjects, and diabetic patients (10-12), but sera from these subjects were unable to produce cytotoxicity to cultured human endothelial cells (12). Recent studies from our laboratory indicate that lipolytic remnants of triglyceride-rich lipoproteins produced *in vitro* are not only one cause of foam cell formation in macrophages but also one cause of necrosis (cell death) in cultured macrophages and endothelial cells, and this necrosis can be inhibited by HDL (13,14). Thus, our results potentially provide a single hypothesis that can explain several events in the atherogenic process.

CYTOTOXICITY OF LIPOLYZED HYPERTRIGLYCERIDEMIC (HTG) SERUM TO CULTURED MACROPHAGES

In vitro incubation of HTG serum with lipoprotein lipase (LpL) (purified from raw bovine milk) for 90 min at 37°C usually resulted in the hydrolysis of 60-85% of the triglycerides in serum and the removal of 60-90% of

Hypercholesterolemia, Hypocholesterolemia, Hypertriglyceridemia
Edited by C.L. Malmendier *et al.*, Plenum Press, New York, 1990

341

VLDL-cholesterol from the VLDL density region and its transfer into the IDL, LDL and HDL density fractions. When control serum (incubated with heat inactivated lipoprotein lipase) or lipolyzed serum was added into serum free culture medium of peritoneal macrophages, the lipolyzed serum was highly cytotoxic to the cells, killing >90% of cells with 5 µg lipolyzed serum per ml culture medium. Control serum at concentrations of up to 60 µg per ml culture medium had no detectable effect on cell viability (Fig. 1).

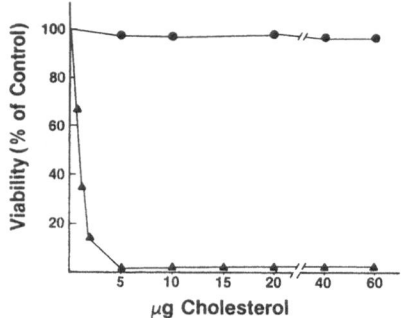

Fig. 1. Viability of macrophages following incubation with control HTG serum (●-●) or lipolyzed HTG serum (▲-▲). Aliquots of the control or lipolyzed serum containing 1-60 µg cholesterol were added to culture dishes of mouse peritoneal macrophages containing 1 ml culture medium. After an 18 h incubation of the dish at 37°C in a cell culture incubator, the viability of the cells was examined by counting the percent of cells excluding trypan blue.

Normolipidemic serum, obtained from a fasting subject and lipolyzed under identical conditions as HTG serum, was not cytotoxic to cultured macrophages. However, when this serum was supplemented with preisolated VLDL to a level similar to that of HTG serum and the supplemented serum was incubated with lipoprotein lipase, the resultant product was cytotoxic to macrophages. Normolipidemic subjects would experience an acute hypertriglyceridemia after the meals (postprandial hypertriglyceridemia); therefore, the cytotoxicity of the lipolyzed samples of fasting serum and postprandial lipemic serum, obtained from the normolipidemic subjects with normal or brisk chylomicron response, to cultured macrophages was further examined. Ingestion of a fatty meal by the normotriglyceridemic subjects resulted in a ~380 mg/dl increase in plasma TG and 10 mg/dl increase in plasma cholesterol 4 hr after the meal in the subject with a brisk chylomicron response but resulted in only a 33 mg/dl increase in TG, with little or no change in plasma cholesterol in the subject with a normal chylomicron response (Fig. 2). The increase in cholesterol was associated exclusively with an increase in cholesterol in the VLDL density region of the plasma (Fig. 2). *In vitro* lipolysis produced little or no cytotoxicity in fasted serum from the normal or brisk chylomicron responder or in postprandial serum from the normal responder (Fig. 3). However, the postprandial serum from the brisk chylomicron responder was cytotoxic to cultured macrophages after *in vitro* lipolysis (Fig. 3).

Light microscopic examination of cultured macrophages showed that cells in the culture dish containing a sublethal dose of lipolyzed serum (2µg cholesterol) contained numerous lipid inclusions, resembling those seen in foam cells, within the cytoplasm of the surviving macrophages; little or no lipid inclusions can be seen in the cells incubated with control serum or with a lethal dose (10 µg cholesterol) of lipolyzed HTG serum (Fig. 4).

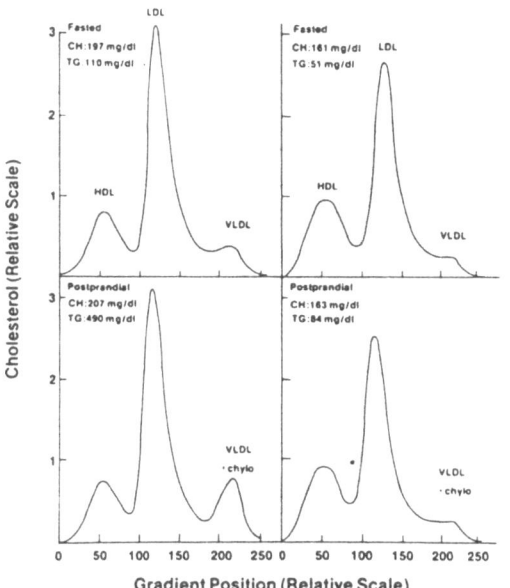

Fig. 2. Lipoprotein cholesterol profiles of fasting and postprandial lipemic serum (obtained 4 h after a fatty meal) from normolipidemic subjects with normal (left) or brisk (right) chylomicron response.

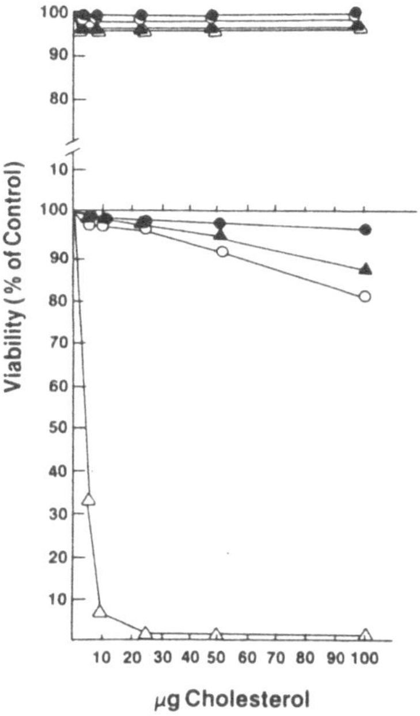

Fig. 3. Viability of macrophages incubated with pre-lipolysis serum (top) or postlipolysis serum (bottom): fasting (○-○) or postprandial lipemic serum (△-△) from the brisk responder and fasting (●-●) or postprandial serum (▲-▲) from the normal responder were lipolyzed *in vitro* with purified milk LpL and incubated with cultured macrophages.

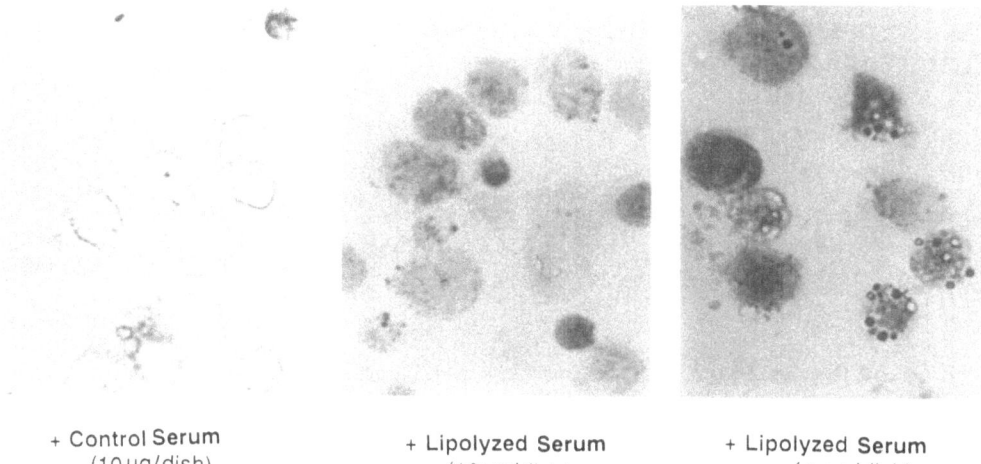

+ Control Serum (10 ug/dish)

+ Lipolyzed Serum (10 ug/dish)

+ Lipolyzed Serum (2 ug/dish)

Fig. 4. Appearance by phase-contrast light microscopy of cultured macrophages after incubation with 10 µg cholesterol of control HTG serum (left), lipolyzed HTG serum (Middle), or 2 µg cholesterol of lipolyzed HTG serum (Right).

Examination of the cytotoxicity of the control or lipolyzed serum following its fractionation into various plasma fractions by single vertical
spin density gradient ultracentrifugation showed that all of the lipoprotein
fractions from *in vitro* lipolyzed serum were cytotoxic and foam cell
inducing; plasma free protein fractions were not cytotoxic but induced a
moderate number of lipid inclusions (Fig. 5). Corresponding control plasma
fractions were neither cytotoxic nor foam-cell inducing (Fig. 5).

When VLDL, LDL and HDL fractions, isolated from control serum were
incubated with LpL in the presence of fatty acid depleted bovine serum albumin, and the cytotoxicity of these LpL treated lipoprotein fractions to cultured macrophages was examined, we found that only the VLDL fraction became
cytotoxic to cultured macrophages after incubation with LpL. The above
results indicate that the cytotoxicity associated with the LDL and HDL
fractions in lipolyzed serum is a result of transfer of cytotoxic remnant
products from the VLDL to the LDL and HDL density fractions.

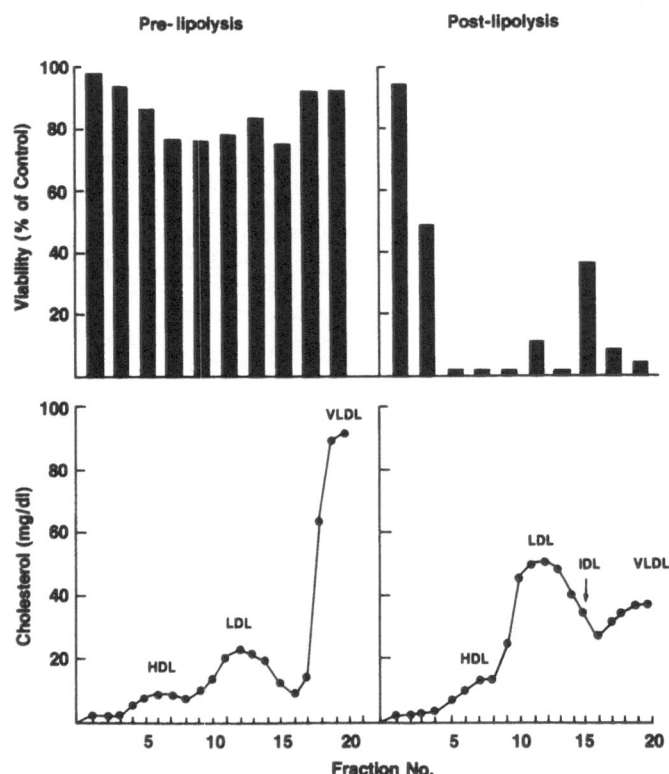

Fig. 5. Cytotoxicity of density gradient fractions of prelipolysis or post-
    lipolysis samples of HTG serum. The pre- and postlipolysis serum
    samples were subjected to single vertical spin density gradient
    ultracentrifugation , and aliquots containing 40 µg cholesterol of
    lipoproteins in prelipolysis sample, 10 µg cholesterol of lipopro-
    teins in postlipolysis sample or 50 ml of plasma free protein frac-
    tions were removed and added to culture dishes of macrophages. Cell
    viability was measured after 18 h incubation.

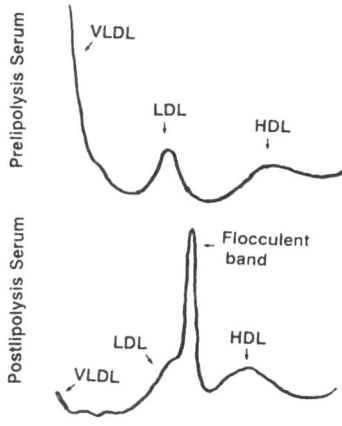

Fig. 6. Absorption (280 nm) profiles of density gradient samples of prelipolysis and postlipolysis serum.

We have identified and characterized a most potent cytotoxic fraction in lipolyzed serum. When lipolyzed HTG serum is subjected to a density gradient ultracentrifugation, and the absorbance (at 280 nm) of the effluent of the samples in the gradient tube is monitored, a large new absorbance peak just below the LDL peak but above the HDL peak appeared; little or no such peak was detected from the control (prelipolysis) serum (Fig. 6). Materials in this peak are very flocculent and can be pelleted free of LDL contaminants by a high speed centrifugation (at 8000 x g for 30 min). The flocculent remnant particles consisted mainly of free fatty acids and were poor in core lipids and rich in unesterified cholesterol (Table 1).

This lipid composition suggests that these remnants may be the surface products of TG-rich lipoproteins. Quantification of the level of flocculent remnants and free fatty acids among various plasma fractions following *in vitro* lipolysis of HTG serum containing a high level of TG (520 mg/dl) showed that flocculent remnants contained about 1.2% of total plasma cholesterol but contained 25.2% of total free fatty acids generated during lipolysis. Cytotoxicity studies of VLDL, LDL and flocculent remnants from lipolyzed sera showed that the flocculent remnants were the most cytotoxic particles in the lipolyzed serum, and the HDL fraction was much more cytotoxic than the VLDL or LDL fractions (Fig. 7).

A number of previous studies have shown that early atherosclerotic lesions in both humans and experimental animals contain numerous liposome-like vesicular particles rich in phospholipid and unesterified cholesterol

Table 1. Composition of flocculent remnants, LDL and HDL isolated from lipolyzed HTG serum.

Lipoproteins	Prot	TG	PL	UC	EC	FFA
			% of total	mass		
LDL	21.6	3.4	21.8	8.8	33.5	10.9
HDL	41.4	4.9	30.8	3.1	11.8	8.0
Flocculent remnants	22.4	4.8	17.6	7.9	2.3	45.0

TG: Triglycerides; PL: Phospholipids; UC: Unesterified cholesterol;
EC: Esterified cholesterol; FFA: Free fatty acids

345

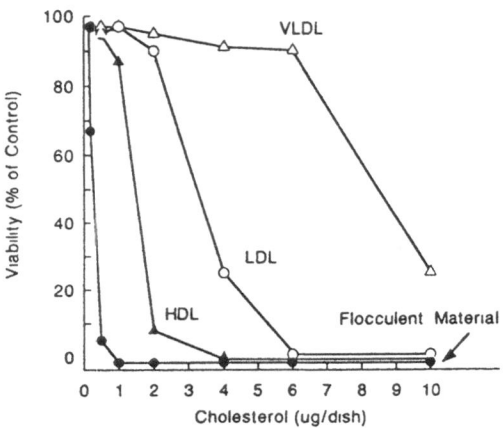

Fig. 7. Viability of macrophages following incubation with VLDL, LDL, HDL, and flocculent remnants from lipolyzed HTG serum.

(15-17). We found that the flocculent remnants isolated from lipolyzed serum have lipid composition, density and morphology similar to that reported in unesterified cholesterol-rich vesicles isolated from human atherosclerotic lesions (18).

EFFECT OF HDL ON CYTOTOXICITY, STRUCTURE AND COMPOSITION OF THE REMNANTS OF TG-RICH LIPOPROTEINS

Because HDL plays a role in circulating blood of accepting lipolytic surface remnants of TG-rich lipoproteins, the effect of the presence of the HDL in culture dishes or in the lipolysis mixtures on the cytotoxicity of the lipolyzed VLDL or the flocculent remnants to cultured macrophages was

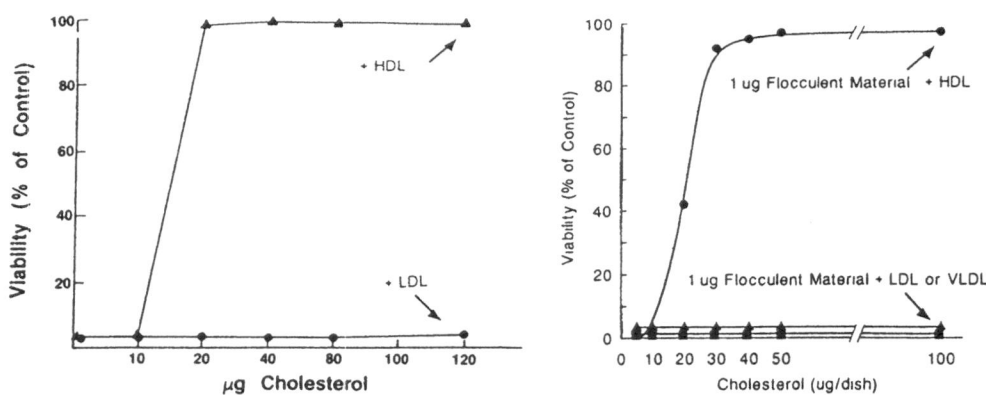

Fig. 8. Effect on macrophage viability of the presence of LDL, HDL and/or VLDL in culture medium containing a cytotoxic level of lipolyzed VLDL (Left) or flocculent remnants (Right). Macrophages viability was determined after incubation of macrophage culture dishes containing 10 μg cholesterol of lipolyzed VLDL or 1 μg cholesterol of flocculent remnants plus indicated amount of control LDL, HDL and/or VLDL for 18 H at 37°C.

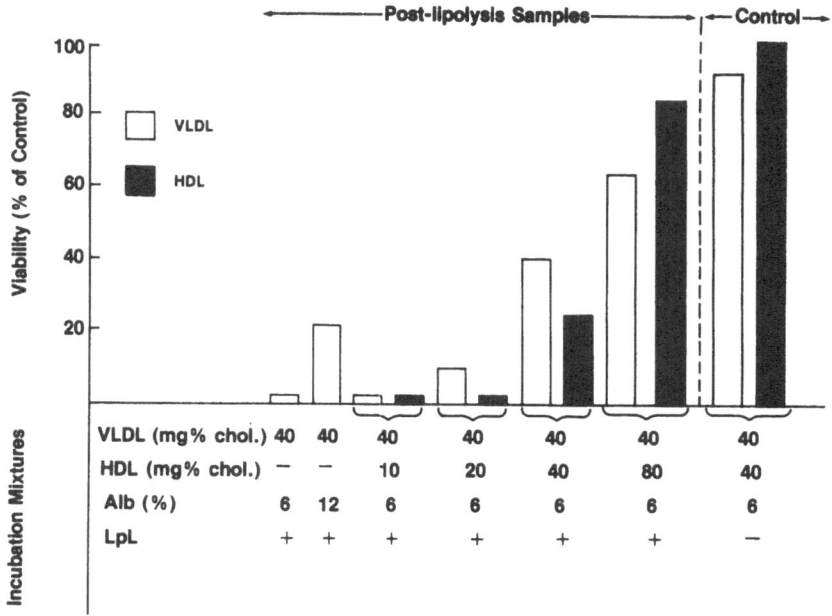

Fig. 9. Effect of the presence of HDL during lipolysis of VLDL on macrophage
viability. VLDL was lipolyzed with LpL in the presence of increasing
amounts of HDL and/or albumin. The VLDL remnants and HDL (10 μg
cholesterol), separated from albumin, were incubated with macrophages, and the viability of macrophages was determined after
incubation for 18 h at 37°C.

further examined. Addition of a 2X excess amount of control HDL (20 μg cholesterol) than the cytotoxic level of lipolyzed VLDL (10 μg cholesterol) into
the culture dishes inhibited the cytotoxicity of lipolyzed VLDL (Fig. 8). In
culture dishes containing a cytotoxic level of flocculent remnants (1 μg
cholesterol), 20X excess amounts of control HDL were required to inhibit the
cytotoxicity of the flocculent remnants (Fig. 8). The addition of VLDL
and/or LDL to the culture dish at up to 5-6 X the effective concentration of
HDL had no measurable effect on remnant cytotoxicity (Fig. 8).

The presence of a 2 X excess amount of HDL than the VLDL in the lipolysis mixture during lipolysis of VLDL can also inhibit the cytotoxicity of
the lipolyzed VLDL (Fig. 9). At HDL to VLDL ratios of 1.0 or lower, both the
remnant VLDL and the HDL isolated from the lipolysis mixture were cytotoxic,
killing 60% or more of the cells (Fig. 9). Additional experiments indicated
that the serum albumin at twice the physiological concentration in the lipolysis mixtures provided some protection (20% viability) from cytotoxicity
(Fig. 9).

Further characterization of cytotoxic remnants of VLDL, which were produced in the absence of HDL, and noncytotoxic remnants, which were produced
in the presence of 2 X excess amount of HDL than the VLDL, showed that the
former consisted of electron-dense spheroidal particles smaller than those
of control VLDL, core depleted spheroidal particles to which the surface
layer was still attached, and numerous liposome-like vesicles, some of which
were larger than control VLDL (Fig. 10-Up). These cytotoxic remnants of VLDL
retained most of the apo Cs of control VLDL (Fig. 10-Down). Noncytotoxic
remnants of VLDL produced in the presence of HDL consisted mostly of

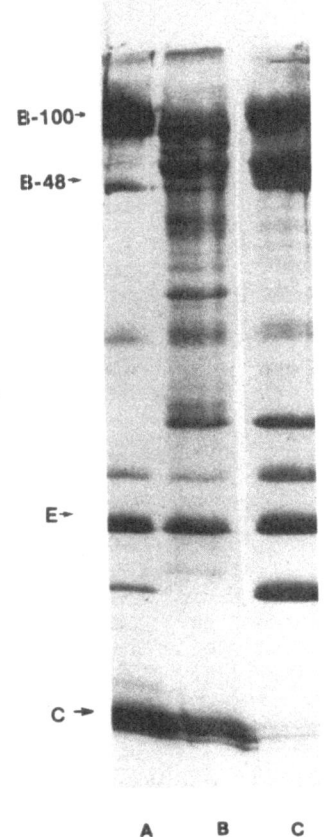

B-100→

B-48→

E→

C→

A   B   C

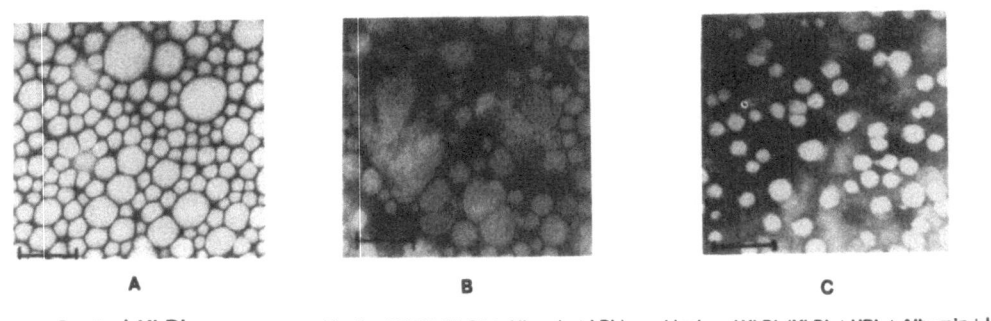

A   B   C

Control VLDL   Lipolized VLDL (VLDL + Albumin + LPL)   Lipolyzed VLDL (VLDL + HDL + Albumin + LPL)

Fig. 10. Electron micrographs (Up) and SDS gradient gel analysis of apolipoproteins (Down) of control VLDL (A), VLDL lipolyzed in the presence
of albumin alone (B) or VLDL lipolyzed in the presence of albumin
and HDL (C).

spheroidal particles smaller than control VLDL and retained little or no apo Cs of control VLDL (Fig. 10).

The presence of apo Cs in the remnant fraction containing liposome-like vesicles and the absence of liposome-like vesicles in the remnant fraction produced in the presence of HDL suggests that liposome-like vesicular particles are most likely the surface remnant products of VLDL, and these particles may be cleaned by HDL. The presence of cytotoxicity in the remnant fraction containing liposome-like vesicular particles, absence of cytotoxicity in the remnant fraction containing no vesicular particles (Fig. 9-10) and the association of most potent cytotoxicity with the remnant particles having free cholesterol, and phospholipid-rich and core lipid-poor particles in the lipolyzed HTG serum (Table 1 and Fig. 7) further suggest that the lipolytic surface remnants of TG-rich lipoproteins may contain the cytotoxic components of lipolyzed HTG serum.

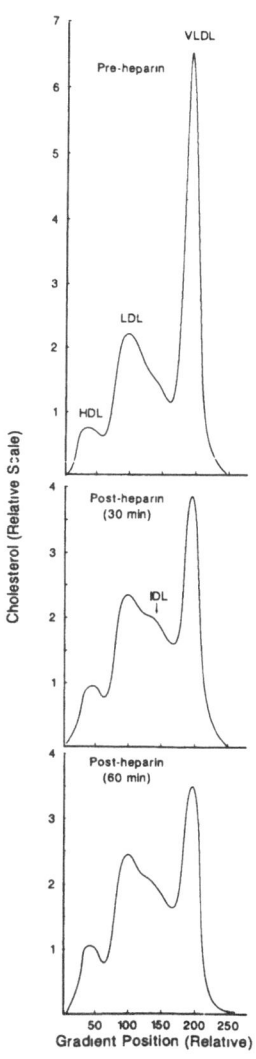

Fig. 11. Lipoprotein cholesterol profiles of preheparin serum and post-heparin serum from a moderately severe HTG subject.

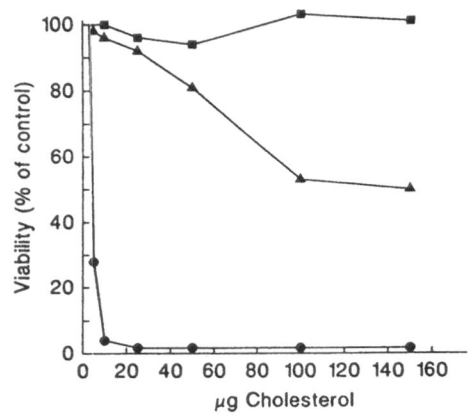

Fig. 12. Viability of macrophages incubated with preheparin serum (■-■),
or postheparin serum obtained at 30 min (●-●) or 60 min (▲-▲)
after an injection of heparin.

CYTOTOXICITY TO MACROPHAGES OF *IN VIVO* LIPOLYZED SERUM

Heparinization induces the lipolysis of TG-rich lipoproteins in circulating blood. Heparin injection into moderately severe HTG subject resulted
in hydrolysis of about 68% and 71% of serum TG within 30 and 60 min. after
injection, respectively. Lipoprotein cholesterol profiles of pre- and post-
heparin serum showed that about 40-50% of VLDL cholesterol in serum were
moved into the intermediate density lipoproteins (IDL), LDL and HDL density
regions (Fig. 11). When pre- or postheparin serum was added to cultured
macrophages, the postheparin serum obtained at 30 min. was highly cytotoxic
and foam-cell inducing; > 90% of the macrophages were killed with 10 µg
(cholesterol) of serum (Fig. 12). The preheparin serum at five times the
concentration of postheparin serum was not cytotoxic to the cells. The
postheparin serum obtained at 60 min was only marginally cytotoxic to the
cells (Fig. 12).

These results suggest that the cytotoxic components of lipolyzed serum
can be produced *in vivo*, but are removed rapidly from the circulation. The
cytotoxicity observed with postheparin serum support an *in vivo* significance
for remnant-associated cytotoxicity.

SUMMARY

None of more widely accepted theories of atherogenesis can explain all
the more pertinent features of atherosclerosis: a) foam cell formation; b)
endothelial cell stress/injury; c) protective effect of HDL; d) atherogenicity of triglyceride-rich lipoproteins; e) the vesicular nature of early
lipid deposits in atherosclerosis, f) dissociation of diet risk from the
risk due to elevation in plasma cholesterol; or g) correlation of postprandial lipemia with CAD risk. The data obtained from our studies provide a new
theory of atherogenesis. This theory is that: a) lipolytic surface remnants
of TG-rich lipoproteins may represent a major class of atherogenic lipoproteins which are exacerbated during postprandial hyperlipidemia; b) clearance
of these surface remnants by HDL *in vivo* may be one important way that HDL
prevents atherosclerosis; c) excess surface remnants may be linked to delayed clearance of potentially atherogenic core remnants, directly linked to
atherogenicity via surface remnant-mediated cytotoxicity to cells of the
artery wall and/or linked to the deposit of unesterified cholesterol-rich
vesicles in early atherosclerosis.

An appealing aspect of this hypothesis is that it can account for several unexplained features of atherosclerosis, such as anatomic differences in susceptibility to atherosclerosis in the vascular tree. the preference of early atherosclerosis in humans to the region of the coronary artery subjected to low hemodynamic shear stresses, and the vesicular nature of lipid deposits in early atherosclerosis.

REFERENCES

1. R. Ross, N. Engl. J. Med., 314:295 (1986).
2. A. Faggiotto, R. Ross, and L. Harker, Arteriosclerosis, 4:323 (1984).
3. R. Ross and L. Harker. Science. 193:1094 (1976).
4. R. Ross. Arteriosclerosis. 1:293 (1981).
5. T. Henriksen. L. A. Evensen. and B. Carlander. J. Lab. Clin. Lab. Invest., 39:361 (1979).
6. S. A. Evensen. K. S. Galdal. and E. Nilsen, Atherosclerosis. 49:23 (1983).
7. J. R. Hessler. D. W. Morel, L. J. Lewis. and G. M. Chisolm, Arteriosclerosis, 3:215 (1983).
8. S. H. Gianturco. S. G. Eskin. L. T. Navarro. C. J. Lahart, L. C. Smith, and A. M. Gotto. Biochim Biophys. Acta, 618:143 (1980).
9. B. W. Arbogast. L. R. Gill. and H. A. Schwertner, Atherosclerosis. 57:75 (1985).
10. A. Szebzeklik and R. J. Gryglewski, Artery. 7:488 (1980).
11. Y. Sata. N. Hotta. N. Sakamoto, S. Matsuoka. N. Ohisi. and K. Yagi, Biochem. Med., 21:104 (1979).
12. R. T. Wall, J. M. Harlan, L. A. Harker, and G. E. Striker,, Thromb. Res., 17:753 (1980).
13. B. H. Chung. J. P. Segrest, K. Smith, F. M. Griffin. and C. G. Brouillette, J. Clin. Invest.. 83:1363 (1989).
14. M. T. Speidel. F. M. Booyse. A. Abrams, M. A. Moore. and B. H. Chung. Thromb. Res., 58:251 (1990).
15. H. S. Kruth, D. L. Fry, Exp. Mol. Pathol., 40:288 (1984).
16. V. Simionescu, E. Vasile. F. Lupu, G. Popescu, and M. Simionescu. Am. J. Pathol., 123:109 (1986).
17. J. S. Frank and A. M. Fogelman. J. Lipid Res., 30:967(1989).
18. F. F. Chao. C. J. Branchette-Mackie, Y. T. Chen, B. F. Dickens, E. Berlin, L. M. Amende, S. I. Skarlotos, and H. S. Kruth. Am. J. Pathol., 136:169 (1990).

THE ROLE OF OXIDIZED LDL IN ATHEROSCLEROSIS

Joseph L. Witztum

Department of Medicine
University of California, San Diego
La Jolla, CA 92093

INTRODUCTION

Atherosclerosis is a complex disease involving a variety of circulating blood components, lipoproteins, and their interactions with the cells and proteins of the extracellular matrix of the artery wall. The end stage lesion, which results in the occlusion of a vessel and/or provides a thrombogenic surface which leads to intravascular thrombosis is frequently an acellular, fibrotic section of tissue reflecting the end stage of a complicated inflammatory process. The atherosclerotic plaque contains the cellular debris of dead and decaying cells and a variety of substances trapped amidst this debris, including cholesterol, chiefly in the form of precipitated crystals[1]. While many pathogenic mechanisms must be involved in transforming normal arterial tissue, which contains only a one- or two-layer thick intimal layer, into this highly thickened intima, it is nevertheless clear that deposition of cholesterol within the intima plays a central role in the pathogenesis of the atherosclerotic plaque. Because arterial wall cholesterol is derived almost exclusively from circulating lipoproteins, chiefly in the form of low density lipoproteins (LDL), research must focus on trying to understand the sequence of events by which the plaque develops and by which LDL accumulates[2-5]. There are many lines of evidence that support the hypothesis that circulating cholesterol, principally in the form of LDL, is central to the atherogenic process and that without "sufficient" plasma LDL levels, the atherogenic process cannot proceed[4,6].

Hypercholesterolemia, Hypocholesterolemia, Hypertriglyceridemia
Edited by C.L. Malmendier *et al.*, Plenum Press, New York, 1990

Trying to understand the pathogenesis of the advanced plaque by studying this lesion alone is analogous to attempts to understand the etiology of end stage renal disease by studying only the shrunken fibrotic kidney found at the end of the disease process. It is unlikely that therapies designed to reverse the end stage atherosclerotic plaque are likely to be as useful as are therapies designed to inhibit the initial steps, or at least those steps involved in the progression of the initial lesion. In order to develop such strategies, however, we must understand the natural history of the lesion and the pathogenetic events from the earliest stages of its development. Pathologists have long recognized that several stages of experimental and human atherosclerosis occur, namely the fatty streak, the transitional lesion and the plaque[5,7]. Fatty streaks are found in the arterial beds of even young individuals in the industrialized world and most investigators believe that these are the precursors of the transitional and more advanced lesions[1,4,5]. If this is true, then a detailed study of the pathogenesis of the fatty streak should give insights that might enable us to develop successful strategies to prevent its occurrence and/or inhibit its progression.

It is beyond the scope of this report to document the early events in atherosclerosis, but there are a number of excellent reviews currently available[5,7]. Most investigators now believe that the earliest event, at least as modeled by cholesterol-fed rabbits and nonhuman primates, is the initial focal accumulation of cholesterol in subintimal sites that are later predestined to develop fatty streaks at a time when other regions within the same arterial tree are more resistant[2-3]. It should be noted that this occurs under an intact endothelium and prior to any known morphologic or even metabolic alteration in the endothelium. Consequently, monocytes begin to attach to the overlying endothelium and to migrate into the subintimal space[8-11] possibly at a greater rate in those sites where the subintimal accumulation of LDL has occurred. This chemoattraction is likely to be produced both by release of chemoattractants directly from the modified LDL found in the subintimal space (as described below), as well as by alterations in the endothelium, which in turn may be produced by products of modified LDL. The surface of the endothelium, which normally repels monocytes and other leukocytes, now contains surface ligands, which serve to specifically bind circulating monocytes[12]. In addition, substances released by activated endothelial cells, such as MCP-1, may serve as a further chemoattractant bringing monocytes into the subintimal space[13]. As monocytes take up residence within the intima they phenotypically modulate into macrophages. Subsequently, they begin to take up LDL and eventually become the classic "foam cell," which is due to enrichment of cholesterol esters[9-11]. As more monocytes enter into these focal accumulations, and more foam cells are generated, their volume increases and the endothelial cells overlying them may be eventually stretched to the point of rupture, now exposing the underlying macrophages to circulating blood elements. At this stage, a variety of new mechanisms may become involved, as originally envisioned in the "response to injury hypothesis[5]." For example, platelets may attach to the exposed monocytes as well as to the exposed structural components of the artery wall. Platelet aggregation and release may then occur leading to release of a variety of growth factors, such as PDGF, which serves to both recruit smooth muscle cells into the intimal layer, as well as to stimulate their proliferation[14]. At this point, events leading to initiation of the proliferative phase of the lesion may have begun.

Atherogenecity of Oxidized LDL

The major focus of the remainder of this discussion will be on the mechanisms by which LDL uptake by the macrophage occurs and the consequences. Several lines of evidence suggest that macrophage uptake of LDL does not occur to any significant extent by way of the classic LDL receptor pathway[15]. Exposure of cultured macrophages to LDL does not lead to foam cell formation, which is probably due to downregulation of LDL receptors, as occurs in other cells. Humans with homozygous familial hypercholesterolemia, as well as the receptor-deficient WHHL rabbit, develop premature and severe atherosclerosis, with marked foam cell formation, despite a total lack of LDL receptor activity. What then accounts for foam cell formation? Brown, Goldstein and coworkers first suggested that it was not native LDL that was taken up by macrophages, but rather a modified or denatured LDL[16]. Reasoning that the primary function of macrophages is to remove denatured proteins, they showed that a chemical modification of LDL, acetylation, led to a negatively charged and modified LDL that now had a rapid uptake in cultured macrophages, occurring by a pathway distinct from that of the LDL receptor. This pathway was saturable, not downregulated, and in fact could promote foam cell formation. This "acetyl LDL receptor" pathway was found on monocyte-derived macrophages, Kupffer cells and some endothelial cells. This pathway also could bind other chemically modified forms of LDL, such as malondialdehyde-modified LDL, which also is negatively charged[17,18]. This strategy has prompted a detailed search by many laboratories to find modifications of LDL that would enhance its atherogenicity. We now recognize a number of such modifications that theoretically enhance the atherogenicity of LDL[4] by somehow modifying native LDL. A paradigm guiding research in this area is shown in Fig. 1. In 1981 Henriksen and coworkers in La Jolla found that when LDL was incubated with cultured endothelial cells, and subsequently reisolated, it was taken up by cultured macrophages 2-4 times more rapidly than native LDL, and this uptake was partially competed for by acetyl LDL[19]. More recent data suggests that a portion of the uptake and degradation of EC-LDL also occurs by way of yet another alternative high affinity receptor[20]. The recent cloning of the acetyl-LDL receptor demonstrates its complex structure and it remains to be demonstrated whether these are one or a family of modified LDL receptors[21] or at least a modification of the acetyl LDL receptor[20]. Subsequently, investigators have shown that smooth muscle cells, monocyte-derived macrophages, and even fibroblasts, under the right conditions, can all effect modification of LDL that will lead to its enhanced uptake by macrophages[4].

It was Steinbrecher *et al.* who first demonstrated that the endothelial cell modified LDL occurred as a result of cell-induced initiation of peroxidation of the polyunsaturated fatty acids present in the LDL[22]. This process could be inhibited by the presence of antioxidants in the medium, such as butylated hydroxytoluene (BHT) or vitamin E. Morrell *et al.*[23] and Heinecke *et al.*[24] showed similar results in human smooth muscle cells. Oxidative modification was dependent on low concentrations of heavy metals in the medium, such as copper, and could be completely inhibited by chelators such as EDTA. During oxidative modification there was extensive conversion of lecithin to lysolecithin, the result of a phospholipase A_2 activity intrinsic to LDL, and it was shown that inhibitors of phospholipase A_2 activity blocked the oxidative modification of LDL as well[25]. In addition, a large number of other as yet uncharacterized polar fatty acid and sterol products are generated that may have important biological consequences.

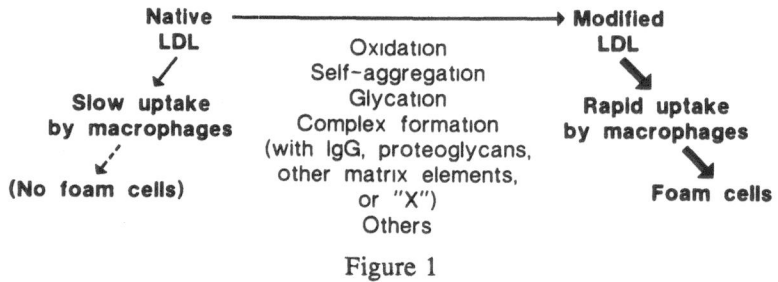

Figure 1

We also know that subsequent to the initiation of the peroxidation of fatty acids, there is the generation and release of a variety of fatty acid oxidation fragments, such as malondialdehyde (MDA) and 4-hydroxynonenal (4-HNE)[26]. These highly reactive fragments can complex with apoprotein B, forming stable lipid-protein adducts. In addition, apoprotein B is cleaved into multiple fragments presumably due to oxidative attack. As a result of these and other changes, the surface charge of the particle becomes more negative, neoepitopes are generated on LDL, and in some manner these changes lead to enhanced recognition by one or more receptors on macrophages.

In addition to having an enhanced recognition and uptake by macrophages, presumably leading to net cholesterol ester deposition over time and the generation of foam cells, oxidized LDL may contribute to atherogenesis by other mechanisms as well. In large part, these other effects relate to the consequences of the variously oxidized fatty acid fragments and sterols formed during lipid peroxidation. These polar products can diffuse out of the LDL into adjacent cells or structures where they may exert important biological effects. For example, Quinn et al.[27] have shown that LysoPC can account in part for the fact that oxidized LDL is a potent chemotactic factor for monocytes, as noted above. Furthermore, once monocytes have phenotypically modulated into macrophages, products of oxidized LDL serve to inhibit cellular movement, thus inhibiting their egress from the lesion. Thus, oxidized LDL serves both to attract monocytes into the lesion, and to trap macrophages within the subintimal space. In addition, the cytotoxicity of oxidized LDL depends on the oxidation of LDL lipids[23,28]. Not only may these products cause direct cellular damage, but they may have other cytotoxic or cytopathic effects on the cells that play an important role in the atherogenic process. For example, polar oxidation products of minimally modified LDL, i.e., LDL having undergone the initial stages of oxidative modification but not yet extensive enough to have led to receptor-mediated uptake in the macrophage can lead to activation of endothelial cells[13], affecting transcriptional events, causing chemotactic and even mitogenic substances to be released by endothelial cells. In addition, products from oxidized LDL, such as LysoPC, have recently been shown to alter endothelial responses to agonists that normally release endothelial-dependent relaxing factors[29]. Thus, oxidized LDL could have profound effects on vasomotor properties of the arteries and provide a mechanism by which changes in plasma LDL levels may have fairly prompt effects on coronary artery hemodynamics. Thus, the presence of oxidized LDL in the artery wall could have profound consequences for the atherogenic process.

Evidence for Presence of Oxidized LDL *in vivo*

The data above suggest a variety of ways in which oxidized LDL could be atherogenic. However, it is important to demonstrate that oxidized LDL occurs *in vivo*. Our group in La Jolla have recently developed a number of lines of evidence that in fact suggest that oxidized LDL does exist in the artery wall in rabbits and in man[4]. First, LDL has been gently eluted from arterial sections taken from WHHL rabbits, as well as from aortic sections obtained from human organ donors[30-31]. The LDL isolated from the atherosclerotic aortic sections, but not from normal arteries, had many of the physical properties found in LDL oxidized *in vitro*. Thus, lesion LDL had increased electrophoretic mobility, contained more free cholesterol, had a higher percentage of LysoPC in the total phospholipids compared with plasma LDL isolated simultaneously, and had extensive fragmentation of apo B. These data are summarized in Table 1.

A second line of evidence comes from the use of immunologic probes. A variety of immunological reagents specific for oxidation-specific epitopes present in

Table 1. Properties of Oxidized LDL, of LDL Isolated from Normal Intima and LDL Isolated from Atherosclerotic Lesions Relative to Those of Native Plasma LDL

	In Vitro Oxidized LDL[a]	Lesion LDL	Normal Intimal LDL
Negative charge[b]	↑↑↑	↑↑	↑
Density	↑	↑	→
Lysolecithin content	↑	↑	→
Free cholesterol content	↑	↑↑	→
Linoleate in cholesterol esters and phospholipids	↓↓	N.D.	↓
Fragmentation of apo B-100	+++	++	+
MDA and 4-HNE epitopes in apo B and its fragments	+	+	-
Chemotactic for monocytes	+	+	–
Degradation in macrophages	↑↑	↑↑	↑
Degradation in fibroblasts	↓	↓	N.D.

N.D. = not determined.

[a] Copper oxidized.

[b] As determined by the electrophoretic mobility in agarose electrophoresis (0.05 M barbital buffer, pH 8.6). (↑ increased, ↓ decreased, → unchanged, - absent)

Reproduced with permission from Ref. 45.

oxidized LDL but not in native LDL were prepared[32]. These reagents were generated using extensively modified forms of LDL as models of oxidized LDL, i.e., MDA-conjugated LDL and 4-HNE-conjugated LDL. The conjugated forms of LDL were injected into homologous species so that antibodies against native LDL were not generated. Thus, MDA-modified guinea pig LDL was injected into guinea pigs, or MDA-modified mouse LDL was injected into mice. Thus, antisera and monoclonal antibodies were produced that were specific to lipid-protein adducts, which we refer to as "oxidation-specific epitopes[32]." Western blot analysis of isolated lesion LDL demonstrated that both intact apo B as well as apo B fragments reacted strongly with these antisera, but LDL isolated from plasma or from normal sections of human intima did not react[30-31].

Third and most importantly, it was shown that LDL isolated from atherosclerotic aortic sections had enhanced uptake in macrophages and that this uptake was competed for by a variety of ligands known to compete with oxidized LDL for macrophage uptake. Furthermore, lesion LDL, but not plasma LDL, could promote cholesterol ester formation in cultured macrophages and was a potent chemotactic substance for monocytes[31].

Fourth, when the antibodies described above were used to immunostain the intact artery, there was extensive localization of the oxidation-specific epitopes to atherosclerotic sections of the aorta but not normal aorta, confirming the *in vivo* presence of these oxidation products. Haberland *et al.*[34] and Boyd *et al.*[35] have also recently provided immunocytochemical evidence for the presence of epitopes of oxidized LDL in rabbit arterial wall.

Fifth, we have shown that autoantibodies, at least to one such epitope of oxidized LDL, MDA lysine adducts, are present in the serum of both rabbits and humans[30]. The logic for looking for such antibodies is derived from our previous demonstration that even minor modifications of autologous LDL renders it highly immunogenic, and in fact was used as the strategy to generate the oxidation-specific antisera and monoclonal antibodies used in the studies described above. Since the various neoepitopes generated during the process of oxidative modification of LDL were indeed present in the artery wall, we reasoned that these might result in the generation of autoantibodies to such modifications. In recently published studies, Rosenfeld *et al.*[33] have used such human autoantibodies to stain sections of rabbit atherosclerotic aorta. They observed very similar patterns of staining to that obtained with the induced MDA-lysine specific antisera prepared in rabbits and mice.

A sixth line of evidence supporting the *in vivo* presence of oxidized LDL comes from studies using an antioxidant compound, probucol. Parthasarathy *et al.*[36] showed several years ago that probucol, a commonly used hypolipidemic agent, was a potent antioxidant. Because it is highly lipophilic and transported in plasma in the core of lipoproteins, chiefly in LDL, it is strategically located to best prevent the oxidation of LDL lipids. When LDL was isolated from human subjects on conventional doses of probucol, their LDL were resistant to cell-induced modification of LDL when tested *in vitro*. To test the ability of probucol to inhibit the atherosclerotic process, Carew *et al.*[37] fed probucol to WHHL rabbits and demonstrated that sufficient concentrations existed in their LDL to protect it from *in vitro* oxidation. When probucol was fed to such rabbits for nine months it was shown conclusively that the probucol-treated animals had a marked reduction in their atherosclerosis that could not be explained by the small degree of cholesterol lowering produced by probucol. A control group, treated with lovastatin, although having greater degrees of cholesterol lowering, had far less protection against atherosclerosis. Kita *et al.*[38] have confirmed these results. In addition, Carew *et al.* clearly showed that the uptake and degradation of LDL was specifically inhibited in atherosclerotic lesions but not in normal portions of the aorta, observations consistent with a specific inhibition of macrophage uptake of oxidized LDL. Of course, it is also possible that probucol has other antiatherosclerotic effects, as recently it has been described to accelerate reverse cholesterol transport[39], as well as to inhibit interleukin I release by macrophages in culture[40].

Therapeutic Implications

As noted in Fig. 1, the new paradigm suggested is that oxidative modification of LDL is one post-secretory modification of LDL that may increase its atherogenicity. However, as noted in this figure, there are likely to be other modifications that are also relevant. For example, during the process of oxidative modification, a variety of neoepitopes are generated that result from oxidative modification of LDL, cross-linking of LDL to other matrix proteins, or even due to nonenzymatic glycosylation. In response to these neoantigens a variety of autoantibodies may form. Such autoantibodies could form immune complexes with their target antigens within the artery wall, fix complement, leading to enhanced uptake of such complexes by macrophages, in part mediated by way of Fc and complement receptor mediated mechanisms. In addition, the presence of immune complexes could iniate a cascade of immunological mechanisms

360

resulting in activation of a variety of immune cells, leading to a classic inflammatory response with attendant tissue destruction, as recently reviewed elsewhere[41]. Such processes, as well as a variety of other kinds of modifications, indicated in Figure 1, all may contribute to alterations of the physical and chemical properties of LDL, changing its immunogenicity, its recognition by macrophages, and thereby enhancing its overall atherogenicity.

It is important for us to learn the details of the mechanisms by which such modifications occur because they offer the potential for therapeutic intervention. For example, inhibiting the inflammatory process may be beneficial for reasons noted above. We also need to understand the mechanism(s) by which cells lead to the initiation of lipid peroxidation in LDL. Recently, Rankin et al.[42] have demonstrated that 15-lipoxygenase activity is crucial for the ability of macrophages to effect the oxidative modification of LDL. Ylä-Herttuala et al.[43] have recently demonstrated, using in situ hybridization and immunocytochemistry, that 15-lipoxygenase mRNA and protein can be demonstrated in the macrophage-rich area of atherosclerotic lesions and colocalizes with the oxidation-specific epitopes described above. This strongly suggests that lipoxygenase activity is involved in the oxidative modification of LDL. If this is so, then interventions to specifically inhibit such lipoxygenase activity, should also result in amelioration of the atherosclerotic process.

Yet another therapeutic strategy would be to inhibit the ability of LDL to be a substrate for lipid peroxidation in the first place. Since polyunsaturated fatty acids are the primary lipids subject to lipid peroxidation, replacing them with monounsaturated fatty acids, which would not sustain propagation of lipid peroxidation, should theoretically result in decreased lipid peroxidation and oxidative modification of LDL. Indeed, we have recently demonstrated that feeding a high oleic acid rich diet to rabbits protects their LDL from in vitro mediated oxidative modification[44]. Similarly, supplementing the other natural antioxidants found in LDL such as vitamin E and various carotenoids, should also serve to protect LDL from oxidative modification.

In summary, we need to continue to accumulate evidence that various postsecretory modifications of LDL enhance its atherogenicity. Specifically, we need to continue to accumulate evidence that oxidative modification is involved in the atherogenic process and we need to continue to learn the mechanisms by which such oxidative modification occurs. Thus, new strategies can be developed to prevent oxidative modification and it can be anticipated that the successful development of such modalities may play an important adjunctive role in the prevention and treatment of atherosclerosis in addition to therapies designed to lower plasma cholesterol.

REFERENCES

1. D.M. Small, Progression and regression of atherosclerotic lesions. Insights from lipid physical biochemistry, Arteriosclerosis 8:103 (1988).

2. D.C. Schwenke DC, and T.E. Carew, Initiation of atherosclerotic lesiona in cholesterol-fed rabbits. I. Focal increases in arterial LDL concentration precede development of fatty streak lesions, Arteriosclerosis 9:895 (1989).

3. D.C. Schwenke, and T.E. Carew, Initiation of atherosclerotic lesions in cholesterol-fed rabbits. II. Selective retention of LDL vs. selective increases in LDL permeability in susceptible sites of arteries, Arteriosclerosis 9:908 (1989).

4. D. Steinberg, S. Parthasarathy, T.E. Carew et al., Beyond cholesterol: Modifications of low density lipoprotein that increase its atherogenicity, New Engl. J. Med. 320:915 (1989).

5. R. Ross, The pathogenesis of atherosclerosis - an update, New Engl. J. Med. 314:418 (1986).

6. J.L. Witztum, Current approaches to drug therapy for the hypercholesterolemic patient, Circulation 80:1101 (1989).

7. A.M. Gown, T. Tsukada, and R. Ross, Human atherosclerosis: Immunocyto-chemical analysis of the cellular composition of human atherosclerotic lesions, Am. J. Pathol. 125:191 (1986).

8. R.G. Gerrity, H.K. Naito, M. Richardson et al., Dietary induced atherogenesis in swine, Am .J. Pathol. 95:775 (1979).

9. A. Faggiotto, R. Ross, and L. Harker, Studies of hypercholesterolemia in the nonhuman primate. I. Changes that lead to fatty streak formation, Arterio-sclerosis 4:323 (1984).

10. N.M. Aqel, R.Y. Ball, H. Waldman et al., Monocytic origin of foam cells in human atherosclerotic plaques, Atherosclerosis 53:265 (1984).

11. M.E. Rosenfeld, T. Tsukada, A.M. Gown et al., Fatty streak initiation in the WHHL and comparably hypercholesterolemic fat-fed rabbits, Arteriosclerosis 1:9 (1987)

12. M.P. Bevilacqua, J.S. Pober, M.E. Wheeler et al., Interleukin 1 acts on cultured human vascular endothelium to increase the adhesion of polymorphonuclear leukocytes, monocytes, and related leukocyte cell lines, J. Clin. Invest. 76:2003 (1985).

13. J.A. Berliner, M.C. Territo, A. Sevanian *et al.*, Minimally modified low density lipoprotein stimulates monocyte endothelial interactions, J. Clin. Invest. 85:1260 (1990).

14. R. Ross, J. Masuda, E.W. Raines *et al.*, Localization of PDGF-B protein in macrophages in all phases of atherogenesis, Science 248:1009 (1990).

15. M.S. Brown, J.L. Goldstein, Lipoprotein metabolism in the macrophage: Implications for cholesterol deposition in atherosclerosis, Ann. Rev. Biochem. 52:223 (1983).

16. J.L. Goldstein, Y.K. Ho, S.K. Basu *et al.*, Binding site on macrophages that mediates uptake and degradation of acetylated low density lipoprotein, producing massive cholesterol deposition, Proc. Natl. Acad. Sci. USA 76:333 (1979).

17. A.M. Fogelman, I. Schechter, J. Seager *et al.*, Malondialdehyde alteration of low density lipoprotein leads to cholesterol accumulation in human monocyte-macrophages, Proc. Natl. Acad. Sci. USA 74:2214 (1980).

18. M.E. Haberland, A.M. Fogelman, and P.A. Edwards, Specificity of receptor-mediated recognition of malondialdehyde-modified low density lipoproteins, Proc. Natl. Acad. Sci. USA 79:1712 (1982).

19. T. Henriksen, E.M. Mahoney, and D. Steinberg, Enhanced macrophage degradation of low density lipoprotein previously incubated with cultured endothelial cells: Recognition by the receptor for acetylated low density lipoproteins, Proc. Natl. Acad. Sci. USA 78:6499 (1981).

20. C.P. Sparrow, S. Parthasarathy, and D. Steinberg, A macrophage receptor that recognizes oxidized LDL but not acetylated LDL, J. Biol. Chem. 264:2599 (1989).

21. T. Kodama, M. Freeman, L. Rohrer *et al.*, Type I macrophage scavenger receptor contains alpha-helical and collagen-like coiled coils. Nature 343:531 (1990).

22. U.P. Steinbrecher, S. Parthasarathy, D.S. Leake *et al.*, Modification of low density lipoprotein by endothelial cells involves lipid peroxidation and degradation of low density lipoprotein phospholipids, Proc. Natl. Acad. Sci. USA 83:3883 (1984).

23. D.W. Morel, P.E. DiCorleto, and G.M. Chisolm, Endothelial and smooth muscle cells alter low density lipoprotein in vitro by free radical oxidation, Arteriosclerosis 4:357 (1984).

24. J.W. Heinecke, H. Rosen, and A. Chait, Iron and copper promote modification of low density lipoprotein by human arterial smooth muscle cells in culture, J. Clin. Invest. 74:1890 (1984).

25. S. Parthasarathy, U.P. Steinbrecher, J. Barnett *et al*, Essential role of phospholipase A_2 activity in endothelial cell-induced modification of low density lipoprotein, Proc. Natl. Acad. Sci. USA 82:3000 (1985).

26. H.F. Hoff, J. O'Neil, G.M. Chisolm 3d *et al.*, Modification of low density lipoprotein with 4-hydroxynonenal induces uptake by macrophages, Arteriosclerosis 9:538 (1989).

27. M.T. Quinn, S. Parthasarathy, L.G. Fong *et al.*, Oxidatively modified low density lipoproteins: A potential role in recruitment and retention of monocyte/macrophages during atherogenesis, Proc. Natl. Acad. Sci. USA 84:2995 (1987).

28. J.R. Hessler, A.L. Robertson, Jr, and G.M. Chisolm, LDL-induced cytotoxicity and its inhibition by HDL in human vascular smooth muscle and endothelial cells in culture, Atherosclerosis 32:213 (1979).

29. K. Kugiyama, S.A. Kerns, J.D. Morrisett *et al.*, Impairment of endothelium-dependent arterial relaxation by lysolecithin in modified low-density lipoproteins, Nature 344:160 (1990).

30. W. Palinski, M.E. Rosenfeld, S. Ylä-Herttuala *et al.*, Low density lipoprotein undergoes oxidative modification in vivo, Proc. Natl. Acad. Sci. USA 86:1372 (1989).

31. S. Ylä-Herttuala, W. Palinski, M.E. Rosenfeld *et al.*, Evidence for the presence of oxidatively modified low density lipoprotein in atherosclerotic lesions of rabbit and man. J Clin Invest 84:1086 (1989).

32. W. Palinski, S. Ylä-Herttuala, M.E. Rosenfeld, Antisera and monoclonal antibodies specific for epitopes generated during the oxidative modification of low density lipoproteins, Arteriosclerosis 10:325 (1990).

33. M.E. Rosenfeld, W. Palinski, S. Ylä-Herttuala *et al.*, Distribution of oxidized proteins and apolipoprotein B in atherosclerotic lesions of varying severity from WHHL rabbits: Immunocytochemical analysis using antibodies generated against modified and native LDL, Arteriosclerosis 10:336 (1990).

34. M.E.Haberland, D. Fong, and L. Cheng, Malondialdehyde-altered protein occurs in atheroma of Watanabe heritable hyperlipidemic rabbits, Science 241:215 (1988).

35. H.C. Boyd, A.M. Gown, G. Wolfbayer *et al.*, Direct evidence for a protein recognized by a monoclonal antibody against oxidatively modified LDL in atherosclerotic lesions from Watanabe Heritable Hyperlipemic rabbit, Am. J. Pathol. 135:1372 (1989).

36. S. Parthasarathy, S.G. Young, J.L. Witztum *et al.*, Probucol inhibits oxidative modification of low density lipoprotein, J. Clin. Invest. 77:641 (1986).

37. T.E. Carew, D.C. Schwenke, and D. Steinberg, Antiatherogenic effect of probucol unrelated to its hypocholesterolemic effect: Evidence that antioxidants in vivo can selectively inhibit low density lipoprotein degradation in macrophage-rich fatty streaks slowing the progression of atherosclerosis in the WHHL rabbit, Proc. Natl. Acad. Sci. USA 84:7725 (1987).

38. T. Kita, Y. Nagano, M. Yokode, Probucol prevents the progression of athero-sclerosis in Watanabe heritable hyperlipidemic rabbit, an animal model for familial hypercholesterolemia, Proc. Natl. Acad. Sci. USA 84:5928 (1987).

39. G. Franceschini, M. Sirtori, V. Vaccarino *et al.*, Mechanisms of HDL reduction after probucol. Changes in HDL subfractions and increased reverse cholesteryl ester transfer. Arteriosclerosis 9:462 (1989).

40. G. Ku, N.S. Doherty, L.F. Schmidt *et al.*, Ex vivo lipopolysaccharide-induced interleukin-1 secretion from murine peritoneal macrophages inhibited by probucol, a hypocholesterolemic agent with antioxidant properties. FASEB J. 4:1645 (1990).

41. G.K. Hansson, L. Jonasson, P.S. Seifert *et al.*, Immune mechanisms in atherosclerosis, Arteriosclerosis 9:567 (1989).

42. S.M. Rankin, S. Parthasarathy, and D. Steinberg, Evidence for a dominant role of lipoxygenase(s) in the oxidation of LDL by mouse peritoneal macrophages. Submitted for publication (1990).

43. S. Ylä-Herttuala, M.E. Rosenfeld, S. Parthasarathy *et al.*, Colocalization of 15-lipoxygenase mRNA and protein with epitopes of oxidized low density lipoprotein in macrophage-rich areas of atherosclerotic lesions, Proc. Natl. Acad. Sci. USA, in press (1990).

44. S. Parthasarathy, J.C. Khoo, E. Miller *et al.*, Low density lipoprotein rich in oleci acid is protected against oxidative modification: Implications for dietary prevention of atherosclerosis, Proc. Natl. Acad. Sci. USA 87:3894, 1990.

45. S. Ylä-Herttuala, W. Palinski, M.E. Rosenfeld *et al.*, Lipoproteins in normal and atherosclerotic aorta, Eur. Heart J. II, Suppl. E, in press (1990).

PROBUCOL AND ITS MECHANISMS FOR REDUCING ATHEROSCLEROSIS

Richard L. Jackson, Roger L. Barnhart and Simon J.T. Mao

Merrell Dow Research Institute
2110 E. Galbraith Road
Cincinnati, OH 45215 USA

INTRODUCTION

Probucol (1) is a marketed lipid-lowering drug used in the treatment of hypercholesterolemia, including forms of familial hypercholesterolemia. In hyperlipidemic subjects, probucol reduces total plasma cholesterol and low density lipoprotein (LDL)-cholesterol by 10 to 20% and high density lipoprotein (HDL)-cholesterol by 20 to 30% (2). The drug has little or no effect on triglyceride concentrations in hypercholesterolemic subjects, although Hattori et al. (3) reported a significant reduction in serum triglycerides with probucol treatment in patients with noninsulin-dependent diabetes mellitus. The mechanism by which probucol reduces plasma cholesterol is not known with certainty. The most convincing evidence in man is that it increases the fractional catabolic rate of LDL clearance (4,5) and enhances bile acid excretion (5,6). The fall in HDL-cholesterol with probucol treatment results from a decrease in the rate of synthesis of apo A-I and apo A-II (5,7), the major protein constituents of HDL. In addition to its lipid-lowering activity probucol is a potent antioxidant, and it is this property which prevents cell and metal ion-mediated oxidative modification of LDL (8). Mao et al. (9) synthesized a number of analogs of probucol that retained the antioxidant property but had little or no lipid-lowering activity. Even though lipid levels were unchanged, these analogs were effective in reducing atherosclerosis in modified Watanabe heritable hyperlipedemic (WHHL) rabbits, suggesting that other mechanisms besides lipid-lowering account for this antiatherosclerotic property. In this report, we review these mechanisms of probucol's action.

ANTIOXIDANT PROPERTIES OF PROBUCOL

Probucol has a structure that is dissimilar to other known lipid-lowering agents (Fig. 1). The molecule is a lipid-soluble, bis-tertiary butyl phenol. Thus, the phenolic-OH groups have easily donatable hydrogens and it is this property of probucol that accounts for its antioxidant activity. As is shown in Fig. 1, a carbon centered radical (R·), peroxyl radical (ROO·), alkoxy radical (RO·) or hydroxy radical (HO·) can abstract a hydrogen from probucol. Since a hydrogen atom has only one electron, this process converts probucol into a free radical itself, probucol-O·. However, since the bulky tertiary butyl groups shield the radical it is poorly reactive with water or other radicals. In addition, the unpaired electron delocalizes into the benzene ring forming a quinone structure (Fig. 1). The net effect of probucol's action, as it relates to its antiatherosclerotic property, is its free radical chain-terminating activity, which results in inhibition of lipid peroxidation.

Hypercholesterolemia, Hypocholesterolemia, Hypertriglyceridemia
Edited by C.L. Malmendier *et al.*, Plenum Press, New York, 1990

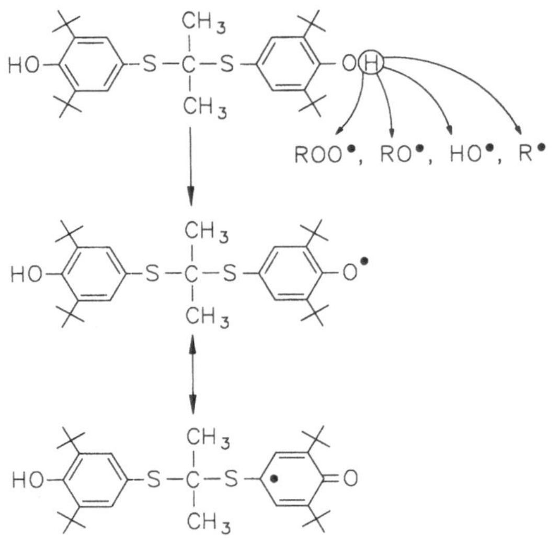

Fig. 1. Structure of probucol and mechanism for its antioxidant activity.

Peroxyl radicals and cytotoxic aldehydes have been implicated in a number of disease states, including atherosclerosis (10-12). Fig. 2 shows a proposed scheme for the free-radical mediated generation of 4-hydroxynonenal (4-OH nonenal), malondialdehyde (MDA) and pentane from arachidonic acid. The free fatty acid, arachidonic acid, is produced by the actions of phospholipase A_2 on membrane or lipoprotein phospholipids or cholesteryl ester hydrolase on cholesteryl archidonate. The key step in lipid peroxidation is the abstraction of a hydrogen atom from the conjugated diene-4-arachidonic acid by the action of 15-lipoxygenase or another free radical, thus forming a carbon-centered lipid radical which then combines with molecular oxygen to yield the lipid hydroperoxide 15-HPETE which further breaks down to 15-hydroxy-eicosatetraenoic acid (15-HETE). As shown, 15-HPETE is further metabolized by 12-lipoxygenase, free radicals or metal ions to yield several end products of lipid

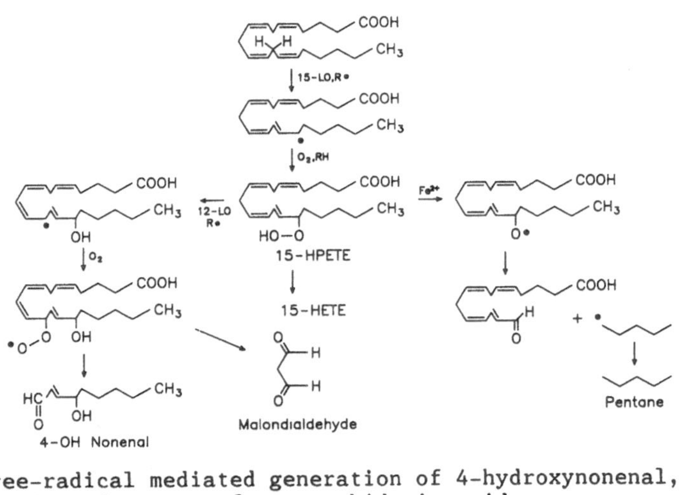

Fig. 2. Free-radical mediated generation of 4-hydroxynonenal, malondial-dehyde and pentane from arachidonic acid.

peroxidation. Some lipid aldehydes produced from this process, e.g., 4-hydroxynonenal, are extremely reactive and rapidly react with amino groups of lysine residues on enzymes and receptors resulting in cell and tissue damage. Because of their reactivity, it is difficult to quantitate free lipid aldehydes in tissue. However, Mao et al. (13) developed an enzyme-linked immunosor- bent assay for the MDA-lysine epitope and reported that the plasma level of MDA-Lys was 313 ± 187 nM in hyperlipidemic subjects; the antibody also detected MDA-protein adducts in atherosclerotic lesions of WHHL rabbits. That these aldehyde adducts may be derived from 15-HPETE (Fig. 2) is consistent with the recent findings of Simon et al. (14). These investigators developed a high performance liquid chromatography method (HPLC) for separating and quantitating all of the arachidonic acid-derived lipid products. The amount of 15-HETE in the aorta of normal rabbits was below the detection level of the method. However, in cholesterol-fed rabbits and WHHL rabbits the levels of 15-HETE ranged from 3 to 14 pmol/cm^2 aorta, suggesting that 15-lipoxygenase activity or free radicals are increased in atherosclerotic lesions.

CELLULAR MECHANISMS OF PROBUCOL'S ACTION

A key feature of the atherosclerotic process is the accumulation of foam cells in the lesion and proliferation of smooth muscle cells. Steinberg et al. (11) have proposed that oxidation of LDL and uptake of this modified lipoprotein by resident macrophages are the key steps in foam cell formation in the arterial wall. In this hypothesis, loss of endothelial cells is not a prerequisite for the initiation of the process. In fact, endothelial cells and smooth muscle cells secrete free radicals and various lipid peroxides of linoleic acid breakdown and these may be the major adducts for modification of LDL (11,15).

Fig. 3 summarizes some of the cellular mechanisms of the atherosclerotic process and sites where probucol may influence this process. LDL are exquisitely sensitive to oxidation (16-19). The major endogenous antioxidant in LDL is α-tocopherol. Lipid peroxidation commences once the lipoprotein particle is depleted of this antioxidant. Since probucol is carried in LDL, the drug protects against lipid peroxidation in vitro and in vivo (8,18-20). Furthermore, Barnhart et al. (16) showed a concentration-dependent relationship between LDL-probucol concentration and the onset of lipid peroxidation. This finding implies that the number of probucol molecules in the LDL particle is an important factor in the prevention of LDL oxidation and modification.

Another mechanism by which LDL may be modified in the arterial wall is by a covalent interaction between the lysine residues of apoliproprotein B with MDA. This product of lipid peroxidation is produced during the catabolism of arachidonic acid by blood platelets at the site of arterial injury (Fig. 3). Since MDA-modified LDL have been shown to bind to the scavenger receptor on macrophages (21), it is not necessary to postulate a free radical-mediated oxidation of LDL-lipids as a prerequisite for foam cell formation. There is no evidence to date suggesting that probucol effects either the production of cellular MDA or its modification of LDL.

A key cell type in the atherosclerotic process is the activated monocyte (Fig. 3). The arterial wall contains a number of chemotactic agents and presumably circulating monocytes migrate into the arterial intima under the influence of one of these factors. Steinberg et al. (11) have also shown that oxidatively modified LDL are potent chemo-attractants for circulating monocytes. The cellular mechanisms for conversion of a newly recruited resident monocyte in the arterial wall to an activated monocyte is not well detailed. As is illustrated in Fig. 3,

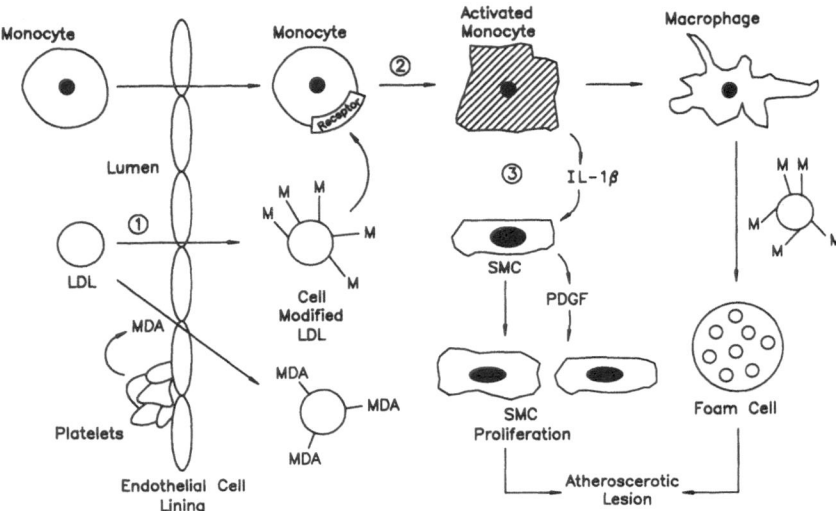

Fig. 3. Mechanisms for inhibition of atherosclerosis by probucol. 1. Inhibition of oxidative modification of LDL. 2. Inhibition of modified LDL-induced differentiation of monocytes into activated monocytes. 3. Inhibition of IL-1β expression by activated monocytes.

cell modified LDL are shown binding to a monocyte receptor. How the putative binding results in monocyte differentiation is only speculative, although Fostegård, et al. (22) have recently shown that Cu²⁺-oxidized LDL stimulate monocytes to differentiate into resident macrophages. In this process, probucol would have an indirect effect on monocyte activation by preventing cell-mediated LDL modification.

The hallmark of the atherosclerotic lesions is smooth muscle cell proliferation and foam cell accumulation (Fig. 3). A number of growth factors, including interleukin 1β (IL-1β) and platelet-derived growth factor (PDGF), are known to stimulate smooth muscle cell proliferation. Wang et al. (23) measured the transcriptional phenotypic expression of these proteins in a normal and an atheroscleroic carotid artery using polymerase chain reaction techniques. The number of molecules of IL-β mRNA per µg of total RNA was 100 in normal tissue as compared to 64000 in the diseased; the values for PDGF-B chain were 22000 and 76000. These findings are of particular interest since probucol inhibits IL-1β secretion from activated macrophages (24,25); one mechanism for this inhibition is a decrease in IL-1β gene transcription.

In conclusion, probucol is a potent antioxidant, and this property prevents oxidative modification of LDL and would account for the inhibition of foam cell formation. In addition, probucol is a potent inhibitor of IL-1β expression, and this activity could lead to a decrease in smooth muscle cell proliferation and inhibition of the atherosclerotic process.

ACKNOWLEDGMENTS

We wish to thank Ms. Mary Lynn Points for the preparation of this manuscript.

REFERENCES

1. J. W. Barnhart, J. A. Sefranka and D. D. McIntosh, Hypocholesterolemic effect of 4,4'-(isopropylidenedithio)-bis(2,6-di-t-butyl-
phenol) (probucol), Am. J. Clin. Nutr. 23:1229 (1970).

2. M. T. Buckley, K. L. Goa, A. H. Price and R. N. Brogden, Probucol,
a reappraisal of its pharmacological properties and therapeutic
use in hypercholesterolaemia, Drugs, 37:761 (1989).

3. M. Hattori, K. Tsuda, T. Taminato, S. Nishi, J. Fujita, K. Tsuji,
T. Kurose, G. Koh, Y. Seino and H. Imura, Effect of probucol on
serum lipids and apoproteins in patients with noninsulin-
dependent diabetes mellitus, Cur. Ther. Res., 42:967 (1987).

4. Y. A. Kesäniemi and S. M. Grundy, Influence of probucol on
cholesterol and lipoprotein metabolism in man, J. Lipid Res.
25:780 (1984).

5. P. J. Nestel and T. Billington, Effects of probucol on low density
lipoprotein removal and high density lipoprotein synthesis,
Atherosclerosis 38:203 (1981).

6. T. A. Miettinen, Mode of action of a new hypocholesteraemic drug
(DH-581) in familial hypercholesterolaemia, Atherosclerosis
15:163 (1972).

7. R. F. Atmeh, J. M. Stewart, D. E. Boag, C. J. Packard, A. R.
Lorimer and J. Shepherd, The hypolipidemic action of probucol: a
study of its effects on high and low density lipoproteins, J.
Lipid Res. 24:588 (1983).

8. S. Parthasarathy, S. G. Young, J. L. Witztum, R. C. Pittman and D.
Steinberg, Probucol inhibits oxidative modification of low
density lipoprotein, J. Clin. Invest. 77:641 (1986).

9. S. J. T. Mao, M. T. Yates, A. E. Rechtin, R. L. Jackson and W. A.
Van Sickle, Probucol and its derivatives prevent atherosclerosis
in modified Watanabe rabbits, (submitted).

10. B. Halliwell, Free radicals, reactive oxygen species and human
disease: a critical evaluation with special reference to
atherosclerosis, Br. J. Exp. Path. 70:737 (1989).

11. D. Steinberg, S. Parthasarathy, T. E. Carew, J. C. Khoo and J. L.
Witztum, Modifications of low-density lipoprotein that increase
its atherogenicity, N. Eng. J. Med. 320:915 (1989).

12. M. D. Stringer, P. G. Görög, A. Freeman, V. V. Kakkar, Lipid
peroxides and atherosclerosis, Br. Med. J. 298:281 (1989).

13. S. J. T. Mao, A. E. Rechtin and R. L. Jackson, Immunochemical
identification and determination of malondialdehyde in human
plasma, Adv. Life Sci. (in press).

14. T. C. Simon, A. N. Makheja and J. M. Bailey, Formation of 15-
hydroxyeicosatetraenoic acid (15-HETE) as the predominant
eicosanoid in aortas from Watanabe heritable hyperlipidemic and
cholesterol-fed rabbits, Atherosclerosis 75:31 (1989).

15. J. W. Heinecke, Free radical modification of low-density
lipoprotein: mechanisms and biological consequences, Free Rad.
Biol. Med. 3:65 (1987).

16. R. L. Barnhart, S. J. Busch and R. L. Jackson, Concentration-
dependent antioxidant activity of probucol in low density
lipoproteins in vitro: probucol degradation precedes lipoprotein
oxidation, J. Lipid Res. 30:1703 (1989).

17. H. Esterbauer, G. Jürgens, O. Quechenberger and E. Koller,
Autoxidation of human low density lipoprotein: loss of
polyunsaturated fatty acids and vitamin E and generation of
aldehydes, J. Lipid Res. 28:495 (1987).

18. L. R. McLean and K. A. Hagaman, Effect of probucol on the physical
properties of low-density lipoproteins oxidized by copper,'
Biochemistry 28:321 (1989).

19. W. Jessup, S. M. Rankin, C. V. De Whalley, J. R. S. Hoult, J. Scott
D. S. Leake, α-Tocopherol consumption during low-density-
lipoprotein oxidation, Biochem. J. 265:399 (1990).

20. D. W. Morel and G. M. Chisolm, Antioxidant treatment of diabetic rats inhibits lipoprotein oxidation and cytotoxicity, J. Lipid Res. 30:1827 (1989).

21. M. E. Haberland, A. M. Fogelman and P. A. Edwards, Specificity of receptor-mediated recognition of malondialdehyde-modified low density lipoproteins, Proc. Natl. Acad. Sci. USA 79:1712 (1982).

22. J. Frostegård, J. Nilsson, A. Haegerstrand, A. Hamsten, H. Wigzell and M. Gidlund, Oxidized low density lipoprotein induces differentiation and adhesion of human monocytes and the monocytic cell line U937, Proc. Natl. Acad. Sci. USA 87:904 (1990).

23. A. M. Wang, M. V. Doyle and D. F. Mark, Quantitation of mRNA by the polymerase chain reaction, Proc. Natl. Acad. Sci. USA 86:9717 (1989).

24. G. Ku, N. S. Doherty, J. A. Wolos and R. L. Jackson, Inhibition by probucol of interleukin 1 secretion and its implication in atherosclerosis, Amer. J. Cardiol. 62:77B (1988).

25. G. Ku, N. S. Doherty, L. F. Schmidt, R. L. Jackson and R. J. Dinerstein, Probucol, a hypocholesterolemic agent with antioxidant properties inhibits ex vivo lipopolysaccharide-induced interleukin 1 (IL-1) secretion from murine peritoneal macrophages FASEB J. 4:1645 (1990).

PROTEOGLYCANS, LIPOPROTEINS, AND ATHEROSCLEROSIS

S.R. Srinivasan[1,2], B. Radhakrishnamurthy[1,2],
P. Vijayagopal[1], and G.S. Berenson[1]

Departments of Medicine[1] and Biochemistry[2]
Louisiana State University School of Medicine
1542 Tulane Avenue, New Orleans, LA 70112

INTRODUCTION

Extracellular matrix proteoglycans play an important role in maintaining structural integrity and normal function of the arterial wall. These macromolecules provide structural links between fibrous and cellular elements, contribute to viscoelastic properties, regulate permeability and retention of plasma components through the matrix, control vascular cell growth, affect hemostasis and interact with lipoproteins (1,2). These functional properties indicate that proteoglycans are clearly involved in the pathogenesis of atherosclerosis.

This discussion will focus on the role of arterial wall proteoglycans in the extracellular and intracellular accumulation of plasma lipoproteins in atherosclerosis.

Nature of arterial wall proteoglycans

Proteoglycans are macromolecules in which sulfated glyosaminoglycan (GAG) chains are covalently linked to a protein core. A striking feature of proteoglycans is their polydispersity and heterogeneity with respect to size, charge density, and chemical characteristics of constituent GAG chains. Arterial wall contains at least two types of proteoglycans, chondroitin sulfate-dermatan sulfate proteoglycans and heparan sulfate proteoglycans. Chondroitin sulfate-dermatan sulfate proteoglycans are the major type in the aorta and can be isolated with varying proportions of dermatan sulfate to chondroitin sulfate. A chondroitin sulfate-dermatan sulfate proteoglycan isolated from bovine aorta contained condroitin 6-sulfate, chondroitin 4-sulfate and dermatan sulfate in a proportion of 8:3:1 (3). This proteoglycan formed large aggregates, a higher level of organization, with hyaluronic acid and link proteins. Heparan sulfate proteoglycans, on the other hand, were found to be firmly associated with elastin and to have potent hemostatic properties.

Proteoglycan changes in atherosclerosis

Earlier studies on alterations in proteoglycans in atherosclerosis were made with respect to their constituent GAG (4). It has been shown consistently that the tissue concentrations of hyaluronic acid and heparan

Hypercholesterolemia, Hypocholesterolemia, Hypertriglyceridemia
Edited by C.L. Malmendier *et al.*, Plenum Press, New York, 1990

373

sulfate tend to decrease, and chondroitin sulfates and dermatan sulfate tend to increase in atherosclerosis. These findings are very much the reverse of those noted during the experimental regression of atherosclerosis in nonhuman primates.

Recent studies on proteoglycan changes in human atherosclerotic aorta showed that proteoglycans form lesions have greater molecular size than the proteoglycans from uninvolved tissue (5). To explore this further, we

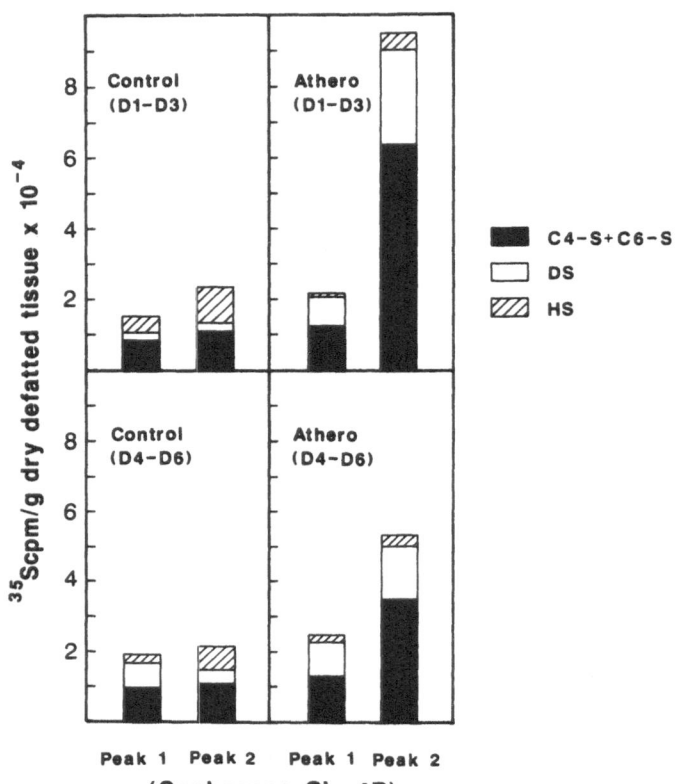

Fig. 1. Changes in ^{35}S-labeled proteoglycans of rabbit aorta in atherosclerosis. C4-S and C6-S, chondroitin 4- and 6-sulfate, respectively; DS, dermatan sulfate; HS, heparan sulfate (With permission, Ref 6).

studied the effect of experimental atherosclerosis on the composition of proteoglycans synthesized by aorta explants from rabbits with diet-induced atherosclerosis (6). ^{35}S-incorporation into proteoglycans, a measure of proteoglycan synthesis, was 2-fold higher in atherosclerotic tissue than in control tissue. This increased synthesis was predominantly reflected in chondroitin sulfate and dermatan sulfate proteoglycans (Figure 1). Further, these · isomeric chondroitin sulfate proteoglycans from atherosclerotic tissues were of larger molecular size than those from control tissue. Thus, proteoglycans undergo both quantitative and qualitative changes in atherosclerosis, reflecting the enhanced activities of smooth muscle cells and endothelial cells.

Unlike other arterial cells, the endothelial cells are in intimate contact with circulating blood, and hence are exposed to higher concentrations of low-density lipoproteins (LDL), especially in hypercholesterolemia and attendant atherosclerosis. To test whether increased concentration of LDL can modulate proteoglycan 'synthesis by confluent vascular endothelial cells, ^{35}S-incorporation into proteoglycans in cultures of human umbilical-vein endothelial cells was studied (7). LDL markedly increased the synthesis of proteoglycans in culture medium and cell fractions (Figure 2). Further, the exposure of cells to LDL resulted in the secretion of a larger isomeric chondroitin sulfate proteoglycan enriched in chondroitin 6-sulfate. Accumulation of such proteoglycans in the subendothelial layer may influence the permeation and retention of LDL in the arterial wall (discussed below).

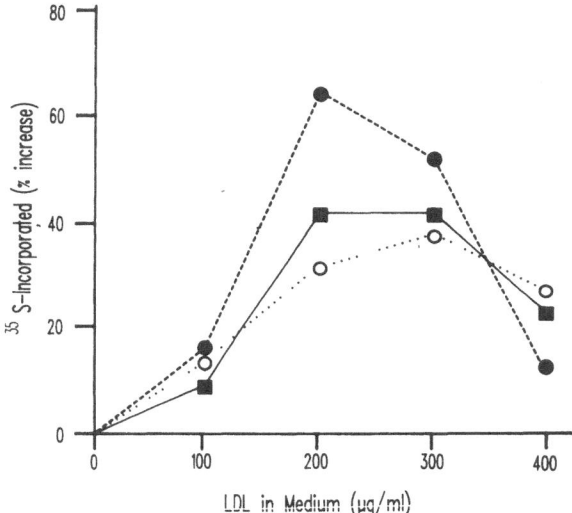

Fig. 2.   Incorporation of ^{35}S into proteoglycans
          in cultures of endothelial cells as a
          function of LDL concentration. Results
          are expressed as the percentage increase
          in incorporation over control, ● , cell
          layer; O , medium; ■ , cell layer + medium.
          (With permission, Ref. 7).

Interaction of proteoglycans with LDL - Extracellular lipid accumulation

Arterial wall proteoglycans and its constituent sulfated GAG chains form both soluble and insoluble complexes with serum apolipoprotein (apo) B - containing lipoproteins (8,9). The fact that proteoglycans form complexes with atherogenic LDL and very low-density lipoprotein (VLDL) but not with antiatherogenic high-density lipoprotein indicate that apoB in VLDL and LDL imparts selectivity in terms of proteoglycan binding. On the other hand, the nature of GAG determines their complexing ability with LDL. For example, unlike heparan sulfate proteoglycan, isomeric chondroitin sulfate proteoglycan forms complex readily with LDL. It is of interest that chondroitin 6-sulfate isomer was the predominant GAG present in LDL-proteoglycan complex isolated from human atherosclerotic lesions (10); in addition, these in vivo complexes contained small amounts of heparin.

Both isomeric chondroitin sulfate proteoglycans and heparin display considerable divergence with respect to LDL binding affinity due to their heterogeneity in size, charge density, and chemical characteristics (11,12). When an isomeric chondroitin sulfate proteoglycan preparation from bovine aorta was fractionated by LDL-affinity chromatography, a minor subfraction (25%) eluted similar to heparin, heretofore considered to be the most strongly binding GAG with LDL (Figure 3). This subfraction (M_r = 1.3 . 10^6) was enriched in chondroitin 6-sulfate isomer, and formed insoluble complex like heparin. The binding data showed two classes of interactions with 0.1 and 0.3 proteoglycan monomer bound per LDL particle characterized by apparent K_d of 4 and 21 nM, respectively. This indicates

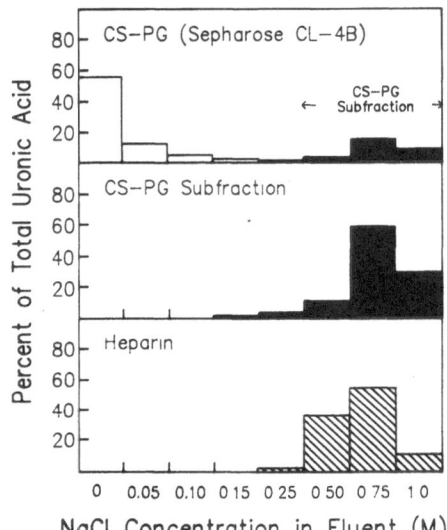

Fig. 3. Affinity chromatography of chondroitin
sulfate proteoglycans (CS-PG) and heparin
on LDL-agarose. Materials eluted between
0.5 and 1.0M NaCl was designated CS-PG
subfraction (Solid bars). (With permission,
Ref 11).

that multiple LDL particles bind to single proteoglycan monomers even at saturation. In contrast, LDL-heparin interactions showed a major component characterized by an apparent K_d of 151 nM and a B_{max} of 9 heparin molecules per LDL particle.

The occurrence of a potent LDL-binding proteoglycan subfraction within the family of arterial isomeric chondroitin sulfate proteoglycans may be of importance in terms of lipid accumulation in atherogenesis. Studies have shown that a non-specific process of injury and repair promotes increased arterial chondroitin sulfate proteoglycan content (2).

It is likely that increased synthesis of high affinity chondroitin sulfate proteoglycan occurs under these conditions, leading to focal lipid accumulation.

Recently we found that heparin also contained a minor (8%) high affinity subfraction; this subfraction was relatively higher in molecular weight (11,000 vs 17,000) and contained more iduronyl sulfate as hexuronic acid (76% vs 86%), N-sulfate ester (0.75 vs 0.96 mole/mole hexosamine), and O-sulfate ester (1.51 vs 1.68 moles/mole hexosamine). Although both heparin preparations formed insoluble complex with LDL quantitatively, the concentrations of NaCl required for 50% reduction in maximal insoluble complex formation was markedly higher with high affinity subfraction (0.55 M vs 0.04 M). The presence of such a high affinity heparin subfraction may favor lipid accumulation in the arterial wall. In fact, mast cells, the major source of heparin, have been identified in the arterial wall, especially in atherosclerotic lesions (13).

Proteoglycan-mediated LDL uptake by macrophages - Foam cell formation

Foam cells of atherosclerotic lesions are derived from blood-borne monocyte-macrophages. The transformation of macrophages to foam cells involves massive accumulation of cholesteryl esters in these cells. It is widely considered that modification of LDL, such as oxidation of LDL, is a prerequisite for macrophage recognition and uptake, leading to massive intracellular cholesteryl ester accumulation (14). Although proteoglycans are integral components of arterial extracellular matrix, and the presence of in vivo LDL-proteoglycan complex in the atherosclerotic lesions has been established for a long time (10), the role of LDL-proteoglycan complex in this process has not been fully appreciated until recently.

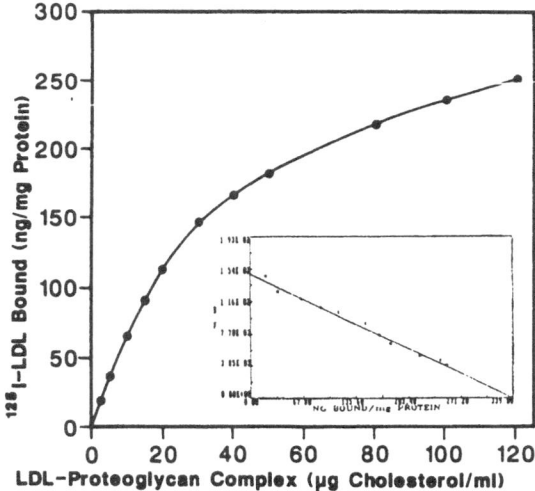

Fig 4. Specific binding of ^{125}I-labeled LDL-proteoglycan complex by macrophages as a function of its concentration. The inset shows Scatchard plot of the specific binding data. (With permission, Ref. 16).

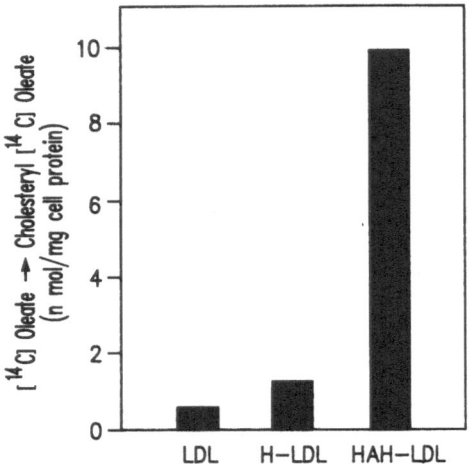

Fig. 5. Effect of interaction of high
affinity heparin (HAH) with LDL
on cholesteryl ester synthesis
in macrophages. H, whole heparin.

We have observed that the interaction of an isomeric chondroitin
sulfate proteoglycan aggregate with LDL lead to the degradation of the
lipoprotein by macrophages (15,16). The proteoglycan-stimulated degradation of LDL produced a marked increase in cholesteryl ester synthesis and
content in macrophages. Further, the binding of the complex was concentration dependent, specific, and saturable (Figure 4). The K_d, calculated
from the Scatchard plot, was 23 μg LDL-protein per ml. At receptor
saturation, the membrane bound complex contained 32,000 LDL particles per
cell. While binding was not inhibited by an excess of native LDL and
β-VLDL, acetyl-LDL and proteoglycan-LDL complex demonstrated only partial
reciprocal competition for binding and degradation. Polyinosinic acid,
fucoidin, and dextran sulfate, known inhibitors of acetyl-LDL binding and
degradation in macrophages, did not affect proteoglycan-LDL binding and
degradation.

Like proteoglycan-LDL complex, high affinity heparin subfraction-LDL
(HAH-LDL) complex produced marked increase in the degradation of
lipoproteins by macrophages as well as cholesteryl ester synthesis (Figure
5) and content (12). Whole heparin, on the other hand, did not produce
such effect. While unlabled HAH-LDL complex produced a dose-dependent
inhibition of the degradation of labeled complex, native unlabeled LDL did
not elicit a similar effect. Although particulate LDL aggregate competed
for 33% of degradation of labeled complex, cytochalasin D, known inhibitor
of phagocytosis, did not effectively inhibit the degradation of labeled
complex. Unlabeled acetyl-LDL produced a partial (33%) inhibition of the
degradation of labeled complex. Macrophages accumulated 2.3-fold more
steady-state cell-associated radioactivity from labeled complex than from
labeled acetyl-LDL, suggesting a slow or incomplete degradation of the
complex due to interference of the complex with the functioning of
endosomal or lysosomal components (Figure 6). Overall, these results
indicate that proteoglycan-LDL and HAH-LDL complexes are taken up by
macrophages by a scavenger receptor-mediated endocytosis, sharing similar
but not identical binding sites with acetyl-LDL. The pathobiologic
significance of these observations to foam cell formation in atherogenesis
is obvious.

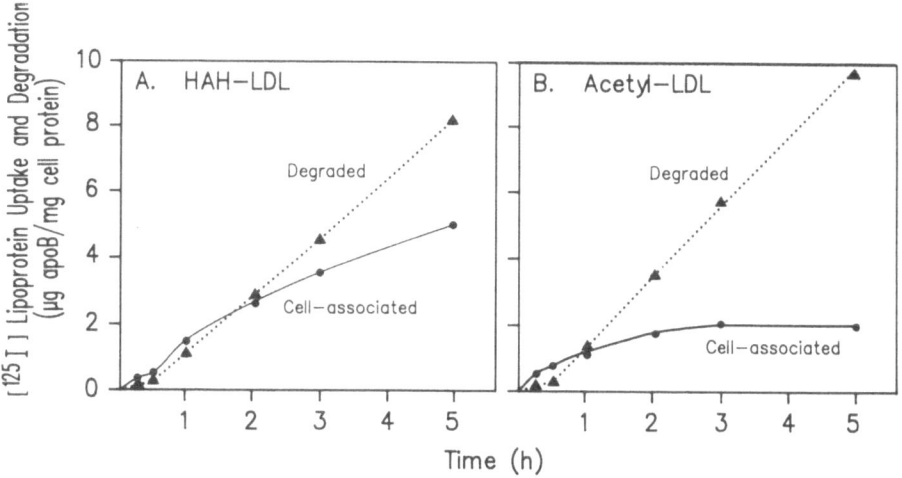

Fig. 6. Time course of macrophage uptake and degradation of high affinity heparin (HAH) - [^{125}I]LDL complex and [^{125}I] acetyl-LDL.

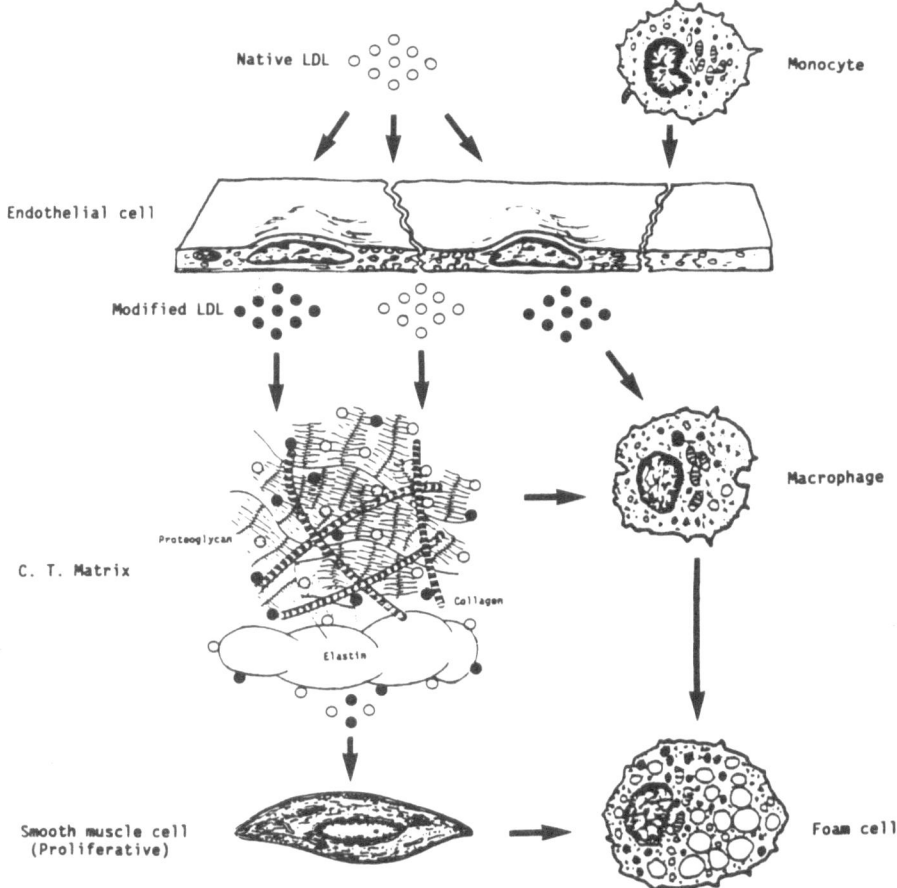

Fig. 7. Arterial wall proteoglycans and lipid accumulation:potential mechanisms.

Summary

   The arterial wall proteoglycans play a crucial role in the
pathogenesis of atherosclerosis as depicted schematically in Figure 7.
Plasma components including lipoproteins cross the endothelium mainly by a
non-specific bulk-phase vesicular transport. A selective interaction of
apoB-containing lipoproteins occurs with proteoglycans of the
subendothelial layer which results in extracellular retention and accumulation of lipoproteins. Such interaction alters the structural and charge
characteristics of LDL particles. These altered LDL are taken up by
monocyte-derived macrophages by scavenger receptor-mediated endocytosis,
leading to cholesteryl ester accumulation and foam cell formation.
Further, retention of LDL by proteoglycans in the extracellular matrix
also increases the chances of oxidative modification of lipoproteins. All
of these changes may be occurring at a marginal level as a normal adaptive
process of the arterial wall. However, focal response to chronic
hemodynamic stress, hyperlipidemia or other forms of injury may function-
ally alter the endothelium, and cause greater influx of lipoproteins and
smooth muscle cell proliferation, resulting in increased synthesis of
proteoglycans with altered characteristics. Enhanced binding of apo-B
containing lipoproteins to proteoglycans under these conditions sets the
stage for the development of athersclerosis.

References

1. Berenson, G. S., Radhakrishnamurthy, B., Srinivasan, S. R.,
    Vijayagopal, P., Dalferes, E.R. Jr., Sharma, C., 1984,
    Carbohydrate-protein macromolecules and arterial wall injury - A role
    in atherogenesis, Exp. Mol. Pathol., 41:267.
2. Wight, T. N., 1989, Cell biology of arterial proteoglycans,
    Arteriosclerosis, 9:1.
3. Radhakrishnamurthy, B., Jeansonne, N., Berenson, G. S., 1986,
    Organization of glycosaminoglycan chains in a chondroitin
    sulfate-dermatan sulfate proteoglycan from bovine aorta, Biochim.
    Biophys. Acta., 822:85.
4. Berenson, G. S., Radhakrishnamurthy, B., Srinivasan, S. R.,
    Vijayagopal, P., Dalferes, E. R., Jr., 1988, Arterial wall injury and
    proteoglycan changes in atherosclerosis Arch. Pathol. Lab. Med.,
    122:1002.
5. Wagner, W. D., 1985 Proteoglycan structure and function as related to
    atherosclerosis, N. Y. Acad. Sci., 454:52.
6. Radhakrishnamurthy, B., Srinivasan, S. R., Eberle, K., Ruiz, H,
    Dalferes, E. R. Jr., Sharma, C., Berenson, G. S., 1988, Composition of
    proteoglycans synthesized by rabbit aortic explants in culture and the
    effect of experimental atherosclerosis, Biochim. Biophys, Acta.,
    964:231.
7. Vijayagopal, P., Srinivasan, S. R., Dalferes, E. R., Jr.,
    Radhakrishnamurthy, B., Berenson, G. S., 1988, Effect of low-density
    lipoproteins on the synthesis and secretion of proteoglycans by human
    endothelial cells in culture, Biochem. J. 255:639.
8. Srinivasan, S. R., Lopez-S, A., Radhakrishnamurthy, B., Berenson, G.
    S., 1970, Complexing of serum pre-β and β-lipoproteins and acid
    mucopolysaccharides, Atherosclerosis 12:321.
9. Vijayagopal, P., Srinivasan, S. R., Radhakrishnamurthy, B., Berenson,
    G. S., 1981, Interaction of serum lipoproteins and a proteoglycan from
    bovine aorta, J. Biol. Chem. 256:8234.
10. Srinivasan, S. R., Dolan, P., Radhakrishnamurthy, B., Pargaonkar,
    P. S., Berenson, G. S. 1975, Lipoprotein-acid mucopolysaccharide
    complexes of human atherosclerotic lesions, Biochim. Biophys. Acta
    388:58.

11. Srinivasan, S. R., Vijayagopal, P., Eberle, K., Radhakrishnamurthy, B., Berenson, G. S., 1989, Low-density lipoprotein binding affinity of arterial wall proteoglycans: Characteristics of a chondroitin sulfate proteoglycan subfraction, <u>Biochim. Biophys. Acta</u>, 1996:159.
12. Srinivasan, S. R., Vijayagopal, P., Eberle, K., Radhakrishnamurthy, B., Berenson, G. S., 1989, Interaction of a heparin subfraction with low density lipoprotein (LDL) promotes cholesteryl ester accumulation in mouse macrophages, <u>Arteriosclerosis</u> 9:764a.
13. Cairns, A., Constantinides, P., 1954, Mast cells in human atherosclerosis, <u>Science</u> 120:31.
14. Steinberg, D., Parthasarathy, S., Carew, T. E., Khoo, J. L., Witztum, J. L., 1989, Beyond cholesterol modification of low-density lipoprotein that increases its atherogenesity, <u>N. Eng. J. Med.</u> 320:915.
15. Vijayagopal, P., Srinivasan, S. R., Jones, K. M., Radhakrishnamurthy, B., Berenson, G. S., 1985, Complexes of low-density lipoprotein and arterial proteoglycan aggregates promote cholesteryl ester accumulation in mouse macrophages, <u>Biochim. Biophys. Acta</u> 837:251.
16. Vijayagopal, P., Srinivasan, S. R., Jones, K. M., Radhakrishnamurthy, B., Berenson G. S., 1988, Metabolism of low-density lipoprotein-proteoglycan complex by macrophages: further evidence for a receptor pathway, <u>Biochim. Biophys. Acta</u> 960:210.

IMMUNE MECHANISMS IN THE PATHOGENESIS OF ATHEROSCLEROSIS

Maria F. Lopes-Virella[†] and G. Virella*

Department of Medicine [†]and Department of Basic and Clinical Immunology and Microbiology*, Medical University of South Carolina and VA Medical Center, Charleston, South Carolina, USA.

INTRODUCTION

It is generally accepted that arteriosclerosis is a multifactorial disease and that several risk factors contribute to its development. However the correlation between the development of arteriosclerosis and the presence of any of the known risk factors or associations of risk factors is not perfect. Furthermore, the precise mechanism by which each risk factor contributes to the pathogenesis of arteriosclerosis is not well understood. Thus, there has been a persistent development of new thoughts and theories concerning risk factors and their relative pathogenic roles. In the past decade an upsurge of interest in the role of immune mechanisms in the development of arteriosclerosis has emerged. In the present review we will analyzed data suggesting that immunological factors may contribute, directly or indirectly, to the sequence of pathological events leading to the development of arteriosclerosis.

One of the more accepted theories to explain the development of arteriosclerosis is the response to injury hypothesis, that was formulated by Ross and Glomset in 1976 (68).The hypothesis in its initial formulation postulated that injury to the endothelium could occur by mechanical, chemical, toxic, viral, or immunologic mechanisms which promoted endothelial denudation, followed by platelet adhesion, aggregation, and release of platelet-derived growth factor (PDGF) - which in turn would stimulate the migration and proliferation of smooth muscle cells within the intima. However, it became apparent that the actual denudation of the endothelium is not a consistent feature of arteriosclerosis, and also that platelet adhesion is not necessary to cause atheromatous plaque formation. Therefore, over the years this hypothesis underwent several modifications: (a) the endothelium can respond to a variety of stimuli, and may undergo either denudation or suffer only subtle functional changes such as the expression of adherence molecules or the release of mediators; (b) smooth muscle proliferation can be induced not only by PDGF but also by PDGF-like molecules which can be secreted by cells other than activated platelets.and (c) monocyte/macrophages are the major cellular components of lesions and these cells are likely to play a major role in the initiation and the evolution of the atheromatous plaque (67).

Recently Steinberg (72) proposed a new theory that combines the response to injury hypothesis with the lipid insudation theory, focusing upon the many areas of overlap between the two theories. Some of these areas of overlap include: (a) elevated LDL levels may damage the endothelial cells (40) and they may enhance platelet and leukocyte aggregation (49), (b) endothelial "injury," by removing the transport barrier and therefore increasing the infiltration of lipoproteins into the intimal area, creates optimal conditions for

Hypercholesterolemia, Hypocholesterolemia, Hypertriglyceridemia
Edited by C.L. Malmendier *et al.*, Plenum Press, New York, 1990

383

lipoprotein modification and subsequent uptake of these lipoproteins by macrophages and foam cell formation, (c) LDL itself has been shown to have mitogenic properties for smooth muscle cells, (d) PDGF has been shown to stimulate LDL receptor activity in an autocrine fashion within smooth muscle cells, fibroblasts, and monocytes (18,85), and (e) stimulation of smooth muscle cell growth is likely to result in deposition of connective tissue matrix materials which may enhance the trapping of LDL within the subintima (11).

Immune mechanisms have been proposed as playing an important role in the pathogenesis of arteriosclerosis. Several observations have been made that suggest a link between immune mechanisms and several stages in the development of the arteriosclerotic process. These observations can be grouped as follows: a) those that recognize the contribution of immune-competent cells to the formation of the arteriosclerotic plaque and their role in its development b) those that stress the effects of cytokines and mediators released by lymphocytes and phagocytic cells in the initiation and progression of arteriosclerosis and finally, c) those that attribute a role to autoantibodies and immune complexes in the development of arteriosclerosis.

ROLE OF IMMUNE COMPETENT CELLS AND CYTOKINES IN THE INITIATION AND PROGRESSION OF THE ARTERIOSCLEROTIC PLAQUE

The first two groups of observations, i.e., the presence of immunecompetent cells in the arteriosclerotic lesion and the effects of cytokines and mediators released by these cells will be discussed jointly. The presence of immunecompetent cells in the arteriosclerotic lesion is well known. In the past decade, the development of monoclonal antibodies recognizing specific cell markers has allowed to demonstrate that immunocompetent cells (lymphocytes and macrophages) are present in the atheromatous lesion. Studies employing monoclonal antibodies against human monocytes carried out by Faggiotto and co-workers (28,29) suggested that the majority of foam cells seen in the atheromatous lesion are of the monocytic lineage. This finding was confirmed by morphological studies carried out by Gerrity (36). Faggioto et al demonstrated also that monocytes from monkeys receiving hypercholesterolemic diets adhered to the endothelium, migrated subendothelially and evolved into foam cells being apparently responsible for the formation of fatty streaks (28.29). Thus, considerable interest has arisen in defining the conditions which lead to leukocyte adherence to endothelial cells, a phenomenon which is believed to have key significance not only in the onset of the atheromatous lesion, but also in the onset of inflammatory processes in general.

Leukocyte adherence to endothelial cells may follow damage of the EC or result from the release of a variety of mediators by monocytes/macrophages that act on undamaged endothelial cells enhancing the adherence of granulocytes, lymphocytes and monocytes to these cells. If the initial EC damage is caused by anti-EC antibodies, as proposed by Cerilli et al (16)., it would depend primarily on the activation of the complement system, which could result in direct cytotoxicity but also in the release of C5a which is a potent chemoattractant for monocytes and neutrophils (7). Besides attracting these cells to the area of endothelial lesion, C5a promotes their adherence to normal human endothelial cell monolayers (25, 79), probably due to increased expression of the leukocyte function-associated antigen 1 (LFA-1). The interaction of neutrophils with cell-bound IC also leads to an enhanced adhesiveness of these cells to healthy endothelial cells, which is followed by endothelial cell detachment and release of [^{3}H]-glucose, indicating that cell damage has occurred (44). The same sequence of events can also take place after EC are damaged by any other mechanism. The adhesion of monocytes non-immunologically damaged to EC may be mediated, at least in part, by an interaction between the monocyte Fc$_\gamma$ receptor and IgG passively absorbed onto cytoskeleton intermediate filaments (38). The fibronectin-bound IgG seems also to be able to activate the complement cascade thus having the potential of inducing the same pathologic events triggered by anti-EC antibodies. Thus, once initial endothelial damage has occurred, several mechanisms of recruitment and activation of phagocytic cells are turned on, contributing to the progression of the pathological process.

Enhanced adherence of leukocytes to endothelial cells can also be induced by a variety of mediators released by immunecompetent cells that are able to affect undamaged endothelial cells. The most important of these mediators are interleukin 1 (IL-1) and tumor necrosis factor α (TNF-α). IL-1 has been shown to activate the expression of a molecule designated as endothelial-leukocyte adhesion molecule-1 (ELAM-1) which promotes interactions with leukocytes independently of those mediated by the LFA-1 molecule (5,6). The expression of LFA-1 molecules on neutrophils, on the other hand, is enhanced by a factor(s) released by IL-1 or TNF-α - activated EC (64). Thus, IL-1, TNF-α, and several other mediators induce the expression of several specific adhesion molecules, which appears to vary according to the stimulus and to specific stages of cell activation (53). The data obtained by many different groups using *in vitro* systems strongly suggests that the expression of leukocyte adhesive proteins on the phagocytic and endothelial cells will promote interactions between these cells. Also, at least in the case of the neutrophil, the adherence of activated neutrophils to EC in vitro is usually followed by EC damage, irrespectively of how the neutrophil has been activated (24,44).

The effects of mediators released by activated monocyte/ macrophages are not limited to inducing the expression of surface molecules on endothelial cells. The monocytes/macrophages are biologically multipotent cells that among other biological activities have been shown to be capable, when activated, of a) secreting growth factors such as fibroblast growth factor and PDGF-like growth factor; b) secreting IL-1 and TNF-α (55,61); c) releasing modulatory substances such as PGE_2 (8,31,69) and α-interferon (74); and d) releasing proteases (59), collagenases (84), oxygen radicals (60), and other compounds capable of damaging the arterial wall.

Several of these mediators have been shown to have effects that could be directly related to the development of arteriosclerosis. PDGF-like, released by monocytes/macrophages, besides playing a role in stimulating smooth muscle cell proliferation, can also increase endocytosis, cholesterol synthesis, and LDL receptor expression in mononuclear cells. IL-1 and TNF-α are known to affect both endothelial and smooth muscle cells. IL-1 has been shown to induce synthesis and cell surface expression of procoagulant activity in endothelial cells (4), to increase vascular permeability (54) to induce IL-1 release from EC by a positive feedback mechanism (83) and to promote the release of increased amount of PAF by endothelial cells which, in turn, activates platelets and neutrophils, enhancing the adhesiveness of the later (10). Also, IL-1 can be indirectly responsible for fibroblast and smooth muscle cell proliferation by inducing the production of PDGF-AA by these cells, activating what appears to be an autocrine growth regulating mechanism (65). However since IL-1 induces equally secretion of prostaglandins by smooth muscle cells (50) which are known to have growth-inhibiting properties, the in vivo effect of IL-1 release in the arterial wall is unclear. TNF-α, which can be produced not only by macrophages but also by smooth muscle cells, induces some responses similar to those of IL-1, such as cell surface expression of procoagulant activity (4) and of adhesive surface proteins in EC leading to increased binding of neutrophils and lymphocytes to EC (14, 34), and induce the production of IL-1 by the EC (62). Another potentially important role of TNF-α is the ability to suppress LPL activity (15). It has been shown that LPL is secreted by macrophages in culture (17) and it has been proposed that macrophages induce lipolysis in the arteriosclerotic plaque by secreting LPL. However recent studies by Jonasson et al (42) were not able to detect immunoreactive LPL in the macrophages of arteriosclerotic lesions and this may reflect local TNF-α-induced inhibition of LPL production. The main conclusion from all these observations is that the monocyte/macrophage can induce and perpetuate EC damage by a variety of mechanisms. Therefore, it is obvious that their activation could play a very significant role in the onset of the sequence of events leading to the formation of an atheromatous lesion.

Another population of immunocompetent cells usually present in the arteriosclerotic lesion are T- lymphocytes. CD8+ T-lymphocytes are found mostly in early plaques(27) and fatty streaks (58), while CD4+ T lymphocytes are seen mostly in complicated arteriosclerotic lesions (43). Although the presence of T lymphocytes in the plaque does not necessarily mean that they are immunological active, studies performed to determine their activation status have shown that one third of the T-cells present in the arteriosclerotic plaque are activated (38). It is still not known, however, which is the role of these cells in the development of the arteriosclerotic process; they may represent an immune response to a specific component of the plaque. It is known that helper T lymphocytes are activated by fragments of antigens bound to MCH class II antigens on the cell surface of antigen-presenting cells (22). Although large vessel EC and smooth muscle cells do not normally express class II antigens they may do so when exposed to interferon γ (IFNγ) (63). Since the release of this lymphokine is a property of activated T lymphocytes, it follows that some other factor may have to account for the initial T lymphocyte activation. Several monocyte/ macrophage mediators, including IL-1, TNF-α, and PAF are known to enhance lymphocyte adherence to EC (13,14, 66). If such adherence is followed by a sufficient degree of T cell activation as to release a sufficient amount of IFNγ to induce the expression of MHC-II molecules and to activate further the macrophages, then the conditions necessary for additional lymphocyte recruitment would be fulfilled since both the expression of MHC-II molecules as well as the release of IL-1 and TNF-α may enhance lymphocyte adhesion to EC (56). The release of IFNγ by activated lymphocytes, besides enhancing lymphocyte adhesion to EC, inhibits endothelial and smooth muscle cell proliferation. However, since it induces IL-1 release which promotes growth, its role *in vivo* is uncertain. As a consequence of efficient signaling of helper T cells, IL-2 is likely to be released. IL-2-activated lymphocytes exhibit increased adherence to normal EC and can cause their lysis, at least in vitro (21). Thus, a complete T cell activation cycle can theoretically be triggered by cytokines released by activated macrophages.

The role of macrophages in arteriosclerosis is extremely complex. The most important components of early atheromatous lesions are "foam cells" which have been considered as an important hallmark of arteriosclerosis. It is well established that foam cells derive mostly from cells of the monocytes/macrophage lineage (28,29). Macrophages express a variety of receptors, some of which have been well characterized, and are capable of mediating lipoprotein uptake through the recognition of specific lipoproteins and lipoprotein complexes. The classical LDL receptor is involved in the uptake of native LDL (nLDL), necessary for the supply of cholesterol to the cell. Initial observations discounted the role of nLDL uptake through the native LDL receptor in the transformation of macrophages into foam cells on the basis of both limited nLDL receptor number and of the stringent regulation of its expression. In fibroblasts, lipid over-accumulation is not observed when the cells are incubated with native LDL and the assumption that the same would be true in any other type of cell, including the monocyte/macrophage, was made without an experimental basis. This assumption was first challenged by Tabas et. al., who showed that the J774 mouse macrophage-like tumor cell line had increased uptake of nLDL and concomitant CE accumulation. These authors postulated that the accumulation of nLDL in the J774 cell line was secondary to the enhanced metabolic activity of the transformed cells (77). Furthermore, the LDL receptor expressed by normal macrophages can bind a variety of ligands, including nLDL, ß-VLDL (26,46), Lp(a) (32,39,48), and chylomicron remnants (47), as well as complexes formed between LDL and LPS (80), and between LDL and anti-LDL antibodies (30,37). Experimentally, incubation of macrophages with several of these lipoproteins and lipoprotein complexes has been shown to lead to their transformation into foam cells, even though binding and uptake occur through the LDL receptor. Therefore, the optimal regulation of the LDL receptor seems to be observed only when native LDL is internalized ; the internalization of other lipoproteins or of LDL complexed with a variety of proteins can lead to intracellular accumulation of cholesteryl esters (CE).

Unquestionably, modified LDL is effective in promoting intracellular accumulation of CE through the well studied "scavenger" receptor, which, in contrast with the native LDL receptor, is not regulated. Accumulating evidence suggests that macrophages may have a "family" of scavenger receptors which are distinct from the "classical" scavenger receptor but participate in the recognition and removal of an overlapping range of modified products (73). Monocyte-derived macrophages may also internalize LDL by ingesting LDL complexes through receptors which recognize the protein or substance complexed with LDL. For example, LDL can interact with extracellular matrix components such as heparin, fibronectin, collagen and proteoglycans (12,41,70,81), and anti-LDL antibodies (37,45). The internalization of LDL included in these complexes through receptors for the non-lipoprotein moiety of the complex seems to lead inevitably to intracellular CE accumulation.

The complexity of the interactions between the various lipoproteins and lipoprotein complexes with macrophages is progressively being understood. The available data shows that there are differences not only in lipoprotein-receptor interactions but also in intracellular processing. The effects of internalization of two different lipoproteins through the same receptor may be strikingly different. For example, both nLDL and ß-VLDL can be internalized via the nLDL receptor. However, the internalization of nLDL usually will result in down regulation of the receptor and the intracellular contents of cholesterol and CE will remain within narrow physiologic limits. In contrast , ß-VLDL internalization has no effect on the expression of the LDL receptor and the cells accumulate cholesterol intracellularly (26,46). On the other hand, when macrophages are activated, the control mechanisms that usually prevent intracellular cholesterol accumulation may cease to be effective. This was well documented in studies performed in our laboratory. We have demonstrated increased uptake of native LDL and subsequent CE accumulation in human macrophages stimulated with microbial or microbial-related products (51,52). Therefore, foam cell formation in human macrophages may occur not only with uptake of modified or complexed lipoproteins but may also depend upon the functional state of the macrophages.

ROLE OF AUTOANTIBODIES AND IMMUNE COMPLEXES IN THE DEVELOPMENT OF ARTERIOSCLEROSIS

The interest of immunologists in arteriosclerosis has been focused on the putative role of autoantibodies and antigen-antibody complexes (immune complexes, IC). The potential role of antibodies to EC has been previously discussed. Although EC damage mediated by anti-EC antibodies remains as one of the likely explanations for accelerated arteriosclerosis following organ transplantation, the putative role of autoantibodies to EC has not been properly substantiated. The interest on IC as possible pathogenic factors in arteriosclerosis was initially suggested by data published by Minick & Murphy (57) demonstrating that the induction of chronic serum sickness in rabbits fed a lipid-rich diet resulted in the formation of vascular lesions similar to those seen in human arteriosclerosis. *In vitro* experiments have suggested that heat aggregated gamma globulin and soluble IC present in the sera of patients with systemic lupus erythematosus or rheumatoid vasculitis may induce tissue factor production by EC and prime EC for neutrophil-mediated damage (9,78). The binding of IC to EC could be mediated by Fc and/or C1q receptors (1), and we have documented using RBC as models that cell-bound IC are very potent activators of neutrophils (82) and monocytes (19). In humans, reports of elevated incidence of circulating IC in patients with clinical manifestations of arteriosclerosis have been published (33,75,76). However, the precise nature of the antigen(s) involved in the formation of arteriosclerosis-related IC has remained controversial.

For several years, data has been available suggesting that LDL-anti-LDL immune complexes (LDL-IC) are associated with striking abnormalities in LDL metabolism. Beaumont et al., showed that spontaneous or immunization-induced anti-LDL antibodies cause decreased clearance of LDL, leading to hyperlipoproteinemia (3). Experimental data suggesting that incubation of cells with LDL-IC significantly disturbed lipoprotein and cholesterol metabolism was first published by Beaumont et al., who showed that the exposure of human fibroblasts to LDL-anti-LDL complexes resulted in excess production of free cholesterol (20). Later on Klimov et al. showed excessive CE accumulation in mouse peritoneal macrophages exposed to LDL-IC (45). Recently we have demonstrated that human macrophages incubated in vitro with insoluble LDL-IC show a marked accumulation of CE evolving into foam cells and have enhanced LDL receptor expression (37). Relatively large concentrations of insoluble LDL-IC were required to elicit the above observations. In order to determine whether minimal amounts of circulating LDL-IC were able to induce similar results we incubated macrophages with small amounts of LDL-IC bound to RBC and we were able to induce the transformation of macrophages into foam cells under these experimental conditions.(Unpublished observations). Thus, it appears as if LDL-IC may play a very important role in the pathogenesis of arteriosclerosis. The interest in this hypothesis was,

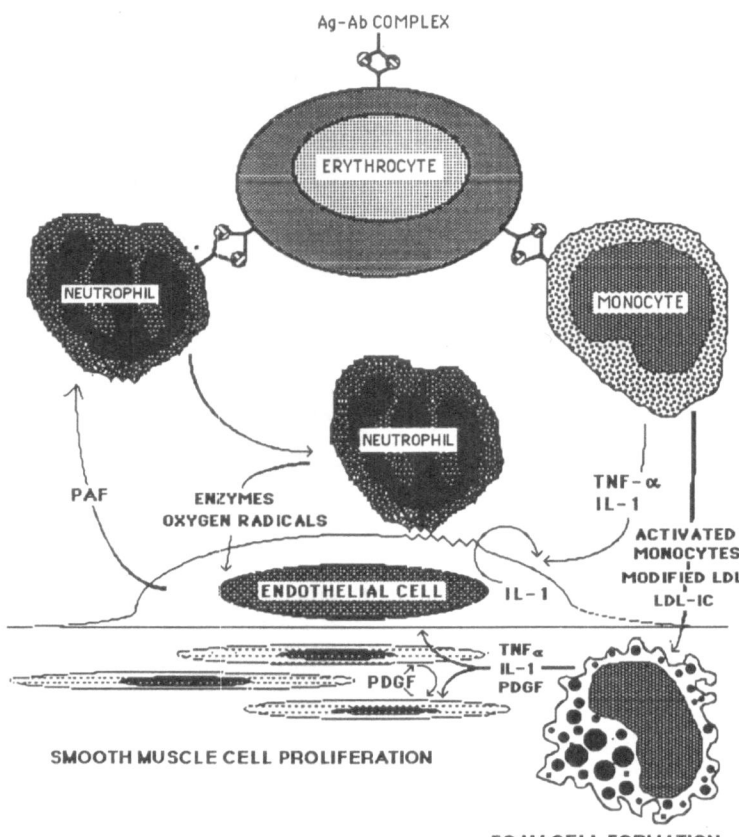

Figure 1. Diagrammatic representation of the role of IC in the pathogenesis of arteriosclerosis. RBC-bound IC (not necessarily involving LDL as antigen) could play the initial role of phagocytic cell activators. Activated monocytes, through the release of monokines, would induce the expression of membrane adhesion molecules on the EC and set the stage for EC-phagocytic cell interactions. As a consequence of such interactions endothelial damage would result, with increased permeability and migration of monocytes to the subendothelial space. LDL, modified by activated phagocytes and endothelial cells, would be taken up by monocytes which would transform into foam cells (particularly if LDL-IC were formed as a consequence of the immune response against modified LDL). Both monocytes and EC would release factors which could trigger smooth muscle cell proliferation.

however, relatively low until data started to accumulate supporting the possibility that anti-LDL antibodies may be formed *in vivo*. For example, it has been shown that various modifications, such as nonenzymatic glycation of Apo-B, enhance the immunogenicity of LDL. Witztum et al. demonstrated the presence of autoantibodies to glucosylated LDL in diabetic patients (86). Furthermore, the presence of circulating anti-LDL antibodies and LDL-IC in patients with coronary artery disease has been described in several independent investigations (3,33,34,71,75,76). A diagrammatic summary of the possible role(s) of IC in the pathogenesis of arteriosclerosis is reproduced in figure 1.

A crucial question that remains to be answered is wether or not foam cell transformation is associated to functional activation. The macrophage, as we have previously mentioned, can release several important mediators which may play significant roles in the pathogenesis of atherosclerosis, including IL-1, TNF-α, and PAF. Another interesting macrophage-derived mediator is the PDGF-like growth factor. It has been shown that this growth factor can not only stimulate smooth muscle cell proliferation but also increase endocytosis, cholesterol synthesis, and LDL receptor expression in mononuclear cells. Preliminary data obtained in our laboratory suggests that human monocyte-derived macrophages which have been transformed into foam cells by uptake of LDL-IC secrete increased amounts of TNF-α. However, at least in the case of the release of TNF-α we have shown that the uptake of modified LDL, which can also lead to foam cell formation, is not associated to its release. There are two significant differences in the presentation of LDL-IC vs. modified LDL. First, when equal concentrations of LDL are presented to human monocyte-derived macrophages as LDL-IC or as modified LDL, the uptake of complexed LDL is significantly greater. Secondly, LDL-IC are ingested predominantly through the Fc receptor, whose cross-linking is known to cause TNF-α release (23), while modified LDL is taken up by the scavenger receptor. Thus, is very possible that wether or not macrophages are activated in the process of becoming foam cells may depend on how are the lipoproteins presented and what receptors are involved in their uptake. Therefore, the formation of LDL-IC involving IgG anti-LDL antibodies would be have considerably more pathogenic potential than the simple chemical modification of LDL since LDL-IC not only appear to deliver LDL to a cell in a way which escapes normal metabolic control, but are also taken up through the Fc receptor, whose occupancy delivers an activating signal to the macrophage which is not delivered as a consequence of the occupancy of the scavenger receptor.

CONCLUSIONS

A variety of immune mechanisms may be involved at different stages of the pathogenesis of arteriosclerosis. A major obstacle to our better understanding of their relative significance lies in the fact that, by necessity, most of the data has been generated in *in vitro* systems. Therefore all our hypothetical schemes have an inherent degree of uncertainty which is difficult to eliminate. Additional progress in our understanding may depend in the design of proper approaches for the clinical evaluation of these mechanisms. This is not a small undertaking and it is not clear at this point wether feasible and informative studies in humans or in adequate animal models can be designed, given the multitude of factors involved. However, the opening of new hypothesis may lead to the reevaluation of available data and shed new understanding into an area in which the questions formulated continue to greatly exceed the answers obtained.

BIBLIOGRAPHY

1. Andrews B.S., Shadforth M., Cunningham P., Davis J.S. IV. Demonstration of a C1q receptor on the surface of human endothelial cells. J. Immunol. 127:1075, 1981.
2. Beaumont J.L. Autoimmune hyperlipidemia. An atherogenic metabolic disease of immune origin. Rev. Eur. Stud. Clin. Biol. 15:1037, 1970.
3. Beaumont J.L., Beaumont V. Immunological Aspects of Atherosclerosis. Atherosclerosis. Reviews 3:133, 1978.
4. Bevilacqua M.P., Pober J.S., Majeau G.R., Cotran R.S., Gimbrone M.A., Jr. Interleukin 1 induces biosynthesis and cell surface expression of procoagulant activity in human vascular endothelial cells. J. Exp. Med. 160:618, 1984.
5. Bevilacqua M.P., Pober J.S., Wheeler M.E., Cotran R.S., Gimbrone M.A. jr. Interleukin 1 acts on cultured human vascular endothelium to increase the adhesion of polymorphonuclear leukocytes, monocytes, and related leukocyte cell lines. J. clin. Invest. 76:2003, 1985.

6. Bevilacqua M.P., Stengelin S., Gimbrone M.A. Jr., Seed B. Endothelial leukocyte adhesion molecule 1: and inducible receptor for neutrophils related to complement regulatory proteins and lectins. Science 243:1160, 1989.

7. Boackle R. The complement system. In "Introduction to Medical Immunology", 2nd Ed. (Virella G., Goust J.M. and Fudenberg H.H., Eds.) M. Dekker, NY 1990, pp.143.

8. Bonney R.J., Humes J.L. Physiological and pharmacological regulation of prostaglandin and leukotriene production by macrophages. J. Leukocyte Biol. 35:1, 1984.

9. Breedveld F.C., Heurkens A.H., Lafeber G.J.M., van Hinsbergh V.W.M., Cats A. Immune complexes in sera from patients with rheumatoid vasculitis induce polymorphonuclear cell-mediated injury to endothelial cells. Clin. Immunol. Immunopath. 48:202, 1988.

10. Breviario F., Bertocchi F., Dejana E., Bussolino F. IL-1 induced adhesion of polymorphonuclear leukocytes to cultured human endothelial cells. Role of platelet-activating factor. J. Immunol. 141:3391, 1988.

11. Brownlee M., Vlassara H., Cerami A. Nonenzymatic glycosylation products on collagen covalently trap low density lipoprotein. Diabetes 34: 938, 1985.

12. Camejo G. The Interactions of Lipids and Lipoproteins with the Intercellular Matrix of Arterial Tissue: It's Possible Role in Atherogenesis. Adv. Lipid Res. 19:1, 1982.

13. Cavender D., Haskard D., Foster N., Ziff M. Superinduction of T lymphocyte-endothelial cell (EC) binding by treatment of EC with interleukin 1 and protein synthesis inhibitors. J. Immunol., 138: 2149, 1987.

14. Cavender D., Saegusa Y., Ziff M. Stimulation of endothelial cell binding of lymphocytes by tumor necrosis factor. J. Immunol., 139: 1855, 1987.

15. Cerami A., Beutler B. The role of cachectin/TNF in endotoxic shock and cachexia. Immunol. Today 9:28, 1988.

16. Cerilli J., Brasile L., Sosa J., Kremer J., Clarke J., Leather R., Shah D. The role of autoantibody to vascular endothelial cell antigens in atherosclerosis and vascular disease. Transplant. Proc. 4 (Suppl.5):47, 1987.

17. Chait A., Iverius P.H., Brunzell J.D.: Lipoprotein lipase secretion by human monocyte-derived macrophages. J. Clin. Invest. 69: 490, 1982.

18. Chait A., Ross R., Albers J.J., Bierman E.L. Platelet-Derived Growth Factor Stimulates Activity of Low Density Lipoprotein Receptors. Proc. Natl. Acad. Sci. U.S.A. 77:4084, 1980.

19. Chou Y.K., Sherwood T., Virella G. Erythrocyte-bound immune complexes trigger the release of interleukin-1 from human monocytes. Cell. Immunol., 91:308, 1985.

20. Dachet C., Bandet M.F., Beaumont, J.L. Cholesterol synthesis by human fibroblasts in the presence of LDL and anti-LDL IgA. Biomedicine 31:80, 1979.

21. Damle N.K., Doyle L.V., Bender J.R., Bradley E.C. Interleukin 2-activated human lymphocytes exhibit enhanced adhesion to normal vascular endothelial cells and cause their lysis. J. Immunol. 138: 1779, 1987.

22. Davis M.M., Bjorkman P.J. T-cell antigen receptor genes and T-cell recognition. Nature 334: 395, 1988.

23. Debets J.M.H., van der Linden C.J., Dieteren I.E.M., Leeuwenberg J.F.M., Buurman W.A.: Fc-receptor cross-linking induces rapid secretion of tumor necrosis factor (cachectin) by human peripheral blood monocytes. J. Immunol. 141:1197, 1988.

24. Diener A., Beatty P.G., Ochs H.D., Harlan J.M. The role of neutrophil membrane glycoprotein 150 (GP-150) in neutrophil-mediated endothelial cell injury in vitro. J. Immunol. 135:537, 1985.

25. Doherty D.E., Haslett C., Tonnensen M.G., Henson P.M. Human monocyte adherence: a primary effect of chemotactic factors on the monocyte to stimulate adherence to human endothelium. J. Immunol. 138:1762, 1987.

26. Ellsworth J.L., Kraemer F.B., Cooper A.D. Transport of ß-very low density lipoproteins and chylomicron remnants by macrophage is mediated by the LDL receptor pathway. J. Biol. Chem. 262:2316, 1987.

27. Emeson E.E., Robertson A.L. T lymphocytes in aortic and coronary intimas. Their potential role in atherogenesis. Am. J. Pathol. 130:369, 1988.

28. Faggiotto A., Ross R. Studies of Hypercholesterolemia in the Nonhuman Primate. II. Fatty Streak Conversion to Fibrous Plaque. Arteriosclerosis 4:341, 1984.

29. Faggiotto A., Ross R., Harker, L. Studies of Hypercholesterolemia in the Nonhuman Primate. I. Changes that Lead to Fatty Streak Formation. Arteriosclerosis 4:323, 1984.

30. Feingold K.R., Castro G.R., Yo I., Fielding P.E., Fielding C.J. Cutaneous Xanthoma in Association with Paraproteinemia in the Absence of Hyperlipidemia. J. Clin. Invest. 83:796, 1989.

31. Ferreri N. R., Howland W. C., Spiegelberg H. L. Release of leukotrienes C4 and B4 and prostaglandin E2 from human monocytes stimulated with aggregated IgG, IgA, and IgE. J. Immunol. 136: 4188, 1986.

32. Floren C.H., Albers J.J., Bierman E.L. Uptake of Lp(a) lipoprotein by cultured fibroblasts. Biochem. Biophys. Res. Commun. 102:636, 1981.

33. Füst G., Szondy E., Szekely J., Nanai I., Gerö S. Studies on the occurrence of circulating immune complexes in vascular disease. Arteriosclerosis 29:181, 1978.

34. Gamble J.R., Harlan J.N., Klebanoff S.J., Vadas M.A. Stimulation of the adherence of neutrophils to umbilical vein endothelium by human recombinant tumor necrosis factor. Proc. Nat. Acad. Sci. U.S.A. 82:8667,1985.

35. Gero S., Szondy E., Mezey Z., Székely J. Immune response against lipoproteins in coronary patients. In "Latent dyslipoproteinemias and arteriosclerosis", edited by DeGennes J. L., Raven Press, New York, 1984, pp.73.

36. Gerrity R.G. The role of the monocyte in atherogenesis. I. Transition of blood-borne monocytes into foam cells in fatty lesions. Am. J. Pathol. 103:181, 1981.

37. Griffith R.L., Virella G.T., Stevenson H.C., Lopes-Virella M.F. Low Density Lipoprotein Metabolism By Human Macrophages Activated With Low Density Lipoprotein Immune Complexes. J. Exp. Med. 168:1041, 1988.

38. Hansson, G., Jonasson L., Seifert P., Stemme S. Immune Mechanisms in Arteriosclerosis. Arteriosclerosis 9; 567, 1989.

39. Havekes L., Vermeer B.J., Brugman T., Emeis J. Binding of Lp(a) to the low density lipoprotein receptor of human fibroblasts. FEBS Lett. 132:169, 1981.

40. Henricksen T., Evensen S.A., Carlander B. Injury to human endothelial cells in culture induced by low density lipoproteins. Scand. J. Clin. Lab. Invest. 39:361, 1979.

41. Hurt E., Camejo G. Effect of Arterial Proteoglycans on the Interaction of LDL with Human Monocyte-derived Macrophages. Atherosclerosis 67:115, 1987.

42. Jonasson L., Bondjers G., Hansson G.K. Lipoprotein lipase in atherosclerosis: its presence in smooth muscle cells and absence from macrophages. J. Lipid Res. 28:437,1987.

43. Jonasson L., Holm J., Skalli O., Bondjers G., Hansson G.K. Regional accumulation of T cells, macrophages and smooth muscle cells in the human atherosclerotic plaque. Arteriosclerosis. 6:131, 1986.

44. Kilpatrick J.M., Hyman B., Virella G. Human endothelial cell damage induced by interactions between polymorphonuclear leukocytes and immune complex-coated erythrocytes. Clin Immunol. Immunopath., 44:335, 1987.

45. Klimov A.N., Denisenko A.D., Popov A.V., Nagornev V.A., Pleskov V.M., Vinogradov A.G., Denisenko T.V., Magracheva E. Y., Kheifes G.M., Kuznetzov A.S. Lipoprotein-antibody immune complexes: Their catabolism and role in foam cell formation. Atherosclerosis 58:1, 1985.

46. Koo C., Wernette-Hammond M. E., Innerarity T. L. Uptake of canine beta-very low density lipoprotein by mouse peritoneal macrophages is mediated by a low lipoprotein receptor. J. Biol. Chem. 261:194, 1986.

47. Koo C., Wernette-Hammond M.E., Garcia Z., Malloy M.J., Uauy R., East C.,.Bilheimer D.W, Mahley R.W., Innerarity T.L. Uptake of Cholesterol-Rich Remnant Lipoproteins by Human Monocyte-Derived Macrophages is Mediated by Low Density Lipoprotein Receptors. J. Clin. Invest. 81:1332, 1988.

48. Krempler F., Kostner G. M., Roscher A., Haslauer F., Bolzano K., Sandhofer F. Studies on the role of specific cell surface receptors in the removal of lipoprotein(a) in man. J. Clin. Invest. 71:1431, 1983.

49. Lechi C., Zatti M., Corradini P. Bonadona G. Arosio E., Pedrolii C., Lechi A. Increased leukocyte aggregation in patients with hypercholesterolaemia. Clin. Chim. Acta 144:11, 1984.

50. Libby P., Warner S.J.C., Friedman G.B. Interleukin 1: a mitogen for human vascular smooth muscle cells that induces the release of growth inhibitory prostanoids. J. Clin. Invest. 81:487, 1988.

51. Lopes-Virella M.F., Virella G. Immunological and Microbiological Factors in the Pathogenesis of Atherosclerosis. Clin. Immunol. Immunopath. 37:377, 1985.

52. Lopes-Virella M.F., Klein, R.L., Stevenson H.C. Low density lipoprotein metabolism in human macrophages stimulated with microbial or microbial-related products. Arteriosclerosis 7:176, 1987.

53. Luscinskas F.W., Brock A.F., Arnaout M.A., Gimbrone M.A. Endothelial-leukocyte adhesion molecule-1-dependent and leukocyte (CD11/CD18)-dependent mechanisms contribute to polymorphonuclear leukocyte adhesion to cytokine-activated human vascular endothelium. J. Immunol. 142:2257, 1989.

54. Martin S., Maruta K., Burkart V., Gillis S., Kolb H. IL-1 and INF-γ increase vascular permeability. Immunology 64:301, 1988.

55. Marx J.L. Cytokines are Two-edged Swords in Disease. Science 239:257, 1988.

56. Masuyama J.-I., Minato N., Kano S. Mechanisms of lymphocyte adhesion to human vascular endothelial cells in culture. J. clin. Invest. 77:1396, 1986.

57. Minick C.R., Murphy G.E. Experimental induction of arteriosclerosis by the synergy of allergic injury to arteries and lipid-rich diet. II. Effect of repeatedly injected foreign protein in rabbits fed a lipid-rich, cholesterol-poor diet. Amer. J. Path. 73:265, 1973.

58. Munro J.M., van der Walt J.D., Munro C.S., Chalmers J.A.C., Cox E.L.: An immunohistochemical analysis of human aortic fatty streaks. Hum. Pathol. 18: 375, 1987.

59. Musson R.A., Shafran H., Henson P.M. Intracellular levels and stimulated release of lysosomal enzymes from human peripheral blood monocytes and monocyte-derived macrophages. J. Reticuloendothelial Soc. 28:249, 1980.

60. Nakagawara A., Nathan C.F., Cohn Z.A. Hydrogen peroxide metabolism in human monocytes during differentiation in vitro. J. Clin. Invest. 68:1243, 1981.

61. Nathan C.F., Murray H.W., Cohn Z.A. Current concepts: the macrophage as an effector cell. N. Eng. J.. Med. 303:622, 1980.

62. Nawroth P.P., Bank I., Hadley D., Cassimeris J., Chess L., Stern D. Tumor necrosis factor/cachectin interacts with endothelial cell receptors to induce release of interleukin 1. J. exp. Med. 165:1363, 1986.

63. Pober J.J., Gimbrone M.A.,Jr., Cotran R.S., Reiss C.S., Burakoff S.J., Fiers W., Ault K.W. Ia expression by vascular endothelium is inducible by activated T cells and human γ interferon. J. exp. Med. 157:1339, 1983.

64. Pohlman T.H., Staness K.A., Beatty, P.G., Ochs, H.D., Harlan J.M. An endothelial cell surface factor(s) induced in vitro by lipopolysaccharide, interleukin 1, and tumor necrosis factor α increases neutrophil adherence by a CDw18-dependent mechanism. J. Immunol., 136:4548, 1986.

65. Raines E.W., Dower S.K., Ross R. Interleukin-1 mitogenic activity for fibroblasts and smooth muscle cells is due to PDGF-AA. Science 243:393, 1989.

66. Renkonen R., Mattila P., Turunen J.-P., Häyry P. Lymphocyte binding and penetration through vascular endothelium is stimulated by platelet-activating factor. Scand. J. Immunol. 30:673, 1989.

67. Ross R. The pathogenesis of atherosclerosis - an update. N. Eng J. Med. 1986.

68. Ross R., Glomset J.A. The pathogenesis of atherosclerosis. N. Eng. J. Med. 295: 369, 1976.

69. Rouzer C.A., Scott W.A., Kempe J., Cohn. Z.A. Prostaglandin synthesis by macrophages require a specific receptor-ligand interaction. Proc. Natl. Acad. Sci. U.S.A. 77:4279, 1980.

70. Salisbury B.G.J., Falcone D.J., Minick C.R. Insoluble low density lipoprotein-proteoglycan complexes enhance cholesteryl ester accumulation in macrophages. Am.J. Path. 120:6, 1985.

71. Smith E.B. The relationship between plasma and tissue lipids in human atherosclerosis. Adv. Lipid Res. 12:1, 1974.

72. Steinberg D. Lipoproteins and Atherosclerosis: Some Unanswered Questions. Am. Heart Journal 113:626, 1987.

73. Steinberg D., Parthasarathy S., Carew T.E., Khoo J.C., Witztum J.L. Beyond Cholesterol: Modifications of Low Density Lipoprotein that Increase it's Atherogenicity. New Eng. J. Med. 320:915, 1989.

74. Stevenson H.C., Dekaban G.A., Miller P.J., Benyajati C., Pearson M.L. Analysis of human blood monocyte activation at the level of gene expression. J. Exp. Med. 161:503, 1985.

75. Szondy E., Lengyel E., Mezey Z., Füst G., Gero S. Occurrence of anti-low density lipoprotein antibodies and circulating immune complexes in aged subjects. Mechanisms of Aging and Development 29:117, 1985.

76. Szondy E., Horvath M., Mezey Z., Szekely J., Lengyel E., Gero S. Free and Complexed Anti-lipoprotein Antibodies in Vascular Disease. Atherosclerosis 49:69, 1983.

77. Tabas I., Weiland D.A., Tall A.R. Unmodified low density lipoprotein causes cholesteryl ester accumulation in J774 macrophages. Proc. Natl. Acad. Sci. U.S.A. 82:416, 1985.

78. Tannenbaum S.H., Finko R., Cines D.B. Antibody and immune complexes induce tissue factor production by human endothelial cells. J. Immunol. 137:1532, 1986.

79. Tonnensen M.G., Anderson D.C., Springer T.A., Knedler A., Avdl N., Henson P.M. Adherence of neutrophils to cultured human microvascular endothelial cells. J. clin. Invest. 83:637, 1989.

80. Van Lenten B.J., Fogelman A. M., Haberland M.E., Edwards P. A. The role of lipoproteins and receptor-mediated endocytosis in the transport of bacterial LPS. Proc. Natl. Acad. Sci. U.S.A. 83: 2704, 1986.

81. Vijayagopal P., Srinivasan S.R., Jones K.M., Radhakrishnamurthy B., Berenson G.S. Complexes of LDL and arterial proteoglycan aggregates promote cholesteryl ester accumulation in mouse macrophage. Biochimica et. Biophysica Acta. 837:251, 1985.

82. Virella G., Lopes-Virella M.F.L., Shuler C., Sherwood T., Espinoza G.A., Winocour P. Colwell, J.A. Release of PAF by human polymorphonuclear leukocytes stimulated by immune complexes bound to sepharose particles and human erythrocytes. Immunology 50:43, 1983.

83. Warner S.J.C., Auger K.R., Libby P. Interleukin 1 induces interleukin . II. Recombinant human interleukin 1 induces interleukin 1 production by adult human vascular endothelial cells. J. Immunol. 139:1911, 1987.

84. Werb Z., Bonda M.J., Jones P.A. Degradation of connective tissue matrices by macrophages: I. Proteolysis of elastin, glycoproteins, and collagens by proteinases isolated from macrophages. J. Exp. Med. 152:1340, 1980.

86. Witztum J.L., Steinbrecher U.P., Kesaniemi Y.A., Fisher M. Autoantibodies to glucosylated proteins in the plasma of patients with diabetes mellitus. Proc. Natl. Acad. Sci. U.S.A. 81:3204, 1984.

85. Witte L.D., Cornicelli J.A. Platelet-Derived Growth Factor Stimulates Low Density Lipoprotein Receptor Activity in Cultured Human Fibroblasts. Proc. Natl. Acad. Sci. U.S.A. 77:5962, 1980.

CHOLESTEROL LEVEL IN CIRCULATING IMMUNE COMPLEXES AS A MARKER OF

CORONARY ATHEROSCLEROSIS

Alexander N. Orekhov, Oleg S. Kalenich,

Vladimir V. Tertov, Il'ya D. Novikov, Elena G. Vorob'eva

USSR Cardiology Research Center, Moscow, Russia

INTRODUCTION

Recently it has been revealed that blood plasma or serum of patients with angiographically assessed coronary atherosclerosis, unlike that of healthy donors, causes lipid accumulation in the cells cultured from human aortic intima.[1,2] This property of serum was termed "atherogenicity" since the accumulation of lipids was paralelled by the stimulation of other atherosclerotic manifestations at the cellular level, namely by increased proliferative activity and synthesis of extracellular matrix.[3] Atherogenicity of plasma is accounted for at least two factors: modified (desialylated) low density lipoproteins (LDL) and autoantibodies to LDL which are formed as a response to the appearance of modified LDL in the blood.[4,5] One may assume that modified LDL and autoantibodies form in blood circulating immune complexes (CIC) underlying atherogenic potential of blood plasma. A simple method was developed to determine the level of LDL component in such complexes. This method is based on the estimation of total cholesterol or apo B in the CIC precipitates obtained as a result of polyeththylene glycol treatment of serum samples.[6,7] Elevated level of CIC cholesterol were more often detected in the serum of patients with angiographically documented coronary atherosclerosis as compared to that of healthy donors.[6,7]

In the present study we attempted to clarify whether CIC cholesterol level correlates with severity of coronary atherosclerosis. The second task was to reveal diagnostic value of CIC cholesterol in comparison with other lipid and lipoprotein parameters used as markers of dislipidaemia associated with atherosclerosis.

MATERIALS AND METHODS

The subjects were 107 ischemic heart disease patients (92 men and 15 women) aged 32-69 years. Coronary atherosclerosis was documented by polypositional coronary angiography. The blood samples were drawn following an overnight fast during first 7 days after coronary angiography.

Serum levels of total cholesterol, triglyceride, HDL-cholesterol,

Hypercholesterolemia, Hypocholesterolemia, Hypertriglyceridemia
Edited by C.L. Malmendier *et al.*, Plenum Press, New York, 1990

apo B and apo A-1 were determined in all the examined samples according to routine procedures. LDL cholesterol was estimated according to Friedenwald et al.[8] CIC cholesterol level was assayed as described elsewhere.[6,7]

The highest value of normal CIC cholesterol level was determined as 15 ug/ml, which was estimated using X^2-test. To determine this value, sensitivity, specificity, accuracy and X^2 were assessed for each CIC cholesterol threshold value in the range of 11 to 19 ul/ml. Normal levels for other serum lipid parameters were similar to those recommended by European Atherosclerosis Society.[9]

Severity of coronary atherosclerosis was assessed as follows: (version A) by the number of stenosed coronary arteries (0, 1, 2, and 3 vessels); (version B) by the degree of stenosis - patients were grouped as exhibiting <75% and >75% stenosis of coronary vessels; (version C) by both degree (0 to 25%, 25 to 50%, 50 to 70%, 70 to 90%, >90%) and localization (proximal, medium, distal part and branches). For version A and B, the correlation between each chemical parameter and the degree of atherosclerotic lesion of coronary vessels was assessed by the method of disperse analysis using ANOVA procedure obtained from SAS 82 system.[8] For version C, significant difference from zero of the correlation between the severity of coronary atherosclerosis and the parameter was estimated using the CORR procedure from SAS 82 system.[10]

RESULTS AND DISCUSSION

The comparison of blood serum parameters in patients with one stenosed coronary artery and in those without angiographically documented stenoses revealed no significant difference in any of the examined chemical parameters (Table 1). In patients with two stenosed vessels significantly elevated CIC cholesterol level was revealed, whereas the other parameters failed to differ significantly from those exhibited in the control group (without stenoses in coronary arteries). Patients with three stenosed vessels showed significantly elevated levels of CIC cholesterol, LDL-cholesterol, apo B, and apo B/apo A-1 ratio (Table 1).

Table 1. Values of serum parameters in atherosclerotic patients

Parameter	Number of stenosed arteries			
	0 (Control)	1	2	3
CIC cholesterol	14+1	20+3	42+8*	28+2*
Total cholesterol	214+7	226+10	225+11	249+7*
Triglycerides	152+20	132+10	163+24	167+11
LDL cholesterol	153+7	169+13	174+12	186+8*
HDL cholesterol	37+2	34+2	33+3	33+1
apo B	128+6	140+7	147+9	150+5*
apo A-1	114+1	105+7	105+10	102+4
apo B/apo A-1	1.2+0.1	1.5+0.1	1.5+0.2	1.6+0.1*

The data presented are mean+SE. *, significant difference from the control group (p<0.05).

Patients were classified into two groups: with normal coronary arteries and with stenoses of any degree (Table 2, version A). It appeared that only CIC cholesterol and apo B/apo A-1 ratio contributed strongly to the discrimination between patients with coronary atherosclerosis and those without stenoses. As to the other serum lipid parameters, no significant correlation between their levels and the presence of atherosclerosis was statistically documented.

Table 2. Correlation between serum parameters and coronary atherosclerosis

Parameter	Version of atherosclerosis estimation		
	A	B	C
CIC cholesterol	0.0001*	0.0020*	0.0104*
Total cholesterol	0.0575	0.0559	0.0525
Triglycerides	0.4852	0.2590	0.2403
LDL cholesterol	0.0591	0.0378*	0.0681
HDL cholesterol	0.2146	0.1070	0.2462
apo B	0.0518	0.1118	0.0605
apo A-1	0.3342	0.1877	0.2768
apo B/apo A-1	0.0162*	0.0235*	0.0164*

The data presented are P values. *, $p < 0.05$. Coronary atherosclerosis was estimated according to version A, B and C as described in Materials and Methods.

Table 2 shows the correlation of the examined chemical parameters with the degree of stenosis (version B) and severity of coronary atherosclerosis (version C). Degree of stenosis (version B) correlates significantly with CIC cholesterol, LDL, and apo B/apo A-1 ratio, whereas CIC cholesterol and apo B/apo A-1 ratio are suitable parameters for the severity of the disease (version C). Thus, it can be concluded that only CIC cholesterol and apo B/apo A-1 ratio correlate significantly with both the presence and severity of atherosclerosis.

Table 3 presents data on the significance of the examined parameters for the discrimination of coronary atherosclerosis. Sensitivity, specificity and accuracy of coronary atherosclerosis diagnosis were assessed for each parameter examined. LDL and total cholesterol turned out highly sensitive, however, they were less specific as compared with CIC cholesterol that proved to be more informative for precise diagnosis of coronary artery disease. HDL-cholesterol and apo B/apo A-1 ratio were less suitable markers as compared to CIC cholesterol. Triglycerides, apo A-1 and apo B appeared to be poor markers that provided only 50% discrimination. Hence, CIC cholesterol is most reliable marker of coronary atherosclerosis as compared to other chemical parameters used in clinical practice.

Table 3. Diagnostic value of serum parameters

Parameter	Treshold value	Sensitivity, %	Specificity, %	Accuracy, %
CIC cholesterol	15 ug/ml	81	70	78
Total cholesterol	200 mg/dL	74	33	69
Triglycerides	200 mg/dL	21	74	31
LDL cholesterol	125 mg/dL	87	11	68
HDL cholesterol	35 mg/dL	63	67	64
apo B	161 mg/dL	30	92	44
apo A-1	91 mg/dL	43	77	50
apo B/apo A-1	1.48	56	69	58

Thus, it can be concluded that CIC cholesterol level and apo B/apo A-1 ratio are most powerful discriminators for either presence or severity of coronary atherosclerosis, the former parameter being more diagnostically valuable. The results obtained in this study enable us to consider CIC cholesterol as a marker of coronary atherosclerosis that may prove to be useful in clinic.

REFERENCES

1. E. I. Chazov, V. V. Tertov, A. N. Orekhov, A. A. Lyakishev, N. V. Perova, Kh. A. Kurdanov, Kh. A. Khashimov, I. D. Novikov and V. N. Smirnov, Atherogenicity of blood serum from patients with coronary heart disease, Lancet 2:595 (1986).
2. A. N. Orekhov, V. V. Tertov, S. N. Pokrovsky, I. Yu. Adamova, O. N. Martsenyuk, A. A. Lyakishev and V. N. Smirnov, Blood serum atherogenicity associated with coronary atherosclerosis. Evidence for nonlipid factor providing atherogenicity of low-density lipoproteins and an approach to its elimination, Circ. Res. 62:421 (1988).
3. A. N. Orekhov, V. V. Tertov, S. A. Kudryashov and V. N. Smirnov, Triggerlike stimulation of cholesterol accumulation and DNA and extracellular matrix synthesis induced by atherogenic serum or low density lipoprotein in cultured cells, Circ. Res. 66:311 (1990).
4. A. N. Orekhov, V. V. Tertov, D. N. Mukhin and I. A. Mikhailenko, Modification of low density lipoprotein by desialylation causes lipid accumulation in cultured cells. Discovery of desialylated lipoprotein with altered cellular metabolism in the blood of atherosclerotic patients, Biochem. Biophys. Res. Commun. 162:206 (1989).
5. A. N. Orekhov, V. V. Tertov, D. N. Mukhin and A. E. Kabakov, Modified (desialylated) low density lipoprotein and autoantibodies against lipoprotein cause atherogenic manifestations in cell culture, in: "Atherosclerosis and Cardiovascular Disease," Volume 4, Part B. G. C. Descovich, A. Gaddi, G.L. Magri and S. Lenzi, eds, Editrice Compositori, Bologna, p. 523 (1989).
6. V. V. Tertov, A. N. Orekhov, A. G. Kacharava, I. A. Sobenin, N. V. Perova and V. N. Smirnov, Low density lipoprotein-containing circulating immune complexes and coronary atherosclerosis, Exp. Mol. Pathol. 52: in press (1990).

7. V. V. Tertov, A. N. Orekhov, Kh. S. Sayadyan, S. G. Serebrennikov, A. G. Kacharava, A. A. Lyakishev and V. N. Smirnov, Correlation between cholesterol content in circulating immune complexes and atherogenic properties of CHD patients' serum manifested in cell culture, Atherosclerosis 81:183 (1990).

8. W. T. Friedenwald, R. I. Levy and D. J. Frederickson, Estimation of the concentration of low-density lipoprotein cholesterol in plasma, without use of the preparative ultracentrifuge, Clin. Chem. 18:499(1972).

9. Study Group of Atherosclerosis Society. Eur. Heart. J. 9:571(1988).

10. "SAS User's Guide: Basic," SAS Institute Inc., Cary (1982).

ANTIBODY-LIKE IMMUNOGLOBULINS G AGAINST LOW DENSITY LIPOPROTEIN THAT

STIMULATE LIPID ACCUMULATION IN CULTURED CELLS

Alexander N. Orekhov and Vladimir V. Tertov

Institute of Experimental Cardiology
USSR Cardiology Research Center
Moscow, Russia

INTRODUCTION

Recently, we have found that low density lipoprotein (LDL) derived from the blood of patients with angiographically documented coronary atherosclerosis has an atherogenic potential which manifests itself in cultures of human aortic intimal cells in the accumulation of cellular lipids (mostly free and esterified cholesterol), elevation of proliferative activity and stimulation of extracellular matrix synthesis.[1-6] LDL of most healthy donors were devoid of such atherogenic potential. We have established that atherogenic LDL circulating in patients' blood is modified lipoprotein differing from nonatherogenic LDL of healthy donors by a low sialic acid content.[7] On the other hand, removal of immunoglobulins G (IgG) fraction from the whole serum of atherosclerotic patients substantially decreased its atherogenic potential.[8] We have assumed that the blood serum of atherosclerotic patients contains autoantibodies against modified (desialylated) LDL since the antibodies against LDL can stimulate the accumulation of intracellular cholesterol.[9,10]

In the present study, IgG fraction was isolated from circulating immune complexes of the blood serum of atherosclerotic patients possessing an atherogenic potential and IgG interacting with LDL were purified by affinity chromatography. We used these anti-LDL IgG to investigate their effect on the LDL-induced accumulation of cellular cholesterol.

MATERIALS AND METHODS

The reagents were purchased from Sigma Chemical Company (St.Louis, MO) if not stated otherwise.

Blood was drawn from the ulnar vein into plastic tubes in the morning before meals after which serum was obtained. We selected 10 patients (males, 35-60-year old) with angiographically documented coronary atherosclerosis and 10 healthy donors (males, 40-62-year old) that were chosen on the basis of the presence or absence in their blood serum of the ability to induce the accumulation of cholesterol in

Hypercholesterolemia, Hypocholesterolemia, Hypertriglyceridemia
Edited by C.L. Malmendier *et al.*, Plenum Press, New York, 1990

cultured cells (atherogenic and nonatherogenic sera, respectively). All the patients' sera selected were atherogenic while all the sera of healthy donors were not. The characteristics of atherosclerotic patients and healthy donors as well as the method for evaluating serum atherogenicity were described in detail elsewhere.[1,5] None of the donors had diabetes mellitus. The mean cholesterol level in both types of sera was similar: 190 ± 6 mg/dL for patients and 186 ± 5 mg/dL for healthy subjects. Healthy donors had no signs of heart disease as determined by epidemiological criteria.

A lipoprotein-deficient fraction was obtained by centrifuging the serum at 300,000g (density 1.250 g/cm^3) for 48 h at $+4^{\circ}$C according to Lindgren[11] and dialyzed as described below. The obtained sera were sterilized by filtration (pore size, 0.45 um).

LDL (density, 1.019-1.063 g/cm^3) was isolated from the plasma obtained from patients and healthy donors according to the conventional method of ultracentrifugation in a stepwise gradient of NaBr[11] as described elsewhere.[1,2] Lipoprotein preparations and the lipoprotein-deficient serum were dialyzed for 24 h against 2,000 volumes of phosphate-buffered saline (PBS), sterilized by filtration and stored at $+4^{\circ}$C. LDL obtained from healthy donors was modified by desialylation.[7] Desialylation of LDL was carried out by agarose-bound neuraminidase (cat. no. N-4883) treatment for 2 h at 37°C. As a result of this procedure, LDL lost 70% of sialic acid. Lipoproteins were utilized within 1-4 days after preparation.

Subendothelial cells for culture were isolated from grossly normal intima by dispersion of human aortic tissue with 0.15% collagenase,[12] suspended in the growth medium containing Medium 199, 10% fetal calf serum, 2 mM L-glutamine, and antibiotics and seeded into 24-well tissue culture plates with a density of 8×10^4 cells per 1 cm^2 of the growth area.[12] The cells were cultured at 37°C in a humidified CO_2-incubator (95% air/5% CO_2). The primary cultures contained a mixed cell population made up primarily by typical and modified smooth muscle cells as defined by the ultrastructural and immunofluorescent features.[12] The medium was changed every day. Starting from the 7th day in primary culture, cells were incubated for 24 h in the media containing serum under testing or LDL and 10% lipoprotein-deficient serum of healthy donors. Lipoprotein and serum were filtered immediately before addition to cultured cells. On the 8th day, the cells were rinsed and cellular cholesterol was determined. Cellular protein was determined according to Lowry at al.[13]

Lipids were extracted from cells with hexane-isopropanol (3:2, v/v) according to Hara and Radin.[14] The total cholesterol content in the lipid extracts was determined using Boehringer Mannheim Monotest^R (Boehringer Mannheim GmbH, Mannheim, FRG). Each determination of cholesterol was done in triplicate.

To isolate immune complexes, blood serum was combined with an equal volume of a 5% polyethylene glycole (PEG) 6000 (E. Merck, Darmstadt, FRG) according to Creighton et al.[15] After a 24-h incubation, immune complexes were sedimented by centrifugation (30 min, 6000 rpm) and washed three times with a 2.5% PEG. In the experiments with cell cultures, serum and PEG-precipitates were dialyzed against 2000 volumes of phosphate buffered saline at 4°C for 24 h and then for 12 h against 1000 volumes of Medium 199.

400

To isolate IgG, PEG-precipitates obtained from a pooled serum of atherosclerotic patients were treated with 1.35 M ammonium sulfate. The IgG precipitate was dissolved in water and precipitated four times with ammonium sulfate. The resulting preparations were dyalyzed against 0.01 M phosphate buffer (pH 6.5) with subsequent purification of IgG by ion-exchange chromatography on DEAE-Sephadex A-50.[16]

IgG having affinity to LDL (anti-LDL) were purified by affinity chromatography on a column with immobilized LDL. LDL immobilization on BrCN-activated Sepharose CL 4B was carried out as described earlier.[1] Twenty to thirty milligrams of the total IgG fraction was applied to a column containing 10 ml LDL-Sepharose. The column was rinsed with 100 ml PBS, containing 1% bovine serum albumin (BSA) and 200 ml 0.5 M NaCl. The IgG bound to the column were eluted with 0.15 M glycine buffer, pH 2.5. After neutralization the eluate was dialyzed against 2000 volumes of PBS overnight and concentrated by reverse dialysis in Ficoll Type 400.

The affinity constant of IgG to LDL was measured using [^{125}I]LDL of known specific activity (1000 cpm/ng apo-B). Serial dilutions of [^{125}I]LDL in PBS containing 1% BSA and 0.05% Tween-20 (0.01-100 ug apo-B/ml) were combined with 1 ug/ml of affinity purified IgG and incubated for 8 h at 20^{0}C. The incubation mixture was transferred to plates pre-coated with goat anti-human IgG antibodies (Organon Teknika Corp., West Chester, PA). The same LDL dilutions incubated in the absence of affinity purified IgG were used as control. To immobilize the immune complex of IgG with LDL formed after an overnight incubation, the plates were rinsed with BSA-Tween-PBS with subsequent determination of radio-activity in each well. The identical procedures were carried out in the presence of 0.001-1 mg/ml unlabeled LDL. The affinity constant was calculated according to Muller[17] proceeding from the known amounts of the radioactive and non-radioactive LDL added as well as the amount of [^{125}I]LDL bound by IgG.

The significance of differences was evaluated by dispersion analysis methods using a BMDP statistical program package.[18]

RESULTS

Blood serum of atherosclerotic patients taken in a 40% concentration caused a 3-fold increase in total cholesterol of human aortic intimal cells within 24 h of cultivation (Table 1). Removal of circulating immune complexes from the serum was performed by treatment with 2.5% PEG 6000. PEG-treated serum lost its atherogenicity by 76% (Table 1). Addition to cultured cells of PEG-precipitate obtained from this serum partially restored its atherogenic properties (Table 1). During cell cultivation in the presence of a 40% lipoprotein-deficient serum of healthy donors and circulating immune complexes (PEG-precipitate) isolated from atherosclerotic patints' serum, the latter brought about a 2-fold rise in cellular cholesterol (Table 1).

From circulating immune complexes precipitated by PEG 6000, IgG fraction was isolated and then purified by affinity chromatography using LDL-Sepharose. Table 2 shows the affinity constants of purified anti-LDL IgG to lipoproteins. The antibody-like IgG show a higher affinity for LDL isolated from the blood of atherosclerotic patients as compared with LDL of healthy donors. The LDL desialylated in vitro by neuraminidase had the highest affinity constant (Table 2).

TABLE 1. Effect of circulating immune complex removal on serum atherogenicity

	Cellular cholesterol, % of control
Lipiprotein-deficient serum (control)	100±5
Inital serum of atherosclerotic patients	293±15*
PEG-treated serum	144±9*
PEG-treated serum + PEG-precipitate	204±12*
Lipoprotein-deficient serum + PEG-precipitate	199±13*

Sera were added to cultured cells in a 40% concentration. Total cholesterol content in control cells cultured with lipoprotein-deficient serum of healthy donors was 38.6±1.8 ug/mg cell protein. *, Significant difference from the control (p<0.05). Values listed are means±SE of 4 determinations.

TABLE 2. Affinity constants of lipoprotein-anti-LDL interaction

	Affinty constant x 10^{-7}, M^{-1}
Healthy donors' LDL	1.9
Atherosclerotic patients' LDL	8.7
Desialylated LDL (neuraminidase treated)	68.0

The affinity constant of anti-LDL was determined using native and modified [^{125}I]LDL of known specific activity (1000 cpm/ng apo B). The known amounts of added radioactive and non-radioactive LDL and the amount of [^{125}I]LDL bound to anti-LDL was used to calculate the affinity constant.

LDL isolated from nonatherogenic plasma of healthy donors failed to stimulate the deposition of intracellular cholesterol (Table 3). In combination with anti-LDL IgG fraction derived from the patients' circulating immune complexes, initialy nonatherogenic LDL caused a 2-fold increase in the cholesterol content of cultured cells (Table 3). LDL isolated from atherogenic patients' serum as well as LDL desialylated by neuraminidase treatment were atherogenic per se, however, anti-LDL IgG enhanced atherogenicity of these LDLs (Table 3).

DISCUSSION

Earlier, the antibodies against lipoproteins or LDL-binding factors were found in patients suffering from different diseases as well as in healthy subjects.[19-22] It was demonstrated that immunoglobulins are major LDL binding proteins in human plasma.[23] The antibody-like IgG isolated in this study from the blood of atherosclerotic patients had a higher affinity to LDL desialylated with neuraminidase and to LDL of atherosclerotic patients as compared with nonatherogenic LDL of healthy donors. Apparently, in the blood of atherosclerotic patients auto-antibodies against LDL are produced in response to the emergence of desialylated lipoproteins. The detection of antibodies with a high affinity to desialylated LDL in patients' blood serum is a strong argument in favour of the presence of desialylated LDL in patients' blood in vivo.

TABLE 3. Effect of anti-LDL IgG on cellular cholesterol accumulation

	Cellular cholesterol, % of control	
	− anti-LDL	+ anti-LDL
Lipoprotein-deficient serum (control)	100+6	−
Healthy donors' LDL	104+7	189+13**
Atherosclerotic patients' LDL	204+15*	328+19**
Desialylated LDL (neuraminidase treated)	308+21*	443+38**

Total cholesterol content in control cells cultured with 10% lipoprotein-deficient serum of healthy donors was 48.4+3.0 ug/mg cell protein. LDL was added to cultured cells in the concentration of 100 ug protein/ml. Anti-LDL IgG were added to cultured cells in the concentration of 50 ug/ml. *, Significant difference from the control ($p<0.05$); **, significant difference from the cells cultured without anti-LDL IgG ($p<0.05$). Other details are the same as in Table 1.

The present study demonstrated that autoantibody-like IgG interacting with modified LDL circulating in the blood of atherosclerotic patients substantially increase the ability of this lipoprotein to induce the deposition of cellular cholesterol. Besides, by interacting with initially nonatherogenic LDL of healthy donors these IgG make them atherogenic, i.e. capable of inducing the accumulation of intracellular cholesterol.

Our data suggest that antibodies against LDL might play the role of a major blood factor responsible for blood plasma atherogenic potential. It can be concluded that the primary factor of atherogenicity present in the blood of atherosclerotic patients is a modified (desialylated) LDL. The second blood factor substantially increasing blood plasma atherogenic potential is represented by autoantibodies produced in response to the emergence of modified LDL.

REFERENCES

1. A. N. Orekhov, V. V. Tertov, S. N. Pokrovsky, I. Yu. Adamova, O. N. Martsenyuk, A. A. Lyakishev and V. N. Smirnov, Blood serum atherogenicity associated with coronary atherosclerosis. Evidence for nonlipid factor providing atherogenicity of low-density lipoproteins and an approach to its elimination, Circ. Res. 62:421 (1988).

2. V. V. Tertov, A. N. Orekhov, O. N. Martsenyuk, N. V. Perova and V. N. Smirnov, Low density lipoproteins isolated from the blood of patients with coronary heart disease induce the accumulation of lipids in human aortic cells, Exp. Mol. Pathol. 50:337 (1989).

3. V. V. Tertov, A. N. Orekhov, H. R. Li and V. N. Smirnov, Intra-cellular cholesterol accumulation is accompanied by enhanced proliferative activity of human aortic intimal cells, Tissue Cell 20:849 (1988).

4. A. N. Orekhov, V. V. Tertov, S. A. Kudryashov and V. N. Smirnov, Triggerlike stimulation of cholesterol accumulation and DNA and extracellular matrix synthesis induced by atherogenic serum or low density lipoprotein in cultured cells, Circ. Res. 66:311 (1990).

5. E. I. Chazov, V. V. Tertov, A. N. Orekhov, A. A. Lyakishev, N. V. Perova, Kh. A. Kurdanov, Kh. A. Khashimov, I. D. Novikov and V. N. Smirnov, Atherogenicity of blood serum from patients with coronary heart disease, Lancet 2:595 (1986).

6. E. I. Chazov, A. N. Orekhov, V. V. Tertov, S. N. Pokrovsky, I. Yu. Adamova, A. A. Lyakishev, N. A. Gratsiansky, A. S. Nechaev, N. V. Perova, Kh. A. Khashimov, Kh. A. Kurdanov, V. V. Kukharchuk and V. N. Smirnov, Atherogenicity of blood plasma from patients with coronary atherosclerosis and its correction, Atherosclerosis Rev. 17:9 (1988).

7. A. N. Orekhov, V. V. Tertov, D. N. Mukhin and I. A. Mikhailenko, Modification of low density lipoprotein by desialylation causes lipid accumulation in cultured cells. Discovery of desialylated lipoprotein with altered cellular metabolism in the blood of atherosclerotic patients, Biochem. Biophys. Res. Commun. 162:206 (1989).

8. V. V. Tertov, A. N. Orekhov, Kh. S. Sayadyan, S. G. Serebrennikov, A. G. Kacharava, A. A. Lyakishev and V. N. Smirnov, Correlation between cholesterol content in circulating immune complexes and atherogenic properties of CHD patients' serum manifested in cell culture, Atherosclerosis 81:183 (1990).

9. A. N. Klimov, A. D. Denisenko, A. G. Vinogradov, V. A. Nagornev, Y. I. Pivovarova, O. D. Sitnikova and V. M. Pleskov, Accumulation of cholesteryl esters in macrophages incubated with human-antibody autoimmune complex, Atherosclerosis 74:41 (1988).

10. A. N. Orekhov, V. V. Tertov, D. N. Mukhin, V. E. Koteliansky, M. A. Glukhova, M. G. Frid, G. K. Sukhova, Kh. A. Khashimov and V.N. Smirnov, Insolubilization of low density lipoprotein induces cholesterol accumulation in cultured subendothelial cells of human aorta, Atherosclerosis 79:59 (1989).

11. F. T. Lindgren, Preparative ultracentrifugal laboratory procedures and suggestions for lipoprotein analysis, in: "Analysis of Lipids and Lipoproteins," E. G. Perkins, ed., American Oil Chemical Society, New York, p. 205 (1975).

12. A. N. Orekhov, V. V. Tertov, I. D. Novikov, A. V. Krushinsky, E. R. Andreeva, V. Z. Lankin and V. N. Smirnov, Lipids in cells of atherosclerotic and uninvolved human aorta. I. Lipid composition of aortic tissue and enzyme isolated and cultured cells, Exp. Mol. Pathol. 42:117 (1985).

13. O. H. Lowry, N. J. Rosenbrough, A. L. Farr, R. J. Randall, Protein measurement with the Folin phenol reagent, J. Biol. Chem. 193:265 (1951).

14. A. Hara, N. S. Radin, Lipid extraction of tissue with a low-toxicity solvent, Anal. Biochem. 90:420 (1978).

15. W. D. Creighton, P. H. Lambert and P.A. Misher, Detection of antibodies and soluble antigen-antibody complexes by precipitation with polyethylene glycol, J. Immunol. 111:1219 (1973).

16. J. Brock, Immunoglobulin isolation, in: "Immunologische Arbeitsmethoden," H. Friemel, ed., VEB Gustav Fischer Verlag, Jena, p. 390 (1984).

17. R. Muller, Calculation of average antibody affinity in anti-hapten sera from data obtained by competitive radioimmunoassay, J. Immunol. Methods 34:345 (1980).

18. W. J. Dixon and M. B. Brown, "Biomedical Computer Programs. P-Series," University of California Press, Berkeley, p. 185 (1977).

19. J.-L. Beaumont, L'hyperlipidemie par auto-anticorps anti-beta-lipoproteine. Une nouvelle entite pathologique, C. R. Acad. Sci. Paris [Ser D] 261:4563 (1965).

20. J. S. Taylor, L. A. Lewis, J. D. Battle, Jr, A. Butkus, A. L. Robertson, S. Deodhar and H. H. Roenigk, Jr, Plane xanthoma and multiple myeloma with lipoprotein-paraprotein complexing. Arch. Dermatol. 114:425 (1978).

21. P. J. Gallagher, B. D. Jones, C. R. Casey and G. P. Sharratt, Circulating immune complexes in cardiac disease, Atherosclerosis 44:241 (1982).

22. E. Szondy, M. Horvath, Z. Mezey, J. Szekely, E. Lengyel, G. Fust and S. Gero, Free and complexed anti-lipoprotein antibodies in vascular diseases, Atherosclerosis 49:69 (1983).

23. B. J. Bauer, K. Blashfield, D. A. Buthala and L. C. Ginsberg, Immunoglobulin as the major low density lipoprotein binding protein in plasma, Atherosclerosis 44:153 (1982).

NUTRITIONAL REGULATION OF APOLIPOPROTEIN GENES: EFFECT OF

DIETARY CARBOHYDRATES AND FATTY ACIDS

Agnès Ribeiro, Marise Mangeney, Philippe Cardot*, Claude Loriette, Jean
Chambaz, Yves Rayssiguier* and Gilbert Béréziat

Laboratoire de biochimie (CNRS URA 1283) CHU Saint-Antoine, Université
Pierre et Marie Curie PARIS & *Laboratoire des maladies métaboliques (INRA)
THEIX FRANCE

SUMMARY

The effect of nutritional factors on apolipoprotein gene expression by rat liver were studied. Dietary carbohydrates or fatty acids regulate the expression of apo E gene, by altering either gene transcription or mRNA stability. Conversely, apo AI regulation occurs at a post transcriptional level. *In vivo* and *in vitro* experiments gave contradictory results concerning apo B gene expression. The more dramatic changes in plasma lipids and apolipoproteins are obtained under dietary fish oil. Hepatocytes from fish oil-fed rats retain for several days modification in fatty acid metabolism, i.e. a shift in oleic acid channeling towards oxidation at the expense of esterification and a reduced ability to synthesize and secrete triacylglycerol. These modifications are paralleled with a decrease in the synthesis and in the secretion of apo Bs. Hepatocytes from fish oil fed rats secrete degradative forms of apo B which might result from either a sluggish VLDL synthesis and secretion or a more specific effect of n-3 long chain polyunsaturated fatty acid peroxidative products. Hepatocytes from fish oil fed rats exhibit a reduced ability to synthesize cholesterol, associated with a decrease in apo AI synthesis and secretion without any modification in apo AI mRNA. In contrast, the hepatocytes exhibit a concomitent decrease in apo E synthesis and secretion and in cellular apo E mRNA levels.

INTRODUCTION

Dietary fatty acids and carbohydrates have a profound effect on the metabolism of lipids and lipoproteins. Diets rich in saturated fats and devoid of essential fatty acids induce a reduction in plasma triacylglycerols and a rise in the hepatic secretion of VLDL[1,2]. The effects on plasma cholesterol are variable[3-5] ; they induce a reduction in the plasma apo E level and a rise in the apo AI level, while the apo B level is virtually unchanged[5]. The rate of synthesis and the composition of the VLDL secreted by the liver are altered by the nature of the carbohydrates in the diet. Fructose stimulates the synthesis and secretion of triacylglycerols, inducing hypertriglyceridaemia[2,6,7] without affecting the plasma cholesterol level[2,8]. These modifications are accompagnied by increased hepatic synthesis of apo E and apo C[9,10].

The introduction of fish oil into the diet induces a reduction in the plasma triacylglycerol level in man. This effect has been attributed to the presence of eicosapentaenoic and docosahexaenoic acids, produced by interconversion of α-linolenic acid[11]. However, the reported effects on plasma cholesterol and plasma apo B and LDL concentrations are contradictory[12]. In monkeys, enrichment of the diet with fatty acids of the n-6 series does not have any effect on the hepatic secretion of apo B or on the messenger RNA level[13]. When the source of lipids is composed of fish oil, the hepatic accumulation of cholesterol decreases without affecting the plasma concentration of apo B[14]. In the rat, fish oil induces a reduction in the synthesis and secretion of triacylglycerols[15] and apolipoproteins by the perfused liver[16].

Hypercholesterolemia, Hypocholesterolemia, Hypertriglyceridemia
Edited by C.L. Malmendier *et al.*, Plenum Press, New York, 1990

407

Hepatocytes obtained from rats fed with fish oil and rat hepatocytes[18] or HepG2 cells[19] cultured in the presence of eicosapentaenoic acid present a reduced capacity for the synthesis and secretion of triacylglycerols. In contrast, the results concerning the synthesis of apolipoproteins by cultured hepatocytes or HepG2 cells are contradictory[19,20].

We re-evaluated all of these effects by measuring, in the weaned rat, the effects of the nature of the fatty acids and carbohydrates in the diet on the plasma and hepatic concentrations of cholesterol and triacylglycerols as well as the plasma apolipoproteins concentrations and the hepatic content of the corresponding messenger RNA. As diets containing fish oil are more effective for reducing plasma cholesterol and plasma triacylglycerols in rats, we investigated their ability to induce persistent modification in the metabolism of lipids and lipoproteins in primary cultures of hepatocytes.

MATERIALS AND METHODS

Weaned male Wistar rats (62 ± 2 g) were kept in individual cages maintained under 12 hours light/dark cycles at 20°C. The diets were prepared according to the recommendations of the American Institute of Nutrition[21]. They contained 70.5% of carbohydrates (weat starch, sucrose or glucose), 20% of casein devoid of essential fatty acids, 3.5% of mineral salts, 1% of a vitamin mixture and 5% of lipids (corn oil, hydrogenated coconut oil or salmon oil). The rats were sacrified during the postprandial period at 9 o'clock in the morning.

Blood was collected by aortic puncture and the livers were washed and immediately frozen at-70°C. The lipids were extracted according to the method proposed by Bligh and Dyer[22] and their plasma and hepatic concentrations were measured by means of BioMerieux kits (France). The plasma lipoproteins were isolated by ultracentrifugation according to Havel's technique[23]. After dialysis and measurement of the proteins according to Lowry's method[24], the samples were submitted to gel electrophoresis on 4% or 12% polyacrylamide gel containing 1% of sodium dodecylsulfate according to Laemmli's technique[25]. The gels were stained with Coomassie blue and the bands were measured by densitometry using an LBK ultroscan.

The hepatocytes were isolated by collagenase perfusion and primary culture performed (5 x 10^6 cells per well) according to a previously described technique[26]. After 4 hours of adhesion, the medium was replaced by a resting medium containing antibiotics as previously indicated[26] and a supplement of 10^{-6} M dexamethasone and 10^{-6} M insulin. After standing overnight, the hepatocytes were incubated for 24 hours in the presence of 0.6 mM oleic acid complexed with 0.15 mM albumin or in the presence of albumin alone.

The cholesterol synthesis was measured by the incorporation of [U-^{14}C] acetate (2 µCi/well, 94 mCi/mole) for 3 hours. The β oxidation of the fatty acids was assessed by measuring the radioactivity derived from [^{14}C] oleic acid (2 µCi per well, 60 mCi/mole) detected in the ketone bodies after 24 hours, as previously described[27]. The synthesis and secretion of triacylglycerols were measured by the incorporation of [U^{14}C] glycerol (2 µCi per well, 120 mCi/mole) into the hepatocytic triacylglycerols or triacylglycerols in the medium. Labelled glycerol was added during the last three hours of incubation. After extraction of the cells and the media, the lipids were separated by thin layer chromatography[26], eluted and their radioactivity was measured.

In order to evaluate the synthesis of apolipoproteins, the hepatocytes were preincubated for 24 hours in the presence or absence of oleic acid, before being incubated for three hours in the presence of a medium devoid of methionine and supplemented with [^{35}S] methionine (60 µCi/well, 1,000 Ci/mole). The medium was then collected and centrifuged at 4,000 r.p.m. for 10 minutes to eliminate cellular debris. The media were immunoprecipitated according to the technique described by Andus et al.[28]. The antisera used were an antitotal lipoproteins antiserum and a specific anti-apo B antiserum and the proteins immunoprecipitated were dissolved as described elsewhere[29]. Gel electrophoresis on 4% or 12% polyacrylamide gel was performed in the presence of 1% sodium dodecylsulfate[25]. The gels were soaked for 30 minutes in a mixture of isopropanol : water : acetic acid (25:65:10 v/v/v) then in a labelling-amplifying reagent (Amersham) before being submitted to autoradiography. The autoradiographs were measured by an LKB ultroscan.

The total RNA of the liver was prepared according to the technique described by Lenich et al.[30] and the hepatocyte RNA was prepared according to the method of Chomcznski and Sacchi[31]. The

RNA prepared from 3 to 6 wells were pooled. The total RNA was quantified by spectrophotometry and its integrity was confirmed by agarose gel electrophoresis. Successive dilutions of RNA (0.5 to 4 μg) were deposited onto nylon membranes. In order to quantify the messenger RNA extracted from the liver and from the hepatocytes, hybridisations were performed using cDNA probes corresponding to apo E mRNA, to the common part of the messenger RNAs for apo B_H and B_L and apo A_I mRNA labelled by nick translation with $[\alpha^{32}P]dCTP$ (specific activity of between 10^8 and 10^9 cpm/mg of cDNA according to the technique described elsewhere[32]. The relative quantity of messenger RNA was calculated by reference to the messenger RNA content of β-actin.

RESULTS

The plasma and hepatocytic concentrations of triacylglycerols, phospholipids and cholesterol are indicated in Table I. As expected, rats fed with sucrose and corn oil had the highest plasma triacylglycerol level and the rats fed with starch and fish oil had the lowest plasma triacylglycerol level. Deprivation of essential fatty acids caused a reduction in the plasma triacylglycerol level. The plasma cholesterol was remarkably stable, slightly increased in the rats receiving saturated fats and lower in the rats receiving fish oil. The hepatocytic lipids were very stable.

Table I . Plasma and liver lipids

	HCO sucrose	CO	CO glucose	CO starch	SO
plasma (mM)					
TAG	1.50 + 0.11	2.30 + 0.32	0.84 + 0.07	1.15 + 0.12	0.60 + 0.03
PL	1.40 + 0.05	1.45 + 0.45	1.52 + 0.07	1.83 + 0.07	1.40 + 0.05
Chol	2.05 + 0.03	1.80 + 0.05	1.80 + 0.10	1.90 + 0.04	1.45 + 0.04
liver (mg/g.liver)					
TAG	16.0 + 0.7	12.0 + 1.3	10.7 + 0.6	14.4 + 0.7	10.3 + 0.8
PL	37.7 + 0.9	35.5 + 1.6	37.5 + 2.0	36.5 + 0.7	34.5 + 0.5
Chol	6.5 + 0.2	5.9 + 0.3	6.1 + 0.2	5.3 + 0.2	5.8 + 0.2

HCO hydrogenated coconut oil, CO corn oil, SO : salmon oil.

Table II . Effect of dietary sucrose or dietary saturated fats
on plama apolipoproteins and liver mRNAs.

	Hydrogenated coconut oil Sucrose	Corn oil	
		Sucrose	Glucose
apo A_I (%)	78.2 + 1.2	62.2 + 0.9	67.5 + 1.2
apo E (%)	6.8 + 0.6	18.2 + 1.2	13.4 + 1.3
apo E mRNA (AU)	0.38 + 0.06	0.60 + 0.10	0.40 + 0.10
apo B μg/ml	184 + 11	198 + 10	168 + 15
apo B mRNA (AU)	1.90 + 0.10	2.20 + 0.12	1.60 + 0.70

Results are expressed as % of total apolipoproteins for apo A_I and apo E and as arbitrary units for mRNA.

The rats receiving diets only containing saturated fatty acids had a lower proportion of plasma apo E, as previously reported[5]. A similar result was obtained when dietary sucrose was replaced by glucose[10]. These variations were correlated with the hepatic concentrations of messenger RNA for apo E (Table II). The effect of carbohydrates may be related to the presence of fructose as previously demonstrated by Strobl et al.[33]. Conversely, the plasma apo AI concentration was increased when the diet contained saturated fatty acids or when sucrose was replaced by glucose. The absence of essential fatty acids in the diet did not modify either the plasma apo B level or the concentration of liver apo B mRNA. Glucose slightly but significantly reduced the plasma apo B concentration, which was accompanied by a reduction in hepatic messenger RNA. Table III presents the results for apolipoproteins in rats receiving salmon oil compared with those receiving corn oil. Fish oil did not significantly modify the proportion of plasma apo B, although its hepatic messenger RNA was decreased. The apo BH concentration was not significantly modified but he BL concentration decreased in the rats fed with fish oil although the messenger RNA for apo B remained unchanged.

As the diet enriched with fish oil exerted the greatest effect on plasma triacylglycerols and plasma cholesterol, we wanted to determine whether it corresponded to a lasting modification of the

Table III . Effect of dietary fish oil or corn oil on plasma apolipoproteins and liver mRNAs.

	Corn oil	Salmon oil
apo AI(%)	69.5 ± 0.6	65.7 ± 1.7
apo E (%)	11.7 ± 0.3	12.7 ± 1.0
apo E mRNA (AU)	1.64	1.13
apo BH/BL ratio	3.3 ± 0.3	6.3 ± 0.3
apo B mRNA (AU)	1.01	1.00

Results are expressed as % of total apolipoproteins for apo AI and apo E, and as arbitrary units for mRNAs.

Table IV . Cholesterol, triacylglycerol and ketone bodies synthesis
in primary culture of hepatocytes from corn oil-fed or fish oil-fed rats.

	Corn oil	Salmon oil %
incorporation of :		
[U14C] acetate into cholesterol	$3.80 + 0.33$	$1.36 + 0.7$ (-65%)
[U14C] glycerol into triacylglycerol	$62.8 + 8.6$	$46.9 + 1.5$ (-26%)
[114C] oleate into acido soluble compounds	$48.9 + 2.5$	$68.3 + 2.5$ (+39%)

Results are expressed as % of the radioactivity taken up by hepatocytes recovered with cholesterol, triacylglycerol or acido soluble compounds.

phenotype. We therefore studied the synthesis and secretion of lipids and apolipoproteins in primary hepatocyte cultures. Table IV shows that the hepatocytes obtained from the livers of rats fed with salmon oil had a lower capacity for synthesis of cholesterol (-65%) and triacylglycerols (-26%) than those hepatocytes obtained from rats fed with corn oil. In contrast, they more effectively transformed oleic acid into ketone bodies (+39%). The decreased synthesis of triacylglycerols was accompanied by a decrease in their secretion, measured either indirectly by incorporation of labelled glycerol into the triacylglycerols secreted or directly by assay of the triacylglycerol concentration in the hepatocytes and in the culture medium (Figure 1).

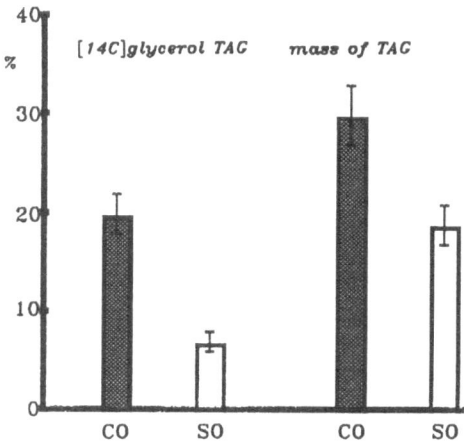

Figure 1. Effect of corn oil and fish oil diets on triacylglycerol secretion by rat hepatocytes

Secreted TAG are estimated by mass determination or by the proportion of [^{14}C] glycerol recovered in secreted TAG

The synthesis and secretion of apolipoproteins were studied in hepatocytes preincubated for 24 hours in the presence or absence of oleic acid then labelled with [^{35}S] methionine for 3 hours. The media were then immunoprecipated with either an anti-total lipoproteins antiserum or a specific anti-apo B antiserum. The proteins were separated by electrophoresis on 4 % and 12% polyacrylamide gel. Figure 2 left shows that the secretion of newly synthesized apo A$_I$ and apo E was reduced in the cultures of hepatocytes derived from rats fed with salmon oil. This reduction was less marked when the hepatocytes were incubated in the presence of oleic acid. Oleic acid also modified the distribution of the isoforms of apo E, as previously reported by Davis et al.[34]. The separation of the

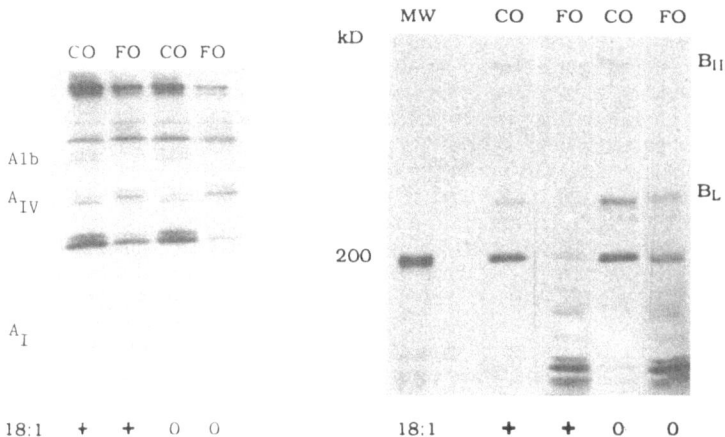

Figure 2. Incorporation of [^{35}S]methionine into secreted apolipoproteins.

Autoradiography of a SDS-PAGE electrophoresis of medium proteins immunoprecipitated with either a total anti rat apolipoprotein antiserum (12% PAGE), left or a specific anti rat apo B antiserum (4% PAGE), right.

immunoprecipitated apo B by the 4% gel revealed that the hepatocytes obtained from rats fed with fish oil secreted less newly synthesized apo B and apo B_H disappeared completely. Numerous bands with an apparent molecular weight less than 200 kD were also observed (Figure 2 right).

The messenger RNA concentration of the cultured hepatocytes is shown in Table V. The apo E mRNA concentration was lower in the hepatocytes derived from rats fed with fish oil, whether they were cultured in the presence or in the absence of oleic acid. A reduction in apo B mRNA was also observed, but it was less marked when the hepatocytes were cultured in the presence of oleic acid. The variations in apo A_I were not correlated with the variations in the corresponding messenger RNA.

Table V . Apo E, B and A_I mRNA concentration in primary culture
of rat hepatocytes from corn oil- or fish oil-fed rats.

oleic acid 0.6 mM	Corn oil		Salmon oil	
	-	+	-	+
apo E mRNA	1.12	1.02	0.75 (-33%)	0.45 (-56%)
apo B mRNA	1.07	1.20	0.70 (-35%)	0.95 (-21%)
apo A_I mRNA	0.70	0.87	0.85 (+21%)	0.65 (+26%)

Results are expressed in arbitrary unit, values between brakets indicate the decrease in salmon oil as compared with corn oil.

DISCUSSION

Dietary conditions, particularly the nature of the dietary carbohydrates and fats, affect the plasma triacylglycerols and cholesterol concentrations and consequently have an important influence on the risk of atherosclerosis[35]. These plasma levels depend on complex processes in which the liver plays a central role. In fact, the liver is the site of remodelling of lipids derived from the uptake of chylomicron, LDL and HDL residues or from the neosynthesis from acetate radicals[36]. The liver is also the site of synthesis of apolipoproteins B, E and A_I which play an important role in the hepatic secretion of VLDL and the catabolism of lipoprotein[37]. The possible nutritional regulation of lipoprotein gene expression is still a subject of debate.

The results presented here demonstrate that nutritional factors regulate the expression of the gene coding for apolipoprotein E. In agreement with Strobl et al.[33], but in contradiction with Kim et al.[38], we showed that sucrose caused an increase in the hepatic levels of messenger RNA for apo E (Table II). Inversely, replacement of essential fatty acids in the diet by saturated fats caused a marked reduction in the messenger RNA for apo E (Table II). In both cases, the proportion of apo E in the plasma was modified accordingly. In contrast, feeding the rats with fish oils decreased the messenger RNA concentration in the liver without affecting the plasma apo E concentration (Table III). However, the hepatocytes prepared from these animals presented both a reduction in the synthesis and secretion of apo E and a reduction in its messenger RNA (Figure 2, Table V) in comparison with hepatocytes from rats fed with corn oil. At the present time, only fasting or a raised glucagon/insulin ratio have been clearly related to expression of the apo E gene in the liver[27,39]. Contradictory results have been obtained from cultured hepatocytes concerning the role of cholesterol in the control of the apo E gene[40-42]. Our results indicate that when cholesterol synthesis is reduced by a diet enriched with fish oil, there is a reduction in the incorporation of [^{35}S] methionine in the apo E secreted together with a reduction in the level of its messenger RNA. However, under nutritional situations in which apo E mRNA is decreased in the liver, no significant variation is observed in the hepatic cholesterol level, which may indicate that although the rate of cholesterol synthesis influences expression of the gene, it involves the participation of particular cell pools. The plasma cholesterol level (Table I) was inversely correlated to the plasma apo A_I level (Tables II and III). This may be explained by a reduction in the synthesis and secretion of apo A_I, as in the case of cultured hepatocytes obtained from rats fed with fish oil (Figure 2). This effect intervenes at a post-transcriptional stage as the messengers are not modified (Table V).

The replacement of glucose by sucrose in the diet increased the plasma triacylglycerol level and the hepatic level of apo B mRNA, but increased the plasma apo B level to a lesser degree. In contrast, the effect of apo B was reinforced by fasting for 24 hours[39]. Saturated fatty acids decreased the plasma triacylglycerol level without affecting either the plasma apo B concentration or the concentration of its messenger RNA in the liver. The most dramatic effect on the plasma triacylglycerol level was obtained with diets enriched with fish oil. The effect on circulating apo B was discrete and essentially affected light apo B, while the hepatic messenger RNA level remained unchanged.

Primary culture of hepatocytes retain the capacity to accumulate triacylglycerol and to secrete lipoproteins into the culture medium when fatty acids are added[26]. Hepatocytes derived from rats fed with fish oil maintained their characteristics for several days, i.e. reduced synthesis and secretion of triacylglycerols determined either in terms of mass or by measuring the incorporation of labelled glycerol (Figure 1) associated with an increase in β-oxidation (Table IV), as has been previously demonstrated[44,45]. In agreement with previous studies on rat hepatocytes[41,46] or chicken hepatocytes[47], the increase in triacylglycerols obtained in the presence of oleic acid was not accompagnied by an increased synthesis and secretion of apo B. This could be due to an increase in the intracellular association of triacylglycerols with apolipoproteins and the secretion of larger particles[10]. The reduction in the incorporation of [^{35}S] methionine into the apo B secreted by hepatocytes derived from rats fed with fish oil compared with those fed with corn oil was accompagnied by a slight decrease in messenger RNA (Table V) which was attenuated when the culture was performed in the presence of oleic acid. The presence on the autoradiograph of several [^{35}S] methionine-labelled proteins immunoprecipitated by the anti-apo B antiserum and presenting a molecular weight equal to or less than 200 kD reinforces the hypothesis of a diet-induced alteration in the stability of apo B. Borchardt and Davis have shown that a large proportion of newly synthesized apo B is broken down in cultured rat hepatocytes[48]. An increased breakdown of apo B could result from a decreased utilisation of newly synthesised apo B by VLDL secretion due to a reduction in the synthesis of triacylglycerols or from a more specific action of the peroxidation derivatives of eicosapentaenoic acid and docosahexaenoic acid[49].

REFERENCES

1 Sinclair, AJ. and Collins, FD. Fatty livers in rats deficient in essential fatty acids. Biochim. Biophys. Acta 152 : 498-510 (1968).

2. Trugnan, G, Thomas, G, Cardot, P, Rayssiguier, Y and Béréziat, G. Short-term essential fatty acid deficiency in rats. Influence of dietary carbohydrates. Lipids 20 : 862-868 (1985).

3. Williams MA, Tinoco, J, Hincenbergs, I and Thomas, B. Increased plasma triglyceride secretion in EFA-deficient rats fed diets with or without saturated fat. Lipids 24 : 448-453 (1989).

4. Sugano, M and Portman OW. Essential fatty acid deficiency and cholesterol esterification activity of plasma and liver in vitro and in vivo. Arch. Biochim. Biophys. 109 : 312-315 (1965).

5. Ney, DM, Ziboh, VA and Schneeman, BO. Reduction in plasma apolipoprotein E and HDL$_1$ levels in rats with essential fatty acid deficiency. J. Nutr. 117 : 2016-2020 (1987).

6. Kannan, R, Baker, N and Bruckdorfer, KR. Secretion and turnover of very low density lipoprotein triacylglycerols in rats fed chronically diets rich in glucose and fructose. J. Nutr. 111 : 1216-1223 (1981).

7. Yamamoto, M, Yamamoto, I, Tanaka, Y and Ontko, JA. Fatty acid metabolism and lipid secretion by perfused livers from rats fed laboratory stock and sucrose-rich diets. J. Lipid Res. 28 : 1156-1165 (1987).

8. Cardot, P, Chambaz, J, Thomas, G, Rayssiguier, Y and Béréziat, G. Essential fatty acid deficiency during pregnancy in the rat : influence of dietary carbohydrates. J. Nutr. 117 : 1504-1513 (1987).

9. Boogaerts, JR, Malone-Mc Neal, M, Archambault-Schexnayder, J and Davis, RA. Dietary carbohydrate induces lipogenenis and very-low-density lipoprotein synthesis. Am. J. Physiol. 246 : E77-E83 (1984).

10. Witztum, JL and Schonfeld, G. Carbohydrate diet-induced changes in very low-density lipoprotein composition and structure. Diabetes 27 : 1215-1229 (1978).

11. Sanders, TAB, Vickers, M and Haines, AP. Effect on blood lipids and haemostasis of a supplement of cold-liver, rich in eicosapentaenoic and docosahexaenoic acid, in healthy young men. Clin. Sci.61 : 317-324 (1981).

12. Harris, WS. Fish oils and plasma lipid and lipoprotein metabolism in humans : a critical review. J. Lipid Res. 30 : 785-807 (1989).

13. Sorci-Thomas, M, Wilson, MD, Johnson, FL, Williams, DL and Rudel, LL. Studies on the expression of genes encoding apolipoproteins B100 and B48 and the low density lipoproteins receptor in nonhuman primates. J. Biol. Chem. 264 : 9039-9045 (1989).

14. Parks, JS, Wilson, MD, Johnson, FL and Rudel, LL. Fish oil decreases hepatic cholesteryl ester secretion but not apo B secretion in african green monkeys. J. Lipid Res. 30 : 1535-1544 (1989).

15. Wong, SH, Nestel, JP, Timble, RP, Storer, GB, Illman, RJ and Topping, DL. The adaptive effects of dietary fish and safflower oil on lipid and lipoprotein metabolism in perfused rat liver. Biochim. Biophys. Acta 792 : 103-109 (1984).

16. Wong, SH and Marsh, JB. Inhibition of apolipoprotein secretion and phosphatidate phosphohydrolase activity by eicosapentaenoic and docosahexaenoic acids in the perfused rat liver. Metabolism 37 : 1177-1181 (1988).

17. Wong, S, Reardon, M and Nestel, PJ. Reduced triglyceride formation from long chain polyenoïc fatty acids in rat hepatocytes. Metabolism 43 : 900-905 (1985).

18. Strum-Odin, R, Adkins-Finke, B, Blake, WL, Phinney, SD and Clarke, SD. Modification of fatty acid composition of membrane phospholipid in hepatocyte monolayer with n-3, n-6 and n-9 fatty acid and its relationship to triacylglycerol production. Biochim. Biophys. Acta 921 : 378-391 (1987).

19. Wong, S and Nestel PJ. Eicosapentaenoic acid inhibits the secretion of triacyl glycerol and of apoprotein B and the binding of LDL in HepG2 cells. Atherosclerosis 64 : 139-146 (1987).

20. Nossen, JO, Rustan, AC, Gloppestad, SH, Malbakken, S and Drevon, CA. Eicosapentaenoic acid inhibits synthesis and secretion of triacylglycerols by cultured rat hepatocytes. Biochim. Biophys. Acta 879 : 56-65 (1986).

21. Bieri, JG, Stoewsand, GS, Briggs, GM, Phillips, RW, Woodard, JC and Knapka, JJ. Report of the american institution of nutrition ad hoc. Committee on standards for nutritional studies. J. Nutr. 107 : 1340-1348 (1977).

22. Bligh, EG and Dyer, WJ. A rapid method of total lipid extraction and purification. Can. J. Biochem. Physiol. 37 : 911-917 (1959).

23. Havel, RJ, Eder, HA and Bragdon JH. The distribution and chemical composition of ultracentrifigally separated lipoproteins in human serum. J. Clin. Invest. 34 : 1345-1447 (1955).

24. Lowry, OH, Rosebrough, NJ, Farr, AL and Randall, RJ. Protein measurement with the Folin phenol reagent. J. Biol. Chem. 193 : 265-275 (1951).

25. Laemmli, UK. Cleavage of structural proteins during the assembly of the head of bacteriophage T4. Nature 227 : 680-685 (1970).

26. Chambaz, J, Guillouzo, A, Cardot, P, Pepin, D and Béréziat, G. Essential fatty acid uptake and esterification in primary culture of rat hepatocytes. Biochim. Biophys. Acta 878 : 310-319 (1986).

27. Mangeney, M, Sire, O, Montagne, J and Nordmann, J. Effect of D-galactosamine in vitro on [U-14C]palmitate oxidation, triacylglycerol synthesis and secretion in isolated hepatocytes. Biochem. Biophys. Acta 833 : 119-127 (1985).

28. Andus, T, Gross, V, Tran-Thi, TA, Schreiber, G, Nagashima, M and Heinrich, PC. The biosynthesis of acute-phase proteins in primary cultures of rat hepatocytes. Eur. J. Biochem. 133 : 561-571 (1983).

29. Ribeiro, A, Mangeney, M, Cardot, P, Loriette, C, Rayssiguier, Y, Chambaz, J and Béréziat, G. Effect of dietary fish oil and corn oil on lipid metabolism and apolipoprotein gene expression by rat liver. Soumis à publication.

30. Lenich, C, Brecher, P, Makrides, S, Chobanian, A and Zannis, VI. Apolipoprotein gene expression in the rabbit : abundance, size and distribution of apolipoprotein mRNA species in different tissues. J. Lipid Res. 29 : 755-764 (1988).

31. Chomczynski, P and Sacchi, N. Single-step method of RNA isolation by acid guanidinium thiocyanate-phenol-chloroform extration. Anal. Biochem. 162 : 156-159 (1987).

32. Mangeney, M, Cardot, P, Lyonnet, S, Coupe, C, Benarous, R, Munnich, A, Girard, J, Chambaz, J and Béréziat, G. Apolipoprotein-E-gene expression in rat liver during development in relation to insulin and glucagon. Eur. J. Biochem. 181 : 225-230 (1989).

33. Strobl, W, Gorder, NL, Fienup, GA, Lin-Lee, YC, Gotto, Jr AM and Patsch, W. Effect of sucrose diet on apolipoprotein biosynthesis in rat liver (increase in apolipoprotein E gene transcription). J. Biol. Chem. 264 : 1190-1194 (1989).

34. Davis, RA, Diuz, SM, Leighton, JK and Brengaze, VA. Increased translatable mRNA and decreased lipogenesis are responsible for the augmented secretion of lipid-deficient apolipoprotein E by hepatocytes from fasted rats. J. Biol. Chem. 264 : 8970-8977 (1989).

35. Cotran RS and Munro JM. Pathogenesis of atherosclerosis. In Recent concept in the role of cholesterol in atherosclerosis. Grundy SM and Bearn AG eds Hanley and Belfus inc. Philadelphia. pp 5-24 (1987).

36. Ruderman NB, Richards KC, De Bourges VN and Jones AL. Regulation of production and release of lipoprotein by the perfused rat liver. J. Lipid Res. 9 : 613-619 (1988).

37. Davis, RA and Boogaerts JR. Intrahepatic assembly of very-low-density lipoproteins : effects of fatty acids on triacylglycerol and apolipoprotein synthesis. J. Biol. Chem. 257 : 10908-10913 (1982).

38. Kim, MH, Nakayama, R, Manos, P, Tomlinson JE, Choi, E, Joseph, D.Ng, and Holten D. Regulation of apolipoprotein E synthesis and mRNA by diet and hormones. J. Lipid Res. 30 : 663-671 (1989).

39. Cardot, P, Ribeiro, A, Lablanquie, C, Mangeney, M, Rayssiguier, Y and Chambaz, J. Effect of dietary sucrose or glucose on apolipoprotein B and E gene expression in rat liver. Influence of fasting and essential fatty acid deficiency. Soumis à publication.

40. Lin-Lee YC, Tanaka, Y, Lin, CT and Chan, L. Effect of an atherogenic diet on apolipoprotein E biosynthesis in the rat. Biochemistry 20 : 6474-6480 (1981).

41. Davis, RA and Malone-Mc Neal M. Dietary cholesterol does not affect the synthesis of apolipoproteins B and E by rat hepatocytes. Biochem. J. 227 : 29-35 (1985).

42. Kosykh, VA, Preobrazhensky, SN, Fuki, IV, Zaikina OE, Tsibulski, VP, Repin, VS and Smirnov, VN. Cholesterol can stimulate secretion of apolipoprotein B by cultured human hepatocytes. Biochim. Biophys. Acta 836 : 385-389 (1985).

43. Apostopoulos, JJ, Howlett GJ and Fidge N. Effect of dietary cholesterol and hypothyroidism on rat apolipoprotein mRNA metabolism. J. Lipid Res. 28 : 642-648 (1987).

44. Bergseth, S, Christiansen, EN and Bremer, J. The effect of feeding fish oils, vegetable oils and clofibrate on the ketogenesis from long chain fatty acids in hepatocytes. Lipids, 21 : 508-514 (1986).

45. Yamasaki, RK, Shen, T and Shade, GB. A diet rich in (n-3) fatty acids increases peroxisomal β-oxidation activity and lowers plasma triacylglycerols without inhibiting glutathione-dependent detoxification activities in the rat liver. Biochim. Biophys. Acta 920 : 62-67 (1987).

46. Patsch, W, Tama, T and Schonfeld, G. Effect of fatty acids on lipid and apoprotein secretion and association in hepatocyte cultures. J. Clin. Invest. 72 : 371-378 (1983).

47. Situa-Manango, P, Janero, DR and Lane MD. Association and assembly of triglyceride and phospholipid with glycosylated and un glycosylated apoproteins of very-low-density lipoprotein in the intact liver cell. J. Biol. Chem. 257 : 11463-11467 (1982).

48. Borchardt, RA and Davis, RA. Intrahepatic assembly of very-low-density lipoproteins (rate of transport out of the endoplasmic reticulum determines rat of secretion). J. Biol. Chem. 262 : 16394-16402 (1987).

49. Van Rollins, M. Cytochrome P-450 metabolites of eicosapentaenoic acids are strong inhibitors of lipid synthesis in rat hepatocytes. Circulation 80 : A2467 (1989).

LIPOPROTEIN STRUCTURE AND METABOLISM DURING PROGRESSION AND REGRESSION OF ATHEROSCLEROSIS IN PIGS FED WITH FISH OIL-DERIVED FATTY ACIDS

A. Van Tol[1], T. Van Gent[1], L.M. Scheek[1], J.E.M. Groener[1], L.M.A. Sassen[2], J.M.J. Lamers[1] and P.D. Verdouw[2]

Department of Biochemistry I[1] and Laboratory for Experimental Cardiology (Thoraxcenter)[2], Faculty of Medicine and Health Sciences Erasmus University Rotterdam, P.O. Box 1738, 3000 DR Rotterdam The Netherlands

INTRODUCTION

Evidence in favor of beneficial effects of dietary fish oil on prevention or regression of atherosclerosis has been obtained from epidemiological studies in man, as well as from experimental studies in pigs (1-5). Both human and animal studies have shown that fish oil-derived fatty acids may have several mechanisms of action, resulting in protection against the atherosclerotic proces. Effects have been described to occur on platelet aggregation (6,7), cytokine synthesis by mononuclear cells (8) and on plasma lipoproteins, especially the very low density lipoproteins (9-13). Some studies show beneficial effects on progression of atherosclerosis even in the absence of significant effects on (extremely elevated) plasma low density lipoprotein (LDL) levels (2,3,14), but a complete analysis of the chemical composition of all lipoprotein classes is lacking. The aim of the present study is to evaluate the concentration and chemical composition of plasma

Hypercholesterolemia, Hypocholesterolemia, Hypertriglyceridemia
Edited by C.L. Malmendier *et al.*, Plenum Press, New York, 1990

417

lipoproteins during progression and regression of atherosclerosis, induced by dietary interventions. Plasma cholesterol metabolism by lecithin:cholesterol acyltransferase (LCAT) is also emphasized. Lecithin:cholesterol acyltransferase (LCAT) plays a major role in the formation of plasma cholesterylesters. Its activity is crucial for normal plasma cholesterol turnover and a specific function in reverse cholesterol transport (e.g. transport of cholesterol out of the arterial wall to the liver) has been proposed (15).

MATERIALS AND METHODS

Protocol

Weanling pigs were put on a diet containing 2% (w/w) cholesterol and 8% lard fat to induce accelerated atherogenesis. After 1 month parts of the abdominal aorta were subjected to balloon abrasion (2,16). 0.5% (w/w) bile acid was added to the cholesterol-rich diet after 3 months. After a total period of 8 months one group of animals was sacrificed (induction group), while the remaining animals were divided into two groups which were fed diets (without cholesterol or bile acid) containing either 10% lard fat or 5% lard fat mixed with 5% fish oil-derived fatty acids (regression groups). The fish oil preparation (OMEPA, cholesterol-free) was obtained from Hastar Enterprises Int. Soest, The Netherlands). The animals fed fish oil received about 0.25 g EPA and 0.17 g DHA per kg body weight. These diets were fed for 4 months. Pigs fed diets containing 10% lard fat (without cholesterol or bile acid) for 8 months served as a control group. The protocol is shown schematically in Fig. 1.

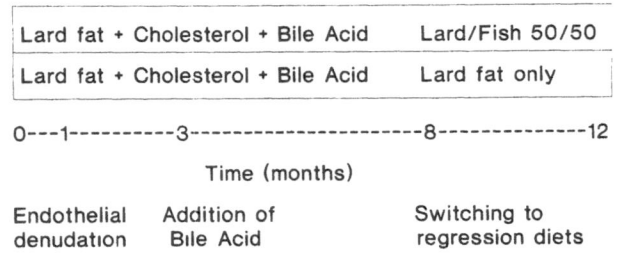

Fig. 1 Protocol for induction/regression of atherosclerosis

Isolation of Bloodplasma

Blood was drawn, following overnight fasting, and collected in tubes containing EDTA (final concentration 1.5 mg/ml). The blood was cooled immediately to 0-4 °C and centrifuged at 3000 rpm for 15 min at 4 °C.

Isolation of Plasma Lipoproteins

Plasma lipoproteins were isolated both by density gradient ultracentrifugation (17) and by automated gel permeation chromatography with preparative fast protein liquid chromatography (18).

Chemical Analyses

Total cholesterol, unesterified cholesterol, triglycerides and phospholipids were measured by enzymatic methods. Cholesterolesters were calculated by substracting unesterified cholesterol from total cholesterol.

Assay of Lecithin:Cholesterol Acyltransferase (LCAT) Activity

LCAT activity levels were measured, using exogenous substrate (19), according to Glomset & Wright (G-W), or endogenous substrate (by measuring the rate of decrease in plasma unesterified cholesterol during incubation of plasma).

RESULTS AND DISCUSSION

As mentioned above, atherosclerosis was induced by feeding the pigs a diet containing 8% lard fat, 2% cholesterol and 0.5% bile acids for 8 months. The extent to which atherosclerosis had developed was determined from the cholesteryl(ester) content of aortic fatty streaks/plaques. Atherosclerosis developed both in abraded and non-abraded parts of the aorta. The diet induced an increase in the plasma levels of all lipoproteins: VLDL, IDL, LDL and HDL. All lipoprotein classes acquired an abnormal composition, e.g. an increased weight % of total cholesterol (TC). The average sizes of LDL and HDL were increased. Plasma TC concentration was 2.3 ± 0.7 mM in pigs fed 10% lard fat and 12.3 ± 1.1 mM in pigs fed 8% lard fat + cholesterol and bile acids.

LCAT levels (G-W) were increased in the severely hypercholesterolemic pigs, if compared to normolipidemic control animals. There was however no increment in the plasma cholesterol esterification rate and, consequently, the fractional esterification rates were very low during severe hypercholesterolemia. This may be related to the observed change in HDL structure (formation of HDLc with increased size and cholesterol content).

The induction period was followed by a regression period of 4 months, during which the pigs were fed diets containing either 10% lard fat, or 5% lard fat + 5% fish oil-derived fatty acids (see Fig. 1). During this period a decrease in cholesteryl(ester) content of aortic plaques (measured in pieces of intima + media) was observed. This regression was most pronounced in pigs receiving fish oil-derived fatty acids. After 1 month in the regression period, plasma TC was reduced from 12.3 ± 1.1 mM to about 2 mM in the pigs receiving 10% lard fat, as well as in those receiving 5% lard fat + 5% fish oil. The effects of these two diets were compared in more detail after 4 months in the regression phase. Plasma TC levels were 2.2 ± 0.3 and 1.7 ± 0.3 mM in animals fed the 10% lard fat diet or the diet with 5% lard fat + 5% fish oil, respectively. By this time the pigs were more than one year old. It was found that plasma LDL-TC levels were identical on the diets with or without fish oil and that both VLDL-TC and HDL-TC levels were decreased after replacement of part of the dietary lard fat by fish oil. The chemical composition of HDL changed in pigs receiving fish oil, e.g. the ratio of free cholesterol/phospholipid (FC/PL) on HDL increased significantly.

Average plasma LCAT levels (G-W) were 27% lower after replacement of part of the lard fat by fish oil. The absolute cholesterol esterification rates were decreased by 10%. The latter change was not statistically significant. The high FC/PL ratio on HDL in the pigs fed fish oil may be caused by several mechanisms, e.g. decreased LCAT activity, increased rate of desorption of cholesterol from plasma membranes or a change in HDL structure.

CONCLUSIONS

1) a high HDL-TC/LDL-TC ratio is not prerequisite for regression of atherosclerosis in fish oil fed pigs. 2) the FC/PL ratio on HDL may be useful for the evaluation of regression of atherosclerosis. 3) Plasma LCAT levels may be relatively low but plasma cholesterol esterification rates are not decreased significantly after replacement of part of the dietary lard fat by fish oil.

REFERENCES

1. Leaf A, Weber PC (1988) New Engl J Med 318:549-557

2. Weiner BH, Ockene IS, Levine PH et al. (1986) New Engl J Med 315:841-846

3. Shimokawa H, Vanhoutte PM (1988) Circulation 78:1421-1430

4. Kim DN, Ho H-T, Lawrence DA, Schmee J, Thomas WA (1989) Atherosclerosis 76:35-54

5. Sassen, LMA, Hartog JM, Lamers JMJ, Klompe M, Van Woerkens LJ, Verdouw PD (1989) Eur Heart J 10:838-846

6. Van Houwelingen AC, Hennissen, AAHM, Verbeek-Schippers F, Simonsen T, Kester ADM, Hornstra G (1988) Thromb Haemostas 59:507-513

7. Hartog JM, Lamers JMJ, Essed CE, Schalkwijk WP, Verdouw PD (1989) Atherosclerosis 76:79-88

8. Enders S, Ghorbani R, Kelley VE et al. (1989) New Engl J Med 320:265-271

9. Harris WS, (1989) J Lipid Res 30:785-807

10. Subbaiah, PV, Davidson MH, Ritter MC, Buchanan W, Bagdade JD (1989) Atherosclerosis 79:157-166

11. Sullivan DR, Sanders TAB, Trayner IM, Thompson GR (1986) Atherosclerosis 61:129-134

12. Van Houwelingen AC, Zevenbergen, JH, Groot, PHE, Hornstra G (1990) Amer J Clin Nutr, in press

13. Groot PHE, Scheek LM, Dubelaar M-L, Verdouw PD, Hartog JM, Lamers JMJ (1989) Atherosclerosis 77:1-6

14. Cahill PD, Sarris GE, Cooper AD et al. (1988) J Vasc Surg 7:108-118

15. Glomset JA (1968) J Lipid Res 9:155-167

16. Sassen LMA, Hartog JM, Lamers JMJ, Klompe M, Van Woerkens, LJ, Verdouw PD (1989) Europ Heart J 10:838-846

17. Redgrave TG, Roberts DCK, West CE (1975) Anal Biochem 65:42-49

18. Van Gent T, Van Tol A (1990) J Chromatography 525:433-441

19. Glomset JA, Wright JL (1964) Biochim Biophys Acta 89:266-276

CONTRIBUTORS

Numbers in parentheses indicate the pages on which the author's
contributions begin

K. AALTO-SETÄLÄ (33)   Institute of Biotechnology, University of
                              Helsinki, SF-00380 Helsinki, Finland

L.P. AGGERBECK (39)   Centre de Génétique Moléculaire, CNRS,
                              F-91190 Gif-sur-Yvette, France

G. AGUIE (93)   SERLIA et INSERM U325, Institut Pasteur,
                              F-59019 Lille, France

G. AILHAUD (85,93)   Centre de Biochimie CNRS, Université de
                              Nice, F-06034 Nice, France

P. ALAUPOVIC (299)   Lipoprotein and Atherosclerosis Resaerch
                              Program, Oklahoma Medical Research
                              Foundation, Oklahoma City, OK 73104, U.S.A.

G.M. ANANTHARAMAIAH (131)   Departments of Medecine,Biochemitry and
                              the Atherosclerosis Unit, UAB Medical
                              Center, Birmingham, AL 35294, U.S.A.

A. ATHIAS (109)   Laboratoire de Biochimie des Lipoprotéines,
                              Hôpital du Bocage, CHRU, F-21034 Dijon,
                              France

G. BAGGIO (329)   Groupe d'Endocrinologie-Métabolisme,
                              Hôpital de la Pitié, F-75651 Paris, France

R. BARBARAS (85)   Centre de Biochimie, CNRS, Université de
                              Nice, F-06034 Nice, France

J-M. BARD (299)   Lipoprotein and Atherosclerosis Research
                              Program, Oklahoma Medical Research
                              Foundation, Oklahoma City, OK 73104, U.S.A.

R.L. BARNHART (65,367)   Merrell Dow Research Institute, Cincinnati,
                              DH 45215, U.S.A.

P.H.R. BARRETT (201)   Center for Bioengineering, FL-20, University
                              of Washington, Seattle, WA 98195, U.S.A.

P.J. BARTER (59)   Baker Medical Research Institute, Prahran 3181
                                 Victoria, Australia

M.W. BAUMSTARK (123)   Medizinische Universitätsklinik, Institüt
                                 für Biophysik, Universität Freiburg, D-7800
                                 Freiburg, F.R.G.

P. BEAU (183)   Service de Gastroentérologie et Assistance
                                 Digestive, CHU, F-83000 Poitiers, France

U. BEISIEGEL (329)   Universitätsklinik, Eppendorf, D-2000
                                 Hamburg 20, F.R.G.

G. BENGTSSON-OLIVECRONA(329)   Medical Biochemistry and Biophysics,
                                 University of Umea, S-901 87 Umea, Sweden

G.S. BERENSON (373)   Department of Medicine, Louisiana State
                                 University School of Medicine, New Orleans.
                                 LA 70112, U.S.A.

G. BEREZIAT (407)   Laboratoire de Biochimie (CNRS, URA 1283)
                                 CHU Saint Antoine, Université Piere et
                                 Marie Curie, F-75571 Paris, France

A. BERG (123)   Mediziniche Universitätsklinik,
                                 Universität Freiburg. D-7800 Freiburg.
                                 F.R.G.

F.BERTHEZENE (147)   INSERM U197, Laboratoire de Métabolisme des
                                 Lipides, Hôpital de l'Antiquaille. F-69321
                                 Lyon, France

E.L.BIERMAN (81)   Division of Metabolism, Endocrinology and
                                 Nutrition, University of Washington,
                                 Seattle, WA 98195, U.S.A.

A. BISHOP (275)   University of Wales College of Medicine,
                                 Heath Park, Cardiff, Wales, U.K.

D.H. BLANKENHORN (299)   Department of Medicine, University of
                                 Southern California School of Medicine, Los
                                 Angeles, CA 90033, U.S.A.

J. BOOGAERTS (131)   Departments of Medicine, Biochemistry and
                                 the Atherosclerosis Unit, UAB Medical
                                 Center, Birmingham, AL 35294, U.S.A.

K. BOSTRÖM (25)   Gladstone Foundation Laboratories for
                                 Cardiovascular Disease, Cardiovascular
                                 Research Institute. University of
                                 California, San Fransisco. CA 94140, U.S.A.

H.B. BREWER (237,329)   Molecular Disease Branch, National Heart.
                                 Lung and Blood Institute, NIH, Bethesda. MD
                                 20892, U.S.A.

C.G. BROUILLETTE (131)   Departments of Medicine, Biochemistry and
                                 the Atherosclerosis Unit, UAB Medical
                                 Center, Birmingham. AL 35294, U.S.A.

J.D. BRUNZELL (329)   University of Washington, Seattle, WA, U.S.A.

S.J. BUSCH (65)   Merrell Dow Research Institute, Cincinnati,
                             OH 45215, U.S.A.

P. CARDOT (407)   Laboratoire des Maladies Métaboliques
                             (INRA), Theix, France

Y.A. CARPENTIER (335)   Clinical Nutrition, Hôpital Saint-Pierre,
                             B-1000 Brussels, Belgium

G. CASTRO (93)   SERLIA et INSERM U325, Institut Pasteur,
                             F-59019 Lille, France

J. CHAMBAZ (407)   Laboratoire de Biochimie (CNRS URA 1283)
                             CHU Saint-Antoine, Université Pierre et
                             Marie Curie, F-75571 Paris, France

J.C.CHAMBERLAIN (275)   Department of Human Genetics & Metabolism,
                             Medical Professorial Unit, St Bartholomew's
                             Hospital, London EC1A 7BE, U.K.

L.B.F.CHANG (59)   Baker Medical Research Institute, Prahran 3181
                             Victoria, Australia

B.H. CHUNG (341)   Department of Medicine, University of Alabama
                             Medical Center, University of Alabama at
                             Birmingham, AL 35294, U.S.A.

C. CLADARAS (1)   Section of Molecular Genetics,
                             Cardiovascular Institute, Boston University
                             Medical Center, Boston. MA 02118. U.S.A.

S.B. CLARK (281)   Department of Biophysics. Boston University
                             School of Medicine, Boston. MA 02118. U.S.A.

V. CLAVEY (85)   Institut Pasteur/SERLIA, F-59019 Lille, France

M.CLERC (161)   Laboratoire de Biochimie Medicale A,
                             Université de Bordeaux II, F-33076 Bordeaux.
                             France

W. DAERR (319)   Medizinische Kernklinik, Universitäts
                             Krankenhaus Eppendorf, D-2000 Hamburg 20,
                             F.R.G.

R.J. DECKELBAUM (335)   Pediatric Gastroenterology, Columbia
                             University, New York, NY 10032, U.S.A.

J.L. DE COEN (141)   Laboratoire de Chimie Générale I . Université
                             Libre de Bruxelles and Research Foundation
                             on Atherosclerosis, B-1000 Brussels, Belgium

J.L. DE GENNES (329)   Groupe d'Endocrinologie-Métabolisme, Hôpital
                             de la Pitié, F-75651 Paris, France

C. DELCROIX (141,223)   Research Foundation on Atherosclerosis and
                             Université Libre de Bruxelles, B-1000
                             Brussels, Belgium

H. DE LOOF (131)   Departments of Medicine, Biochemistry and the
                                    Atherosclerosis Unit, UAB Medical Center,
                                    Birmingham, AL 35294, U.S.A.

A.DERKSEN (281)   Department of Biophysics, Boston, MA 02118,
                                    U.S.A.

M. DJAVAHERI (39)   Centre de Génétique Moléculaire, CNRS, F-91190
                                    Gif-sur Yvette, France

P.J.DOLPHIN (71)   Department of Biochemitry, Dalhousie
                                    University, Halifax, Nova Scotia, B3H 4H7
                                    Canada

D.Y. DUBOIS (123,223)   Research Foundation on Atherosclerosis,
                                    B-1000 Brussels, Belgium

P. DUCHATEAU (93)   SERLIA et INSERM U325, Institut Pasteur,
                                    F-59019 Lille, France

M-F. DUMON (161)   Laboratoire de Biochimie Médicale A,
                                    Université de Bordeaux II, F-33076 Bordeaux,
                                    France

N. DUVERGER (93)   SERLIA et INSERM U325, Institut Pasteur,
                                    F-59019 Lille, France

J.A. ENGLER (131)   Departments of Medicine, Biochemistry and the
                                    Atherosclerosis Unit, UAB Medical Center,
                                    Birmingham, AL 35294, U.S.A.

J. FEREZOU (183)   Laboratoire de Physiologie de la Nutrition,
                                    Université Paris-Sud, F-91405 Orsay, France

M.A. FLANAGAN (65)   Merrell Dow Research Institute, Cincinnati,
                                    OH 45215, U.S.A.

G. FLESS (295)   The University of Chicago, Departments of
                                    Medicine, Biochemistry and Molecular
                                    Biology, Chicago, IL 60637, U.S.A.

S.S FOJO (329)   Molecular Disease Branch, National Heart, Lung
                                    and Blood Institute, NIH, Bethesda, MD 20892,
                                    U.S.A.

D.M. FOSTER (201)   Center for Bioengineering, FL-20, University
                                    of Washington, Seattle, WA 98195, U.S.A.

O.L. FRANCONE (325)   INSERM U.177, Physiopathologie de la
                                    Nutrition, F-75270 Paris, France

M. FRENEIX-CLERC (161)   Laboratoire de Biochimie Médicale A,
                                    Université de Bordeaux II, F-33076 Bordeaux,
                                    France

J-C. FRUCHART (85,93)   Institut Pasteur/SERLIA, F-59019 Lille, France

D.J. GALTON (275)   Department of Human Genetics & Metabolism,
                                    Medical Professorial Unit, St Bartholomew's
                                    Hospital, London EC1A 7BE, U.K.

D. GANTZ (281)   Department of Biophysics, Boston University School of Medicine, Boston, MA 02118, U.S.A.

P. GAMBERT (109)   Laboratoire de Biochimie des Lipoprotéines, Hôpital du Bocage, CHRU, F-21034 Dijon, France

N. GHALIM (85,93)   Institut Pasteur/SERLIA, F-59019 Lille, France

P.C. GREIF (189)   Molecular Disease Branch, National Heart, Lung and Blood Institute, NIH, Bethesda, MD 20892, U.S.A.

H.GRETEN (319)   Medizinische Kernklinik, Universitäts-Krankenhaus Eppendorf, D-2000 hamburg 20, F.R.G.

G. GRIFFATON (325)   INSERM U.177, Physiopathologie de la Nutrition, F-75270 Paris, France

S. GRIGLIO (311)   INSERM U.177, Physiopathologie de la Nutrition, F-75270 Paris, France

J.E.M. GROENER (417)   Department of Biochemistry I, Erasmus University Rotterdam, 3000 DR Rotterdam, The Netherlands

S.M. GRUNDY (213)   Center for Human Nutrition, University of Texas Southwestern Medical Center at Dallas, Dallas. TX 75235, U.S.A.

M. HADZOPOULOU-CLADARAS (1)   Section of Molecular Genetics, Cardiovascular Institute, Boston University Medical Center, Boston, MA 02118, U.S.A.

T. HAJRI (183)   Physiologie de la Nutrition, Université Paris-Sud, F-91405 Orsay, France

R.J. HAVEL (245)   Cardiovascular Research Institute and Department of Medicine, University of California. CA 94143, U.S.A.

J.M. HOEG (237)   Molecular Disease Branch, National Heart, Lung and Blood Institute, NIH, Bethesda, MD 20892. U.S.A.

M. HULTIN (335)   Medical Biochemistry and Biophysics, University of Umea, S-901 87 Umea, Sweden

T.L. INNERARITY (25)   Gladstone Foundation Laboratories for Cardiovascular Disease, Cardiovascular Research Institute. University of California. San Fransisco, CA 94140, U.S.A.

S. JÄCKLE (319)   Mediziniche Kernklinik, Universitäts-Krankenhaus Eppendorf, D-2000 Hamburg 20, F.R.G.

R.L. JACKSON (65,367)   Merrell Dow Research Institute. Cincinnati, OH 45215, U.S.A.

R.W. JAMES (101) Division de Diabétologie, Département de Médecine, Hôpital Cantonal Universitaire, 1211 Genève 4, Suisse

M. JAUHIAINEN (71) Department of Biochemistry, National Public Health Institute, Helsinki, Finland

O.S. KALENICH (393) Institute of Experimental Cardiology, USSR Cardiology Research Center, Moscow 121552, USSR

A.D. KALOPISSIS (325) INSERM U.177, Physiopathologie de la Nutrition, F-75270 Paris, France

D. KARDASSIS (1) Section of Molecular Genetics, Cardiovascular Institute, Boston University Medical Center, Boston, MA 02118, U.S.A.

M.L. KASHYAP (233) Cholesterol Center, University of California at Irvine, V.A. Medical Center, Long Beach, CA 90822, U.S.A.

J. KEUL (123) Mediziniche Universitätsklinik, Universität Freiburg, D-7800 Freiburg, F.R.G.

C. KNIGHT-GIBSON (299) Lipoprotein and Atherosclerosis Research Program, Oklahoma Medical Research Foundation, Oklahoma City, OK 73104, U.S.A.

P. KOCHER (141) Laboratoire de Chimie Générale I, U.L.B. and Research Foundation on Atherosclerosis, B-1000 Brussels, Belgium

K. KONTULA (33) Institute of Biotechnology, University of Helsinki, SF-00380 Helsinki, Finland

W. KREUTZ (123) Institut für Biophysik, Universität Freiburg, D-7800 Freiburg, F.R.G.

D. LAGRANGE (311) INSERM U. 177, Physiopathologie de la Nutrition, F-75270 Paris, France

L. LAGROST (109) Laboratoire de Biochimie des Lipoprotéines, Hôpital du Bocage, CHRU, F-21034 Dijon, France

C. LALLEMANT (109) Laboratoire de Biochimie des Lipoprotéines, Hôpital du Bocage, CHRU, F-21034 Dijon, France

J.M.J. LAMERS (417) Department of Biochemistry I, Erasmus University Rotterdam, 3000 DR Rotterdam, The Netherlands

D.M. LEE (49) Lipoprotein and Atherosclerosis Research Program, Oklahoma Medical Research Foundation, Oklahoma City, OK 73104, U.S.A.

B. LEVKAU (319) Medizinische Kernklinik, Universitäts-Krankenhaus Eppendorf, D-2000 Hamburg 20, F.R.G.

J.F. LONTIE (141,173,223)   Research Foundation on Atherosclerosis,
                                B-1000 Brussels, Belgium

M.F. LOPES-VIRELLA (383)   Department of Medicine, Medical University of
                                South Carolina and VA Medical Center,
                                Charleston, SC 29425, U.S.A.

T. LORENZEN (319)   Medizinische Kernklinik, Universitäts-
                                Krankenhaus Eppendorf, D-2000 Hamburg 20,
                                F.R.G.

C. LORIETTE (407)   Laboratoire de Biochimie (CNRS URA 1283) CHU
                                Saint-Antoine, Université Pierre et Marie
                                Curie, F-75571 Paris, France

C. LUTTON (183,257)   Laboratoire de Physiologie de la Nutrition,
                                Université de Paris-Sud, F-91405 Orsay, France

T. MAGOT (183,257)   Laboratoire de Physiologie de la Nutrition,
                                Université de Paris-Sud, F-91405 Orsay, France

C.L. MALMENDIER(141,173,223)   Research Foundation on Atherosclerosis,
                                B-1000 Brussels, Belgium

M.MANGENEY (407)   Laboratoire de Biochimie (CNRS URA 1283) CHU
                                Saint-Antoine, Université Pierre et Marie
                                Curie, F-75571 Paris, France

A. MANN (237)   Molecular Disease Branch National Heart, Lung
                                and Blood Institute, NIH, Bethesda,
                                MD 20892, U.S.A.

S.J.T. MAO (367)   Merrell Dow Research Institute, Cincinnati,
                                OH 45215, U.S.A.

Y.L. MARCEL (77)   Laboratoire du Métabolisme des Lipoprotéines,
                                Institut de Recherches Cliniques de Montréal,
                                Montréal, Quebec H2W 1R7, Canada

G.A. MARTIN (65)   Merrell Dow Research Institute, Cincinnati,
                                OH 45215, U.S.A.

C. MATUCHANSKY (183)   Service de Gastroenthérologie et Assistance
                                Digestive, CHU, F-86000, Poitiers, France

M-J. MAVIEL (161)   Laboratoire de Biochimie Médicale A,
                                Université de Bordeaux II, F-33076 Bordeaux,
                                France

B.J.MCKEONE (289)   Department of Medicine, Baylor College of
                                Medicine and The Methodist Hospital,
                                Houston, TX 77030, U.S.A.

R. MCPHERSON (77)   Lipid Research Laboratory, Royal Victoria
                                Hospital, McGill University, Montreal,
                                Quebec, Canada

A. MENDEZ (81)

Division of Metabolism, Endocrinology and Nutrition, University of Washington, Seattle, WA 98195, U.S.A.

R.W. MILNE (77)

Clinical Research Institute of Montreal, Montreal, Quebec, Canada

R. MORGAN (275)

University of Wales College of Medicine, Heath Park, Cardiff, Wales, U.K.

I.D. NOVIKOV (393)

Institute of Experimental Cardiology, USSR Cardiology Research Center, Moscow 121552, USSR

K. OGAMI (1)

Section of Molecular Genetics, Cardiovascular Institute, Boston University Medical Center, Boston, MA 02118, U.S.A.

K. OKA (275)

Medlantic Research Foundation, Washington DC, U.S.A.

T. OLIVECRONA (335)

Medical Biochemistry and Biophysics, University of Umea, S-901 87 Umea, Sweden

J. ORAM (81)

Division of Metabolism, Endocrinology and Nutrition, University of Washington. Seattle, WA 98195, U.S.A.

A.N. OREKHOV (393,399)

Institute of Experimental Cardiology, USSR Cardiology Research Center, Moscow 121552. USSR

K.OUGUERRAM (257)

Laboratoire de Physiologie de la Nutrition. Université de Paris-Sud, F-914015 Orsay,France

M. PARQUET (183)

Laboratoire de Physiologie de la Nutrition. Université de Paris-Sud, F-91405 Orsay, France

J. PATSCH (335)

Department of Medicine, University of Innsbruck, Austria

J.R. PATSCH (289)

Department of Medicine, Baylor College of Medicine and the Methodist Hospital. Houston, TX 77030, U.S.A.

J. PETERSON (335)

Medical Biochemistry and Biophysics, University of Umea, S-901 87 Umea, Sweden

D. POMETTA (101)

Division de Diabétologie, Département de Médecine, Hôpital Cantonal Universitaire. 1211 Genève 4, Suisse

G. PONSIN (147)

INSERM U 197. Laboratoire de Métabolisme des Lipides, Hôpital de l'Antiquaille, F-69231 Lyon, France

H.J. POWNALL (289)

Department of Medicine, Baylor College of Medicine and the Methodist Hospital, Houston, TX 77030, U.S.A.

A. PRADINES-FIGUERES (85)   Centre de Biochimie CNRS, Université de
                                   Nice, F-06034 Nice, France

P.PUCHOIS (85)   Institut Pasteur/SERLIA, F-59019 Lille, France

D.J. RADER (189,237)   Molecular Disease Branch National Heart, Lung
                                   and Blood Institute, NHI, Bethesda, MD 20892
                                   U.S.A.

B. RADHAKRISHNAMURTHY (373)   Departments of Medicine and Biochemistry,
                                   Louisiana State University School of
                                   Medicine, New Orleans, LA 70112, U.S.A.

O.V. RAJARAM (59)   Baker Medical Research Institute, Prahran 3181
                                   Victoria, Australia

Y. RAYSSIGUIER (407)   Laboratoire des Maladies Métaboliques (INRA),
                                   Theix, France

A. REES (275)   University of Wales College of Medicine, Heath
                                   Park, Cardiff, Wales, U.K.

A. RIBEIRO (407)   Laboratoire de Biochimie (CNRS URA 1283) CHU
                                   Saint-Antoine, Université Pierre et Marie
                                   Curie, F-75571 Paris, France

F. RINNINGER (319)   Mediziniche Kernklinik, Universitäts-
                                   Krankenhaus Eppendorf, D-2000 Hamburg 20,
                                   F.R.G.

K. SAKU (233)   Cholesterol Center, University of California
                                   at Irvine, V.A. Medical Center, Long Beach,
                                   CA 90822, U.S.A.

L.M.A. SASSEN (417)   Laboratory for Experimental Cardiology
                                   (Thoraxcenter), Erasmus University Rotterdam,
                                   3000 DR Rotterdam, The Netherlands

A.M. SCANU (295)   The University of Chicago, Departments of
                                   Medicine, Biochemistry and Molecular
                                   Biology, Chicago, IL 60637, U.S.A.

L.M. SCHEEK (417)   Department of Biochemistry I, Erasmus
                                   University Rotterdam, 3000 DR Rotterdam, The
                                   Netherlands

S. SEGREST (131,341)   Department of Medicine, Biochemistry and the
                                   Atherosclerosis Unit, UAB Medical Center,
                                   University of Alabama at Birmingham,
                                   AL 35294, U.S.A.

D. SEIDEL (155)   Institute for Clinical Chemistry, Klinikum
                                   Großhadern, D-8000 München 70, F.R.G.

D.M. SHAMES (245)   Cardiovascular Research Institute and
                                   Department of Medicine, University of
                                   California, San Francisco, CA 60637, U.S.A.

D.M. SMALL (281)   Department of Biophysics, Boston University
                                   School of Medicine, Boston, MA 02118, U.S.A.

431

S.R. SRINIVASAN (373)   Department of Medicine and Biochemistry, Louisiana State University School of Medicine, New Orleans, LA 70112, U.S.A.

A.F.H. STALENHOEF (329)   Universitäts Klinik-Eppendorf, D-2000 Hamburg 20, F.R.G.

O. STEIN (117)   Department of Experimental Medicine and Cancer Research, Hebrew University Hadassah Medical School, Jerusalem, Israel

Y. STEIN (117)   Lipid Research Laboratory, Department of Medicine B, Hadassah University Hospital, Jerusalem, Israel

J. STEINER (281)   Department of Biophysics, Boston, University School of Medicine, Boston MA 02118, U.S.A.

A. STEINMETZ (85)   Institut Pasteur/SERLIA, F-59019 Lille, France

J.STOCKS (275)   Department of Human Genetics & Metabolism, Medical Professorial Unit, St Bartholomew's Hospital, London EC1A 7BE, U.K.

F. SULTAN (311)   INSERM U.177, Physiopathologie de la Nutrition F-75270 Paris, France

A.R. TALL (77)   Department of Medicine, Columbia University College of Physicans and Surgeons, New York, NY 10032, U.S.A.

M. TAVELLA (299)   Lipoprotein and Atherosclerosis Research Program, Oklahoma Medical Research Foundation, Oklahoma City, OK 73104, U.S.A.

G. TENNYSON (237)   Molecular Disease Branch, National Heart, Lung and Blood Institute, NIH, Bethesda, MD 20892, U.S.A.

A. TERCYAK (281)   Department of Biophysics, Boston University School of Medicine, Boston, MA 02118, U.S.A.

V.V. TERTOV (393,399)   Institute of Experimental Cardiology, USSR Cardiology Research Center, Moscow 121552, USSR

N. THÉRET (93)   SERLIA et INSERM U325, Institut Pasteur, F-59019 Lille, France

J.A. THORN (275)   Department of Human Genetics & Metabolism, Medical Professorial Unit, St Bartholomew's Hospital, London EC1A 7BE, U.K.

T. VAN GENT (417)   Department of Biochemistry I, Erasmus University Rotterdam, 3000 DR Rotterdam, The Netherlands

A. VAN TOL (417)   Department of Biochemistry I, Erasmus University Rotterdam, 3000 DR Rotterdam, The Netherlands

G.L. VEGA (213)   Center for Human Nutrition, University of
                                    Texas Southwestern Medical Center, Dallas,
                                    TX 75235, U.S.A.

Y.V. VENKATACHALAPATHY (131) Departments of Medicine, Biochemistry and the
                                    Atherosclerosis Unit, UAB Medical Center,
                                    Birmingham, AL 35294, U.S.A.

P.D. VERDOUW (417)   Laboratory for Experimental Cardiology
                                    (Thoraxcenter), Erasmus University Rotterdam,
                                    3000 DR Rotterdam, The Netherlands

C. VIALLE-VALENTIN (147)   INSERM U197, Laboratoire de Métabolisme des
                                    Lipides, Hôpital de l'Antiquaille, F-69321
                                    Lyon, France

P. VIJAYAGOPAL (373)   Department of Medicine, Louisiana State
                                    University School of Medicine, New Orleans,
                                    LA 70112, U.S.A.

S. VILARO (335)   Cell Biology, University of Barcelona, Spain

G. VIRELLA (383)   Department of Basic and Clinical Immunology
                                    and Microbiology, Medical University of South
                                    Carolina and VA Medical Center, Charleston,
                                    SC 29425, U.S.A.

E.G. VOROB'EVA (393)   Institute of Experimental Cardiology, USSR
                                    Cardiology Research Center, Moscow 121552,
                                    USSR

E. WINDLER (319)   Medizinche Kernklinik Universitäts-
                                    Krankenhaus Eppendorf, D-2000 Hamburg 20,
                                    F.R.G.

J.L. WITZTUM (353)   Department of Medicine, University of
                                    California, San Diego, La Jolla, CA 92093,
                                    U.S.A.

V.I. ZANNIS (1)   Section of Molecular Genetics, Cardiovascular
                                    Institute, Boston University Medical Center,
                                    Boston, MA 02118, U.S.A.

L.A. ZECH (189)   Molecular Disease Branch, National Heart, Lung
                                    and Blood Institute, NIH, Bethesda, MD 20892,
                                    U.S.A.

INDEX